D1127034

The Annotated *Origin*

The Annotated *Origin*

A Facsimile of the First Edition of *On the Origin of Species*

Charles Darwin

ANNOTATED BY JAMES T. COSTA

The Belknap Press of Harvard University Press

Cambridge, Massachusetts, and London, England • *2009*

Printed in the United States of America

Library of Congress Cataloging-in-Publication Data

Darwin, Charles, 1809–1882.
[On the origin of species]
The annotated Origin : a facsimile of the first edition of On the origin of species / Charles Darwin; annotated by James T. Costa.
p. cm.
Includes bibliographical references and index.
ISBN 978-0-674-03281-1 (alk. paper)
1. Evolution (Biology). 2. Natural selection. 3. Darwin, Charles, 1809–1882. On the origin of species. I. Costa, James T., 1963– II. Title.
QH365.O2 2009
576.8′2--dc22 2008043895

For Leslie, who loves ginkgo leaves best

Contents

Introduction

Darwin's *Origin of Species* is a living document that has, in many respects, even greater currency today than when it was first published in 1859. It is at once a founding treatise of a major scientific discipline, a philosophical argument for a novel worldview, and a masterly piece of science writing. A close reading throws open a window on a time and place, giving us insight into the cultural context in which its ideas were fermented and debated. That window also reveals much about scientific pursuit in the nineteenth century, a time when professional science was in its infancy and scholars began refining ideas about science as a way of knowing the natural world. As for the book as science literature, we have the renowned naturalist Alfred Russel Wallace, co-discoverer of the principle of natural selection, to thank for derailing Darwin's planned mega-tome on the subject and giving us instead this remarkable book, *On the Origin of Species by Means of Natural Selection.* Darwin apologetically called the *Origin* an "abstract," which never fails to elicit a chuckle from students braving its 490-odd pages; but indeed it is an abbreviated version of the treatise he intended, lacking proper footnotes and the many argument-buttressing examples Darwin marshaled in that larger work. As a result, the *Origin* is written in a style far more accessible than the "big species book" might have been. The *Origin* has its dense passages, but in places the book is nothing short of lyrical. Its arguments are backed by case studies, data, and observations, yet they are presented in a narrative style that was unusual for serious scientific books at the time, and that had the welcome effect of broadening the book's audience.

To a greater extent than perhaps any other watershed scientific document, the *Origin* rippled through society: its implications were felt in literature, philosophy, politics, and art; the book was midwife to the Modern period to come in the next century. Yet despite its initial impact, today readers see the *Origin* as little more than a presentation of Darwin's ideas of evolution and natural selection; in fact, it is probably more *mis*understood than understood by the general public. Even within the biological sciences few venture far beneath its cover, and fewer still read the book in its entirety. There are several reasons for this inattention, chief among them the nature of the scientific pursuit. Today the extent and pace of scientific research are unprecedented, and the sheer quantity of new knowledge is so great that, in a discipline that emphasizes the latest cutting-edge findings, not much room is left in biology curricula for reading the old literature. Another problem is the distance between Darwin's time and our own. People, places, and many of the ideas discussed in the *Origin* mean little to a modern reader, and for many people Darwin's Victorian prose poses its own stumbling block.

The annotations presented here aim to remedy this situation. It is my hope that by highlighting guideposts along the path of Darwin's "one long argument," as he described the book—transporting the reader to Darwin's time and place by fleshing out such details as biographical references, natural history observations, and the intent or meaning of arguments that are obscure from a modern perspective—a new generation of students will be inspired to read Darwin. This book can also be seen as a guide to the making of the *Origin,* the behind-the-scenes ferment revealed in Darwin's correspondence, notebooks, and diaries, and the writings of his contemporaries. My aim, finally, is to show readers the breathtaking sweep of Darwin's method, in hopes that more educators and others will take a page from his playbook in making the case for biological evolution in school curricula and beyond. Indeed, it is my hope that the very notion of having to defend biological evolution and its teaching in a court of law will become a thing of the past, a phenomenon seen as a historical peculiarity of twentieth- and early twenty-first–century American society, symptomatic of a particularly pernicious brand of scientific illiteracy.

As the late astronomer Carl Sagan pointed out in his book of essays *Demon Haunted World,* ours is an age of contradictions. American society in particular has been a powerhouse of scientific innovation and advance made possible by the unrivalled collective brainpower of its homegrown and immigrant scientists and prosperous economic conditions. Yet this is true even as a significant proportion of this same society embraces mysticism and the supernatural, and evinces profound scientific illiteracy (see, e.g., Gross 2006). This national cognitive dissonance can only spell disaster if science and the understanding and innovation it brings become eclipsed by ignorance, fear, superstition, and religious zealotry. Scientific knowledge of and by itself is no panacea, to be sure. But the importance of thoughtfully considering and debating the implications of scientific pursuits, and the technological innovations stemming from them, cannot be stressed enough. Science as a way of knowing is a potent way forward for the understanding of both ourselves and the world we live in, as it has been since the time of Francis Bacon. In its modest way, this book seeks to help readers better understand Darwin's *Origin of Species,* and thereby better understand the scientific pursuit as a highly successful mode of inquiry into the natural world.

Before analyzing the *Origin* itself, however, we must understand where the book came from. By focusing on Darwin's personal intellectual odyssey and how that was incorporated into the *Origin*'s very structure, we set the stage for our own odyssey through Darwin's *one long argument.*

The Road to the *Origin*

Alfred Russel Wallace knew. As the only other person to have stood, with Darwin, gazing upon the grand sweep of life's diversity with understanding, the brilliant naturalist knew that nothing would be the same upon publication of *On the Origin of Species.* He declared that "Mr. Darwin has given the world a new science, and his name should, in my opinion, stand above that of every philosopher of ancient or modern times. The force of admiration can no further go!!!" (Berry 2002). Wallace received his copy of the book in early 1860, while still far off in southeast Asia. He was not to return from his eight-year adventure in the Malay Archipelago for another two years, and so his adulation marks what was almost certainly the farthest-flung, most remote reverberation made by the *Origin* when it was published. Now, here we are reading the book 150 years later, and the reverberations continue still.

Charles Robert Darwin was an unlikely revolutionary. An indifferent student in youth, a medical school dropout, and a sometime theology student, this mild-mannered, good-tempered young man really just wanted to ride, hunt, and naturalize. By his own account he had the common child's penchant for collecting—shells, coins, rocks, and minerals—the kind of passionate hobby that "leads a man to be a systematic naturalist, a virtuoso, or a miser," he wrote in his autobiography. He never really grew out of it. Instead, he just discovered new things to collect. Beetle-collecting was nearly a blood sport with Darwin and many of his fellow students at Cambridge, and after discovering the joys of botanizing with his professor John Stevens Henslow, he threw himself into learning the plants of Cambridgeshire with zeal.

Despite initial doubts, he fully intended to take Holy Orders after finishing at Cambridge. But not before a taste of adventure: Darwin and his fellow naturalists planned a jaunt to Tenerife, in the Canary Islands. In preparation for this "Canary scheme," as he put it, Darwin accompanied another Cambridge don, Rev. Adam Sedgwick, on a three-week geology expedition to Wales in August 1831. But the Canaries were not to be. Upon returning home from his field excursion with Sedgwick, Darwin found a letter awaiting him with the offer of a far grander trip—no less than a multiyear voyage around the world! He could barely contain his excitement.

With the support of his Uncle Josiah, Darwin convinced his no-nonsense father to permit him to take the journey, complete with promises to settle down as a respectable country parson on his return—and assurances that his rather liberal spending habits developed at Cambridge would inevitably be curbed in the confines of a small ship plying the world's oceans. "I should be deuced clever to spend more than my allowance whilst on board the *Beagle,*" Darwin said solicitously. "But they tell me you are very clever," came the shrewd reply from his father.

Darwin sailed from England in December 1831, companion-naturalist to Captain Robert Fitzroy, Commander of HMS *Beagle.* Orthodox in his religious views, gentlemanly, well trained in several branches of natural history, amiable, earnest, curious, and, most of the time, seasick, Darwin struck out to see the world. Upon his return five years later his father exclaimed, "Why, the shape of his head is quite altered." It probably appeared so owing to his now-receding hairline. What was inside that head was altered more profoundly, but not quite in the manner commonly supposed. Indeed, it is often mistakenly assumed that Darwin experienced a revelation of sorts while on the *Beagle* voyage—a sudden stroke of insight that immediately led him to abandon his belief in the immutability of species. But there was no "eureka" moment on the voyage. How, then, *did* this young naturalist, so recently and earnestly able to recite from William Paley's *Evidences of Christianity,* and so fully orthodox with respect to the question of species immutability and the veracity of the Bible, come to embrace the heretical notion of transmutation? Knowing something of Darwin's intellectual odyssey is highly instructive on several levels: it expresses the essence of the scientific pursuit and underscores the very humanity of Darwin and his contemporaries. Understanding just how Darwin first glimpsed the reality of transmutation and doggedly pursued the idea for some two decades is essential to appreciating Wallace's prophetic words. Darwin's intellectual odyssey also gives us insight into the *Origin* itself. Darwin described the book as "one long argument," but what is the nature of that argument? First things first.

Darwin had a sharp eye for the natural world. We see what are perhaps his earliest musings on the grand philosophical questions pertaining to species and centers of creation in a diary entry from January 1836, toward the end of the voyage. In New South Wales, Australia, he recorded these thoughts:

> I had been lying on a sunny bank & was reflecting on the strange character of the Animals of this country as compared to the rest of the world. An unbeliever in everything beyond his own reason, might exclaim "Surely two distinct Creators must have been [at] work; their object however has been the same & certainly the end in each case is complete".— Whilst thus thinking, I observed the conical pitfall of a Lion-Ant:— A fly fell in & immediately disappeared; then came a large but unwary Ant; his struggles to escape being very violent, the little jets of sand described by Kirby (Vol. I p. 425) were promptly directed against him.— His fate however was better than that of the poor fly's:— Without a doubt this predacious Larva belongs to the same genus, but to a different species from the Europaean one.— Now what would the *Dis*believer say to this? Would any two workmen ever hit on so beautiful, so simple & yet so artificial a contrivance? It cannot be thought so.— The one hand has surely worked throughout the universe." (Keynes 2001, pp. 402–403)

Here we see young Darwin tentatively testing the waters then pulling back. The ideas were gestating and would continue to do so for more than a year as he struggled retrospectively to make sense of his collections and observations. The data he cited as forming the foundation for his subsequent evolutionary thinking—the relationships between extinct and living species of South America, and the curiosities of the Galápagos Archipelago and their species relationships among those islands as well as with mainland species—all came together for him later, back in England (see Sulloway 1982a,b; 1984). Writing up his notes on the final leg of the *Beagle*'s return journey, Darwin seemed to be on the scent, musing,

> when I see these Islands [Galápagos] in sight of each other and possessed of but a scanty stock of animals, tenanted by these [mockingbirds] but slightly differing in structure & filling the same place in Nature, I must suspect they are only varieties. The only fact of a similar kind of which I am aware is the constant as-

serted difference between the wolf-like Fox of East & West Falkland Isds.— If there is the slightest foundation for these remarks, the Zoology of Archipelagoes will be well worth examining; for such facts would undermine the stability of species. (Barlow 1963)

The "stability of species" was a matter that struck at the very heart of understanding creation. These remarks, penned sometime between mid-June and August 1836 (Sulloway 1982b), include reference to the mockingbirds (*Nesomimus*) of the Galápagos Islands, a group of birds that exhibits three species, each endemic to one of the southernmost islands in the archipelago, plus one species that is now regarded as a complex of six subspecies, each also confined to one or more of the main islands. Darwin seemed to see the significance of these birds, but he did not yet realize that the finches of those equatorial islands were even more remarkable, nor that the tortoises told a similar story. Nonetheless, crucial data came to Darwin almost as a revelation when, in London and Cambridge, specialists impressed upon him the curious nature of his collections from South America and the Galápagos.

Winter and spring of 1837 was a watershed period in Darwin's thinking. The *Beagle* arrived home on October 2, 1836, and Darwin wasted no time seeing family and friends and making arrangements to distribute his biological collections for study. He took up residence in Cambridge in mid-December, on Fitzwilliam Street, where he stayed through March. He twice trekked back to London during this time to attend scientific meetings and confer with friends and colleagues. Much happened in those heady months, particularly in connection with his fossil discoveries from South America and the Galápagos avifauna.

Darwin delivered his fossil mammal specimens to Richard Owen in late December or early January, and within a month the great anatomist knew that they were remarkable. Owen said as much in a letter to Charles Lyell dated January 23, in which he revealed that Darwin had discovered no fewer than five extinct relatives of mammals found in South America today—and moreover, that these extinct species were gigantic forms: a massive armadillo, a giant sloth, a tank-like armored sloth relative, a rodent with a skull more than two feet long, dwarfing all living rodents; and a llama so large that Darwin had tentatively labeled its bones those of a mastodon! (Herbert 2005, pp. 320–324). Something profound was happening, and Owen knew it. Darwin's finds confirmed the "law of succession," which described the replacement of so-called types by related, though distinct, forms through geological time. Lyell was so impressed by Owen's conclusions that he devoted his Presidential Address to the Geological Society to the subject. He pointedly invited Darwin to hear his address, given on February 17, 1837: "These fossils . . . establish the fact that the peculiar type of organization which is now characteristic of the South American mammalia has been developed on that continent for a long period." Darwin was impressed, but he was not yet a transmutationist.

Even while Lyell enthused and Darwin marveled over the South American fossils, Darwin's bird collection was being scrutinized by the respected ornithologist John Gould at the Zoological Society. Gould received the birds on January 4, 1837, and began analyzing them soon after. He reported his results at successive meetings of the society: the curious Galápagos finches on January 10 (finding some dozen unique *Geospiza* species in three subgenera); the raptorial birds on January 24 (including the remarkable Galápagos hawk, which he cited as a "beautiful intervening link" between the genus *Buteo* and the mainland caracara genus *Polyborus*); the distinctive mockingbirds of the Galápagos on February 28 (three island-specific species yet so like mainland ones); and the South American rheas, a large species and a smaller relative, on March 14. Gould's analysis appears to have been the final catalyst precipitating Darwin's conversion to transmutation. Darwin met with Gould between March 7 and 12, 1837 (Sulloway 1982b), and was given a summary of the ornithologist's findings: more than two-thirds of Darwin's Galápagos birds were new species! His finds on the continent were no less interesting. Gould named the small or "petisse" rhea in Darwin's honor (though the name does not stand today, as discussed on p. 349). Darwin found it curious that the common large rhea should be geographically replaced by its diminutive cousin in the south of the continent with no obvious geographical feature demarking the boundary. Why two rheas where surely one might suffice?

Darwin's first evolutionary entry in the then-recently reopened notebook, now called the Red notebook, considers just this point: "Speculate on neutral ground of 2. Ostriches; bigger one encroaches on smaller.—change not progressif: produced at one blow . . . Yet new creation affected by Halo of neighboring continent" (Barrett et al. 1987, p. 61). Darwin was increasingly convinced that species change, but how? As this and other early passages suggest, he first toyed with the idea that transmutation occurs quickly, perhaps instantaneously. Two pages later he muses again about what secret these ostriches hold, and he relates their juxtaposition in space to the temporal juxtaposition of extinct and living mammals of South America. "Not gradual change or degeneration," he writes on p. 130. "If one species does change into another it must be per saltum [fast, by sudden transitions]" (Barrett et al. 1987, p. 63).

The important point about Gould's analysis was not his taxonomic assessments per se but what they meant for an understanding of species and varieties. As Darwin later wrote in the *Zoology* of the *Beagle* voyage (1841, pp. 63–64), referring to the mockingbirds: "I may observe, that as some naturalists may be inclined to attribute these differences [among island forms of the mockingbirds] to local varieties; that if birds so different as *O. trifasciatus* and *O. parvulus,* can be considered as varieties of one species, then the experience of all the best ornithologists must be given up, and whole genera must be blended into one species." In that book and in the *Journal of Researches* (1839), Darwin recounted Gould's remarkable assessment: "In my collections from these islands, Mr. Gould considers that there are twenty-six different species of land birds. With the exception of one, all probably are undescribed kinds, which inhabit this archipelago, and no other part of the world." He goes on to discuss each group, noting, for example, that the finches are "the most singular of any [birds] in the archipelago," that "it is very remarkable that a nearly perfect gradation of structure in this one group can be traced in the form of the beak, from one exceeding in dimensions that of the largest gros-beak, to another differing but little from that of a warbler." Darwin is not tipping his hand; he offers only this suggestive comment:

> I will not here attempt to come to any definite conclusions, as the species have not been accurately examined; but we may infer, that, with the exception of a few wanderers, the organic beings found on this archipelago are peculiar to it; and yet that their general form strongly partakes of an American character . . . The circumstances would be explained, according to the views of some authors, by saying that the creative power had acted according to the same law over a wide area. (Darwin 1839, p. 474)

Darwin soon opened another notebook, in July 1837, dedicated to transmutation. This is the notebook he was referring to when he wrote in his diary: "In July opened first note Book on 'transmutation of Species'—Had been greatly struck from about month of previous March—on character of S. American fossils—& species on Galapagos Archipelago. These facts origin (especially latter) of all my views." Dubbed the B notebook (after the A notebook on geological matters), running through February 1839, it reflects a mind expanding at warp speed—a riot of ideas, questions, suggestions for experiments, and observations. That summer we find him grappling with the rate of transmutation; he had yet to see change in gradualistic terms. We also see him pondering islands, his remarkable Galápagos finds fresh in his mind: "According to this view animals, on separate islands, ought to become different if kept long enough . . . Now Galapagos Tortoises, Mocking birds" (Barrett et al. 1987, p. 172).

Darwin explored evolutionary ideas and their ramifications over the next months and years. There were doubts, false starts, and blind alleys. He soon gave up the notion of sudden (saltational) evolution, but what was driving gradual change? How necessary was isolation? What was the relationship between geological change and species change? Were the formation and extinction of species somehow tied together, with more or less constrained lifespans? The essential elements of his theory as ultimately published came together over a period of time, some much later than others. In the B notebook (Barrett et al. 1987, pp. 177, 180) we find the first expressions of a genealogical view of species: first a tentative branching "coral of life," then, a few pages later, a bold ramifying tree-like sketch with species or species groups branching from a common trunk; this beautiful diagram is prefaced with the words "I think"—a vision that clearly set Darwin's conception of transmutation apart from all earlier attempts.

In October 1838 Darwin gained insight from re-reading the Rev. Thomas Malthus's *Essay on Population.* Malthus taught him that the continual "struggle for existence" would lead to differential survival, and that those individuals with constitutions best suited to the demands of the struggle would be the most likely to make it through. Malthus was writing about people, but the applicability of his ideas to the natural world immediately struck Darwin, giving him "a theory by which to work," as he wrote in his diary. That same year Darwin apparently realized that domesticated varieties provide a case study, a natural experiment on species and varieties. What is the extent of variability? How are variants propagated and not lost by crossing? His reading of the great agricultural breeders of his day—among them Robert Baker, John Sebright (who used the term "selection" in describing agricultural breeding), and William Youatt—underscored how methodical "picking and choosing" gradually bends plant and animal breeds to human whims. Darwin coined the phrase "natural selection" to contrast with this process; the earliest appearance of the phrase is in a marginal note he made in Youatt's book on horses in March 1840 (Evans 1984).

The essential elements of the theory were worked out by 1842, when Darwin sat down to write a "pencil sketch." Here we see the domestication analogy, the mechanism of natural selection based upon the postulates of heritable variation and struggle for existence, and comments on the wide applicability of the theory in explaining patterns in diverse departments of natural history. The *Sketch,* as it is now called, was greatly expanded in 1844 to an *Essay* of some 230 pages. Darwin sealed a copy in an envelope with a letter to his wife, Emma, asking that she have it published immediately in the event of his sudden death. He knew that the naturalists of the world would wish to know about this theory. But of course Darwin did not die prematurely; despite chronic ill health he continued steadily to work on the theory as well as on side projects relevant to it. The *Essay* and the *Sketch* were later published by his son Francis as *The Foundations of the Origin of Species* (1909).

Why Darwin did not publish his theory as early as the mid-1840s has long been a matter of speculation. The time between more or less complete formulation of the theory and its publication is sometimes called the "delay years," but this period probably did not represent a delay at all. Darwin was determined to make a thoroughly researched and documented case for his theory, knowing full well the scientific world's extreme skepticism to transmutationist ideas. The very year his *Essay* was penned also saw the pillorying of the then-anonymous author of the sensational book *Vestiges of the Natural History of Creation,* a long transmutationist reverie that embraced an evolutionary vision of the universe, humanity, and everything in between (Secord 2000). Darwin was no metaphysical transmutationist; he aimed to present a scientific case so well argued and supported that it would at least have to be given a fair hearing. Books like *Vestiges* and even his own grandfather's *Zoonomia,* with its unique brand of fanciful transmutationist speculations, were criticized by Darwin and his circle for their unsubstantiated notions of transmutation, but owing to their popularity with a wide audience, these books paved the way for the broader acceptance of the later idea of organic change. At the time, though, with transmutationism represented by speculative philosophers on the one hand and the more sober but perhaps equally unsubstantiated theories of the brilliant French naturalist Jean-Baptiste Lamarck on the other, the elite English natural philosophers of the mid-nineteenth century would have nothing to do with such radical ideas.

During this time, Darwin tentatively reached out to a few friends and other correspondents with a view to sharing his ideas. He tested the waters for a certain openness of mind to cast an unprejudiced eye on his theory. Lyell passed the test. So did the botanists Asa Gray and Joseph Dalton Hooker. These naturalists became friendly critics and sounding boards as well as sources of invaluable experience and information as Darwin continued to develop his theory and struggle with its difficulties. In the mid-1840s Hooker commented to Darwin that anyone making pronouncements about the nature of species and varieties ought to become expert in some group to show he really knew what he was talking about—pay his dues, so to speak. Darwin took this advice to heart and initiated a monographic study of barnacles, a group rich in species and varieties, and conveniently amenable to analysis in the quiet of his study at Down House, Kent, where he had moved his family. Eight years and four monographs later (the last was published in 1854), Darwin could indeed claim intimate knowledge of the nature of species and varieties as manifest in this diverse group, yet the barnacles taught him much more besides: he was struck by the extreme sexual dimorphism of some species (with males so diminutive they were first thought to be tiny parasites adhering to the female), and the varied modifications of anatomical structures.

By the 1850s Darwin had amassed a vast amount of information. He had long since become something of a homebody, frequently indisposed by illness but patriarch of a happy and bustling household. He was far from isolated in his reluctance to travel, though; despite his bouts of illness the Darwins entertained frequently, and he often wrote several letters a day to friends, family, scientific correspondents, and others. The significance of these letters cannot be overstated: Darwin was at the center of a correspondence maelstrom, an almost frenetic give-and-take of information, specimens, books, and critiques of his ideas with hundreds of correspondents. Cambridge University Press has made these available in a magisterial set of annotated volumes, and more recently has put them online via the Darwin Correspondence Project (*www.darwinproject.ac.uk/*).

Darwin thrived in the tranquility of Down House, but the family was not insulated from the maladies of the times. Tragedy struck in 1851; the loss of the Darwins' ten-year-old daughter, Annie, to illness seemed to have taken an especially heavy toll on him, and some authors have suggested that this event extinguished whatever slim remnant of a belief in a personal god Darwin may have retained. Bitter with the loss, he wrote a loving memorial to her and buried himself in his work. Still deep in his barnacle project at the time, Darwin had a flash of insight in 1852, when his "principle of divergence" came to him—a critical piece of the puzzle that explained not just transmutation but diversification (Browne 1980). Darwin later wrote that he always remembered exactly where he was, riding in his carriage, when this insight came.

Into the 1850s there were endless difficulties to work out and applications of his theory to explore: the nature of sterility and interfertility, long-distance dispersal, patterns of species appearance and extinction in the fossil record, instinct and its relationship to habit, and on and on. He experimented and collected field data whenever he could, exploring question after question: How long could seeds remain viable in sea water? Could aquatic invertebrates be transported on the feet of waterfowl? What diversity of plant species can be found in a plot of turf of a given size? What is the rate of seedling mortality? Can simpler versions of even the most complex organs and behaviors be found? Darwin took up pigeon breeding in the mid-1850s, using the range of morphological and behavioral variation in this domesticated group to probe the limits of variability, developmental expression of traits, and of course to better understand the pigeon breeder's art and the power of cumulative selection. He was interested in plants, too, initiating in 1854 a study of the numbers of species in large vs. small plant genera, a project that mushroomed into an exhaustive analysis of the ratio of species to varieties in large and small genera based on no fewer than a dozen botanical manuals of flora around the world (Stauffer 1977, Browne 1980).

During the intervening years, Charles Lyell and Joseph Hooker had become Darwin's closest friends and confidants. Lyell himself was keenly interested in what was called the "species question." Wallace's 1855 paper "On the law which has regulated the introduction of new species," written from his base in Sarawak, Borneo, prompted Lyell to do two things: initiate a species notebook of his own (see Wilson 1970); and urge Darwin not to delay publishing his own ideas on species change. Lyell recognized the significance of Wallace's keen insights underpinning his Sarawak Law: "Every species has come into existence coincident both in time and space with a pre-existing closely allied species," Wallace wrote, something that "connects together and renders intelligible a vast number of independent and hitherto unexplained facts." This statement veritably screams transmutation to a modern reader with the benefit of hindsight, but Darwin was unfazed. The law did not mention natural selection nor expound a principle of divergence. Without a mechanism, Darwin felt that Wallace had little to nothing to share. But he had seriously underestimated Wallace's curiosity and creativity.

Darwin was deep into his species book (initiated at last in 1856 at Lyell's urging) when in June 1858 he opened a fateful package from Wallace with a letter and a manuscript entitled "On the tendency of varieties to depart indefinitely from the parental type." Posted from the tropical southeast Asian island Ternate in the Moluccas some months before, the manuscript had slowly wended its way to Darwin's study, where the previously unfazed Darwin was very fazed indeed. "Your words have come true with a vengeance that I [should] be forestalled," he wrote Lyell in a restrained but anguished letter. "So all my originality, whatever it may amount to, will be smashed." These events and the subsequent "delicate arrangement" that Lyell and Hooker orchestrated—a joint presentation to the Linnean Society of some of Darwin's priority-establishing writings together with Wallace's Ternate essay—all unfolded while Wallace remained in blissful (or tortured?) ignorance in the field. Their papers were read on July 1, 1858; Wallace received word much later and was elated to find himself held in high esteem by the eminent scientific men of London (Berry 2002).

Darwin's work was cut out for him. He could not possibly quickly complete the encyclopedic work over which he had been laboring, so he decided to abstract much of what he had written already and treat the remaining topics in an abbreviated manner. This was no simple task; by the time Wallace's package arrived Darwin had written a hefty manuscript of ten and a half chapters, a proper scientific treatise brimming with example after example supporting his arguments, backed by tables of data and numerous citations. In his Introduction to the draft "big species book," published as *Charles Darwin's Natural Selection*, Stauffer (1975) estimated that, on the basis of the word count of the surviving manuscript of eight and a half chapters (some of the original was later cannibalized for *Variation of Animals and Plants Under Domestication*), the final book would have had a word count of about 375,000. It pained Darwin to have to cut it down, in part because an abstract deviated from standard presentation in scientific books: "My work is now nearly finished," he wrote in the Introduction to the *Origin,* "but as it will take me two or three more years to complete it, and as my health is far from strong, I have been urged to publish this Abstract." He continues: "I can here give only the general conclusions at which I have arrived, with a few facts in illustration, but which, I hope, in most cases will suffice. No one can feel more sensible than I do of the necessity of hereafter publishing in detail all the facts, with references, on which my conclusions have been grounded; and I hope in a future work to do this."

Darwin wanted to make it abundantly clear to readers that he was presenting a mere abstract, and in fact his first title for the work was *An Abstract of an Essay on the Origin of Species and Varieties Through Natural Selection.* John Murray, his publisher, voiced concerns to Lyell about this reluctant-sounding title. Darwin acquiesced: "I am sorry about Murray objecting to term abstract as I look at it as only possible apology for not giving References & facts in full.—but I will defer to him & you" (*Correspondence* 7: 272). The manuscript was completed in nine months. One year and four months later, this "abstract" of 490 pages finally appeared, bearing the full title *On the Origin of Species by Means of Natural Selection, or the Preservation of Favoured Races in the Struggle for Life.* Darwin was more anxious than an expectant father, and he dreaded the coming attention and notoriety. As Janet Browne so compellingly describes in *The Power of Place* (2002), he had retreated to a hydropathic spa in Yorkshire, as much to escape the public eye as to seek treatment for his frazzled nerves, skin rashes, and churning stomach. He fired off letter after disarming letter to friends, colleagues, and would-be critics, hoping to soften the blow, blunt their resistance: "I fear . . . that you will not approve of your pupil . . . If you are in *ever so slight a degree* staggered (which I hardly expect) on the immutability of species, then I am convinced with further reflexion you will become more & more staggered, for this has been the process through which my mind has gone," he wrote to his old Cambridge mentor Henslow. "I know there will be much in it, which you object to . . . I am very far from expecting to convert you to many of my heresies," he told Thomas Henry Huxley in another letter; "my book will horrify & disgust you," he fretted to Thomas Eyton. His grimmest expression may have been sent to Hugh Falconer: "Lord how savage you will be . . . and how you will long to crucify me alive! I fear it will produce no other effect on you" (*Correspondence* 7: 350, 368, 370; Browne 2002, p. 84).

Darwin steeled himself and resolutely stood by his theory. A letter written to his American confidant Asa Gray nicely expresses this stand, as well as his look to the future: "I fully admit there are very many difficulties . . . but I cannot possibly believe that a false theory would explain so many classes of facts as I think it certainly does explain.— On these grounds I drop my anchor, & believe that the difficulties will slowly disappear" (*Correspondence* 7: 369). A copy of the book meanwhile was coursing its way to southeast Asia, arriving at Amboina, in the Moluccas, in February 1860. "God knows what the public will think," Darwin said in an accompanying letter. The public was in for a transmutation of its own. Though few realized it at the time, their world would never be the same.

Alfred Russel Wallace knew.

The *Origin* Revealed

At first glance the *Origin*'s narrative structure can be confusing —a seeming hodgepodge of topics. How can a single argument-cum-narrative unite the entire book? Historians and philosophers of science see consistency in the way Darwin presented his case, beginning with the 1842 *Sketch,* continuing through the 1844 *Essay, Natural Selection,* and the *Origin* itself (Hodge 1977, Ruse 1979, Waters 2003). There is a consensus that Darwin largely followed Herschelian logic in his scientific reasoning, the tenets for establishing *verae causae,* true causes, put forth by the eminent astronomer and mathematician Sir John Herschel. Herschel was something of a polymath who made contributions to astronomy, light theory, and the development of photography. He was conversant in geology, zoology, and botany as well, and achieved renown as a philosopher of science. His important philosophical work *Preliminary Discourse on the Study of Natural Philosophy* was published in 1831, the same year he was knighted. (Coincidentally, that was also the year the *Beagle* sailed from England with the eager young Darwin aboard; Darwin met Herschel while on the voyage, far from home in South Africa in 1836.)

In Herschel's philosophy of science, the case for a *vera causa* is best made if we can (1) identify the *existence* of a mechanism, (2) demonstrate (or persuasively argue for) its *adequacy* or *competence* in effecting the phenomenon of interest, and (3) show that this mechanism has wide explanatory power—*responsibility* for diverse observations. Herschelian responsibility is akin to the *vera causa* of another prominent philosopher of science in Darwin's day: William Whewell's *consilience of inductions,* a "jumping together" of inductive inferences from diverse and unrelated fields (Ruse 1979). According to the philosophical precepts of the day, demonstration of responsibility should be independent of the cases for existence and adequacy of the mechanism (Hodge 1977, 1992; Hull 2003; Waters 2003). These ideas provide a road map to the *Origin.* Darwin opens the book with a chapter on domestication, for viewing domestic varieties as an analog to natural varieties helps argue the existence and adequacy of cases for transmutation by natural selection. Darwin sees selection as the causal agent of change both under domestication and in nature: varieties of domesticated species like dogs, say, are genealogically related, developed by the cumulative effects of selective breeding (artificial selection) in a way that parallels the action of natural selection in promoting the formation of varieties (and ultimately species) in nature.

While chapter I presents the domestication analogy, chapters II–V set forth the logical argument for natural selection based on naturally occurring, heritable variation and differential survival and reproduction linked to this variation. These complete the existence and adequacy cases. Much of the rest of the book is aimed at addressing potential problems (chapters VI–VIII) and showing the wide applicability of the theory in explaining sets of observations from paleontology to biogeography to embryology (chapters IX–XIII), corresponding to the Herschelian responsibility case.

The *Origin* is not the first place Darwin pursued this logical structure. Scholars are essentially agreed that he did so from the earliest formulations of his theory. If we read Darwin's *Sketch* of 1842, *Essay* of 1844, and the *Origin* with an eye toward discerning the layout and presentation of these ideas, a general bipartite structure is evident in all these works. Francis Darwin divided his edition of the *Sketch* into Parts I and II, following his father's mention of two divisions: the domestication analogy is immediately followed by a case for natural variability, struggle, and natural selection. The remainder is largely devoted to applying the idea of transmutation mediated by natural selection to diverse observations. This structure parallels the two overarching arguments Darwin makes: natural selection as agent of change and the reality of transmutation in general.

If we delve a bit deeper, it becomes clear that a tripartite structure reflects even better the way Darwin conceptualized his argument. He describes a three-part layout in the Introduction to his 1868 book *Variation of Animals and Plants Under Domestication.* Significantly, that work was intended as the first of three books Darwin planned post-*Origin,* each of which represents a component of the logical argument. To be clear: having been forestalled in completing his big species book and publishing instead an abstracted form of this work as the *Origin,* Darwin was determined to publish more lengthy and thoroughly documented versions of his views. He decided this would best be done by presenting each of the three components of his conceptual argument as a book in its own right. Accordingly, the first two chapters of the big species book were expanded into the two *Variation* volumes. In the Introduction to *Variation* he states that in a second work he will expand his case for variation in nature, the struggle for existence, natural selection, and other difficulties. "In a third work," he continues, "I shall try the principle of natural selection by seeing how far it will give a fair explanation of the several classes of facts just alluded to": the responsibility case (Darwin 1868b, pp. 8, 9). Darwin never realized his plan for the two additional works—other books seemed to get in the way—so we have no such concise trio reflecting the way he conceptualized his theory and its presentation. Nonetheless, the three-part structure is a useful map to Darwin's argument and clearly reflects how Darwin thought about his theory from the earliest days. It is supported by an early outline, thought to be written about 1842, found later by his son Francis (see F. Darwin 1909, p. xviii):

I. The Principles of Var. in domestic organisms.
II. The possible and probable application of these same principles to wild animals and consequently the possible and probable production of wild races, analogous to the domestic ones of plants and animals.
III. The reasons for and against believing that such races really have been produced, forming what are called species.

This outline may have been intended as the layout for the 1842 *Sketch.* Comparing it with the three-part structure Darwin described in *Variation* some twenty-five years later, it is clear that the basic logical structure of his argument did not change appreciably over time. In some respects it had to change, of course, as Darwin's understanding of the process became more nuanced. What is remarkable, though, is the clarity of Darwin's earliest insight into the evolutionary process, and the fact that the conceptual structure he formulated soon after his return from the *Beagle* voyage has stood the test of time.

It is unfortunate, though, that Darwin did not expressly stick to this three-part scheme in all major presentations of his work. He had some tough decisions to make, in particular where and how to address difficulties. From early on (1842) Darwin anticipated and offered arguments to defuse certain problems that he knew would loom large in the minds of his critics. He decided to take these on immediately following the case for natural selection and before moving on to the chapters on applications of his theory. In the *Origin,* these "difficulties" treatments correspond to chapters VI–VIII.

Most authors take Darwin at his word when, in the opening paragraph of chapter VI ("Difficulties on Theory"), he identifies four issues he considered the most important explanatory obstacles faced by his theory and states that he will address the first two in that chapter and the next two in the subsequent chapters, VII ("Instinct") and VIII ("Hybridism"), respectively. In fact he goes beyond this. Chapter IX, "On the Imperfection of the Geological Record," is a treatment of yet another set of difficulties, albeit not those associated with the efficacy of natural selection. Chapter X on patterns of the fossil record ("On the Geological Succession of Organic Beings") can be seen as paired with IX, presenting the geological evidence favoring his theory to balance IX's geological evidence apparently contradicting his theory. I believe this is evident in the two biogeographical chapters as well, though Darwin is not explicit on this point. Chapter XI can be seen as addressing a difficulty in the geographical distribution of species; namely, by what process did they become distributed in different areas across the globe if not created there? Darwin then posits mechanisms facilitating the migration of species, all turning on such means of dispersal as wind, water, and animals, and the geological and climatic cycles that open up migration corridors. In chapter XII he shifts gears, more explicitly presenting his case for empirical observations of species distributions that support transmutation: for example, relationships of species on the same continent both with each other and with extinct forms on that continent, the fact that geography and not habitat or climate best predicts species relationships, and the striking "worlds within worlds" pattern that oceanic archipelagos manifest with their unique species that show affinities with species of the nearest mainland.

Comparing these with the other difficulties chapters, we can discern a consistency of approach: present the problem, then offer solutions. We see this in the treatments of lack of transitional forms and organs of extreme perfection (chapter VI), highly complex instincts and the special conundrum posed by sterile insect castes (chapter VII), and the problem of hybrid sterility and interfertility (chapter VIII). In each case Darwin's approach is to show that the difficulty is really a non-issue on closer inspection, or can be satisfactorily explained. In this light, chapters VI–XII fit into the general scheme. Admittedly, however, my scenario of Darwin's alternating difficulties with solutions seems to break down in chapter XIII, which consists almost entirely of positive arguments in favor of his theory. But this is unsurprising: Darwin saved his most potent arguments and observations for last, and this entire chapter has a triumphal momentum to it. Indeed, it is no coincidence that he ends his case with embryology and morphology, fields whose facts *alone,* he declares, would convince him of the correctness of his theory "even if it were unsupported by other facts or arguments" (*Origin,* p. 458). Now *that's* confidence.

Regardless of precisely how the *Origin* is philosophically dissected, Darwin's portrayal of the book as "one long argument" is certainly borne out. Whether seen as a unitary work or a work with bi- or tripartite arguments (and it is all of these), it stands as a masterly narrative, bold in its vision and ringing with passion.

Marcel Proust is credited with saying that "the real voyage of discovery consists not in seeking new landscapes but in having new eyes." Darwin managed both. His words in the concluding paragraphs of the *Origin* proved prophetic: "I look with confidence to the future, to young and rising naturalists," a new generation freer of the prejudices of education (!) and preconceived notions. His following did in fact increase with each generation, albeit often with an irritating (to Darwin) tendency to downplay natural selection or to invoke ill-defined metaphysical or supernatural notions like "tendency to complexity" or species senescence. Darwin's commitment to gradualistic transmutation mediated primarily, though not exclusively, by natural selection remained firm to the end. Time and again he would appeal to natural law in explaining species birth, death, and of course transmutation. We see this philosophy in the epigraphs Darwin chose to stand opposite his title page: quotes from Whewell and Bacon speaking, respectively, to the Divine acting through natural law, and to the wisdom of reading the *works* of the creator—the natural world—as attentively as the *words* of the creator.

By the close of Darwin's century, the concept of evolution by natural selection was inextricably tied to his name. According to the *Oxford English Dictionary* the word "Darwinism" first appeared in 1856, but in a novel, and in reference to Charles's remarkable grandfather Erasmus. In 1864 Huxley used the word to describe a new philosophy, that of Charles himself, and it has been linked to evolution by natural selection ever since. We honor the sagacious Alfred Russel Wallace for his own profound contributions to the field, especially his fully independent formulation of the theory, but the fact that Wallace chose to entitle his own definitive statement on the subject *Darwinism* (1889) shows that even he lionized his retiring colleague for his insights (and, by the way, gives lie to the persistent conspiracy theories that he was given short shrift by Darwin and his friends). A decade earlier Gray had set the precedent for eponymous titles honoring Darwin, titling his evolutionary book *Darwiniana: Essays and Reviews Pertaining to Darwinism* (1876).

The *Origin of Species* is epochal. Darwin's identification of a naturalistic explanation for species origins—for *our* origin—does not mean that his ideas are inherently atheistic, as legions of spiritually minded biologists from Gray on can attest. Yet make no mistake: his ideas are fundamentally incompatible with any literal reading of biblical scripture, and indeed with the creation narratives of any religion. Nowadays, biblical literalists (particularly but not exclusively in the United States) constitute a vocal and politically active minority, but the rest of us must be vigilant lest young-earth creationists and the neo-creationist "intelligent design" propagandists manage to legislate their way into science classrooms. Society should not countenance peddling such ideas under the guise of science—a common tactic recently and decisively repudiated by U.S. District Judge John E. Jones III in the Dover, Pennsylvania, intelligent design case.

In the grand journey of self-discovery that started with comprehension of the nature of our planet and star in the seventeenth century, Darwin's *Origin of Species* is one of humanity's crowning achievements. This living document opened windows on grand vistas extending back in Deep Time and sweeping forward to the uncertain future of the ever-ramifying tree of life—a profound step in self-awareness and self-understanding for a remarkable little primate. I share Darwin's exultation that "there is grandeur in this view of life." Carl Sagan once described humans as "star stuff come alive," the very universe aware of itself. That awareness underwent a quantum leap on November 24, 1859.

Galápagos Islands and
Quito, Ecuador
March 2008

The Annotated *Origin*

ON THE

ORIGIN OF SPECIES.

[1] "But with regard to the material world, we can at least go so far as this—we can perceive that events are brought about not by insulated interpositions of Divine power, exerted in each particular case, but by the establishment of general laws."

W. Whewell : *Bridgewater Treatise.*

[2] "To conclude, therefore, let no man out of a weak conceit of sobriety, or an ill-applied moderation, think or maintain, that a man can search too far or be too well studied in the book of God's word, or in the book of God's works; divinity or philosophy; but rather let men endeavour an endless progress or proficience in both."

Bacon : *Advancement of Learning.*

Down, Bromley, Kent,
October 1*st*, 1859.

1 William Whewell (1794–1866) authored the third of nine Bridgewater Treatises, entitling his work *On Astronomy and General Physics Considered with Reference to Natural Theology* (1833). Support for the treatises was established by the Rev. Francis Henry Egerton (1756–1829), the 8th Earl of Bridgewater. In his will the earl commissioned the Royal Society to oversee works "On the Power, Wisdom, and Goodness of God, as manifested in the Creation." Darwin read Whewell's treatise at least twice, the second time in 1840. The passage he quotes here declares forcefully that the creator acts through natural law—something Darwin argues for continually in the *Origin*—as opposed to constant miraculous interposition, such as through repeated creations of suites of species over time. Whewell may have felt this way about the laws of physics, but he was unwilling to apply the principle to species: to Darwin's disappointment, he remained opposed to his theory.

2 The second epigraph comes from Book One of Sir Francis Bacon's *Advancement of Learning* (1605). Bacon (1561–1626) advocated a new system of knowledge based on empirical and inductive principles that is sometimes now called the Baconian method. This passage does not so much echo as complement that of Whewell in embracing empiricism—the need to study the natural world firsthand—as on a par with studying scripture.

Beginning with the second edition of January 1860, Darwin added a third epigraph, taken from *Analogy of Religion, Natural and Revealed* (1736), by Bishop Joseph Butler (1692–1752): "The only distinct meaning of the word 'natural' is stated, fixed, or settled; since what is natural as much requires and presupposes an intelligent agent to render it so, i.e., to effect it continually or at stated times, as what is supernatural or miraculous does to effect it for once." Some readers took this as an admission of supernatural agency in establishing the world.

ON

THE ORIGIN OF SPECIES

BY MEANS OF NATURAL SELECTION,

OR THE

PRESERVATION OF FAVOURED RACES IN THE STRUGGLE FOR LIFE.

By CHARLES DARWIN, M.A.,

FELLOW OF THE ROYAL, GEOLOGICAL, LINNÆAN, ETC., SOCIETIES;
AUTHOR OF 'JOURNAL OF RESEARCHES DURING H. M. S. BEAGLE'S VOYAGE ROUND THE WORLD.'

LONDON:
JOHN MURRAY, ALBEMARLE STREET.
1859.

1 Darwin dropped the word "On" from the title in the sixth, final edition of 1872. Because that edition (more precisely, the corrected 1878 version) became the one most often republished after his death, many people are unaware of the original title. The reasons for the change are unclear; perhaps Darwin came to prefer the more declarative tone of *The Origin of Species.*

2 John Murray became Darwin's publisher on Charles Lyell's recommendation. Murray, an amateur geologist, was the publisher of Lyell's lucrative title *Principles of Geology,* among others, and had acquired the rights to publish Darwin's *Voyage of the Beagle* in 1845. Murray agreed to publish Darwin's work sight unseen, confident that it would be a hit. Darwin accepted his terms of publication, but also left Murray the option of rescinding his offer after seeing some of the manuscript. When the first few chapters arrived in April 1859, Murray did have concerns. Neither he nor his advisers understood Darwin, and they scrambled to change the book's course. Murray turned to the Rev. Whitwell Elwin, editor of the *Quarterly,* which Murray published. After talking with Lyell about the book, Elwin wrote to Darwin suggesting that, on Lyell's advice, he write a book about pigeons instead, and delay his species work until later! "This appears to me an admirable suggestion. Everybody is interested in pigeons," Elwin coaxed (*Correspondence* 7: 288–291). Darwin politely dismissed the suggestion, and fortunately Murray accepted his response and never looked back. Shrewd decision, as the *Origin* was to become quite a profitable book. It is often remarked that the book "sold out" its first printing on November 26, 1859, but the print run was only 1,250 copies—modest even by mid-nineteenth-century standards. Murray more than doubled that number to 3,000 copies the following month, with the publication of the second edition (which differed from the first only in having some corrections). For the next thirty years the book sold steadily, with new print runs every other year on average (Peckham 1959).

LONDON: PRINTED BY W. CLOWES AND SONS, STAMFORD STREET, AND CHARING CROSS.

CONTENTS.

CHAPTER I.

VARIATION UNDER DOMESTICATION.

CHAPTER II.

VARIATION UNDER NATURE.

CHAPTER III.

Struggle for Existence.

CHAPTER IV.

Natural Selection.

CHAPTER V.

Laws of Variation.

CHAPTER VI.

Difficulties on Theory.

CHAPTER VII. ◀1

Instinct.

CHAPTER VIII.

Hybridism.

1 Darwin made many modifications to the *Origin* in subsequent editions (see Peckham 1959 and the Coda), but the most significant structural change was the insertion of a new chapter here in the sixth edition of 1872. The new chapter VII, entitled "Miscellaneous Objections to the Theory of Natural Selection," assembled in one place Darwin's answers to the many different criticisms his theory received over time. "Instinct" and subsequent chapters were renumbered accordingly.

CHAPTER IX.

On the Imperfection of the Geological Record.

CHAPTER X.

On the Geological Succession of Organic Beings.

CHAPTER XI.

Geographical Distribution.

CHAPTER XII.

GEOGRAPHICAL DISTRIBUTION—*continued.*

CHAPTER XIII.

MUTUAL AFFINITIES OF ORGANIC BEINGS: MORPHOLOGY: EMBRYOLOGY: RUDIMENTARY ORGANS.

CHAPTER XIV.

RECAPITULATION AND CONCLUSION.

ON THE ORIGIN OF SPECIES.

INTRODUCTION.

WHEN on board H.M.S. 'Beagle,' as naturalist, I was much struck with certain facts in the distribution of the inhabitants of South America, and in the geological relations of the present to the past inhabitants of that continent. These facts seemed to me to throw some light on the origin of species—that mystery of mysteries, as it has been called by one of our greatest philosophers. On my return home, it occurred to me, in 1837, that something might perhaps be made out on this question by patiently accumulating and reflecting on all sorts of facts which could possibly have any bearing on it. After five years' work I allowed myself to speculate on the subject, and drew up some short notes; these I enlarged in 1844 into a sketch of the conclusions, which then seemed to me probable: from that period to the present day I have steadily pursued the same object. I hope that I may be excused for entering on these personal details, as I give them to show that I have not been hasty in coming to a decision.

My work is now nearly finished; but as it will take me two or three more years to complete it, and as my health is far from strong, I have been urged to publish this Abstract. I have more especially been induced to do this, as Mr. Wallace, who is now studying the

B

1 This brief introduction changed little over time. Beginning with the third edition (1861) it was prefaced with a new, longer section entitled "An Historical Sketch of the Recent Progress of Opinion on the Origin of Species," into which some of the material from the Introduction was moved. There Darwin traced the history of the idea of transmutation. He cited naturalists who had argued in one form or another for evolutionary change, gradual modification, or the unity of life, as well as those who had argued for the importance of invoking natural law in formulating scientific explanations or phenomena. Perhaps most important, Darwin concluded by citing contemporaries such as Hooker and Huxley who had publicly embraced his views.

2 At the time, Darwin was most struck by his fossil finds in South America—several extinct giant mammal species—and by the existence of two closely related rhea species with adjacent ranges. These and other observations all came together for him in the opening months of 1837; by mid-March he was convinced of the reality of transmutation. It was the astronomer Sir John Herschel who, in a letter to Charles Lyell written in 1836, referred to species origins as the "mystery of mysteries" (Cannon 1961).

3 In his autobiography Darwin wrote that his October 1838 reading of Malthus gave him the sudden insight into how natural selection works: "Here then I had at last got a theory by which to work," he wrote, "but I was so anxious to avoid prejudice, that I determined not for some time to write even the briefest sketch of it." The brief sketch was written in early 1842, following the five years' work he mentions here.

1 It is a remarkable quirk of history that Wallace sent his paper to Darwin first rather than to Lyell directly for publication. The matter-of-fact tone of this passage belies the shock and dismay Darwin must have felt upon reading Wallace's manuscript: "Your words have come true with a vengeance that I [should] be forestalled," he wrote to Lyell soon afterward. "I never saw a more striking coincidence. [If] Wallace had my M.S. sketch written out in 1842 he could not have made a better short abstract! Even his terms now stand as Heads of my Chapters" (*Correspondence* 7: 107).

Conspiracy theories abound that Darwin actually cribbed material from Wallace's manuscript before forwarding it to Lyell. It is likely that he did not immediately forward the manuscript as he said he did, and probably fretted over it for as long as two weeks, but it is abundantly clear from the documentary evidence (notebooks, diaries, letters, draft manuscripts) that Darwin had fully worked out the theory in all of its detail and nuance by that point and "stole" nothing from Wallace. For his part Wallace did not feel wronged in any way; on the contrary, he was delighted to see his paper presented alongside Darwin's "brief extracts" of priority-establishing writings.

2 Recall that by the time Wallace's manuscript arrived Darwin had written about ten and a half chapters of the book that was to be called *Natural Selection*. Too much remained to be covered for quick completion following the joint presentation of his and Wallace's Linnean Society papers on July 1, 1858, leading Darwin to "abstract" what he had written to that point and hastily draft brief treatments of the final chapters.

natural history of the Malay archipelago, has arrived at almost exactly the same general conclusions that I have on the origin of species. Last year he sent to me a memoir on this subject, with a request that I would forward it to Sir Charles Lyell, who sent it to the Linnean Society, and it is published in the third volume of the Journal of that Society. Sir C. Lyell and Dr. Hooker, who both knew of my work—the latter having read my sketch of 1844—honoured me by thinking it advisable to publish, with Mr. Wallace's excellent memoir, some brief extracts from my manuscripts.

This Abstract, which I now publish, must necessarily be imperfect. I cannot here give references and authorities for my several statements; and I must trust to the reader reposing some confidence in my accuracy. No doubt errors will have crept in, though I hope I have always been cautious in trusting to good authorities alone. I can here give only the general conclusions at which I have arrived, with a few facts in illustration, but which, I hope, in most cases will suffice. No one can feel more sensible than I do of the necessity of hereafter publishing in detail all the facts, with references, on which my conclusions have been grounded; and I hope in a future work to do this. For I am well aware that scarcely a single point is discussed in this volume on which facts cannot be adduced, often apparently leading to conclusions directly opposite to those at which I have arrived. A fair result can be obtained only by fully stating and balancing the facts and arguments on both sides of each question; and this cannot possibly be here done.

I much regret that want of space prevents my having the satisfaction of acknowledging the generous assistance which I have received from very many naturalists, some of them personally unknown to me. I cannot, however, let this opportunity pass without expressing my deep obligations to Dr. Hooker, who for the last fifteen years has aided me in every possible way by his large stores of knowledge and his excellent judgment.

In considering the Origin of Species, it is quite conceivable that a naturalist, reflecting on the mutual affinities of organic beings, on their embryological relations, their geographical distribution, geological succession, and other such facts, might come to the conclusion that each species had not been independently created, but had descended, like varieties, from other species. Nevertheless, such a conclusion, even if well founded, would be unsatisfactory, until it could be shown how the innumerable species inhabiting this world have been modified, so as to acquire that perfection of structure and coadaptation which most justly excites our admiration. Naturalists continually refer to external conditions, such as climate, food, &c., as the only possible cause of variation. In one very limited sense, as we shall hereafter see, this may be true; but it is preposterous to attribute to mere external conditions, the structure, for instance, of the woodpecker, with its feet, tail, beak, and tongue, so admirably adapted to catch insects under the bark of trees. In the case of the misseltoe, which draws its nourishment from certain trees, which has seeds that must be transported by certain birds, and which has flowers with separate sexes absolutely requiring the agency of certain insects to bring pollen from one flower to the other, it is equally preposterous to account for the structure of this parasite, with its relations to several distinct organic beings, by the effects of external conditions, or of habit, or of the volition of the plant itself. [1]

The author of the 'Vestiges of Creation' would, I presume, say that, after a certain unknown number of [2]

1 Ernst Mayr (1991) pointed out that Darwin's theory of species origins departed from those of his predecessors in that it was a theory of *common descent*, an arborescent pattern of ancestor-descendant relationship. Other models, like Lamarck's, included branching within limits but were in essence linear modes of change. Common descent at once made sense of the lines of evidence such as those listed here. "It was evident," Darwin wrote in his autobiography, "that such facts as these, as well as many others, could only be explained on the supposition that species gradually become modified; and the subject haunted me."

2 *Vestiges of the Natural History of Creation* was published anonymously in 1844. The author turned out to be the Scottish publisher Robert Chambers, but this did not come to light until 1884, well after his death. The wildly speculative brand of universal transmutation Chambers espoused was deemed scandalous; its reception, perhaps not coincidentally in the very year Darwin penned his lengthy species *Essay*, was likely one of the reasons Darwin decided to hold off on publishing his theory until he could marshal a thoroughly convincing case in accordance with the precepts of good science of the day. It had a rather different effect on Alfred Russel Wallace: his reading of *Vestiges* in 1845 immediately convinced him that species change. Writing to kindred spirit Henry Walter Bates, Wallace declared it "an ingenious hypothesis strongly supported by some striking facts and analogies, but which remains to be proved by more facts and the additional light which more research may throw upon the problem. It furnishes a subject for every observer of nature to attend to" (Berry 2002).

1 The brand of transmutationism put forth in *Vestiges* was saltational, with sudden transitions from one species to another. Very soon after his conversion to transmutation Darwin himself speculated that change was saltational. He soon gave up that idea for gradualism, and here he seems to suggest that the idea of one species springing forth from another, "produced perfect as we now see them," as if by fiat, is unacceptable as an explanation or scientific hypothesis. Saltational change had its adherents even in the post-*Origin* years, particularly in those who felt there was not time enough for gradualistic evolution to give rise to the diversity of life. Saltationism's most prominent exponent in the twentieth century was the German geneticist Richard Goldschmidt. He coined the term "hopeful monster" in his book *The Material Basis of Evolution* (1940) to describe the first individuals of a new species formed instantaneously, he posited, by mutations. The modern view holds that most evolutionary change is gradual, but that rates can vary considerably.

2 Domestication provided Darwin with his earliest insights into the mechanism of evolutionary change, so it is not surprising that he here asserts that domestication is the "best and safest clue" to shed light on the means of modification. His comment that the value of studying domestication has been "very commonly neglected by naturalists" is something of an understatement. Indeed, the fact that no *new* species have been produced by domestication was cited even by some of Darwin's friends and supporters as arguing against the ability of selection to effect species change.

3 From this statement it is clear why Darwin opened the *Origin* with a chapter on domestication. Domesticated varieties at once illustrate the point that virtually all characters exhibit hereditary variation, and steady accumulative selection can effect dramatic change. Domesticated species and varieties represent in microcosm what goes on in nature.

generations, some bird had given birth to a woodpecker, and some plant to the misseltoe, and that these had been produced perfect as we now see them; but this [1] assumption seems to me to be no explanation, for it leaves the case of the coadaptations of organic beings to each other and to their physical conditions of life, untouched and unexplained.

It is, therefore, of the highest importance to gain a clear insight into the means of modification and coadaptation. At the commencement of my observations it seemed to me probable that a careful study of domesticated animals and of cultivated plants would offer the best chance of making out this obscure problem. Nor have I been disappointed; in this and in all other perplexing cases I have invariably found that our [2] knowledge, imperfect though it be, of variation under domestication, afforded the best and safest clue. I may venture to express my conviction of the high value of such studies, although they have been very commonly neglected by naturalists.

[3] From these considerations, I shall devote the first chapter of this Abstract to Variation under Domestication. We shall thus see that a large amount of hereditary modification is at least possible; and, what is equally or more important, we shall see how great is the power of man in accumulating by his Selection successive slight variations. I will then pass on to the variability of species in a state of nature; but I shall, unfortunately, be compelled to treat this subject far too briefly, as it can be treated properly only by giving long catalogues of facts. We shall, however, be enabled to discuss what circumstances are most favourable to variation. In the next chapter the Struggle for Existence amongst all organic beings throughout the world, which inevitably follows from their high geometrical powers of

increase, will be treated of. This is the doctrine of Malthus, applied to the whole animal and vegetable kingdoms. As many more individuals of each species are born than can possibly survive; and as, consequently, there is a frequently recurring struggle for existence, it follows that any being, if it vary however slightly in any manner profitable to itself, under the complex and sometimes varying conditions of life, will have a better chance of surviving, and thus be *naturally selected.* From the strong principle of inheritance, any selected variety will tend to propagate its new and modified form.

This fundamental subject of Natural Selection will be treated at some length in the fourth chapter; and we shall then see how Natural Selection almost inevitably causes much Extinction of the less improved forms of life, and induces what I have called Divergence of Character. In the next chapter I shall discuss the complex and little known laws of variation and of correlation of growth. In the four succeeding chapters, the most apparent and gravest difficulties on the theory will be given: namely, first, the difficulties of transitions, or in understanding how a simple being or a simple organ can be changed and perfected into a highly developed being or elaborately constructed organ; secondly, the subject of Instinct, or the mental powers of animals; thirdly, Hybridism, or the infertility of species and the fertility of varieties when intercrossed; and fourthly, the imperfection of the Geological Record. In the next chapter I shall consider the geological succession of organic beings throughout time; in the eleventh and twelfth, their geographical distribution throughout space; in the thirteenth, their classification or mutual affinities, both when mature and in an embryonic condition. In the last chapter I shall give a

1 A *natural* process of selection: natural selection is the second of Darwin's two great insights in the *Origin.* His use of the word "natural" reflects his desire to underscore the thoroughly mechanistic, material nature of the process. The word "selection" caused him more difficulties, however, as we shall see in chapter IV. Although Darwin chose a term to contrast with the human artifice of effecting change by selective breeding (hence "artificial selection"), his use of the word "selection" in the context of nature nonetheless bore overtones of conscious agency, for which he was criticized.

1 "Mutual relations" speaks to ecology, to ecological interactions. Darwin is hinting here that an appreciation of the myriad subtle and not-so-subtle interactions between organisms is essential for an understanding of the nature of species and varieties: it is their adaptations that determine their "present welfare" and prospects for "future success and modification."

2 It may be difficult for the modern reader to appreciate the importance of this seemingly straightforward expression of Darwin's conviction that species are not immutable. Yet to deny species constancy was to deny the veracity of scripture and the social institutions that depended upon it. No wonder Darwin felt out prospective friends and confidants so gingerly. Writing to Joseph Hooker in January 1844, he seems to exhibit classic approach-avoidance behavior as he tentatively reveals where his thinking has tended:

> I was so struck with distribution of Galapagos organisms . . . & with the character of the American fossil mammifers . . . that I determined to collect blindly every sort of fact, which [could] bear any way on what are species.— I have read heaps of agricultural & horticultural books, & have never ceased collecting facts—At last gleams of light have come, & I am almost convinced (quite contrary to opinion I started with) that species are not (it is like confessing a murder) immutable. (*Correspondence* 3: 1)

3 Darwin resolutely defended natural selection despite the misgivings of even some of his supporters that it could do the job. (Recall that the *Origin* succeeded in converting most naturalists to transmutation in short order, but the mechanism of transmutation remained much debated.) He later accepted a limited role for other processes but was indignant when accused of maintaining that natural selection was the *sole* evolutionary process. Darwin pointed out that he was not absolutist about the role of natural selection, and this is clear from the careful wording of his last sentence. In the fifth *Origin* edition of 1869 he amended the wording to state that natural selection "has been the most important, but not the exclusive, means of modification" (Peckham 1959, p. 75). The additional comma draws attention to the phrase "but not the exclusive," emphasizing this point.

brief recapitulation of the whole work, and a few concluding remarks.

No one ought to feel surprise at much remaining as yet unexplained in regard to the origin of species and varieties, if he makes due allowance for our profound
[1] ignorance in regard to the mutual relations of all the beings which live around us. Who can explain why one species ranges widely and is very numerous, and why another allied species has a narrow range and is rare? Yet these relations are of the highest importance, for they determine the present welfare, and, as I believe, the future success and modification of every inhabitant of this world. Still less do we know of the mutual relations of the innumerable inhabitants of the world during the many past geological epochs in its history. Although much remains obscure, and will long remain obscure, I can entertain no doubt, after the most deliberate study and dispassionate judgment of which I am capable, that the view which most naturalists entertain, and which I formerly entertained—namely, that each species has been independently created—is erro-
[2] neous. I am fully convinced that species are not immutable; but that those belonging to what are called the same genera are lineal descendants of some other and generally extinct species, in the same manner as the acknowledged varieties of any one species are the descendants of that species. Furthermore, I am con-
[3] vinced that Natural Selection has been the main but not exclusive means of modification.

CHAPTER I.

VARIATION UNDER DOMESTICATION.

Causes of Variability—Effects of Habit—Correlation of Growth—Inheritance—Character of Domestic Varieties—Difficulty of distinguishing between Varieties and Species—Origin of Domestic Varieties from one or more Species—Domestic Pigeons, their Differences and Origin—Principle of Selection anciently followed, its Effects—Methodical and Unconscious Selection—Unknown Origin of our Domestic Productions—Circumstances favourable to Man's power of Selection.

WHEN we look to the individuals of the same variety or [1] sub-variety of our older cultivated plants and animals, one of the first points which strikes us, is, that they generally differ much more from each other, than do the individuals of any one species or variety in a state of nature. When we reflect on the vast diversity of the plants and animals which have been cultivated, and which have varied during all ages under the most different climates and treatment, I think we are driven to conclude that this greater variability is simply due to [2] our domestic productions having been raised under conditions of life not so uniform as, and somewhat different from, those to which the parent-species have been exposed under nature. There is, also, I think, some probability in the view propounded by Andrew Knight, that this [3] variability may be partly connected with excess of food. It seems pretty clear that organic beings must be exposed during several generations to the new conditions of life to cause any appreciable amount of variation; and that when the organisation has once begun to vary, [4] it generally continues to vary for many generations.

Darwin's main objective in this chapter is to argue for a genealogical pattern of relationship among domesticated varieties of a given group of plants or animals, and against the view that such varieties simply represent a like number of wild ancestors. Under the genealogical view, domesticated groups provide a powerful demonstration of what accumulative selection can achieve—a process analogous to, on a human scale, the action of natural selection in generating species diversity. Darwin termed the process of selective breeding in domestication "artificial" selection to distinguish it from the more "natural"—nonhuman–influenced—process that occurs in the wild.

The chapter also presents a case for variation sufficient to account for the remarkable diversification of organisms. Contrary to the view that organisms vary only to a certain degree, Darwin argues for continual variation in all points of structure and physiology, variation that is then added up over time.

---

1 In this curious opening sentence, Darwin asserts, counterintuitively, that individuals of certain long-domesticated varieties show greater variability than individuals of naturally occurring varieties. He is setting up a central question: *why* do cultivated varieties differ from natural ones in this way? This leads to an exploration of causes of variation, and then the significance of variation for Darwin's model of transmutation by natural selection.

2 Darwin invokes a role for environment in generating or influencing variation. Note the following statements that an excess of food, as well as exposure to new "conditions of life," plays a role in variation.

3 Knight was a leading English botanist and horticulturist, as well as a longtime president of the Horticultural Society of London. The idea referred to here concerns the role of nutrition in growth and variation in plants, the heritability of which is of vital concern to Darwin.

4 This "ripple effect" model suggests that variation may itself help engender further variation, again by contributing to the continued modification of the organism's "inner environment," generating variations by perturbing the reproductive system.

No case is on record of a variable being ceasing to be variable under cultivation. Our oldest cultivated plants, such as wheat, still often yield new varieties : our oldest domesticated animals are still capable of rapid improvement or modification.

It has been disputed at what period of life the causes of variability, whatever they may be, generally act; whether during the early or late period of development of the embryo, or at the instant of conception. Geoffroy [1] St. Hilaire's experiments show that unnatural treatment of the embryo causes monstrosities; and monstrosities cannot be separated by any clear line of distinction [2] from mere variations. But I am strongly inclined to suspect that the most frequent cause of variability may be attributed to the male and female reproductive elements having been affected prior to the act of conception. Several reasons make me believe in this; but the chief one is the remarkable effect which confinement or cultivation has on the functions of the reproductive system; this system appearing to be far more susceptible than any other part of the organisation, to the action of any change in the conditions of life. Nothing is more easy than to tame an animal, and few things more difficult than to get it to breed freely under confinement, even in the many cases when the male and female unite. How many animals there are which will not breed, though living long under not very close confinement in their native country! This is generally attributed to vitiated instincts; but how many cultivated plants display the utmost vigour, and yet rarely or never seed! In some few such cases it has been found out that very trifling changes, such as a little more or less water at some particular period of growth, will determine whether or not the plant sets a seed. I cannot here enter on the copious details which I have collected on

1 Darwin read the work of the celebrated French comparative anatomist Étienne Geoffroy St. Hilaire and his son Isidore. Both naturalists did extensive work on comparative anatomy, with Isidore specializing more in development. Here Darwin is probably referring to information in the *Histoire des Anomalies,* a three-volume treatise by Isidore published between 1832 and 1837, which Darwin recorded in his reading notebook in 1844. This work is also mentioned on p. 11.

2 This is a clear statement of Darwin's view on the origin of variability. Note again his assertion that the act of domestication ("confinement or cultivation") affects the reproductive system, which, Darwin contends in lines 20–22, is *far* more susceptible than *any* other part to the action of *any* change. This line of thought is pursued over the next couple of pages.

this curious subject; but to show how singular the laws are which determine the reproduction of animals under confinement, I may just mention that carnivorous animals, even from the tropics, breed in this country pretty freely under confinement, with the exception of the plantigrades or bear family; whereas, carnivorous birds, with the rarest exceptions, hardly ever lay fertile eggs. Many exotic plants have pollen utterly worthless, in the same exact condition as in the most sterile hybrids. When, on the one hand, we see domesticated animals and plants, though often weak and sickly, yet breeding quite freely under confinement; and when, on the other hand, we see individuals, though taken young from a state of nature, perfectly tamed, long-lived, and healthy (of which I could give numerous instances), yet having their reproductive system so seriously affected by unperceived causes as to fail in acting, we need not be surprised at this system, when it does act under confinement, acting not quite regularly, and producing offspring not perfectly like their parents or variable. 1

Sterility has been said to be the bane of horticulture; but on this view we owe variability to the same cause which produces sterility; and variability is the source of all the choicest productions of the garden. I may add, that as some organisms will breed most freely under the most unnatural conditions (for instance, the rabbit and ferret kept in hutches), showing that their reproductive system has not been thus affected; so will some animals and plants withstand domestication or cultivation, and vary very slightly—perhaps hardly more than in a state of nature.

A long list could easily be given of "sporting plants;" 2
by this term gardeners mean a single bud or offset, which suddenly assumes a new and sometimes very different character from that of the rest of the plant.

B 3

1 Here Darwin stresses that variants result from a perturbed reproductive system. Note, however, that he does not speculate as to how the reproductive elements are affected. He continually struggled with this question, and even his later pangenesis theory fell short of providing a satisfactory model.

2 In this paragraph Darwin asserts that "sports" in plants, which we now understand as spontaneous mutations in meristematic cells, are more common in cultivated species than in natural species. This notion is doubtful from our perspective today, but it was not uncommon in Darwin's time. In the E notebook, passages 13 and 142 (see Barrett et al. 1987), Darwin summarizes an 1820 paper on dahlias by Joseph Sabine in which the author suggests that a period of cultivation was necessary before the plants began to vary (sport) and produce the lovely cultivars we enjoy today. Why is this important to Darwin? If true, the theory would support his suggestion that the act of domestication somehow provokes the reproductive system of plants to generate variations.

1 The German physiologist Johannes Peter Müller authored an important physiological treatise translated as *Elements of Physiology,* a work that Darwin heavily annotated. Darwin discussed these same observations in *Variation of Animals and Plants under Domestication,* in the context of blending inheritance: "The dissimilarity of brothers or sisters of the same family, and of seedlings from the same capsule, may be in part accounted for by the unequal blending of the characters of the two parents" (Darwin 1868b, vol. II, p. 239).

2 Environmental conditions may play a small role in *directly* generating variation, in Darwin's view, but this statement shows that direct action of the environment takes a backseat to "the laws of reproduction . . . growth . . . and inheritance."

3 The "recent experiments" Darwin mentions here refer to James Buckman's investigations into the effects of abiotic factors (fertilizer, light, etc.) on plant growth. Buckman believed that varieties of agricultural plants were a result of differences in such factors during plant growth. As such, varieties would be completely changeable through alteration of growth conditions. Darwin fretted in his notes that Buckman's experiments were "hostile to that part of my theory which attributes so much to selection."

Such buds can be propagated by grafting, &c., and sometimes by seed. These "sports" are extremely rare under nature, but far from rare under cultivation; and in this case we see that the treatment of the parent has affected a bud or offset, and not the ovules or pollen. But it is the opinion of most physiologists that there is no essential difference between a bud and an ovule in their earliest stages of formation; so that, in fact, "sports" support my view, that variability may be largely attributed to the ovules or pollen, or to both, having been affected by the treatment of the parent prior to the act of conception. These cases anyhow show that variation is not necessarily connected, as some authors have supposed, with the act of generation.

Seedlings from the same fruit, and the young of the same litter, sometimes differ considerably from each other, though both the young and the parents, as Müller [1] has remarked, have apparently been exposed to exactly the same conditions of life; [2] and this shows how unimportant the direct effects of the conditions of life are in comparison with the laws of reproduction, and of growth, and of inheritance; for had the action of the conditions been direct, if any of the young had varied, all would probably have varied in the same manner. To judge how much, in the case of any variation, we should attribute to the direct action of heat, moisture, light, food, &c., is most difficult: my impression is, that with animals such agencies have produced very little direct effect, though apparently more in the case of plants. Under this point of view, Mr. Buckman's recent experiments [3] on plants seem extremely valuable. When all or nearly all the individuals exposed to certain conditions are affected in the same way, the change at first appears to be directly due to such conditions; but in some cases it can be shown that quite opposite conditions produce similar changes of structure. Nevertheless some slight amount of change may, I think, be attributed to the direct action of the conditions of life—as, in some cases, increased size from amount of food, colour from particular kinds of food and from light, and perhaps the thickness of fur from climate.

Habit also has a decided influence, as in the period of flowering with plants when transported from one climate to another. In animals it has a more marked effect; for instance, I find in the domestic duck that the bones of the wing weigh less and the bones of the leg more, in proportion to the whole skeleton, than do the same bones in the wild-duck; and I presume that this change may be safely attributed to the domestic duck flying much less, and walking more, than its wild parent. The great and inherited development of the udders in cows and goats in countries where they are habitually milked, in comparison with the state of these organs in other countries, is another instance of the effect of use. Not a single domestic animal can be named which has not in some country drooping ears; and the view suggested by some authors, that the drooping is due to the disuse of the muscles of the ear, from the animals not being much alarmed by danger, seems probable.

There are many laws regulating variation, some few of which can be dimly seen, and will be hereafter briefly mentioned. I will here only allude to what may be called correlation of growth. Any change in the embryo or larva will almost certainly entail changes in the mature animal. In monstrosities, the correlations between quite distinct parts are very curious; and many instances are given in Isidore Geoffroy St. Hilaire's great work on this subject. Breeders believe that long limbs are almost always accompanied by an elongated head. Some instances of correlation are quite whimsical: thus

1 Here the Lamarckian mechanism of use and disuse is invoked as a source of variation. Despite the ridicule Lamarck received from Lyell and others, Darwin did allow for a Lamarckian contribution to variation. It is easy to imagine that different rates of use or disuse of certain structures among individuals lead to differences in the appearance or extent of development of those structures (compare a weightlifter's musculature with that of a person who does not lift weights). The question is whether or to what extent such differences become heritable; in Darwin's day the idea that some bodily modifications could be passed on to offspring was still very much alive. In principle, then, a Lamarckian process of use and disuse could lead to heritable variation.

2 Correlation of traits—as when a modification in one part of the organism seems to be accompanied by certain modifications in another, seemingly unrelated, part—was a mysterious concept in Darwin's time, but it was duly registered as another factor underlying variation. The "great work" of Isidore Geoffroy St. Hilaire is the *Histoire des Anomalies* (Paris, 1832–1837).

1 Here Darwin tries to convey a sense of the magnitude of variation found in organisms: their differences in "endless points in structure and constitution" are the vital fuel for natural selection, his mechanism of evolution. Here we see, too, an example of Darwin's shrewd use of the published literature, a sort of appeal to authority as he invites the reader to inspect treatises on old cultivated plants. Why old plants? All the more time for their varieties to differ in myriad ways.

2 This sentence is important. Darwin may not know the actual basis for variations, but he realizes that the heritability of variation is critical for his mechanism of natural selection. Any variation that cannot be passed on to offspring cannot be selected for or against. Note the reiteration of the "endless" nature of variations in the second sentence.

3 Darwin is referring to the *Traité Philosophique et Physiologique de L'Hérédité Naturelle,* published by the French physician Prosper Lucas in 1847 and 1850.

cats with blue eyes are invariably deaf; colour and constitutional peculiarities go together, of which many remarkable cases could be given amongst animals and plants. From the facts collected by Heusinger, it appears that white sheep and pigs are differently affected from coloured individuals by certain vegetable poisons. Hairless dogs have imperfect teeth; long-haired and coarse-haired animals are apt to have, as is asserted, long or many horns; pigeons with feathered feet have skin between their outer toes; pigeons with short beaks have small feet, and those with long beaks large feet. Hence, if man goes on selecting, and thus augmenting, any peculiarity, he will almost certainly unconsciously modify other parts of the structure, owing to the mysterious laws of the correlation of growth.

▶1 The result of the various, quite unknown, or dimly seen laws of variation is infinitely complex and diversified. It is well worth while carefully to study the several treatises published on some of our old cultivated plants, as on the hyacinth, potato, even the dahlia, &c.; and it is really surprising to note the endless points in structure and constitution in which the varieties and subvarieties differ slightly from each other. The whole organisation seems to have become plastic, and tends to depart in some small degree from that of the parental type.

▶2 Any variation which is not inherited is unimportant for us. But the number and diversity of inheritable deviations of structure, both those of slight and those of considerable physiological importance, is endless. Dr.
▶3 Prosper Lucas's treatise, in two large volumes, is the fullest and the best on this subject. No breeder doubts how strong is the tendency to inheritance: like produces like is his fundamental belief: doubts have been thrown on this principle by theoretical writers alone. When a

deviation appears not unfrequently, and we see it in the father and child, we cannot tell whether it may not be due to the same original cause acting on both; but when amongst individuals, apparently exposed to the same conditions, any very rare deviation, due to some extraordinary combination of circumstances, appears in the parent—say, once amongst several million individuals—and it reappears in the child, the mere doctrine of chances almost compels us to attribute its reappearance to inheritance. Every one must have heard of cases of albinism, prickly skin, hairy bodies, &c., appearing in several members of the same family. If strange and rare deviations of structure are truly inherited, less strange and commoner deviations may be freely admitted to be inheritable. Perhaps the correct way of viewing the whole subject, would be, to look at the inheritance of every character whatever as the rule, and non-inheritance as the anomaly.

The laws governing inheritance are quite unknown; no one can say why the same peculiarity in different individuals of the same species, and in individuals of different species, is sometimes inherited and sometimes not so; why the child often reverts in certain characters to its grandfather or grandmother or other much more remote ancestor; why a peculiarity is often transmitted from one sex to both sexes, or to one sex alone, more commonly but not exclusively to the like sex. It is a fact of some little importance to us, that peculiarities appearing in the males of our domestic breeds are often transmitted either exclusively, or in a much greater degree, to males alone. A much more important rule, which I think may be trusted, is that, at whatever period of life a peculiarity first appears, it tends to appear in the offspring at a corresponding age, though sometimes earlier. In many cases this could

1 Having argued for a high degree of variability, and then pointing out that only heritable variation is important, Darwin now links the two by suggesting that most variation is heritable.

2 This is an understatement: the laws governing inheritance were a complete mystery to Darwin and his contemporaries. Mendel's ideas would not become generally known for over forty years, and his "particulate" model of inheritance would not be extended to complex traits (i.e., the polygenic model) until even later. In the meantime, some version of blending inheritance was the standard model, and Darwin put forth a rather comprehensive (and incorrect) version of his own particulate model (pangenesis) in *The Variation of Animals and Plants under Domestication* (1868).

3 "Reversion" is a confusing phenomenon under a blending inheritance model. If traits are blended in the reproductive process each generation, how can an earlier configuration reemerge?

4 This is another puzzling observation, but one that Darwin senses is significant. We can understand this phenomenon today in terms of our model of development as an unfolding stepwise process, like a software routine.

1 Again, Darwin perceives that this observation is significant—that when "peculiarities" appear in offspring they do so at the same developmental stage at which they appeared in the parent. Hence he considers this "rule" to be of the "highest importance" in embryology.

2 This paragraph, referring once again to reversion, is worth noting with respect to Alfred Russel Wallace's discussion of the subject in the opening paragraph of his 1858 Ternate paper on evolution. In that paper, Wallace criticizes the view that "reversion to parental type" (i.e., the tendency of domesticated varieties to revert to a more generalized ancestral form when they become feral) argues for constancy of species. To Wallace, domesticated organisms and their putative reversion are a red herring with respect to mutability or non-mutability of species in nature. He suggests that we cannot learn about species change in nature from domesticated organisms precisely because they are unnatural and unlikely even to survive in nature. Does Darwin have Wallace in mind here when he makes the similar statement that many of the most strongly differentiated varieties would have a hard time surviving in a state of nature? Darwin differs from Wallace, by contrast, in arguing that we can learn something from reversion in domesticated organisms: see how he makes use of this phenomenon in regard to pigeon crosses on p. 25.

not be otherwise: thus the inherited peculiarities in the horns of cattle could appear only in the offspring when nearly mature; peculiarities in the silkworm are known to appear at the corresponding caterpillar or cocoon stage. But hereditary diseases and some other facts make me believe that the rule has a wider extension, and that when there is no apparent reason why a peculiarity should appear at any particular age, yet that it does tend to appear in the offspring at the same period
1 at which it first appeared in the parent. I believe this rule to be of the highest importance in explaining the laws of embryology. These remarks are of course confined to the first *appearance* of the peculiarity, and not to its primary cause, which may have acted on the ovules or male element; in nearly the same manner as in the crossed offspring from a short-horned cow by a long-horned bull, the greater length of horn, though appearing late in life, is clearly due to the male element.

Having alluded to the subject of reversion, I may
2 here refer to a statement often made by naturalists—namely, that our domestic varieties, when run wild, gradually but certainly revert in character to their aboriginal stocks. Hence it has been argued that no deductions can be drawn from domestic races to species in a state of nature. I have in vain endeavoured to discover on what decisive facts the above statement has so often and so boldly been made. There would be great difficulty in proving its truth: we may safely conclude that very many of the most strongly-marked domestic varieties could not possibly live in a wild state. In many cases we do not know what the aboriginal stock was, and so could not tell whether or not nearly perfect reversion had ensued. It would be quite necessary, in order to prevent the effects of intercrossing, that only a

single variety should be turned loose in its new home. Nevertheless, as our varieties certainly do occasionally revert in some of their characters to ancestral forms, it seems to me not improbable, that if we could succeed in naturalising, or were to cultivate, during many generations, the several races, for instance, of the cabbage, in very poor soil (in which case, however, some effect would have to be attributed to the direct action of the poor soil), that they would to a large extent, or even wholly, revert to the wild aboriginal stock. Whether or not the experiment would succeed, is not of great importance for our line of argument; for by the experiment itself the conditions of life are changed. If it could be shown that our domestic varieties manifested a strong tendency to reversion,—that is, to lose their acquired characters, whilst kept under unchanged conditions, and whilst kept in a considerable body, so that free intercrossing might check, by blending together, any slight deviations of structure, in such case, I grant that we could deduce nothing from domestic varieties in regard to species. But there is not a shadow of evidence in favour of this view: to assert that we could not breed our cart and race-horses, long and short-horned cattle, and poultry of various breeds, and esculent vegetables, for an almost infinite number of generations, would be opposed to all experience. I may add, that when under nature the conditions of life do change, variations and reversions of character probably do occur; but natural selection, as will hereafter be explained, will determine how far the new characters thus arising shall be preserved.

When we look to the hereditary varieties or races of our domestic animals and plants, and compare them with species closely allied together, we generally perceive in each domestic race, as already remarked, less uniformity of character than in true species. Domestic races of

1 In this paragraph the significance (indeed the general occurrence) of reversion is challenged. Note that Darwin sets a high standard for rigorously showing reversion: the organisms must be kept under unchanging conditions, in large enough numbers to allow free intercrossing. Why does he devote so much space to this topic? Largely to dismiss the importance of reversion as a process in and of itself. Darwin tellingly mentions natural selection: it is natural selection that governs the fate of variants, as we will come to see, and if and when reversion occurs it is in response to that process.

1 The "one part" in which varieties often differ is—most tellingly—precisely that trait which has been selected . . .

2 . . . and such varieties present an analogy with congeneric species in nature.

3 This analogy provides a nice segue to "doubtful forms" and the case for an insensible gradation between species and varieties.

the same species, also, often have a somewhat monstrous character; by which I mean, that, although differing from each other, and from the other species of the same genus, in several trifling respects, they often differ in an extreme degree in some one part, both when compared one with another, and more especially when compared with all the species in nature to which they are nearest allied. With these exceptions (and with that of the perfect fertility of varieties when crossed,—a subject hereafter to be discussed), domestic races of the same species differ from each other in the same manner as, only in most cases in a lesser degree than, do closely-allied species of the same genus in a state of nature. I think this must be admitted, when we find that there are hardly any domestic races, either amongst animals or plants, which have not been ranked by some competent judges as mere varieties, and by other competent judges as the descendants of aboriginally distinct species. If any marked distinction existed between domestic races and species, this source of doubt could not so perpetually recur. It has often been stated that domestic races do not differ from each other in characters of generic value. I think it could be shown that this statement is hardly correct; but naturalists differ most widely in determining what characters are of generic value; all such valuations being at present empirical. Moreover, on the view of the origin of genera which I shall presently give, we have no right to expect often to meet with generic differences in our domesticated productions.

When we attempt to estimate the amount of structural difference between the domestic races of the same species, we are soon involved in doubt, from not knowing whether they have descended from one or several parent-species. This point, if it could be cleared up, would be interesting; if, for instance, it could be shown that the greyhound, bloodhound, terrier, spaniel, and bull-dog, which we all know propagate their kind so truly, were the offspring of any single species, then such facts would have great weight in making us doubt about the immutability of the many very closely allied and natural species—for instance, of the many foxes—inhabiting different quarters of the world. I do not believe, as we shall presently see, that all our dogs have descended from any one wild species; but, in the case of some other domestic races, there is presumptive, or even strong, evidence in favour of this view.

It has often been assumed that man has chosen for domestication animals and plants having an extraordinary inherent tendency to vary, and likewise to withstand diverse climates. I do not dispute that these capacities have added largely to the value of most of our domesticated productions; but how could a savage possibly know, when he first tamed an animal, whether it would vary in succeeding generations, and whether it would endure other climates? Has the little variability of the ass or guinea-fowl, or the small power of endurance of warmth by the rein-deer, or of cold by the common camel, prevented their domestication? I cannot doubt that if other animals and plants, equal in number to our domesticated productions, and belonging to equally diverse classes and countries, were taken from a state of nature, and could be made to breed for an equal number of generations under domestication, they would vary on an average as largely as the parent species of our existing domesticated productions have varied.

In the case of most of our anciently domesticated animals and plants, I do not think it is possible to come to any definite conclusion, whether they have descended from one or several species. The argument mainly relied on by those who believe in the multiple origin

1 Here is the analogy in a nutshell: if, Darwin rhetorically asserts, these various canine varieties could be shown to descend from a common ancestral canine species, we would have to doubt that species cannot change. Much of the remainder of this chapter will argue that (1) so-called doubtful forms are more the rule than the exception, and that (2) domesticated varieties of a given type are related by common descent, presenting a potent analogy for relationships between varieties and species, and species and genera, in a state of nature.

of our domestic animals is, that we find in the most ancient records, more especially on the monuments of Egypt, much diversity in the breeds; and that some of the breeds closely resemble, perhaps are identical with, those still existing. Even if this latter fact were found more strictly and generally true than seems to me to be the case, what does it show, but that some of our breeds originated there, four or five thousand years ago? But Mr. Horner's researches have rendered it in some degree probable that man sufficiently civilized to have manufactured pottery existed in the valley of the Nile thirteen or fourteen thousand years ago; and who will pretend to say how long before these ancient periods, savages, like those of Tierra del Fuego or Australia, who possess a semi-domestic dog, may not have existed in Egypt?

The whole subject must, I think, remain vague; neverthelsss, I may, without here entering on any details, state that, from geographical and other considerations, I think it highly probable that our domestic dogs have descended from several wild species. In regard to sheep and goats I can form no opinion. I should think, from facts communicated to me by Mr. Blyth, on the habits, voice, and constitution, &c., of the humped Indian cattle, that these had descended from a different aboriginal stock from our European cattle; and several competent judges believe that these latter have had more than one wild parent. With respect to horses, from reasons which I cannot give here, I am doubtfully inclined to believe, in opposition to several authors, that all the races have descended from one wild stock. Mr. Blyth, whose opinion, from his large and varied stores of knowledge, I should value more than that of almost any one, thinks that all the breeds of poultry have proceeded from the common wild

1 Leonard Horner (1785–1864) was a Scottish geologist and science education advocate. In 1855 he authored a paper on the geology of the Nile for which Darwin was referee.

2 Blyth, who curated the museum of the Asiatic Society of Bengal in Calcutta, India, furnished Darwin with a great deal of information about Indian flora and fauna. *Gallus bankiva* (more recently *Gallus gallus bankiva*) is the red jungle fowl of southeast Asia, the species from which the domestic chicken, *Gallus gallus domesticus*, is derived. The genus is taken from the Latin for chickens: *gallus* and *gallina* mean cock and hen, respectively.

Indian fowl (Gallus bankiva). In regard to ducks and rabbits, the breeds of which differ considerably from each other in structure, I do not doubt that they all have descended from the common wild duck and rabbit.

The doctrine of the origin of our several domestic races from several aboriginal stocks, has been carried to an absurd extreme by some authors. They believe that every race which breeds true, let the distinctive characters be ever so slight, has had its wild prototype. At this rate there must have existed at least a score of species of wild cattle, as many sheep, and several goats in Europe alone, and several even within Great Britain. One author believes that there formerly existed in Great Britain eleven wild species of sheep peculiar to it! When we bear in mind that Britain has now hardly one peculiar mammal, and France but few distinct from those of Germany and conversely, and so with Hungary, Spain, &c., but that each of these kingdoms possesses several peculiar breeds of cattle, sheep, &c., we must admit that many domestic breeds have originated in Europe; for whence could they have been derived, as these several countries do not possess a number of peculiar species as distinct parent-stocks? So it is in India. Even in the case of the domestic dogs of the whole world, which I fully admit have probably descended from several wild species, I cannot doubt that there has been an immense amount of inherited variation. Who can believe that animals closely resembling the Italian greyhound, the bloodhound, the bull-dog, or Blenheim spaniel, &c.—so unlike all wild Canidæ—ever existed freely in a state of nature? It has often been loosely said that all our races of dogs have been produced by the crossing of a few aboriginal species; but by crossing we can get only forms in some degree intermediate between their parents; and if we

1 While acknowledging that some groups of domesticated organisms probably have more than one progenitor, Darwin argues that many have descended from just one. Only as exemplars of his model of descent with modification do domesticated organisms provide the potent analogy that Darwin needs. Some scholars have suggested that Darwin overstated the claim that domestic varieties were typically viewed as having been derived from so many progenitor species. Many pigeon fanciers may have subscribed to this notion, but it was not the general opinion of naturalists. Buffon, in the eighteenth century, first proposed common ancestry for pigeon varieties. This became the accepted scientific view, even to the point that eminent naturalists like Yarrell and Temminck did not bother treating pigeon varieties in their works on ornithology because they viewed them as so many degenerate versions of the Rock Dove (Secord 1981).

2 Here is the crux of the issue: generating varieties by crossing only goes so far toward producing novel varieties. Something more, another process, is necessary to generate the wide range of varieties observed in many domesticated groups.

account for our several domestic races by this process, we must admit the former existence of the most extreme forms, as the Italian greyhound, bloodhound, bull-dog, &c., in the wild state. Moreover, the possibility of making distinct races by crossing has been greatly exaggerated. There can be no doubt that a race may be modified by occasional crosses, if aided by the careful selection of those individual mongrels, which present any desired character; but that a race could be obtained nearly intermediate between two extremely different races or speceies, I can hardly believe. Sir J. Sebright expressly experimentised for this object, and failed. The [1] offspring from the first cross between two pure breeds is tolerably and sometimes (as I have found with pigeons) extremely uniform, and everything seems simple enough; but when these mongrels are crossed one with another for several generations, hardly two of them will be alike, and then the extreme difficulty, or rather utter hopelessness, of the task becomes apparent. Certainly, a breed intermediate between *two very distinct* breeds could not be got without extreme care and long-continued selection; nor can I find a single case on record of a permanent race having been thus formed.

[2] *On the Breeds of the Domestic Pigeon.*—Believing that it is always best to study some special group, I have, after deliberation, taken up domestic pigeons. I have kept every breed which I could purchase or obtain, and have been most kindly favoured with skins from several quarters of the world, more especially by the Hon. W. Elliot from India, and by the Hon. C. Murray from Persia. Many treatises in different languages have been published on pigeons, and some of them are very important, as being of considerable antiquity. I have associated with several eminent fanciers, and have been permitted to join two

1 This cross seems to describe what we now call a classic Mendelian cross: two (distinct) parentals yield rather uniform F1s, which, when themselves crossed, yield highly variable F2s.

2 Pigeons provided a microcosm of Darwin's model of selection, as well as valuable data on development, correlation of traits, and reversion. Between 1855 and 1858 Darwin raised numerous varieties (peaking at ninety birds at one point) and obtained skins and skeletons of many others (Secord 1981). He joined several London pigeon clubs, including the exclusive Philoperisteron Society, which met in the Freemason's Tavern, and the Apiarian Society. The individuals mentioned here, Walter Elliot and Charles Augustus Murray, were especially helpful in procuring numerous specimens. Darwin found their variety astounding; Elliot sent him specimens of an Indian breed called the Lotan, or ground tumbler, for example, which when shaken a bit and placed on the ground commences tumbling around! Darwin commented in *Variation of Plants and Animals under Domestication* that this breed "present[s] one of the most remarkable inherited habits or instincts which have ever been recorded" (Darwin 1868b, vol. I, p. 150).

of the London Pigeon Clubs. The diversity of the breeds is something astonishing. Compare the English carrier and the short-faced tumbler, and see the wonderful difference in their beaks, entailing corresponding differences in their skulls. The carrier, more especially the male bird, is also remarkable from the wonderful development of the carunculated skin about the head, and this is accompanied by greatly elongated eyelids, very large external orifices to the nostrils, and a wide gape of mouth. The short-faced tumbler has a beak in outline almost like that of a finch; and the common tumbler has the singular and strictly inherited habit of flying at a great height in a compact flock, and tumbling in the air head over heels. The runt is a bird of great size, with long, massive beak and large feet; some of the sub-breeds of runts have very long necks, others very long wings and tails, others singularly short tails. The barb is allied to the carrier, but, instead of a very long beak, has a very short and very broad one. The pouter has a much elongated body, wings, and legs; and its enormously developed crop, which it glories in inflating, may well excite astonishment and even laughter. The turbit has a very short and conical beak, with a line of reversed feathers down the breast; and it has the habit of continually expanding slightly the upper part of the œsophagus. The Jacobin has the feathers so much reversed along the back of the neck that they form a hood, and it has, proportionally to its size, much elongated wing and tail feathers. The trumpeter and laugher, as their names express, utter a very different coo from the other breeds. The fantail has thirty or even forty tail-feathers, instead of twelve or fourteen, the normal number in all members of the great pigeon family; and these feathers are kept expanded, and are carried so erect that in good birds the head and tail

1 More than 200 pigeon varieties are recognized today, often bearing such fanciful names as Archangel, Show Homer, Nun, Jacobin, Pouter, Tumbler, and the Runt—oddly named, given that it is the largest domestic variety of pigeon, weighing in at up to 2.5 pounds! (Blechman 2006).

touch; the oil-gland is quite aborted. Several other less distinct breeds might have been specified.

1 In the skeletons of the several breeds, the development of the bones of the face in length and breadth and curvature differs enormously. The shape, as well as the breadth and length of the ramus of the lower jaw, varies in a highly remarkable manner. The number of the caudal and sacral vertebræ vary; as does the number of the ribs, together with their relative breadth and the presence of processes. The size and shape of the apertures in the sternum are highly variable; so is the degree of divergence and relative size of the two arms of the furcula. The proportional width of the gape of mouth, the proportional length of the eyelids, of the orifice of the nostrils, of the tongue (not always in strict correlation with the length of beak), the size of the crop and of the upper part of the œsophagus; the development and abortion of the oil-gland; the number of the primary wing and caudal feathers; the relative length of wing and tail to each other and to the body; the relative length of leg and of the feet; the number of scutellæ on the toes, the development of skin between the toes, are all points of structure which are variable. The period at which the perfect plumage is acquired varies, as does the state of the down with which the nestling birds are clothed when hatched. The shape and size of the eggs vary. The manner of flight differs remarkably; as does in some breeds the voice and disposition. Lastly, in certain breeds, the males and females have come to differ to a slight degree from each other.

2 Altogether at least a score of pigeons might be chosen, which if shown to an ornithologist, and he were told that they were wild birds, would certainly, I think, be ranked by him as well-defined species. Moreover, I do not believe that any ornithologist would place

1 This lengthy, detailed accounting of anatomical difference underscores Darwin's point about the ubiquity and degree of variation: virtually every anatomical detail varies.

2 This statement emphasizes the remarkable degree of morphological divergence among pigeon breeds: the power of artificial selection.

the English carrier, the short-faced tumbler, the runt, the barb, pouter, and fantail in the same genus; more especially as in each of these breeds several truly-inherited sub-breeds, or species as he might have called them, could be shown him.

Great as the differences are between the breeds of [1] pigeons, I am fully convinced that the common opinion of naturalists is correct, namely, that all have descended from the rock-pigeon (Columba livia), including under this term several geographical races or sub-species, which differ from each other in the most trifling respects. As several of the reasons which have led me to this belief are in some degree applicable in other cases, I will here briefly give them. If the several breeds are not varieties, [2] and have not proceeded from the rock-pigeon, they must have descended from at least seven or eight aboriginal stocks; for it is impossible to make the present domestic breeds by the crossing of any lesser number: how, for instance, could a pouter be produced by crossing two breeds unless one of the parent-stocks possessed the characteristic enormous crop? The supposed aboriginal stocks must all have been rock-pigeons, that is, not breeding or willingly perching on trees. But besides C. livia, with its geographical sub-species, only two or three other species of rock-pigeons are known; and these have not any of the characters of the domestic breeds. Hence the supposed aboriginal stocks must either still exist in the countries where they were originally domesticated, and yet be unknown to ornithologists; and this, considering their size, habits, and remarkable characters, seems very improbable; or they must have become extinct in the wild state. But birds breeding on precipices, and good fliers, are unlikely to be exterminated; and the common rock-pigeon, which has the same habits with the domestic breeds, has not been exterminated

1 This statement is important: pigeon divergence by human agency is an epitome of the natural transmutation process. Note, however, that the pattern of descent Darwin mentions here could mean that multiple independent varieties have all derived from the rock pigeon, or that a tree-like pattern of ancestor-descendant relationship exists among the varieties. He does not distinguish between these scenarios with respect to pigeons, but later it will become clear that he thinks natural selection drives a tree-like pattern of divergence. Here, Darwin is merely setting forth the idea that artificial selection has remarkable transformative powers.

2 The remainder of this page and the next are dedicated to arguing against the alternative hypothesis that the various pigeon varieties are each derived from a like number of wild progenitor species. Recall that most naturalists of Darwin's time agreed with him—Secord (1981) points out, for example, that pigeon varieties were of interest to naturalists insofar as they showed that the rock dove could be domesticated. Consider this comment from Darwin's correspondent in India, Edward Blyth: "The more I see of domestic Pigeons, the more obvious becomes the conclusion, according to my judgement, of their having all descended from the livia (or barely separable intermedia, &c); and I think the voice attests this, quite as decidedly as with domestic fowls. Some difference of voice there undoubtedly is, among the races; but what is that amount of difference, compared with the distinct coo of any really different species? It is indeed surprising what varieties of cooing there are among the various wild or true species of Columbiae; in all cases an unmistakeable Pigeon's voice, yet how different one species from another, & comparatively how very similar the coos of the different domestic races! Then their pairing so indiscriminately, however different the races!" (*Correspondence* 6: 67). The view was quite different in the world of fanciers. "My code of natural historical faith is this," wrote the Reverend Edmund Saul Dixon in his treatise on pigeon breeding, "that the domestic races of birds and animals are not developments, but creations" (Dixon 1851, pp. 78–79).

even on several of the smaller British islets, or on the shores of the Mediterranean. Hence the supposed extermination of so many species having similar habits with the rock-pigeon seems to me a very rash assumption. Moreover, the several above-named domesticated breeds have been transported to all parts of the world, and, therefore, some of them must have been carried back again into their native country; but not one has ever become wild or feral, though the dovecot-pigeon, which is the rock-pigeon in a very slightly altered state, has become feral in several places. Again, all recent experience shows that it is most difficult to get any wild animal to breed freely under domestication; yet on the hypothesis of the multiple origin of our pigeons, it must be assumed that at least seven or eight species were so thoroughly domesticated in ancient times by half-civilized man, as to be quite prolific under confinement.

An argument, as it seems to me, of great weight, and applicable in several other cases, is, that the above-specified breeds, though agreeing generally in constitution, habits, voice, colouring, and in most parts of their structure, with the wild rock-pigeon, yet are certainly highly abnormal in other parts of their structure: we may look in vain throughout the whole great family of Columbidæ for a beak like that of the English carrier, or that of the short-faced tumbler, or barb; for reversed feathers like those of the jacobin; for a crop like that of the pouter; for tail-feathers like those of the fantail. Hence it must be assumed not only that half-civilized man succeeded in thoroughly domesticating several species, but that he intentionally or by chance picked out extraordinarily abnormal species; and further, that these very species have since all become extinct or unknown. So many strange contingencies seem to me improbable in the highest degree.

Some facts in regard to the colouring of pigeons well deserve consideration. The rock-pigeon is of a slaty-blue, and has a white rump (the Indian sub-species, C. intermedia of Strickland, having it bluish) ; [1] the tail has a terminal dark bar, with the bases of the outer feathers externally edged with white ; the wings have two black bars ; some semi-domestic breeds and some apparently truly wild breeds have, besides the two black bars, the wings chequered with black. These several marks do not occur together in any other species of the whole family. Now, in every one of the domestic breeds, taking thoroughly well-bred birds, all the above marks, even to the white edging of the outer tail-feathers, sometimes concur perfectly developed. Moreover, when two birds belonging to two distinct breeds are crossed, neither of which is blue or has any of the above-specified marks, the mongrel offspring are very apt suddenly to acquire these characters ; for instance, I crossed some uniformly white fantails with some [2] uniformly black barbs, and they produced mottled brown and black birds ; these I again crossed together, and one grandchild of the pure white fantail and pure black barb was of as beautiful a blue colour, with the white rump, double black wing-bar, and barred and white-edged tail-feathers, as any wild rock-pigeon ! We can understand these facts, on the well-known principle of reversion to ancestral characters, if all the domestic breeds have descended from the rock-pigeon. But if we deny this, we must make one of the two following highly improbable suppositions. Either, firstly, that all the several imagined aboriginal stocks were coloured and marked like the rock-pigeon, although no other existing species is thus coloured and marked, so that in each separate breed there might be a tendency to revert to the very same colours and markings. Or, secondly,

1 Hugh Strickland (1811–1853) was a geologist and zoologist. In 1844 he named a new Indian pigeon species, *Columba intermedia,* which some naturalists believed was simply a variety of the common rock dove, *C. livia.* Strickland's species is recognized as a subspecies today: *C. livia intermedia.* Known as the blue rock pigeon, this subspecies is found in the lower Himalaya in northern India.

2 This is a classic Mendelian cross of parentals to obtain F1s, and cross of F1s to obtain F2s. Note the expected uniformity of F1s. There is no mention here of the range of variation in the F2s, but one resembles the rock pigeon in coloration. This apparent "reversion" to the coloration of its putative distant ancestor is taken as evidence of decent of the parental varieties from a common ancestor.

1 Here Darwin argues that ready interfertility is further evidence for the close affinity of the different pigeon breeds—making it even more unlikely that they were derived simply by crossing distinct species.

that each breed, even the purest, has within a dozen or, at most, within a score of generations, been crossed by the rock-pigeon: I say within a dozen or twenty generations, for we know of no fact countenancing the belief that the child ever reverts to some one ancestor, removed by a greater number of generations. In a breed which has been crossed only once with some distinct breed, the tendency to reversion to any character derived from such cross will naturally become less and less, as in each succeeding generation there will be less of the foreign blood; but when there has been no cross with a distinct breed, and there is a tendency in both parents to revert to a character, which has been lost during some former generation, this tendency, for all that we can see to the contrary, may be transmitted undiminished for an indefinite number of generations. These two distinct cases are often confounded in treatises on inheritance.

1 Lastly, the hybrids or mongrels from between all the domestic breeds of pigeons are perfectly fertile. I can state this from my own observations, purposely made on the most distinct breeds. Now, it is difficult, perhaps impossible, to bring forward one case of the hybrid offspring of two animals *clearly distinct* being themselves perfectly fertile. Some authors believe that long-continued domestication eliminates this strong tendency to sterility: from the history of the dog I think there is some probability in this hypothesis, if applied to species closely related together, though it is unsupported by a single experiment. But to extend the hypothesis so far as to suppose that species, aboriginally as distinct as carriers, tumblers, pouters, and fantails now are, should yield offspring perfectly fertile, *inter se*, seems to me rash in the extreme.

From these several reasons, namely, the improbability of man having formerly got seven or eight supposed species of pigeons to breed freely under domestication; these supposed species being quite unknown in a wild state, and their becoming nowhere feral; these species having very abnormal characters in certain respects, as compared with all other Columbidæ, though so like in most other respects to the rock-pigeon; the blue colour and various marks occasionally appearing in all the breeds, both when kept pure and when crossed; the mongrel offspring being perfectly fertile;—from these several reasons, taken together, I can feel no doubt that all our domestic breeds have descended from the Columba livia with its geographical sub-species.

In favour of this view, I may add, firstly, that C. livia, or the rock-pigeon, has been found capable of domestication in Europe and in India; and that it agrees in habits and in a great number of points of structure with all the domestic breeds. Secondly, although an English carrier or short-faced tumbler differs immensely in certain characters from the rock-pigeon, yet by comparing the several sub-breeds of these breeds, more especially those brought from distant countries, we can make an almost perfect series between the extremes of structure. Thirdly, those characters which are mainly distinctive of each breed, for instance the wattle and length of beak of the carrier, the shortness of that of the tumbler, and the number of tail-feathers in the fantail, are in each breed eminently variable; and the explanation of this fact will be obvious when we come to treat of selection. Fourthly, pigeons have been watched, and tended with the utmost care, and loved by many people. They have been domesticated for thousands of years in several quarters of the world; the earliest known record of pigeons is in the fifth Ægyptian dynasty, about 3000 B.C., as was pointed out to me by Professor Lepsius; but Mr. Birch informs me that pigeons are given in a bill

1 Here and in the following paragraph, Darwin reiterates his argument that pigeon varieties could not have arisen from multiple, separate, domestication events.

2 Karl Richard Lepsius (1810–1884) was a well-known Prussian Egyptologist and author; Samuel Birch (1813–1885) was an English archaeologist and keeper (curator) of Asian, British, and medieval antiquities in the British Museum.

of fare in the previous dynasty. In the time of the [1] Romans, as we hear from Pliny, immense prices were given for pigeons; "nay, they are come to this pass, that they can reckon up their pedigree and race." Pigeons were much valued by Akber Khan in India, about the year 1600; never less than 20,000 pigeons were taken [2] with the court. "The monarchs of Iran and Turan sent him some very rare birds;" and, continues the courtly historian, "His Majesty by crossing the breeds, which method was never practised before, has improved them astonishingly." About this same period the Dutch were as eager about pigeons as were the old Romans. The paramount importance of these considerations in explaining the immense amount of variation which pigeons have undergone, will be obvious when we treat of Selection. We shall then, also, see how it is that the breeds so often have a somewhat monstrous character. It is also a most favourable circumstance for the production of distinct breeds, that male and female pigeons can be easily mated for life; and thus different breeds can be kept together in the same aviary.

I have discussed the probable origin of domestic pigeons at some, yet quite insufficient, length; because when I first kept pigeons and watched the several kinds, [3] knowing well how true they bred, I felt fully as much difficulty in believing that they could ever have descended from a common parent, as any naturalist could in coming to a similar conclusion in regard to the many species of finches, or other large groups of birds, in nature. One circumstance has struck me much; namely, that all the breeders of the various domestic animals and the cultivators of plants, with whom I have ever conversed, or whose treatises I have read, are firmly convinced that the several breeds to which each has attended, are descended from so many aboriginally distinct species.

1 Darwin refers here to Pliny the Elder, Gaius Plinius Secundus, 23–79 CE, whose *Naturalis Historia* (Natural History) was for centuries the authoritative encyclopedia on the natural world. The following is Pliny's account of the value people placed on pigeons, from volume 10 of Philemon Holland's 1601 translation, pp. 270–309 (the same translation that Darwin read): "many men are growne now to cast a speciall affection and love to these birds: they build turrets above the tops of their houses for dovecotes. Nay they are come to this passe, that they can reckon up their pedigree and race, yea they can tell the verie places from whence this or that pigeon first came. And indeed one old example they follow of L. Axius a gentleman sometime of Rome, who before the civill warre with Pompey, sold every paire of pigeons for foure hundred deniers, as M. Varro doth report. True it is, that there goeth a great name of certaine countries where some of these pigeons are bred: for Campanie is voiced to yeeld the greatest and fairest bodied of all other places. To conclude, their manner of flying induceth and traineth me to thinke and write of the flight of other foules."

2 Darwin is quoting volume I of *The Ayeen Akbery, or, The Institutes of the Emperor Akber,* translated in 1783 by Francis Gladwin. This same quote is also given in *Variation,* but in that work Darwin adds that "Akber Khan possessed seventeen distinct kinds [of pigeons], eight of which were valuable for beauty alone" (Darwin 1868b, chapter VI).

3 Note this comparison, made in passing, of a complex of varieties to a complex of species in the same genus. This is, of course, where the argument is going.

Ask, as I have asked, a celebrated raiser of Hereford cattle, whether his cattle might not have descended from long-horns, and he will laugh you to scorn. I have never met a pigeon, or poultry, or duck, or rabbit fancier, who was not fully convinced that each main breed was descended from a distinct species. Van Mons, in his treatise on pears and apples, shows how utterly he disbelieves that the several sorts, for instance a Ribston-pippin or Codlin-apple, could ever have proceeded from the seeds of the same tree. Innumerable other examples could be given. The explanation, I think, is simple: from long-continued study they are strongly impressed with the differences between the several races; and though they well know that each race varies slightly, for they win their prizes by selecting such slight differences, yet they ignore all general arguments, and refuse to sum up in their minds slight differences accumulated during many successive generations. May not those naturalists who, knowing far less of the laws of inheritance than does the breeder, and knowing no more than he does of the intermediate links in the long lines of descent, yet admit that many of our domestic races have descended from the same parents—may they not learn a lesson of caution, when they deride the idea of species in a state of nature being lineal descendants of other species?

Selection.—Let us now briefly consider the steps by which domestic races have been produced, either from one or from several allied species. Some little effect may, perhaps, be attributed to the direct action of the external conditions of life, and some little to habit; but he would be a bold man who would account by such agencies for the differences of a dray and race horse, a greyhound and bloodhound, a carrier and tumbler pigeon. One of the most remarkable features in our domesticated races

1 Jean Baptiste Van Mons (1765–1842) was a Belgian pharmacist and physician who achieved renown as a pomologist. Darwin refers here to Van Mons's 1836 treatise *Arbres Fruitiers ou Pomologie Belge Expérimentale et Raisonnée.*

2 Darwin seems frustrated, even exasperated, at breeders' seemingly blinkered view of their breeds. They are fixated on the differences their breeding efforts yield, yet they cannot seem to recall the innumerable baby steps the breeds took in diverging during the long selective-breeding process. By extension, Darwin is frustrated with naturalists, too. He admonishes them: once they accept the idea that varieties are related by common descent, they will be duly chastened and will think twice before deriding his idea that species can descend from other species.

3 Having argued for the idea of common descent of breeds, Darwin turns to the nuts-and-bolts mechanism in this section. The concluding sentence of this paragraph, on the next page, is a lucid statement of artificial selection.

is that we see in them adaptation, not indeed to the animal's or plant's own good, but to man's use or fancy. Some variations useful to him have probably arisen suddenly, or by one step; many botanists, for instance, believe that the fuller's teazle, with its hooks, which cannot be rivalled by any mechanical contrivance, is only a variety of the wild Dipsacus; and this amount of change may have suddenly arisen in a seedling. So it has probably been with the turnspit dog; and this is known to have been the case with the ancon sheep. But when we compare the dray-horse and race-horse, the dromedary and camel, the various breeds of sheep fitted either for cultivated land or mountain pasture, with the wool of one breed good for one purpose, and that of another breed for another purpose; when we compare the many breeds of dogs, each good for man in very different ways; when we compare the game-cock, so pertinacious in battle, with other breeds so little quarrelsome, with "everlasting layers" which never desire to sit, and with the bantam so small and elegant; when we compare the host of agricultural, culinary, orchard, and flower-garden races of plants, most useful to man at different seasons and for different purposes, or so beautiful in his eyes, we must, I think, look further than to mere variability. We cannot suppose that all the breeds were suddenly produced as perfect and as useful as we now see them; indeed, in several cases, we know that this has not been their history. The key is man's power of accumulative selection: nature gives successive variations; man adds them up in certain directions useful to him. In this sense he may be said to make for himself useful breeds.

The great power of this principle of selection is not hypothetical. It is certain that several of our eminent breeders have, even within a single lifetime, modified to

1 Teasels (or Teazles, in the nineteenth century) are tall Old World herbaceous plants of the genus *Dipsacus*, family Dipsacaceae. The common Wild Teasel *(Dipsacus sylvestris)* is a weedy plant now found worldwide. Fuller's Teasel *(Dipsacus fullonum sativus)* was highly useful in the textile trade ("fuller" refers, of course, to one who fulls cloth). Teasel flowers are borne in dense and rather spiny heads, and the exceptionally well developed flower head hooks of fuller's teasel were used as a natural comb in textile processing throughout the nineteenth century, until they were replaced by metal combs. It is possible that this trait arose suddenly, as a sport.

a large extent some breeds of cattle and sheep. In order fully to realise what they have done, it is almost necessary to read several of the many treatises devoted to this subject, and to inspect the animals. Breeders habitually speak of an animal's organisation as something quite plastic, which they can model almost as they please. If I had space I could quote numerous passages to this effect from highly competent authorities. Youatt, who was probably better acquainted with the works of agriculturists than almost any other individual, and who was himself a very good judge of an animal, speaks of the principle of selection as "that which enables the agriculturist, not only to modify the character of his flock, but to change it altogether. It is the magician's wand, by means of which he may summon into life whatever form and mould he pleases." Lord Somerville, speaking of what breeders have done for sheep, says:—"It would seem as if they had chalked out upon a wall a form perfect in itself, and then had given it existence." That most skilful breeder, Sir John Sebright, used to say, with respect to pigeons, that "he would produce any given feather in three years, but it would take him six years to obtain head and beak." In Saxony the importance of the principle of selection in regard to merino sheep is so fully recognised, that men follow it as a trade: the sheep are placed on a table and are studied, like a picture by a connoisseur; this is done three times at intervals of months, and the sheep are each time marked and classed, so that the very best may ultimately be selected for breeding.

What English breeders have actually effected is proved by the enormous prices given for animals with a good pedigree; and these have now been exported to almost every quarter of the world. The improvement is by no means generally due to crossing different breeds;

1 Darwin read extensively in animal husbandry and breeding. Here he cites some eminent authorities on the subject: William Youatt (1776–1847) was a London veterinarian and author of several books on animal breeding and husbandry. Darwin heavily annotated a copy of Youatt's book *Cattle: Their Breeds, Management, and Diseases* (1834). Lord John Southey Somerville (1765–1819) was a pioneer sheep breeder known especially for the Merino breed. Sir John Sebright (1767–1846) was a renowned breeder of pigeons.

all the best breeders are strongly opposed to this practice, except sometimes amongst closely allied sub-breeds. And when a cross has been made, the closest selection is far more indispensable even than in ordinary cases. If selection consisted merely in separating some very distinct variety, and breeding from it, the principle would be so obvious as hardly to be worth notice; but its importance consists in the great effect produced by the accumulation in one direction, during successive generations, of differences absolutely inappreciable by an uneducated eye—differences which I for one have vainly attempted to appreciate. Not one man in a thousand has accuracy of eye and judgment sufficient to become an eminent breeder. If gifted with these qualities, and he studies his subject for years, and devotes his lifetime to it with indomitable perseverance, he will succeed, and may make great improvements; if he wants any of these qualities, he will assuredly fail. Few would readily believe in the natural capacity and years of practice requisite to become even a skilful pigeon-fancier.

The same principles are followed by horticulturists; but the variations are here often more abrupt. No one supposes that our choicest productions have been produced by a single variation from the aboriginal stock. We have proofs that this is not so in some cases, in which exact records have been kept; thus, to give a very [1] trifling instance, the steadily-increasing size of the common gooseberry may be quoted. We see an astonishing improvement in many florists' flowers, when the flowers of the present day are compared with drawings made only twenty or thirty years ago. When a race of plants is once pretty well established, the seed-raisers do not pick out the best plants, but merely go over their seed-beds, and pull up the "rogues," as they call the plants that deviate from the proper standard. With animals this

1 Gooseberries, genus *Ribes* (family Grossulariaceae) are widely distributed throughout the northern hemisphere. The horticultural gooseberry is *Ribes uva-crispa.* Horticulturists D. L. Barney and K. E. Hummer (2005, p. 20) write of this species: "For centuries, gooseberry breeding efforts were primarily devoted to increasing fruit size. Beginning with an average fruit weight of about 0.3 ounces (7.1 g) in the wild species, by 1852 gooseberry enthusiasts in England had achieved a record-setting berry weight of 1.9 ounces (53.9 g). This works out to an increase in fruit size of nearly 800 percent and a berry almost two inches (5 cm) in diameter."

kind of selection is, in fact, also followed; for hardly any one is so careless as to allow his worst animals to breed.

In regard to plants, there is another means of observing the accumulated effects of selection—namely, by comparing the diversity of flowers in the different varieties of the same species in the flower-garden; the diversity of leaves, pods, or tubers, or whatever part is valued, in the kitchen-garden, in comparison with the flowers of the same varieties; and the diversity of fruit of the same species in the orchard, in comparison with the leaves and flowers of the same set of varieties. See how different the leaves of the cabbage are, and how extremely alike the flowers; how unlike the flowers of the heartsease are, and how alike the leaves; how much the fruit of the different kinds of gooseberries differ in size, colour, shape, and hairiness, and yet the flowers present very slight differences. It is not that the varieties which differ largely in some one point do not differ at all in other points; this is hardly ever, perhaps never, the case. The laws of correlation of growth, the importance of which should never be overlooked, will ensure some differences; but, as a general rule, I cannot doubt that the continued selection of slight variations, either in the leaves, the flowers, or the fruit, will produce races differing from each other chiefly in these characters.

It may be objected that the principle of selection has been reduced to methodical practice for scarcely more than three-quarters of a century; it has certainly been more attended to of late years, and many treatises have been published on the subject; and the result, I may add, has been, in a corresponding degree, rapid and important. But it is very far from true that the principle is a modern discovery. I could give several references to the full acknowledgment of the importance of the principle in works of high antiquity. In rude and

1 The diversity of precisely those plant structures or features that are "valued" or of use to us speaks to the power of artificial selection. Apple *fruits* of different varieties, say, may differ considerably from one another, while the *leaves* of those same varieties are highly similar. Selective breeding has focused on the former, not the latter.

barbarous periods of English history choice animals were often imported, and laws were passed to prevent their exportation: the destruction of horses under a certain size was ordered, and this may be compared to the "roguing" of plants by nurserymen. The principle of selection I find distinctly given in an ancient Chinese encyclopædia. Explicit rules are laid down by some of the Roman classical writers. From passages in Genesis, it is clear that the colour of domestic animals was at that early period attended to. Savages now sometimes cross their dogs with wild canine animals, to improve the breed, and they formerly did so, as is attested by passages in Pliny. The savages in South Africa match their draught cattle by colour, as do some of the Esquimaux their teams of dogs. Livingstone shows how much good domestic breeds are valued by the negroes of the interior of Africa who have not associated with Europeans. Some of these facts do not show actual selection, but they show that the breeding of domestic animals was carefully attended to in ancient times, and is now attended to by the lowest savages. It would, indeed, have been a strange fact, had attention not been paid to breeding, for the inheritance of good and bad qualities is so obvious.

At the present time, eminent breeders try by methodical selection, with a distinct object in view, to make a new strain or sub-breed, superior to anything existing in the country. But, for our purpose, a kind of Selection, which may be called Unconscious, and which results from every one trying to possess and breed from the best individual animals, is more important. Thus, a man who intends keeping pointers naturally tries to get as good dogs as he can, and afterwards breeds from his own best dogs, but he has no wish or expectation of permanently altering the breed. Nevertheless I cannot

1 Observations and advice on the proper breeding of animals are found throughout classical literature. Here Darwin mentions only generically the Bible and ancient Chinese and Roman sources, but in *Variation* he dedicates four and a half pages to the subject, citing Plato, Virgil, Homer, Alexander, Tacitus, Charlemagne, Akbar Khan, Jesuitic works on ancient China, and more.

2 Scottish missionary David Livingstone (1813–1873) achieved renown for his explorations of the African interior. In his reading journal, Darwin declared Livingstone's *Missionary Travels and Researches in South Africa* (1857) "the best travels I ever read" (Vorzimmer 1977).

doubt that this process, continued during centuries, would improve and modify any breed, in the same way as Bakewell, Collins, &c., by this very same process, only carried on more methodically, did greatly modify, even during their own lifetimes, the forms and qualities of their cattle. Slow and insensible changes of this kind could never be recognised unless actual measurements or careful drawings of the breeds in question had been made long ago, which might serve for comparison. In some cases, however, unchanged or but little changed individuals of the same breed may be found in less civilised districts, where the breed has been less improved. There is reason to believe that King Charles's spaniel has been unconsciously modified to a large extent since the time of that monarch. Some highly competent authorities are convinced that the setter is directly derived from the spaniel, and has probably been slowly altered from it. It is known that the English pointer has been greatly changed within the last century, and in this case the change has, it is believed, been chiefly effected by crosses with the fox-hound; but what concerns us is, that the change has been effected unconsciously and gradually, and yet so effectually, that, though the old Spanish pointer certainly came from Spain, Mr. Borrow has not seen, as I am informed by him, any native dog in Spain like our pointer.

By a similar process of selection, and by careful training, the whole body of English racehorses have come to surpass in fleetness and size the parent Arab stock, so that the latter, by the regulations for the Goodwood Races, are favoured in the weights they carry. Lord Spencer and others have shown how the cattle of England have increased in weight and in early maturity, compared with the stock formerly kept in this country. By comparing the accounts given in old pigeon treatises of carriers

1 Robert Bakewell (1725–1795) was a leading English cattle and sheep breeder responsible for developing or improving several important breeds. Darwin perceived that Bakewell's revolutionary breeding method involved careful within-breed crossing and selection, a departure from the widespread practice of attempting to improve breeds by crossing females with males of a different breed (Wood 1973). The "Collins" Darwin mentions here has been something of a mystery for many years. A recent study (Ogawa 2001) argues that Darwin made an error that persisted through all six editions of the *Origin:* the person in question is actually Charles Colling (1751–1836), a well-known breeder of shorthorn cattle. Interestingly, though, the agriculturist William Marshall refers to a sheep breeder named "Mr. Collins" in his two-volume *Rural Economy of Yorkshire,* a book that Darwin read.

2 The key point in this section can be summarized in the phrase "slow and insensible changes." The methodical selection exerted by the "eminent breeders" Darwin refers to change the breeds imperceptively slowly—insensibly.

3 King Charles's spaniel is an English dog breed named in honor of King Charles II, who nearly always traveled in the company of several "toy" spaniels. Did the astronomer Sir Edmond Halley have this in mind when in 1725 he named the brightest star in the hunting dog constellation Canes Venatici "Cor Caroli," or Charles's heart, in memory of Charles? Legend has it that the star seemed especially brilliant on the eve of the king's restoration to the throne on May 29, 1660 (Allen 1963). To all who knew him, his heart certainly seemed to lie with his cherished dogs.

4 George Henry Borrow (1803–1881) was a traveler and writer famed for his books on Romany (Gypsy) culture and language (though Darwin was not very favorably impressed with Borrow's 1851 book *Lavengro; the Scholar, the Gypsy, the Priest,* which in his reading notebooks he called "rubbish yet amusing").

5 The celebrated Goodwood horseracing track, or course, in West Sussex, England, was established by the Duke of Richmond on his estate in 1802 and is still an important racetrack today.

and tumblers with these breeds as now existing in Britain, India, and Persia, we can, I think, clearly trace the stages through which they have insensibly passed, and come to differ so greatly from the rock-pigeon.

[1] Youatt gives an excellent illustration of the effects of a course of selection, which may be considered as unconsciously followed, in so far that the breeders could never have expected or even have wished to have produced the result which ensued—namely, the production of two distinct strains. The two flocks of Leicester sheep kept by Mr. Buckley and Mr. Burgess, as Mr. Youatt remarks, "have been purely bred from the original stock of Mr. Bakewell for upwards of fifty years. There is not a suspicion existing in the mind of any one at all acquainted with the subject that the owner of either of them has deviated in any one instance from the pure blood of Mr. Bakewell's flock, and yet the difference between the sheep possessed by these two gentlemen is so great that they have the appearance of being quite different varieties."

If there exist savages so barbarous as never to think of the inherited character of the offspring of their domestic animals, yet any one animal particularly useful to them, for any special purpose, would be carefully preserved during famines and other accidents, to which savages are so liable, and such choice animals would thus generally leave more offspring than the inferior ones; so that in this case there would be a kind of unconscious selection going on. [2] We see the value set on animals even by the barbarians of Tierra del Fuego, by their killing and devouring their old women, in times of dearth, as of less value than their dogs.

In plants the same gradual process of improvement, through the occasional preservation of the best individuals, whether or not sufficiently distinct to be ranked

1 A telling observation: the Buckley and Burgess flocks are demonstrably different from their known Bakewell common ancestor—in just fifty years, independent selection of these lineages had resulted in their divergence. The quotation is taken from p. 315 of Youatt's book *On Sheep* (London, 1838).

2 In chapter X of *Voyage of the Beagle,* Darwin recounted chilling stories of cannibalism by the Fuegians: "it is certainly true, that when pressed in winter by hunger they kill and devour their old women before they kill their dogs: [a native] boy, being asked by Mr. Low why they did this, answered, 'Doggies catch otters, old women no.' This boy described the manner in which they are killed by being held over smoke and thus choked; he imitated their screams as a joke, and described the parts of their bodies which are considered best to eat. Horrid as such a death by the hands of their friends and relatives must be, the fears of the old women, when hunger begins to press, are more painful to think of; we were told that they then often run away into the mountains, but that they are pursued by the men and brought back to the slaughter-house at their own firesides!" Later scholarship suggests that the early reports of cannibalism by the Fuegians are bogus; see discussion in Hazlewood (2000).

at their first appearance as distinct varieties, and whether or not two or more species or races have become blended together by crossing, may plainly be recognised in the increased size and beauty which we now see in the varieties of the heartsease, rose, pelargonium, dahlia, and other plants, when compared with the older varieties or with their parent-stocks. No one would ever expect to get a first-rate heartsease or dahlia from the seed of a wild plant. No one would expect to raise a first-rate melting pear from the seed of the wild pear, though he might succeed from a poor seedling growing wild, if it had come from a garden-stock. The pear, though cultivated in classical times, appears, from Pliny's description, to have been a fruit of very inferior quality. I have seen great surprise expressed in horticultural works at the wonderful skill of gardeners, in having produced such splendid results from such poor materials; but the art, I cannot doubt, has been simple, and, as far as the final result is concerned, has been followed almost unconsciously. It has consisted in always cultivating the best known variety, sowing its seeds, and, when a slightly better variety has chanced to appear, selecting it, and so onwards. But the gardeners of the classical period, who cultivated the best pear they could procure, never thought what splendid fruit we should eat; though we owe our excellent fruit, in some small degree, to their having naturally chosen and preserved the best varieties they could anywhere find.

A large amount of change in our cultivated plants, thus slowly and unconsciously accumulated, explains, as I believe, the well-known fact, that in a vast number of cases we cannot recognise, and therefore do not know, the wild parent-stocks of the plants which have been longest cultivated in our flower and kitchen gardens. If it has taken centuries or thousands of years to improve

1 Pears (*Pyrus communis,* Rosaceae) have been cultivated for thousands of years. A rather inedible group of pear varieties called choke-pears—or Pliny's pears—are thought to be the fruit described by Pliny in *Natural History.* He referred to this pear as "ampullaceum," which likely reflects its shape (an ampulla is a big-bellied jar or flask). Horticulturist Philip Miller's *Gardener's Dictionary* of 1754 describes 80 pear varieties, but that is just the beginning: "There are many other Sorts of Pears which are still continued in some old Gardens," Miller writes, "but as those here mentioned are the best Sorts known at present, it would be needless to enumerate a great Quantity of ordinary Fruit; since every one who intends to plant Fruits, will rather choose those which are the most valu'd, the Expense and Trouble being the same for a bad Sort of Fruit as a good one." A turn-of-the century manual called *A Practical Guide to Garden Plants,* by John Weathers (London, 1901), reports that between 700 and 800 pear varieties have been identified at one time or another, but that "very few . . . are worthy of cultivation." Well over 3,000 varieties are now recognized worldwide.

or modify most of our plants up to their present standard [1] of usefulness to man, we can understand how it is that neither Australia, the Cape of Good Hope, nor any other region inhabited by quite uncivilised man, has afforded us a single plant worth culture. It is not that these countries, so rich in species, do not by a strange chance possess the aboriginal stocks of any useful plants, but that the native plants have not been improved by continued selection up to a standard of perfection comparable with that given to the plants in countries anciently civilised.

[2] In regard to the domestic animals kept by uncivilised man, it should not be overlooked that they almost always have to struggle for their own food, at least during certain seasons. And in two countries very differently circumstanced, individuals of the same species, having slightly different constitutions or structure, would often succeed better in the one country than in the other, and thus by a process of "natural selection," as will hereafter be more fully explained, two sub-breeds might be formed. This, perhaps, partly explains what has been remarked by some authors, namely, that the varieties kept by savages have more of the character of species than the varieties kept in civilised countries.

On the view here given of the all-important part which selection by man has played, it becomes at once obvious, how it is that our domestic races show adaptation in their structure or in their habits to man's wants or fancies. [3] We can, I think, further understand the frequently abnormal character of our domestic races, and likewise their differences being so great in external characters and relatively so slight in internal parts or organs. Man can hardly select, or only with much difficulty, any deviation of structure excepting such as is externally visible; and indeed he rarely cares for what is internal. He can never act by selection, excepting on variations

1 One wonders if Darwin is including the native peoples of Central and South America in this generalization—ironically, several of the most important crops, including corn, tomatoes, and potatoes, were domesticated by the ancient inhabitants of these regions. Similarly, yams and peanuts were domesticated by native African peoples.

2 Here, and elsewhere, Darwin reveals his Victorian English biases. Progressive-thinking liberals for their day, the Darwin and Wedgwood families were vehemently antislavery and promoted social reforms to aid and educate the underclass. Moreover, Darwin's personal interactions with the indigenous peoples he met in his travels were characterized by humanity and amity. Nonetheless, Darwin is a product of his time, and there is no question in his mind about the inherent superiority of European society. He often makes unabashed reference to non-European native people as "uncivilized" and "savages," terms that only recently came to be appreciated as pejorative.

3 Consistent with his argument, Darwin points out here that the traits that differ among varieties of a given group are outwardly functional ones that we (breeders) have favored. The varieties resemble one another much more closely *inwardly,* as the breeder "rarely cares for what is internal" and so has not modified these traits.

which are first given to him in some slight degree by nature. No man would ever try to make a fantail, till he saw a pigeon with a tail developed in some slight degree in an unusual manner, or a pouter till he saw a pigeon with a crop of somewhat unusual size; and the more abnormal or unusual any character was when it first appeared, the more likely it would be to catch his attention. But to use such an expression as trying to make [1] a fantail, is, I have no doubt, in most cases, utterly incorrect. The man who first selected a pigeon with a slightly larger tail, never dreamed what the descendants of that pigeon would become through long-continued, partly unconscious and partly methodical selection. Perhaps the parent bird of all fantails had only fourteen tail-feathers somewhat expanded, like the present Java fantail, or like individuals of other and distinct breeds, in which as many as seventeen tail-feathers have been counted. Perhaps the first pouter-pigeon did not inflate its crop much more than the turbit now does the upper part of its œsophagus,—a habit which is disregarded by all fanciers, as it is not one of the points of the breed.

Nor let it be thought that some great deviation of [2] structure would be necessary to catch the fancier's eye: he perceives extremely small differences, and it is in human nature to value any novelty, however slight, in one's own possession. Nor must the value which would formerly be set on any slight differences in the individuals of the same species, be judged of by the value which would now be set on them, after several breeds have once fairly been established. Many slight differences might, and indeed do now, arise amongst pigeons, which are rejected as faults or deviations from the standard of perfection of each breed. The common goose has not given rise to any marked varieties; hence the Thoulouse and the common breed, which differ only in colour, that

1 Darwin's point is a good one: breeders did not set out initially to create any particular breed or to develop any particular trait. Rather, natural variations likely hinted at possibilities that subsequently caught the eye of a breeder. Such breeders effected changes stepwise, by and large never seeing beyond the next slight step and not planning out the numerous steps that would eventually result in considerable modification. Much later, when plant and animal breeding became an industry—and especially when the principles of heredity and selection became known—breeders could deliberately attempt to move a breed in a particular direction.

2 Again, the mantra is that slight differences suffice for the breeder to act upon. This sets up Darwin's later transition to natural selection, when he argues that virtually every trait inside and out varies among individuals in myriad slight ways.

1 This is an evocative way for Darwin to express the gradual evolution of varieties: like the dialects of a language diverging to the point where they are themselves distinct languages, varieties slowly become more and more distinct until they are no longer merely varieties but distinct species. As dialects diverge, the languages they become do not have a distinct point of origin in space and time, and neither do species formed by ever-diverging varieties.

most fleeting of characters, have lately been exhibited as distinct at our poultry-shows.

I think these views further explain what has sometimes been noticed—namely, that we know nothing about the origin or history of any of our domestic breeds. But, in fact, a breed, like a dialect of a language, can hardly be said to have had a definite origin. A man preserves and breeds from an individual with some slight deviation of structure, or takes more care than usual in matching his best animals and thus improves them, and the improved individuals slowly spread in the immediate neighbourhood. But as yet they will hardly have a distinct name, and from being only slightly valued, their history will be disregarded. When further improved by the same slow and gradual process, they will spread more widely, and will get recognised as something distinct and valuable, and will then probably first receive a provincial name. In semi-civilised countries, with little free communication, the spreading and knowledge of any new sub-breed will be a slow process. As soon as the points of value of the new sub-breed are once fully acknowledged, the principle, as I have called it, of unconscious selection will always tend,—perhaps more at one period than at another, as the breed rises or falls in fashion,—perhaps more in one district than in another, according to the state of civilisation of the inhabitants,—slowly to add to the characteristic features of the breed, whatever they may be. But the chance will be infinitely small of any record having been preserved of such slow, varying, and insensible changes.

I must now say a few words on the circumstances, favourable, or the reverse, to man's power of selection. A high degree of variability is obviously favourable, as freely giving the materials for selection to work on; not that mere individual differences are not amply sufficient, with extreme care, to allow of the accumulation of a large amount of modification in almost any desired direction. But as variations manifestly useful or pleasing to man appear only occasionally, the chance of their appearance will be much increased by a large number of individuals being kept; and hence this comes to be of the highest importance to success. On this principle Marshall has remarked, with respect to the sheep of parts of Yorkshire, that " as they generally belong to poor people, and are mostly *in small lots*, they never can be improved." On the other hand, nurserymen, from raising large stocks of the same plants, are generally far more successful than amateurs in getting new and valuable varieties. The keeping of a large number of individuals of a species in any country requires that the species should be placed under favourable conditions of life, so as to breed freely in that country. When the individuals of any species are scanty, all the individuals, whatever their quality may be, will generally be allowed to breed, and this will effectually prevent selection. But probably the most important point of all, is, that the animal or plant should be so highly useful to man, or so much valued by him, that the closest attention should be paid to even the slightest deviation in the qualities or structure of each individual. Unless such attention be paid nothing can be effected. I have seen it gravely remarked, that it was most fortunate that the strawberry began to vary just when gardeners began to attend closely to this plant. No doubt the strawberry had always varied since it was cultivated, but the slight varieties had been neglected. As soon, however, as gardeners picked out individual plants with slightly larger, earlier, or better fruit, and raised seedlings from them, and again picked out the best seedlings and bred from them, then, there appeared (aided by some

1 This is an astute observation—as population increases, the chances for novel variants to arise also increases. In modern terms, larger numbers of individuals afford more opportunities for mutation. Darwin is quoting the agriculturist William Marshall's *Rural Economy of Yorkshire*, a popular manual first published in 1788 that eventually went through eleven editions.

2 William Marshall wrote: "Formerly, it was the practice, in the improvement of cattle, to *cross* with other breeds; but modern breeders, who have brought the art to a high degree of perfection, pursue a different method: they pick out the fairest, of the particular breed or variety they want to improve, and prosecute the improvement with the *selected individuals*." (Marshall 1796, p. 9, emphasis in original). The point about the remarkable efficacy of careful attention to breeding is well taken, though Darwin's statement that without it "nothing can be effected" seems contrary to his earlier claim (e.g., p. 37) that much change is unconsciously effected.

3 Darwin's gentle mocking here underscores his point that every species varies. Of course the strawberry did not just happen to begin varying just when horticulturists started paying attention to it. He is serious, however, about the tendency of some people to come to just such nonsensical conclusions. The passage recalls another, from the D notebook (entry 49, p. 347 in Barrett et al. 1987), in which Darwin expresses shock at the philosopher William Whewell's suggestion that day length is adapted to the sleep patterns of man, and that this is proof of divine benevolence. Darwin dismisses the notion as an "instance of arrogance!!"

1 Another astute observation: islands are genetically closed systems where novel variants might be expected to spread (i.e., become common in the population); the Manx cat comes to mind, a tailless breed that arose on the Isle of Man centuries ago.

crossing with distinct species) those many admirable varieties of the strawberry which have been raised during the last thirty or forty years.

In the case of animals with separate sexes, facility in preventing crosses is an important element of success in the formation of new races,—at least, in a country which is already stocked with other races. In this respect enclosure of the land plays a part. Wandering savages or the inhabitants of open plains rarely possess more than one breed of the same species. Pigeons can be mated for life, and this is a great convenience to the fancier, for thus many races may be kept true, though mingled in the same aviary; and this circumstance must have largely favoured the improvement and formation of new breeds. Pigeons, I may add, can be propagated in great numbers and at a very quick rate, and inferior birds may be freely rejected, as when killed they serve for food. On the other hand, cats, from their nocturnal rambling habits, cannot be matched, and, although so much valued by women and children, we hardly ever see a distinct breed kept up; such breeds as we do sometimes see are almost always imported from some
1 other country, often from islands. Although I do not doubt that some domestic animals vary less than others, yet the rarity or absence of distinct breeds of the cat, the donkey, peacock, goose, &c., may be attributed in main part to selection not having been brought into play: in cats, from the difficulty in pairing them; in donkeys, from only a few being kept by poor people, and little attention paid to their breeding; in peacocks, from not being very easily reared and a large stock not kept; in geese, from being valuable only for two purposes, food and feathers, and more especially from no pleasure having been felt in the display of distinct breeds.

To sum up on the origin of our Domestic Races of animals and plants. I believe that the conditions of life, from their action on the reproductive system, are so far of the highest importance as causing variability. I do not believe that variability is an inherent and necessary contingency, under all circumstances, with all organic beings, as some authors have thought. The effects of variability are modified by various degrees of inheritance and of reversion. Variability is governed by many unknown laws, more especially by that of correlation of growth. Something may be attributed to the direct action of the conditions of life. Something must be attributed to use and disuse. The final result is thus rendered infinitely complex. In some cases, I do not doubt that the intercrossing of species, aboriginally distinct, has played an important part in the origin of our domestic productions. When in any country several domestic breeds have once been established, their occasional intercrossing, with the aid of selection, has, no doubt, largely aided in the formation of new sub-breeds; but the importance of the crossing of varieties has, I believe, been greatly exaggerated, both in regard to animals and to those plants which are propagated by seed. In plants which are temporarily propagated by cuttings, buds, &c., the importance of the crossing both of distinct species and of varieties is immense; for the cultivator here quite disregards the extreme variability both of hybrids and mongrels, and the frequent sterility of hybrids; but the cases of plants not propagated by seed are of little importance to us, for their endurance is only temporary. Over all these causes of Change I am convinced that the accumulative action of Selection, whether applied methodically and more quickly, or unconsciously and more slowly, but more efficiently, is by far the predominant Power.

1 This final, lengthy paragraph provides a succinct summary of Darwin's arguments in this chapter. The key point is the power of cumulative selection to diversify the progenitors of domesticated species into distinct varieties—varieties that are therefore *genealogically* related. Darwin thus stresses an important fact: selection is remarkably effective as a tool and as a model of common descent for highly divergent groups of organisms.

2 Note the capitalization of the words "Change," "Selection," and "Power." These three words are, of course, central: *change* refers to transmutation, *selection* to the causal mechanism of transmutation, and *power* to selection as the primary process effecting change.

In this chapter Darwin takes great pains to argue for variation in virtually every point of anatomy and physiology. Prodigious amounts of variation are necessary to convince his readers that, given enough time, natural selection is capable of generating the diversity of life that we see. This chapter is the first of three making a case for the theory of natural selection, the central mechanism of transmutation. This mechanism follows from the ready availability of heritable variations (present chapter) and differential survival and reproduction linked to that variation (next chapter). Another, related point in this chapter is the way in which natural variation is distributed in species and varieties.

---

1 The chapter starts off simply enough, but Darwin soon launches into definitions of such key terms as "species," "variety," and "variation." It may seem odd that, in a book on the origin of species, the author is suggesting that naturalists are not entirely clear on what, exactly, a species is.

CHAPTER II.

VARIATION UNDER NATURE.

Variability — Individual differences — Doubtful species — Wide ranging, much diffused, and common species vary most—Species of the larger genera in any country vary more than the species of the smaller genera—Many of the species of the larger genera resemble varieties in being very closely, but unequally, related to each other, and in having restricted ranges.

1 BEFORE applying the principles arrived at in the last chapter to organic beings in a state of nature, we must briefly discuss whether these latter are subject to any variation. To treat this subject at all properly, a long catalogue of dry facts should be given; but these I shall reserve for my future work. Nor shall I here discuss the various definitions which have been given of the term species. No one definition has as yet satisfied all naturalists; yet every naturalist knows vaguely what he means when he speaks of a species. Generally the term includes the unknown element of a distinct act of creation. The term "variety" is almost equally difficult to define; but here community of descent is almost universally implied, though it can rarely be proved. We have also what are called monstrosities; but they graduate into varieties. By a monstrosity I presume is meant some considerable deviation of structure in one part, either injurious to or not useful to the species, and not generally propagated. Some authors use the term "variation" in a technical sense, as implying a modification directly due to the physical conditions of life; and "variations" in this sense are supposed not to be inherited: but who can say that the dwarfed condition of shells in the brackish waters of the Baltic, or dwarfed

plants on Alpine summits, or the thicker fur of an animal from far northwards, would not in some cases be inherited for at least some few generations? and in this case I presume that the form would be called a variety.

Again, we have many slight differences which may be called individual differences, such as are known frequently to appear in the offspring from the same parents, or which may be presumed to have thus arisen, from being frequently observed in the individuals of the same species inhabiting the same confined locality. No one supposes that all the individuals of the same species are cast in the very same mould. These individual differences are highly important for us, as they afford materials for natural selection to accumulate, in the same manner as man can accumulate in any given direction individual differences in his domesticated productions. These individual differences generally affect what naturalists consider unimportant parts; but I could show by a long catalogue of facts, that parts which must be called important, whether viewed under a physiological or classificatory point of view, sometimes vary in the individuals of the same species. I am convinced that the most experienced naturalist would be surprised at the number of the cases of variability, even in important parts of structure, which he could collect on good authority, as I have collected, during a course of years. It should be remembered that systematists are far from pleased at finding variability in important characters, and that there are not many men who will laboriously examine internal and important organs, and compare them in many specimens of the same species. I should never have expected that the branching of the main nerves close to the great central ganglion of an insect would have been variable in the same species; I should have expected that changes of this nature could have been effected only

1 This is the first clear statement of the significance of variation in providing the raw material fueling transmutation—"accumulating" change. The parallel with domesticated organisms is clearly spelled out.

2 Such laborious comparisons, what the evolutionary biologist Ernst Mayr termed "populational thinking," are necessary to appreciate the degree and significance of individual variation.

[1] by slow degrees: yet quite recently Mr. Lubbock has shown a degree of variability in these main nerves in Coccus, which may almost be compared to the irregular [2] branching of the stem of a tree. This philosophical naturalist, I may add, has also quite recently shown that the muscles in the larvæ of certain insects are very far from uniform. Authors sometimes argue in a circle when they state that important organs never vary; for these same authors practically rank that character as important (as some few naturalists have honestly confessed) which does not vary; and, under this point of view, no instance of an important part varying will ever be found: but under any other point of view many instances assuredly can be given.

There is one point connected with individual differences, which seems to me extremely perplexing: I refer to those genera which have sometimes been called "protean" or "polymorphic," in which the species present an inordinate amount of variation; and hardly two naturalists can agree which forms to rank as species and [3] which as varieties. We may instance Rubus, Rosa, and Hieracium amongst plants, several genera of insects, and several genera of Brachiopod shells. In most polymorphic genera some of the species have fixed and definite characters. Genera which are polymorphic in one country seem to be, with some few exceptions, polymorphic in other countries, and likewise, judging from Brachiopod [4] shells, at former periods of time. These facts seem to be very perplexing, for they seem to show that this kind of variability is independent of the conditions of life. I am inclined to suspect that we see in these polymorphic genera variations in points of structure which are of no service or disservice to the species, and which consequently have not been seized on and rendered definite by natural selection, as hereafter will be explained.

1 Sir John Lubbock was a notable politician, banker, and entomologist—and good friend of Darwin's. In 1858 he undertook a study of the nervous and digestive systems of the soft brown scale, *Coccus hesperidum* (family Coccidae), a pest of citrus trees. He sent Darwin detailed information on the nervous ganglia of this and related species in a letter dated June 10, 1858 (*Correspondence* 7: 105; by then he knew about the Darwin-Wallace papers), and in November of that year he read a paper on the subject to the Royal Society (Lubbock 1858). "In some specimens of both species I have seen the nerve a rise directly from the ganglion and not from the nerve," he wrote to Darwin. Darwin took this as evidence that even highly important organs are subject to variation (Somkin 1962).

2 This is a term of respect and esteem; to be a philosophical naturalist was to ponder and investigate the underlying meaning of the patterns presented by nature. Natural philosophy meant, to the Victorians, what we term "science" (William Whewell coined the term "scientist" in 1833). Hence Darwin's nickname of "Philos" on the voyage of the *Beagle*—he was the ship's natural philosopher, or scientist. His brother Erasmus shared this nickname as a young man, when he and Charles "experimentised" in their chemistry laboratory.

3 *Rubus* is the blackberry and raspberry genus, and *Rosa* is the rose genus (both family Rosaceae); *Hieracium* is the large genus of hawkweeds, which are daisy relatives (Asteraceae). Brachiopods (see p. 316) are marine invertebrates that superficially resemble clams.

4 This observation would be perplexing to one wedded to the idea that environmental conditions somehow engender variation. Why, in the same environment, should some related species vary a great deal and others vary little?

Those forms which possess in some considerable degree the character of species, but which are so closely similar to some other forms, or are so closely linked to them by intermediate gradations, that naturalists do not like to rank them as distinct species, are in several respects the most important for us. We have every reason to believe that many of these doubtful and closely-allied forms have permanently retained their characters in their own country for a long time; for as long, as far as we know, as have good and true species. Practically, when a naturalist can unite two forms together by others having intermediate characters, he treats the one as a variety of the other, ranking the most common, but sometimes the one first described, as the species, and the other as the variety. But cases of great difficulty, which I will not here enumerate, sometimes occur in deciding whether or not to rank one form as a variety of another, even when they are closely connected by intermediate links; nor will the commonly-assumed hybrid nature of the intermediate links always remove the difficulty. In very many cases, however, one form is ranked as a variety of another, not because the intermediate links have actually been found, but because analogy leads the observer to suppose either that they do now somewhere exist, or may formerly have existed; and here a wide door for the entry of doubt and conjecture is opened.

Hence, in determining whether a form should be ranked as a species or a variety, the opinion of naturalists having sound judgment and wide experience seems the only guide to follow. We must, however, in many cases, decide by a majority of naturalists, for few well-marked and well-known varieties can be named which have not been ranked as species by at least some competent judges.

1 By "doubtful," Darwin means difficult to classify definitively as either good (full or bona fide) species or mere varieties. Individual specimens often seem intermediate. Darwin will argue that such intermediates are the rule and not the exception; as such, they support his theory of transmutation.

2 This is the first statement of a central point in this chapter and several to come: such intermediate forms are "most important" because they illustrate the ill-defined boundary between species and varieties.

3 This is a rhetorical device worthy of notice: Darwin suggests that we look to naturalists of sound judgment and wide experience for guidance on what should be considered a species and what a variety. Yet these very naturalists often come to differing conclusions. Darwin here exploits their disagreement to make the point that species and varieties intergrade; there *is* no real line of demarcation between them.

1 That varieties of this doubtful nature are far from
uncommon cannot be disputed. Compare the several
floras of Great Britain, of France or of the United
States, drawn up by different botanists, and see what a
surprising number of forms have been ranked by one
botanist as good species, and by another as mere
2 varieties. Mr. H. C. Watson, to whom I lie under
deep obligation for assistance of all kinds, has marked
for me 182 British plants, which are generally considered as varieties, but which have all been ranked
by botanists as species; and in making this list he
has omitted many trifling varieties, but which nevertheless have been ranked by some botanists as species,
and he has entirely omitted several highly polymorphic
3 genera. Under genera, including the most polymorphic
forms, Mr. Babington gives 251 species, whereas Mr.
Bentham gives only 112,—a difference of 139 doubtful
forms! Amongst animals which unite for each birth,
and which are highly locomotive, doubtful forms, ranked
by one zoologist as a species and by another as a variety,
can rarely be found within the same country, but are
common in separated areas. How many of those birds
and insects in North America and Europe, which differ
very slightly from each other, have been ranked by
one eminent naturalist as undoubted species, and by
another as varieties, or, as they are often called, as
4 geographical races! Many years ago, when comparing,
and seeing others compare, the birds from the separate islands of the Galapagos Archipelago, both one
with another, and with those from the American mainland, I was much struck how entirely vague and arbitrary is the distinction between species and varieties.
5 On the islets of the little Madeira group there are
many insects which are characterized as varieties in
Mr. Wollaston's admirable work, but which it cannot

1 In support of his argument that "doubtful" species or varieties are common, Darwin provides data from well-studied flora.

2 Hewett Cottrell Watson made many contributions to the study of plant biogeography. He had a long and extensive correspondence with Darwin on botanical matters, as is evident from this passage. He was also an avid phrenologist, much to Darwin's disdain. In a letter to Hooker, Darwin wrote: "Here is a good joke: [H. C.] Watson (who I fancy & hope is going to review new Edit. of Origin) says that in first 4 paragraph of the Introduction, the words 'I' 'me' 'my' occur 43 times! I was dimly conscious of this accursed fact.—He says it can be explained phrenologically which I suppose civilly means that I am the most egotistically self-sufficient man alive,—perhaps so—" (*Correspondence* 9: 70).

3 Darwin's approach is simple yet effective: comparing the assessments of two well-regarded botanists, he allows the discrepancies between them to speak to the ill-defined line between species and varieties. Charles Cardale Babington was a botanist, entomologist, and archaeologist. Darwin knew Babington from his Cambridge days. Nicknamed "Beetles" Babington as a student, he and Darwin had a fierce rivalry when it came to beetle collecting. Babington became known for his botanical work and authored the *Manual of British Botany* (London, 1843), which provided Darwin with the data cited here. George Bentham was a prolific botanist, among whose many contributions was a manual of British plants (*Handbook of the British Flora*, London, 1858). Bentham worked closely with Hooker, through whom Darwin often sent him requests for botanical information.

4 In the draft of his "big species book," *Natural Selection*, Darwin wrote that he "was much struck how entirely arbitrary the distinction is between species & varieties, when I witnessed different naturalists comparing the organic productions which I brought home from the islands, off the coast of S. America" (Stauffer 1975, p. 115). While in the Galápagos Islands, however, Darwin paid little attention to the varieties of island-specific mockingbirds or to any other bird group. Later the ornithologist John Gould instructed him on the curious diversity and distri-

(continued)

be doubted would be ranked as distinct species by many entomologists. Even Ireland has a few animals, now generally regarded as varieties, but which have been ranked as species by some zoologists. Several most experienced ornithologists consider our British red grouse as only a strongly-marked race of a Norwegian species, whereas the greater number rank it as an undoubted species peculiar to Great Britain. A wide distance between the homes of two doubtful forms leads many naturalists to rank both as distinct species; but what distance, it has been well asked, will suffice? if that between America and Europe is ample, will that between the Continent and the Azores, or Madeira, or the Canaries, or Ireland, be sufficient? It must be admitted that many forms, considered by highly-competent judges as varieties, have so perfectly the character of species that they are ranked by other highly competent judges as good and true species. But to discuss whether they are rightly called species or varieties, before any definition of these terms has been generally accepted, is vainly to beat the air.

Many of the cases of strongly-marked varieties or doubtful species well deserve consideration; for several interesting lines of argument, from geographical distribution, analogical variation, hybridism, &c., have been brought to bear on the attempt to determine their rank. I will here give only a single instance,—the well-known one of the primrose and cowslip, or Primula veris and elatior. These plants differ considerably in appearance; they have a different flavour and emit a different odour; they flower at slightly different periods; they grow in somewhat different stations; they ascend mountains to different heights; they have different geographical ranges; and lastly, according to very numerous experiments made during several years by

(continued)
bution of the mockingbirds as well as of the group of birds now known as Darwin's finches (see editor's Introduction).

5 Thomas Vernon Wollaston was an entomologist who studied the insects (especially beetles) of Madeira. The "admirable work" to which Darwin is likely referring is Wollaston's book *On the Variation of Species, with Especial Reference to the Insecta* (1856), which Darwin recorded in his reading notebook (Vorzimmer 1977).

1 Primroses, genus *Primula*, family Primulaceae, are common European wildflowers. They were a sort of botanical "lab rat" in the nineteenth century, subject to numerous studies of cross-pollination, hybridization, and heterostyly (variation in pistil length). Darwin experimented with them (e.g., Darwin 1868a) and gave a lengthy treatment in the first two chapters of his book *The Different Forms of Flowers on Plants of the Same Species* (Darwin 1877). Karl Friedrich Gärtner (1772–1850), a leading authority on plant hybridization, also studied these primroses. He is cited later in this paragraph.

In this passage Darwin mixes up his species: he refers to primrose and cowslip as *Primula veris* and *P. eliator*, respectively, but he means *P. vulgaris* and *P. veris*. The error was corrected in the second edition of *Origin*. The observation he reports here comes from Gärtner's book on hybridization, *Versuche und Beobachtungen uber die Bastarderzeugung im Pflanzenreich* (1849). He summarized Gärtner's findings in chapter II of *Different Forms of Flowers:* "The cowslip and primrose, when intercrossed, behave like distinct species, for they are far from being mutually fertile. Gartner crossed 27 flowers of P. vulgaris [another name for *P. eliator*] with pollen of P. veris, and obtained 16 capsules; but these did not contain any good seed . . . He also crossed 21 flowers of P. veris with pollen of P. vulgaris; and now he got only five capsules, containing seed in a still less perfect condition." Such well-studied and apparently distinct species presented excellent case studies for Darwin to illustrate his point about the "doubtful" status of some species.

that most careful observer Gärtner, they can be crossed only with much difficulty. We could hardly wish for better evidence of the two forms being specifically distinct. On the other hand, they are united by many intermediate links, and it is very doubtful whether these links are hybrids; and there is, as it seems to me, an overwhelming amount of experimental evidence, showing that they descend from common parents, and consequently must be ranked as varieties.

Close investigation, in most cases, will bring naturalists to an agreement how to rank doubtful forms. Yet it must be confessed, that it is in the best-known countries that we find the greatest number of forms of doubtful value. I have been struck with the fact, that if any animal or plant in a state of nature be highly useful to man, or from any cause closely attract his attention, varieties of it will almost universally be found recorded. These varieties, moreover, will be often ranked by some authors as species. Look at the common oak, how closely it has been studied; yet a German author makes more than a dozen species out of forms, which are very generally considered as varieties; and in this country the highest botanical authorities and practical men can be quoted to show that the sessile and pedunculated oaks are either good and distinct species or mere varieties.

When a young naturalist commences the study of a group of organisms quite unknown to him, he is at first much perplexed to determine what differences to consider as specific, and what as varieties; for he knows nothing of the amount and kind of variation to which the group is subject; and this shows, at least, how very generally there is some variation. But if he confine his attention to one class within one country, he will soon make up his mind how to rank most of the doubtful forms. His

1 The oak is an iconic tree to the English. "What is the tree, which ought to be best known in Britain? assuredly the oak," wrote Darwin in *Natural Selection.* Yet, he observed, some authorities such as a Dr. Lindley and Sir James Smith recognize these two oaks as distinct species—*Quercus robur* and *Quercus sessiliflora*—while Babington, Hooker, and other authorities treat them as varieties of one species, *Q. robur* (Stauffer 1975, p. 118). The two are recognized as distinct species today.

2 The "young naturalist" may be Darwin himself; is this paragraph an autobiographical sketch of Darwin's own trial-and-error education in the subtleties of distinguishing species and varieties? Darwin undertook his barnacle studies between 1846 and 1854, producing four monographs. This monumental project was apparently undertaken at least partly in response to a comment made by Joseph Hooker, judging from remarks Darwin made in an 1845 letter to Hooker: "How painfully (to me) true is your remark that no one has hardly a right to examine the question of species who has not minutely described many" (*Correspondence* 3: 252). Hooker had by then been taken into Darwin's confidence and knew of his "species theory." Darwin initiated his barnacle project about a year later, in October 1846.

general tendency will be to make many species, for he will become impressed, just like the pigeon or poultry-fancier before alluded to, with the amount of difference in the forms which he is continually studying; and he has little general knowledge of analogical variation in other groups and in other countries, by which to correct his first impressions. As he extends the range of his observations, he will meet with more cases of difficulty; for he will encounter a greater number of closely-allied forms. But if his observations be widely extended, he will in the end generally be enabled to make up his own mind which to call varieties and which species; but he will succeed in this at the expense of admitting much variation,—and the truth of this admission will often be disputed by other naturalists. When, moreover, he comes to study allied forms brought from countries not now continuous, in which case he can hardly hope to find the intermediate links between his doubtful forms, he will have to trust almost entirely to analogy, and his difficulties will rise to a climax.

Certainly no clear line of demarcation has as yet been drawn between species and sub-species—that is, the forms which in the opinion of some naturalists come very near to, but do not quite arrive at the rank of species; or, again, between sub-species and well-marked varieties, or between lesser varieties and individual differences. These differences blend into each other in an insensible series; and a series impresses the mind with the idea of an actual passage.

Hence I look at individual differences, though of small interest to the systematist, as of high importance for us, as being the first step towards such slight varieties as are barely thought worth recording in works on natural history. And I look at varieties which are in any degree more distinct and permanent, as steps leading to more

1 Note that the young naturalist is first impressed with *differences* between forms (hence naming many species), but eventually grows more cognizant of uniting *similarities* and doubtful forms—i.e., intermediates.

2 This is the first clear statement of what Darwin has been driving at: the intermediates represent a series or gradation; the appearance of passage from one form to another—transmutation—is more than mere appearance. The following paragraph is important.

1 Varieties become more and more "strongly marked"—distinct—until they become permanent and well established. These in turn further differentiate to the point of being termed subspecies; still more differentiation renders them distinct enough to be considered separate species. The final sentence of this paragraph nicely sums up the point.

2 Keep in mind that the nature of species and varieties became a critical line of inquiry beginning in the eighteenth century and continuing well into the nineteenth. It was long generally supposed, recall, that species were created more or less as we see them. Later, the renowned French comparative anatomist Cuvier grouped animals into *embranchments,* four basic body plans: Radiata, Vertebrata, Articulata, and Mollusca. All the great naturalists of the late eighteenth and nineteenth centuries discussed species and their varieties, biogeography, and in some radical cases origins. The young Alfred Russel Wallace was motivated to undertake travels in the tropics in part by the species question, and even the eminent geologist Charles Lyell kept notebooks devoted to the topic. The Astronomer Royal Sir John Herschel, writing to Lyell from the Cape of Good Hope at the southern tip of Africa, referred to the question of species origins as the "mystery of mysteries."

strongly marked and more permanent varieties; and at these latter, as leading to sub-species, and to species. The passage from one stage of difference to another and higher stage may be, in some cases, due merely to the long-continued action of different physical conditions in two different regions; but I have not much faith in this view; and I attribute the passage of a variety, from a state in which it differs very slightly from its parent to one in which it differs more, to the action of natural selection in accumulating (as will hereafter be more fully explained) differences of structure in certain definite directions. Hence I believe a well-marked variety may be justly called an incipient species; but whether this belief be justifiable must be judged of by the general weight of the several facts and views given throughout this work.

It need not be supposed that all varieties or incipient species necessarily attain the rank of species. They may whilst in this incipient state become extinct, or they may endure as varieties for very long periods, as has been shown to be the case by Mr. Wollaston with the varieties of certain fossil land-shells in Madeira. If a variety were to flourish so as to exceed in numbers the parent species, it would then rank as the species, and the species as the variety; or it might come to supplant and exterminate the parent species; or both might co-exist, and both rank as independent species. But we shall hereafter have to return to this subject.

From these remarks it will be seen that I look at the term species, as one arbitrarily given for the sake of convenience to a set of individuals closely resembling each other, and that it does not essentially differ from the term variety, which is given to less distinct and more fluctuating forms. The term variety, again, in comparison with mere individual differences, is also applied arbitrarily, and for mere convenience sake.

Guided by theoretical considerations, I thought that some interesting results might be obtained in regard to the nature and relations of the species which vary most, by tabulating all the varieties in several well-worked floras. At first this seemed a simple task; but Mr. H. C. Watson, to whom I am much indebted for valuable advice and assistance on this subject, soon convinced me that there were many difficulties, as did subsequently Dr. Hooker, even in stronger terms. I shall reserve for my future work the discussion of these difficulties, and the tables themselves of the proportional numbers of the varying species. Dr. Hooker permits me to add, that after having carefully read my manuscript, and examined the tables, he thinks that the following statements are fairly well established. The whole subject, however, treated as it necessarily here is with much brevity, is rather perplexing, and allusions cannot be avoided to the "struggle for existence," "divergence of character," and other questions, hereafter to be discussed. ◀1

Alph. De Candolle and others have shown that plants which have very wide ranges generally present varieties; and this might have been expected, as they become exposed to diverse physical conditions, and as they come into competition (which, as we shall hereafter see, is a far more important circumstance) with different sets of organic beings. But my tables further show that, in any limited country, the species which are most common, that is abound most in individuals, and the species which are most widely diffused within their own country (and this is a different consideration from wide range, and to a certain extent from commonness), often give rise to varieties sufficiently well-marked to have been recorded in botanical works. Hence it is the most flourishing, or, as they may be called, the dominant species,— ◀2

1 This paragraph prefaces an argument that occupies the remainder of the chapter. Darwin employs a form of "botanical arithmetic" to reveal what he believes to be a significant underlying pattern in the abundance and distribution of species and varieties. "Botanical arithmetic" was the term given to an approach to plant biogeography first developed by the English botanist Robert Brown. This was the first attempt to quantify species differences among geographical regions using taxonomic ratios. Brown observed a latitudinal gradient in genus:species ratios, and the explorer Alexander von Humboldt later applied Brown's approach to altitudinal gradients in the Andes (Browne 1983). The "tables" that Darwin mentions are given in *Natural Selection*—more than six pages of them—in which he crunches data from more than a dozen botanical manuals (see Stauffer 1975, pp. 138–164).

2 Darwin is citing Alphonse de Candolle's important work *Géographie Botanique Raisonnée* (Paris, 1855).

1 Here is the main thrust of Darwin's botanical arithmetic: Dominant (large, widespread) genera contain species that are disproportionately variable compared with species in small genera. Dominant genera also tend to contain dominant, widespread species. This ties in with Darwin's view of how variation occurs: exposure to varying environmental conditions. Dominant, widespread species vary more than do narrowly restricted species as a result of exposure to a wider range of conditions and by virtue of being numerically abundant, and hence with more individuals potentially varying.

2 Darwin had been poring over botanical manuals (floras), counting up species in larger and smaller genera, when he was saved from a computational blunder by his friend John Lubbock, whom he thanked profusely in a letter dated July 14, 1857: "My dear Lubbock, You have done me the greatest possible service in helping me to clarify my Brains. If I am as muzzy on all subjects as I am on proportions & chance,—what a Book I shall produce!" (*Correspondence* 6: 430). Straight counts will nearly always give more varieties in larger genera. Lubbock had suggested that he compare the *proportion* of species with varieties in large genera with the proportion of such species in smaller genera (see Browne 1980).

1 those which range widely over the world, are the most diffused in their own country, and are the most numerous in individuals,—which oftenest produce well-marked varieties, or, as I consider them, incipient species. And this, perhaps, might have been anticipated; for, as varieties, in order to become in any degree permanent, necessarily have to struggle with the other inhabitants of the country, the species which are already dominant will be the most likely to yield offspring which, though in some slight degree modified, will still inherit those advantages that enabled their parents to become dominant over their compatriots.

2 If the plants inhabiting a country and described in any Flora be divided into two equal masses, all those in the larger genera being placed on one side, and all those in the smaller genera on the other side, a somewhat larger number of the very common and much diffused or dominant species will be found on the side of the larger genera. This, again, might have been anticipated; for the mere fact of many species of the same genus inhabiting any country, shows that there is something in the organic or inorganic conditions of that country favourable to the genus; and, consequently, we might have expected to have found in the larger genera, or those including many species, a large proportional number of dominant species. But so many causes tend to obscure this result, that I am surprised that my tables show even a small majority on the side of the larger genera. I will here allude to only two causes of obscurity. Fresh-water and salt-loving plants have generally very wide ranges and are much diffused, but this seems to be connected with the nature of the stations inhabited by them, and has little or no relation to the size of the genera to which the species belong. Again, plants low in the scale of organisation are

generally much more widely diffused than plants higher in the scale; and here again there is no close relation to the size of the genera. The cause of lowly-organised plants ranging widely will be discussed in our chapter on geographical distribution.

From looking at species as only strongly-marked and well-defined varieties, I was led to anticipate that the species of the larger genera in each country would oftener present varieties, than the species of the smaller genera; for wherever many closely related species (*i. e.* species of [1] the same genus) have been formed, many varieties or incipient species ought, as a general rule, to be now forming. Where many large trees grow, we expect to find saplings. Where many species of a genus have been formed through variation, circumstances have been favourable for variation; and hence we might expect that the circumstances would generally be still favourable to variation. On the other hand, if we look at [2] each species as a special act of creation, there is no apparent reason why more varieties should occur in a group having many species, than in one having few.

To test the truth of this anticipation I have arranged [3] the plants of twelve countries, and the coleopterous insects of two districts, into two nearly equal masses, the species of the larger genera on one side, and those of the smaller genera on the other side, and it has invariably proved to be the case that a larger proportion of the species on the side of the larger genera present varieties, than on the side of the smaller genera. Moreover, the species of the large genera which present any varieties, invariably present a larger average number of varieties than do the species of the small genera. Both these results follow when another division is made, and when all the smallest genera, with from only one to four species, are absolutely excluded from the tables. These

1 Darwin's statement here indicates his belief that if transmutation has been occurring in the recent past, it is likely still occurring. See his explicit statement of this belief on p. 56.

2 This is the first instance of many in which Darwin explicitly contrasts his view of transmutation/common descent with that of special creation with respect to observed patterns. The point is reasonable: if we treat each created entity as separate and distinct from every other, the number of varieties should be about the same on average whether the species falls in a large or a small genus. That the number varies according to the size of the genus indicates that something interesting is going on.

3 Wherever possible Darwin marshals data to support his arguments. He was very excited about the patterns that emerged using a "rule-of-three" proportional approach, and here he reports his findings with respect to plants and beetles. In *Natural Selection* he presents lengthy tables of proportional data from a sizable stack of taxonomic manuals. The first entry, for example, is for Babington's flora of Great Britain (Stauffer 1975, p. 149). Darwin divides this flora into genera with 5 or more species, and those with 4 or fewer species. He then notes how many species occur in each category, and the number of these "presenting varieties." Babington's flora gives 663 total species for all genera with 5 or more species each, and 745 species for those with 4 or fewer species each. The varieties found in the larger genera number 101; proportionally, then, one would expect 113 varieties in the smaller genera: 663:745:101:113. In fact, only 89 varieties are found among the species of the smaller genera—there are fewer than expected, or more than expected in the larger genera. The higher proportion of species with varieties in larger genera versus smaller genera is very significant to Darwin, as is the higher average number of varieties per species in the larger genera. Such patterns make no sense if species are specially created and immutable.

1 We would not necessarily agree with this statement today; genus size is perhaps irrelevant to whether member species are common or rare, widespread or restricted. However, it makes sense that more widespread species, if abundant, would experience a greater absolute number of mutations owing to large population size, and depending on mobility could differentiate into geographic varieties.

facts are of plain signification on the view that species are only strongly marked and permanent varieties; for wherever many species of the same genus have been formed, or where, if we may use the expression, the manufactory of species has been active, we ought generally to find the manufactory still in action, more especially as we have every reason to believe the process of manufacturing new species to be a slow one. And this certainly is the case, if varieties be looked at as incipient species; for my tables clearly show as a general rule that, wherever many species of a genus have been formed, the species of that genus present a number of varieties, that is of incipient species, beyond the average. It is not that all large genera are now varying much, and are thus increasing in the number of their species, or that no small genera are now varying and increasing; for if this had been so, it would have been fatal to my theory; inasmuch as geology plainly tells us that small genera have in the lapse of time often increased greatly in size; and that large genera have often come to their maxima, declined, and disappeared. All that we want to show is, that where many species of a genus have been formed, on an average many are still forming; and this holds good.

There are other relations between the species of large genera and their recorded varieties which deserve notice. We have seen that there is no infallible criterion by which to distinguish species and well-marked varieties; and in those cases in which intermediate links have not been found between doubtful forms, naturalists are compelled to come to a determination by the amount of difference between them, judging by analogy whether or not the amount suffices to raise one or both to the rank of species. Hence the amount of difference is one very important criterion in settling whether two forms should be ranked as species or varieties. Now Fries has remarked in regard to plants, and Westwood in regard to insects, that in large genera the amount of difference between the species is often exceedingly small. I have endeavoured to test this numerically by averages, and, as far as my imperfect results go, they always confirm the view. I have also consulted some sagacious and most experienced observers, and, after deliberation, they concur in this view. In this respect, therefore, the species of the larger genera resemble varieties, more than do the species of the smaller genera. Or the case may be put in another way, and it may be said, that in the larger genera, in which a number of varieties or incipient species greater than the average are now manufacturing, many of the species already manufactured still to a certain extent resemble varieties, for they differ from each other by a less than usual amount of difference. ◀1

Moreover, the species of the large genera are related to each other, in the same manner as the varieties of any one species are related to each other. No naturalist pretends that all the species of a genus are equally distinct from each other; they may generally be divided into sub-genera, or sections, or lesser groups. As Fries has well remarked, little groups of species are generally clustered like satellites around certain other species. And what are varieties but groups of forms, unequally related to each other, and clustered round certain forms—that is, round their parent-species? Undoubtedly there is one most important point of difference between varieties and species; namely, that the amount of difference between varieties, when compared with each other or with their parent-species, is much less than that between the species of the same genus. But when we come to discuss the principle, as I call it, of Divergence of Character, ◀2

1 Note Darwin's appeal to authority in underscoring his point that there is more differentiation among species of larger genera—they resemble varieties more than do species of smaller genera. Darwin asserts that this is the case because the newly formed species of larger genera, freshly minted, so to speak, still resemble the varieties they were not so long before. Perhaps in a large group with active differentiation (speciation) one is more likely to find new species representing all stages and degrees of differentiation, including a higher incidence of very recently differentiated forms.

2 Here Darwin hints at a key argument to come: divergence of character, presented in chapter IV. Note in this passage that he underscores the symmetry of relationship among varieties, species, and beyond: varieties are related to their parent species in the same way that species are related in parent subgenera, subgenera related in parent genera, etc. Note the nice imagery of groups of species clustered like satellites about other species. This comes from the Swedish botanist Elias Magnus Fries's paper "A monograph of the Hieracia," which was published in the *Botanical Gazette* in 1850. Darwin took detailed notes on this paper: "In genera containing many species, the individual species stand much closer together than in poor genera; hence it is well in the former case to collect them around certain types or principal species, about which, as around a centre, the others arrange themselves, as satellites" (Stauffer 1975, p. 93).

we shall see how this may be explained, and how the lesser differences between varieties will tend to increase into the greater differences between species.

[1] There is one other point which seems to me worth notice. Varieties generally have much restricted ranges: this statement is indeed scarcely more than a truism, for if a variety were found to have a wider range than that of its supposed parent-species, their denominations ought to be reversed. But there is also reason to believe, that those species which are very closely allied to other species, and in so far resemble varieties, often have much restricted ranges. For instance, Mr. H. C. [2] Watson has marked for me in the well-sifted London Catalogue of plants (4th edition) 63 plants which are therein ranked as species, but which he considers as so closely allied to other species as to be of doubtful value: these 63 reputed species range on an average over 6·9 of the provinces into which Mr. Watson has divided Great Britain. Now, in this same catalogue, 53 acknowledged varieties are recorded, and these range over 7·7 provinces; whereas, the species to which these varieties belong range over 14·3 provinces. So that the acknowledged varieties have very nearly the same restricted average range, as have those very closely allied forms, marked for me by Mr. Watson as doubtful species, but which are almost universally ranked by British botanists as good and true species.

Finally, then, varieties have the same general characters as species, for they cannot be distinguished from species,—except, firstly, by the discovery of intermediate linking forms, and the occurrence of such links cannot affect the actual characters of the forms which they connect; and except, secondly, by a certain amount of

1 Here Darwin presents one final bit of botanical data concerning doubtful species. Closely related species in a group often resemble varieties, for example, in having restricted ranges. These species, in other words, are very variety-like, and one might therefore infer that they are indeed newly derived from related varieties.

2 The book Darwin cites here is *The London Catalogue of British Plants,* which was edited by Hewett Cottrell Watson from 1844 to 1874.

difference, for two forms, if differing very little, are generally ranked as varieties, notwithstanding that intermediate linking forms have not been discovered; but the amount of difference considered necessary to give to two forms the rank of species is quite indefinite. In genera having more than the average number of species in any country, the species of these genera have more than the average number of varieties. In large genera the species are apt to be closely, but unequally, allied together, forming little clusters round certain species. Species very closely allied to other species apparently have restricted ranges. In all these several respects the species of large genera present a strong analogy with varieties. And we can clearly understand these analogies, if species have once existed as varieties, and have thus originated: whereas, these analogies are utterly inexplicable if each species has been independently created.

We have, also, seen that it is the most flourishing and dominant species of the larger genera which on an average vary most; and varieties, as we shall hereafter see, tend to become converted into new and distinct species. The larger genera thus tend to become larger; and throughout nature the forms of life which are now dominant tend to become still more dominant by leaving many modified and dominant descendants. But by steps hereafter to be explained, the larger genera also tend to break up into smaller genera. And thus, tne forms of life throughout the universe become divided into groups subordinate to groups. ◀1

1 "Groups subordinate to groups" constitutes a nested hierarchy: the Linnaean classification! As we shall soon see, Darwin is keenly aware that his model of transmutation handily explains the nested taxonomic hierarchy that works so well in classification. In this chapter Darwin called attention to the blurred boundaries between species and varieties, making the point that the prevalence of intermediates—those of doubtful status—is consistent with transmutation. His use of botanical arithmetic underscores this point. The underlying pattern in the geographic and taxonomic distribution of varieties only makes sense, he argues, if species are derived from varieties. Continued divergence over time gives us groups subordinate to groups.

Darwin has a difficult task in this chapter. He must impress upon his readers, who are steeped in the tradition of Natural Theology and accustomed to seeing the natural world as a balanced, harmonious place well designed by a benevolent creator, that death, destruction, and competition rage about them at all times. Even more challenging, he must make the case that this destructive process is akin to a purifying fire, the engine of creation for all those exquisite adaptations attributed to the creator. Success in the struggle for existence, when based on the variation presented in the previous chapter, constitutes natural selection. The phrase "struggle for existence" that serves as this chapter's title did not originate with Darwin; the Rev. Thomas Robert Malthus used the phrase in chapter three of his *Essay on Population,* in reference to groups of people ("and the frequent contests with tribes in the same circumstances with themselves, would be so many struggles for existence"). The idea of a struggle was certainly appreciated among naturalists. Augustin De Candolle wrote that "all the plants of a given country are at war one with another" in his important 1820 work *Essai Élémentaire de Géographie Botanique,* and was quoted by Lyell in *Principles of Geology* (1833; vol. II, p. 131). Darwin, perhaps after reading Lyell, referred to "De Candolle's war of nature" in several places. Malthus's expression of the struggle for existence gave both Darwin and Wallace key insights into their independent formulation of the principle of natural selection; both then used the phrase in their subsequent expositions, Darwin first in his 1844 *Essay* (see Darwin 1909) and Wallace in his 1858 transmutation paper (writing, "The life of wild animals is a struggle for existence").

1 Explaining exquisite adaptations is perhaps Darwin's greatest challenge, simply because most of his readers understood such adaptations as evidence of design. This beautifully written paragraph captures our wonder at complex and intricate structures in nature.

CHAPTER III.

STRUGGLE FOR EXISTENCE.

Bears on natural selection—The term used in a wide sense—Geometrical powers of increase — Rapid increase of naturalised animals and plants—Nature of the checks to increase—Competition universal — Effects of climate — Protection from the number of individuals—Complex relations of all animals and plants throughout nature—Struggle for life most severe between individuals and varieties of the same species ; often severe between species of the same genus—The relation of organism to organism the most important of all relations.

BEFORE entering on the subject of this chapter, I must make a few preliminary remarks, to show how the struggle for existence bears on Natural Selection. It has been seen in the last chapter that amongst organic beings in a state of nature there is some individual variability ; indeed I am not aware that this has ever been disputed. It is immaterial for us whether a multitude of doubtful forms be called species or sub-species or varieties ; what rank, for instance, the two or three hundred doubtful forms of British plants are entitled to hold, if the existence of any well-marked varieties be admitted. But the mere existence of individual variability and of some few well-marked varieties, though necessary as the foundation for the work, helps us but little in understanding how species arise in nature. How have ▶1 all those exquisite adaptations of one part of the organisation to another part, and to the conditions of life, and of one distinct organic being to another being, been perfected? We see these beautiful co-adaptations most plainly in the woodpecker and missletoe ; and only a little less plainly in the humblest parasite which clings to the hairs of a quadruped or feathers of a bird; in the structure of the beetle which dives through the water; in the plumed seed which is wafted by the gentlest breeze; in short, we see beautiful adaptations everywhere and in every part of the organic world.

Again, it may be asked, how is it that varieties, which I have called incipient species, become ultimately converted into good and distinct species, which in most cases obviously differ from each other far more than do the varieties of the same species? How do those groups of species, which constitute what are called distinct genera, and which differ from each other more than do the species of the same genus, arise? All these results, as we shall more fully see in the next chapter, follow inevitably from the struggle for life. Owing to this struggle for life, any variation, however slight and from whatever cause proceeding, if it be in any degree profitable to an individual of any species, in its infinitely complex relations to other organic beings and to external nature, will tend to the preservation of that individual, and will generally be inherited by its offspring. The offspring, also, will thus have a better chance of surviving, for, of the many individuals of any species which are periodically born, but a small number can survive. I have called this principle, by which each ◀1 slight variation, if useful, is preserved, by the term of Natural Selection, in order to mark its relation to man's power of selection. We have seen that man by selection can certainly produce great results, and can adapt organic beings to his own uses, through the accumulation of slight but useful variations, given to him by the hand of Nature. But Natural Selection, as we shall ◀2 hereafter see, is a power incessantly ready for action, and is as immeasurably superior to man's feeble efforts, as the works of Nature are to those of Art.

1 Here Darwin is explicit about his choice of the term "natural selection." The term can be semantically problematic in that the word "selection" implies volition—a selector—while the process envisioned by Darwin (and accepted as correct today) is purely mechanistic. Problems with the term led Darwin to regret having coined it; in a letter to Lyell in September 1860, he wrote, "if I had to commence *de novo,* I would have used 'natural preservation'" (*Correspondence* 8: 396). Wallace preferred the phrase "survival of the fittest," which he adopted from the philosopher Herbert Spencer, whom he admired very much. In fact, Wallace was so taken with the phrase that he crossed out "natural selection" through much of his copy of the *Origin* and scribbled over it "survival of the fittest." Darwin somewhat reluctantly followed suit; beginning with the fifth edition of the *Origin* in 1869, he added the sentence, "But the expression often used by Mr. Herbert Spencer of the Survival of the Fittest is more accurate, and is sometimes equally convenient." "Survival of the fittest" was even added to the title of chapter IV of that edition (Peckham 1959, pp. 145, 163). Unfortunately, Darwin's use of this phrase linked his work directly with Spencer's social philosophy, giving rise to social Darwinism and the eugenics movement. That philosophy would be used to justify atrocities in the century to come; see discussion by Paul (1988).

2 Note the use of superlatives and intensifiers in this chapter as Darwin emphasizes the power of natural selection.

1 Here Darwin refers to the passage in Lyell quoted at the head of this chapter, part of a longer discussion on struggle and competition in the plant world.

2 In *Natural Selection* Darwin often cited Rev. William Herbert (1778–1847), clergyman and well-known lily specialist. Here he alludes to an 1846 paper entitled "Local habitations and wants of plants," in which Herbert describes the struggle for existence. Darwin wrote: "Philosophical writers, such as Lyell, Hooker, Herbert, &c. have most ably endeavoured to make others appreciate the struggle & equilibrium of life, as clearly as they do themselves; & I should not have discussed this subject at length, had it not been in many ways of great importance for us; & had I not occasionally met with good observers of nature, who . . . show, as it seems to me, an entire ignorance of the real state of nature" (Stauffer 1975, p. 208).

3 This is a clever opening: we see nature as cheerful and harmonious and forget the constant destruction of some life to support other life. Darwin strives to make his readers conscious of this reality: what does it take to support the populations of those birds whose singing we so uncomprehendingly enjoy?

4 Struggle can be obvious, as with predation or competition, but the challenges posed by the physical environment can be just as deadly.

We will now discuss in a little more detail the struggle for existence. In my future work this subject shall be treated, as it well deserves, at much greater length. [1] The elder De Candolle and Lyell have largely and philosophically shown that all organic beings are exposed to severe competition. In regard to plants, no one has treated this subject with more spirit and ability than [2] W. Herbert, Dean of Manchester, evidently the result of his great horticultural knowledge. Nothing is easier than to admit in words the truth of the universal struggle for life, or more difficult—at least I have found it so—than constantly to bear this conclusion in mind. Yet unless it be thoroughly engrained in the mind, I am convinced that the whole economy of nature, with every fact on distribution, rarity, abundance, extinction, and variation, will be dimly seen or quite misunderstood. [3] We behold the face of nature bright with gladness, we often see superabundance of food; we do not see, or we forget, that the birds which are idly singing round us mostly live on insects or seeds, and are thus constantly destroying life; or we forget how largely these songsters, or their eggs, or their nestlings, are destroyed by birds and beasts of prey; we do not always bear in mind, that though food may be now superabundant, it is not so at all seasons of each recurring year.

I should premise that I use the term Struggle for Existence in a large and metaphorical sense, including dependence of one being on another, and including (which is more important) not only the life of the individual, but success in leaving progeny. Two canine animals in a time of dearth, may be truly said to struggle with each other which shall get food and live. [4] But a plant on the edge of a desert is said to struggle for life against the drought, though more properly it should be said to be dependent on the moisture. A plant which annually produces a thousand seeds, of which on an average only one comes to maturity, may be more truly said to struggle with the plants of the same and other kinds which already clothe the ground. The missletoe is dependent on the apple and a few other trees, but can only in a far-fetched sense be said to struggle with these trees, for if too many of these parasites grow on the same tree, it will languish and die. But several seedling missletoes, growing close together on the same branch, may more truly be said to struggle with each other. As the missletoe is disseminated by birds, its existence depends on birds; and it may metaphorically be said to struggle with other fruit-bearing plants, in order to tempt birds to devour and thus disseminate its seeds rather than those of other plants. In these several senses, which pass into each other, I use for convenience sake the general term of struggle for existence.

A struggle for existence inevitably follows from the high rate at which all organic beings tend to increase. Every being, which during its natural lifetime produces several eggs or seeds, must suffer destruction during some period of its life, and during some season or occasional year, otherwise, on the principle of geometrical increase, its numbers would quickly become so inordinately great that no country could support the product. Hence, as more individuals are produced than can possibly survive, there must in every case be a struggle for existence, either one individual with another of the same species, or with the individuals of distinct species, or with the physical conditions of life. It is the doctrine of Malthus applied with manifold force to the whole animal and vegetable kingdoms; for in this case there can be no artificial increase of food, and no prudential restraint from marriage. Although some species may

1 Mistletoes, to use the modern spelling, are semi-parasitic epiphytic plants. This common name includes members of several families worldwide, but Darwin would have been most familiar with the European mistletoe, *Viscum album* (Santalaceae). Its Latin name reflects how the seeds are dispersed: the pulp of the berries (which the epithet *album* refers to) is very sticky (hence *Viscum*), so the seeds adhere to birds' beaks as they dine on the fruits. When a bird attempts to dislodge the sticky mass by wiping its beak on a branch (ideally of a different tree), the seeds are then stuck to a new host. The most common New World mistletoe is genus *Phoradendron,* "carried to trees."

2 The Malthusian formulation. The Rev. Thomas Robert Malthus's *Essay on Population* provided this key insight to both Darwin and Wallace. Writing about human society, Malthus argues that population growth always outstrips sustenance, and he concludes with a pessimistic outlook on prospects for human happiness. Darwin and Wallace saw that this principle must operate in nature, too, but with "manifold force." Malthus, who lived from 1766 to 1834, wrote his essay as a contribution to the social discourse of the day concerning the perfectibility of mankind and society; in the wake of the Revolution in France, there was utopian optimism about improving the condition of the common people—the prevalence of misery and inequity of the masses being an important factor in precipitating the Revolution. Malthus was sympathetic to the desirability of improving the lot of the masses, but he saw population pressure as a major obstacle to realizing that goal. Darwin and Wallace immediately saw the applicability of these population concerns to species in general.

be now increasing, more or less rapidly, in numbers, all cannot do so, for the world would not hold them.

There is no exception to the rule that every organic being naturally increases at so high a rate, that if not destroyed, the earth would soon be covered by the progeny of a single pair. Even slow-breeding man has doubled in twenty-five years, and at this rate, in a few thousand years, there would literally not be standing room for his progeny. Linnæus has calculated that if an annual plant produced only two seeds—and there is no plant so unproductive as this—and their seedlings next year produced two, and so on, then in twenty years there would be a million plants. [1] The elephant is reckoned to be the slowest breeder of all known animals, and I have taken some pains to estimate its probable minimum rate of natural increase: it will be under the mark to assume that it breeds when thirty years old, and goes on breeding till ninety years old, bringing forth three pair of young in this interval; if this be so, at the end of the fifth century there would be alive fifteen million elephants, descended from the first pair.

But we have better evidence on this subject than mere theoretical calculations, namely, the numerous recorded cases of the astonishingly rapid increase of various animals in a state of nature, when circumstances have been favourable to them during two or three following seasons. Still more striking is the evidence from our domestic animals of many kinds which have run wild in several parts of the world: if the statements of the rate of increase of slow-breeding cattle and horses in South-America, and latterly in Australia, had not been well authenticated, they would have been quite incredible. So it is with plants: cases could be given of introduced plants which have become common throughout whole islands in a period of less than ten years. Several

1 Examples of slow-breeding species underscore the point: if unchecked, even they will soon have sizable populations. Expressed in simple mathematical form—worked out well after Darwin's time—a population (N_0) will grow exponentially if unchecked at a rate dependent on the intrinsic rate of increase, *r*, also called the Malthusian parameter. This in turn is the difference between the birth and death rates in the population. With constant rates of birth and death, at time *t* the population can be expressed as $N_t = N_0 e^{rt}$ where e = natural log base (known as Euler's number after the Swiss mathematician Leonhard Euler).

An anonymous blogger has pointed out that the algorithm and starting assumptions that Darwin used to come up with his estimate of 15 million elephants are something of a mystery. The sex ratio of the offspring makes a big difference to any such calculation. If a 50:50 offspring sex ratio is assumed for elephants producing 1 offspring every 10 years for 60 years, we come up with only about 1.8 million elephants at the end of the 500-year period Darwin gives. Curiously, using an algorithm in which females produce 1 offspring every 10 years from the age of 30 through 90, each giving birth to 4 daughters, a son, and then another daughter, after 500 years the population is 14,999,058—almost exactly Darwin's figure! Did Darwin assume a biased sex ratio? The blogger speculates that if he did, someone may have objected, for in the fifth edition this calculation is altered to reflect an outcome based on an even sex ratio: "it will be safest to assume that [the elephant] begins breeding when 30 years old, and goes on breeding till 90 years old, bringing forth six young in the interval, and surviving till one hundred years old; if this be so then, after a period of from 740 to 750 years there would be nearly nineteen million elephants alive descended from the first pair" (Peckham 1959, p. 148).

of the plants now most numerous over the wide plains of La Plata, clothing square leagues of surface almost to the exclusion of all other plants, have been introduced from Europe; and there are plants which now range in India, as I hear from Dr. Falconer, from Cape Comorin to the Himalaya, which have been imported from America since its discovery. In such cases, and endless instances could be given, no one supposes that the fertility of these animals or plants has been suddenly and temporarily increased in any sensible degree. The obvious explanation is that the conditions of life have been very favourable, and that there has consequently been less destruction of the old and young, and that nearly all the young have been enabled to breed. In such cases the geometrical ratio of increase, the result of which never fails to be surprising, simply explains the extraordinarily rapid increase and wide diffusion of naturalised productions in their new homes.

In a state of nature almost every plant produces seed, and amongst animals there are very few which do not annually pair. Hence we may confidently assert, that all plants and animals are tending to increase at a geometrical ratio, that all would most rapidly stock every station in which they could any how exist, and that the geometrical tendency to increase must be checked by destruction at some period of life. Our familiarity with the larger domestic animals tends, I think, to mislead us: we see no great destruction falling on them, and we forget that thousands are annually slaughtered for food, and that in a state of nature an equal number would have somehow to be disposed of.

The only difference between organisms which annually produce eggs or seeds by the thousand, and those which produce extremely few, is, that the slow-breeders would require a few more years to people, under favourable

1 Darwin is referring to the plains surrounding the Rio de la Plata, which translates as Silver River but is sometimes called the River Plate in English ("plate" being a synonym for gold and silver). Darwin visited this area in the summer of 1832 while on the voyage of the *Beagle.* He commented on the "open slightly-undulating country, covered by one uniform layer of fine green turf, on which countless herds of cattle, sheep, and horses graze." Traveling inland about seventy miles, he found things no better: "the country wore the same aspect, till at last the fine green turf became more wearisome than a dusty turnpike road." Converting the area to rangeland for cattle grazing would have resulted in a high degree of disturbance, conditions under which weedy exotic species thrive.

2 Another way of demonstrating the effects of checks on population increase is to look at how populations explode when such checks are removed. This is often the case when species are introduced into novel environments, where, in the absence of their usual predators, pathogens, and competitors, their populations swell dramatically. Think rabbits in Australia or kudzu in southeastern North America.

3 Here Darwin presents another interesting angle to impress upon readers the omnipresent destruction that keeps populations in check: *we* are the check on populations of domesticated organisms.

conditions, a whole district, let it be ever so large. The condor lays a couple of eggs and the ostrich a score, and yet in the same country the condor may be the more numerous of the two: the Fulmar petrel lays but one egg, yet it is believed to be the most numerous bird in the world. One fly deposits hundreds of eggs, and another, like the hippobosca, a single one; but this difference does not determine how many individuals of the two species can be supported in a district. A large number of eggs is of some importance to those species, which depend on a rapidly fluctuating amount of food, for it allows them rapidly to increase in number. But the real importance of a large number of eggs or seeds is to make up for much destruction at some period of life; and this period in the great majority of cases is an early one. If an animal can in any way protect its own eggs or young, a small number may be produced, and yet the average stock be fully kept up; but if many eggs or young are destroyed, many must be produced, or the species will become extinct. It would suffice to keep up the full number of a tree, which lived on an average for a thousand years, if a single seed were produced once in a thousand years, supposing that this seed were never destroyed, and could be ensured to germinate in a fitting place. So that in all cases, the average number of any animal or plant depends only indirectly on the number of its eggs or seeds.

In looking at Nature, it is most necessary to keep the foregoing considerations always in mind—never to forget that every single organic being around us may be said to be striving to the utmost to increase in numbers; that each lives by a struggle at some period of its life; that heavy destruction inevitably falls either on the young or old, during each generation or at recurrent intervals. Lighten any check, mitigate the

1 The "hippobosca" fly refers to louse flies, also known as keds, family Hippoboscidae. Like true lice, louse flies are external parasites of birds and mammals—one of several fly families to adopt this parasitic lifestyle. Their peculiar morphology enables them to hang on to their host: they are often wingless and flattened, with long powerful legs bearing stout claws splayed out to better grasp hairs or feather shafts. Darwin collected several hippoboscids on the voyage of the *Beagle,* including a specimen of *Ornithomyia intertropica* from a Galápagos hawk (Smith 1987). The name of this family is derived from the common species that parasitizes horses: *hippobosca,* conferred by Linnaeus, is literally derived from the Greek "hippo" (horse) and "bosc" (to feed). They do not invariably deposit a single egg, contrary to Darwin's statement here—in fact, these remarkable flies complete their larval development largely while still unborn, so no eggs are laid per se. Females "larviposit" on a host rather than lay eggs, and the maggotlike larva pupates almost immediately into an adult. Nonetheless, Darwin's point is well taken: they are certainly at the low end of the spectrum for flies when it comes to young produced per reproductive episode. There must be a strong selective pressure not to overwhelm a host with parasites; hundreds of them would soon kill the host, taking the flies with it.

Darwin might have been dismayed to learn that another fly ectoparasite has been discovered infesting the Galápagos finches that now bear his name. The larvae of *Philornis downsi* (Muscidae) were first found to be obligate blood-feeding parasites of Galápagos passerines in 1997. Closer scrutiny found them in an astonishing 97 percent of the finch nests studied, with an average of more than twenty-three parasites per nestling, resulting in a 27 percent nestling mortality (Fessl and Tebbich 2002; Fessl et al. 2006).

2 Darwin periodically stresses this point. Note his choice of terms: *every single* organic being is striving to the *utmost* to increase, but *heavy* destruction *inevitably* falls . . .

destruction ever so little, and the number of the species will almost instantaneously increase to any amount. The face of Nature may be compared to a yielding surface, with ten thousand sharp wedges packed close together and driven inwards by incessant blows, sometimes one wedge being struck, and then another with greater force.

What checks the natural tendency of each species to increase in number is most obscure. Look at the most vigorous species; by as much as it swarms in numbers, by so much will its tendency to increase be still further increased. We know not exactly what the checks are in even one single instance. Nor will this surprise any one who reflects how ignorant we are on this head, even in regard to mankind, so incomparably better known than any other animal. This subject has been ably treated by several authors, and I shall, in my future work, discuss some of the checks at considerable length, more especially in regard to the feral animals of South America. Here I will make only a few remarks, just to recall to the reader's mind some of the chief points. Eggs or very young animals seem generally to suffer most, but this is not invariably the case. With plants there is a vast destruction of seeds, but, from some observations which I have made, I believe that it is the seedlings which suffer most from germinating in ground already thickly stocked with other plants. Seedlings, also, are destroyed in vast numbers by various enemies; for instance, on a piece of ground three feet long and two wide, dug and cleared, and where there could be no choking from other plants, I marked all the seedlings of our native weeds as they came up, and out of the 357 no less than 295 were destroyed, chiefly by slugs and insects. If turf which has long been mown, and the case would be the same with turf closely browsed by quadrupeds, be let to grow,

1 This is Darwin's famous wedges metaphor. The wedges represent competing forms or species, all of which are hammered by selection to secure a space in the "face of nature"—i.e., to exist. Note the central importance of competition in this metaphor: a new wedge can gain a place in the economy of nature only by forcing another out. Here Darwin is not so explicit about forcing others out, but in the D notebook he wrote: "One may say there is a force like a hundred thousand wedges trying [to] force every kind of adapted structure into the gaps in the economy of nature, or rather forming gaps by thrusting out weaker ones" (Barrett et al. 1987, p. 375). Similarly, in *Natural Selection* he wrote of "ten-thousand sharp wedges . . . one driven deeply in forcing out others." In an expression of ecological interconnectedness, he continues: "with the jar & shock often transmitted very far to other wedges in many lines of direction" (Stauffer 1975, p. 208).

For reasons that are unclear, though the wedge metaphor appears in Darwin's earliest evolutionary writings (the D notebook dating to 1838, and his 1842 and 1844 essays) and makes it to this point, the first edition of the *Origin*, it was dropped from all subsequent editions of the *Origin*.

2 This homespun empirical demonstration makes the subtle point that struggle is everywhere, always—even in our very lawns. In January 1857 Darwin cleared this small plot of perennials, and from March through the following August he made daily records of seedings as they appeared and were lost to mortality. "I am amusing myself with several little experiments; I have now got a little weed garden & am marking each seedling as it appears, to see at what time of life they suffer most," he wrote to Hooker in late March 1857 (*Correspondence* 6: 363).

1 It is important to realize that checks on growth come from abiotic as well as biotic factors; most or all of these factors are in operation simultaneously, but their relative importance varies a great deal both spatially and temporally. In extreme environments (e.g., deserts or the arctic) abiotic factors can be the primary agents of struggle. The severity of the winter of 1854–1855 was legendary. The Crimean War, of which Britain was a part, was raging at the time, and troop mortality from cold and disease reached record levels that winter.

the more vigorous plants gradually kill the less vigorous, though fully grown, plants: thus out of twenty species growing on a little plot of turf (three feet by four) nine species perished from the other species being allowed to grow up freely.

The amount of food for each species of course gives the extreme limit to which each can increase; but very frequently it is not the obtaining food, but the serving as prey to other animals, which determines the average numbers of a species. Thus, there seems to be little doubt that the stock of partridges, grouse, and hares on any large estate depends chiefly on the destruction of vermin. If not one head of game were shot during the next twenty years in England, and, at the same time, if no vermin were destroyed, there would, in all probability, be less game than at present, although hundreds of thousands of game animals are now annually killed. On the other hand, in some cases, as with the elephant and rhinoceros, none are destroyed by beasts of prey: even the tiger in India most rarely dares to attack a young elephant protected by its dam.

1 Climate plays an important part in determining the average numbers of a species, and periodical seasons of extreme cold or drought, I believe to be the most effective of all checks. I estimated that the winter of 1854–55 destroyed four-fifths of the birds in my own grounds; and this is a tremendous destruction, when we remember that ten per cent. is an extraordinarily severe mortality from epidemics with man. The action of climate seems at first sight to be quite independent of the struggle for existence; but in so far as climate chiefly acts in reducing food, it brings on the most severe struggle between the individuals, whether of the same or of distinct species, which subsist on the same kind of food. Even when climate, for instance extreme

cold, acts directly, it will be the least vigorous, or those which have got least food through the advancing winter, which will suffer most. When we travel from south to north, or from a damp region to a dry, we invariably see some species gradually getting rarer and rarer, and finally disappearing; and the change of climate being conspicuous, we are tempted to attribute the whole effect to its direct action. But this is a very false view: we forget that each species, even where it most abounds, is constantly suffering enormous destruction at some period of its life, from enemies or from competitors for the same place and food; and if these enemies or competitors be in the least degree favoured by any slight change of climate, they will increase in numbers, and, as each area is already fully stocked with inhabitants, the other species will decrease. When we travel southward and see a species decreasing in numbers, we may feel sure that the cause lies quite as much in other species being favoured, as in this one being hurt. So it is when we travel northward, but in a somewhat lesser degree, for the number of species of all kinds, and therefore of competitors, decreases northwards; hence in going northward, or in ascending a mountain, we far oftener meet with stunted forms, due to the *directly* injurious action of climate, than we do in proceeding southwards or in descending a mountain. When we reach the Arctic regions, or snow-capped summits, or absolute deserts, the struggle for life is almost exclusively with the elements.

That climate acts in main part indirectly by favouring other species, we may clearly see in the prodigious number of plants in our gardens which can perfectly well endure our climate, but which never become naturalised, for they cannot compete with our native plants, nor resist destruction by our native animals.

1 A point worth noting: our direct experience with climate leads us to conclude that it is the most important factor underlying species shifts of the kind Darwin describes. His statement about the enormous destruction constantly visited on all species echoes page 62.

2 The relative importance of competitors/predators and climate in shaping biological communities varies geographically; the more severe the climate, the fewer the species that can survive, let alone compete with each other. Hence, under the most severe conditions such as those that prevail in places like the arctic, arid deserts, or at timberline, Darwin says, organisms struggle primarily with the environment, not each other. The distinctive stunted vegetation of high-elevation exposures is called *krummholz,* a term coined in the early twentieth century from the German for "crooked wood." Poet Alan Sullivan's lovely poem *Krummholz* opens with the lines

Hunched like an anchorite behind its boulder,
a treeline pine weathers the winter storms.
Its knotty branches shrink as nights turn colder.
Caught in its tufts, a fluted snowdrift forms.

1 Population can be checked as effectively by disease and parasites as by the elements or limited food supplies. People have a good understanding of the importance of disease in this regard, but among nonbiologists there is often a profound underappreciation of the power of parasitism. Some of the most bizarre and elaborate strategies and life histories are found among parasitic organisms, some reminiscent of the most chilling creations of science fiction (no coincidence). In fact, the lifestyle of some parasites is cited later by Darwin as evidence against special creation.

2 Poor crop productivity under conditions of low population density is attributable to inefficient pollination. Darwin's point here is that population must remain above some threshold level ("large stocks," as he puts it) in order to remain viable. This is an important principle in conservation biology: minimum viable population (MVP) analysis is used to estimate the extinction risk of populations (see Shaffer 1981).

1 When a species, owing to highly favourable circumstances, increases inordinately in numbers in a small tract, epidemics—at least, this seems generally to occur with our game animals—often ensue: and here we have a limiting check independent of the struggle for life. But even some of these so-called epidemics appear to be due to parasitic worms, which have from some cause, possibly in part through facility of diffusion amongst the crowded animals, been disproportionably favoured: and here comes in a sort of struggle between the parasite and its prey.

On the other hand, in many cases, a large stock of individuals of the same species, relatively to the numbers of its enemies, is absolutely necessary for its preservation. Thus we can easily raise plenty of corn and rape-seed, &c., in our fields, because the seeds are in great excess compared with the number of birds which feed on them; nor can the birds, though having a superabundance of food at this one season, increase in number proportionally to the supply of seed, as their numbers
2 are checked during winter: but any one who has tried, knows how troublesome it is to get seed from a few wheat or other such plants in a garden; I have in this case lost every single seed. This view of the necessity of a large stock of the same species for its preservation, explains, I believe, some singular facts in nature, such as that of very rare plants being sometimes extremely abundant in the few spots where they do occur; and that of some social plants being social, that is, abounding in individuals, even on the extreme confines of their range. For in such cases, we may believe, that a plant could exist only where the conditions of its life were so favourable that many could exist together, and thus save each other from utter destruction. I should add that the good effects of frequent intercrossing, and the ill effects

of close interbreeding, probably come into play in some of these cases; but on this intricate subject I will not here enlarge.

Many cases are on record showing how complex and unexpected are the checks and relations between organic beings, which have to struggle together in the same country. I will give only a single instance, which, though a simple one, has interested me. In Staffordshire, on the estate of a relation where I had ample means of investigation, there was a large and extremely barren heath, which had never been touched by the hand of man; but several hundred acres of exactly the same nature had been enclosed twenty-five years previously and planted with Scotch fir. The change in the native vegetation of the planted part of the heath was most remarkable, more than is generally seen in passing from one quite different soil to another: not only the proportional numbers of the heath-plants were wholly changed, but twelve species of plants (not counting grasses and carices) flourished in the plantations, which could not be found on the heath. The effect on the insects must have been still greater, for six insectivorous birds were very common in the plantations, which were not to be seen on the heath; and the heath was frequented by two or three distinct insectivorous birds. Here we see how potent has been the effect of the introduction of a single tree, nothing whatever else having been done, with the exception that the land had been enclosed, so that cattle could not enter. But how important an element enclosure is, I plainly saw near Farnham, in Surrey. Here there are extensive heaths, with a few clumps of old Scotch firs on the distant hill-tops: within the last ten years large spaces have been enclosed, and self-sown firs are now springing up in multitudes, so close together that all cannot live.

1 This Staffordshire estate is Maer Hall, the grand home of the Wedgwoods. Josiah Wedgwood II (1769–1843), son of the renowned pottery manufacturer, was Darwin's uncle and father-in-law, and the Darwin and Wedgwood families were always very close. No wonder Darwin had "ample means of investigation" there.

2 A single introduced species (in this case, a tree) has measurable effects on other organisms—in short, on the very ecology of the enclosure. The vegetation changes affect the birds, which in turn affect the insects.

3 To treat his mysterious ailments, Darwin periodically stayed at Moor Park, a hydropathic treatment establishment near Farnham, Surrey, run by the physician Edward W. Lane. Even while undergoing treatment, Darwin had ample time to explore his surroundings. This rolling heath is near Crooksbury Hill, not far from Moor Park, and Darwin kept detailed notes on its plants and the effects of enclosure. Some of these observations are found in the *Natural Selection* manuscript (see Stauffer 1975, pp. 570–571). After one passage dated May 5, 1857, he appended a note to himself: "N.B. I have been again all over Farnham common: *part enclosed & part unenclosed & the case is very curious.*— Enclosed part studded with trees" (emphasis in original).

When I ascertained that these young trees had not been sown or planted, I was so much surprised at their numbers that I went to several points of view, whence I could examine hundreds of acres of the unenclosed heath, and literally I could not see a single Scotch fir, except the old planted clumps. But on looking closely between the stems of the heath, I found a multitude of seedlings and little trees, which had been perpetually browsed down by the cattle. In one square yard, at a point some hundred yards distant from one of the old clumps, I counted thirty-two little trees; and one of them, judging from the rings of growth, had during twenty-six years tried to raise its head above the stems of the heath, and had failed. No wonder that, as soon as the land was enclosed, it became thickly clothed with vigorously growing young firs. Yet the heath was so extremely barren and so extensive that no one would ever have imagined that cattle would have so closely and effectually searched it for food.

Here we see that cattle absolutely determine the existence of the Scotch fir; but in several parts of the world insects determine the existence of cattle. Perhaps Paraguay offers the most curious instance of this; for here neither cattle nor horses nor dogs have ever run wild, though they swarm southward and northward in a feral state; and Azara and Rengger have shown that this is caused by the greater number in Paraguay of a certain fly, which lays its eggs in the navels of these animals when first born. The increase of these flies, numerous as they are, must be habitually checked by some means, probably by birds. Hence, if certain insectivorous birds (whose numbers are probably regulated by hawks or beasts of prey) were to increase in Paraguay, the flies would decrease—then cattle and horses would become feral, and this would certainly greatly alter (as

1 Darwin's observations on the effects of enclosure at the Staffordshire and Surrey sites show not only his use of data to support his arguments but also his knack for seeing informative patterns in seemingly mundane places.

2 This is the tip of the iceberg of interconnectedness; here and over the next several pages Darwin is alluding to ecological interaction. Such chains of interaction are food chains, or one strand in a complex food web; indeed, on the next page Darwin refers to animals bound together by "web[s] of complex relations." Cause and effect are not simply linear, however—they can be circular in the form of a feedback loop. The term "ecology" *(oekologie)* was coined by the German zoologist Ernst Haeckel seven years later, in 1866, and thus gave expression to what biologists had long loosely described as the "economy of nature." See Stauffer (1960) for a discussion of Darwin's ecological thinking.

For the account of feral cattle and horse populations checked by parasitic flies, Darwin consulted Félix d'Azara's *Voyages dans l'Amérique mériodionale* (Paris, 1809) and *Essais sur l'histoire naturelle des quadrupèdes de la Province de Paraguay* (Paris, 1801) and Johann Rudolf Rengger's *Naturgeschichte der Säugethiere von Paraguay* (Basel, 1830). Feral livestock populations increased dramatically in La Plata once the colony there was abandoned in 1537, yet very few were found to run wild in Paraguay, apparently owing to this one pestilential fly species. The fly in question could be the New World screw-worm fly *Cochliomyia hominivorax* (family Calliphoridae), obligate parasite of mammals. Female flies are attracted to open wounds or sores, or the navel of newborns, where they lay several hundred eggs. The larvae burrow into body tissue, a condition called myiasis.

must here have gone on during long centuries, each annually scattering its seeds by the thousand; what war between insect and insect—between insects, snails, and other animals with birds and beasts of prey—all striving to increase, and all feeding on each other or on the trees or their seeds and seedlings, or on the other plants which first clothed the ground and thus checked the growth of the trees! Throw up a handful of feathers, and all must fall to the ground according to definite laws; but how simple is this problem compared to the action and reaction of the innumerable plants and animals which have determined, in the course of centuries, the proportional numbers and kinds of trees now growing on the old Indian ruins! ◀1

The dependency of one organic being on another, as of a parasite on its prey, lies generally between beings remote in the scale of nature. This is often the case with those which may strictly be said to struggle with each other for existence, as in the case of locusts and grass-feeding quadrupeds. But the struggle almost invariably will be most severe between the individuals of the same species, for they frequent the same districts, require the same food, and are exposed to the same dangers. In the case of varieties of the same species, the struggle will generally be almost equally severe, and we sometimes see the contest soon decided: for instance, if several varieties of wheat be sown together, and the mixed seed be resown, some of the varieties which best suit the soil or climate, or are naturally the most fertile, will beat the others and so yield more seed, and will consequently in a few years quite supplant the other varieties. To keep up a mixed stock of even such extremely close varieties as the variously coloured sweet-peas, they must be each year harvested separately, and the seed then mixed in due propor- ◀2

1 Darwin seems most poetic when impassioned with his argument, and this is nowhere more evident than when he is exhorting his readers to see things in a new light. In this passage, incessant struggle—wars upon wars—is the rule rather than the exception: predicting the trajectory of a handful of tossed feathers fluttering to the ground is easier than unraveling the complexities of interaction that shape populations. The idea of epic struggle in nature was not original to Darwin; recall de Candolle (1820): "All the plants of a given country are at war one with another."

In the plant world these battles are fought in slow motion, so to speak, over vast stretches of time. Imagine the still forest as a plant battleground, with flanking movements, frontal attacks, and penetration of enemy lines accomplished inch by inch over years, decades, centuries. This passage illustrates the tough row Darwin must hoe: his readers will be struck by the (apparently) peaceful tranquility of the woodland and repulsed by a worldview that will have them see the natural world as so many battlefields, massacre sites. "Lord, how savage you will be . . . and how you will long to crucify me alive!" Darwin wrote to Hugh Falconer when the *Origin* was published (*Correspondence* 7: 368).

2 A key idea! This passage presages Darwin's theory of the mechanism behind divergence of character, which will be discussed at some length in the next chapter. The concept anticipates the Principle of Competitive Exclusion, also known as Gause's Law, a tenet of modern ecological theory developed by the Soviet biologist Georgii Frantsevitch Gause in 1934, through his pioneering experiments with the protozoan *Paramecium.*

tion, otherwise the weaker kinds will steadily decrease in numbers and disappear. So again with the varieties of sheep: it has been asserted that certain mountain-varieties will starve out other mountain-varieties, so that they cannot be kept together. The same result has followed from keeping together different varieties of the medicinal leech. It may even be doubted whether the varieties of any one of our domestic plants or animals have so exactly the same strength, habits, and constitution, that the original proportions of a mixed stock could be kept up for half a dozen generations, if they were allowed to struggle together, like beings in a state of nature, and if the seed or young were not annually sorted.

[1] As species of the same genus have usually, though by no means invariably, some similarity in habits and constitution, and always in structure, the struggle will generally be more severe between species of the same genus, when they come into competition with each other, than between species of distinct genera. We see this in the recent extension over parts of the United States of one species of swallow having caused the decrease of an-
[2] other species. The recent increase of the missel-thrush in parts of Scotland has caused the decrease of the song-thrush. How frequently we hear of one species of rat taking the place of another species under the most different climates! In Russia the small Asiatic cockroach has everywhere driven before it its great congener. One species of charlock will supplant another, and so in other cases. We can dimly see why the competition should be most severe between allied forms, which fill nearly the same place in the economy of nature; but probably in no one case could we precisely say why one species has been victorious over another in the great battle of life.

1 This is a restatement of the main point behind competitive exclusion: individuals from two closely related species will tend to have similar ecological requirements (food and space needs, and other resources), and thus come into direct competition more readily than will those from less closely related species.

2 The "missel-thrush" or mistle-thrush, *Turdus viscivorus,* is a large European thrush that feeds on mistletoe berries. Recall from p. 63 that the European mistletoe is *Viscum album*—this thrush that specializes on its fruits is thus "viscivorus." Darwin got his information about the increased population of this thrush at the expense of the song-thrush in Scotland from an article by I. Edwards in the *Zoologist,* v. 13–14, 1855–1856. He was trying to recall the reference in response to a correspondent, Alfred Newton, who doubted the assertion in a letter written in March 1874. "Since my boyhood, now about 50 years, I feel sure that Missel-thrush has much increased," Darwin replied. "I remember my astonishment when I saw the first which appeared in my Father's grounds at Shrewsbury." After penning his reply, Darwin found the original reference and supplied it in a postscript (De Beer 1959).

A corollary of the highest importance may be deduced from the foregoing remarks, namely, that the structure of every organic being is related, in the most essential yet often hidden manner, to that of all other organic beings, with which it comes into competition for food or residence, or from which it has to escape, or on which it preys. This is obvious in the structure of the teeth and talons of the tiger; and in that of the legs and claws of the parasite which clings to the hair on the tiger's body. But in the beautifully plumed seed of the dandelion, and in the flattened and fringed legs of the water-beetle, the relation seems at first confined to the elements of air and water. Yet the advantage of plumed seeds no ◀1 doubt stands in the closest relation to the land being already thickly clothed by other plants; so that the seeds may be widely distributed and fall on unoccupied ground. In the water-beetle, the structure of its legs, so well adapted for diving, allows it to compete with other aquatic insects, to hunt for its own prey, and to escape serving as prey to other animals.

The store of nutriment laid up within the seeds of many plants seems at first sight to have no sort of relation to other plants. But from the strong growth of young plants produced from such seeds (as peas and beans), when sown in the midst of long grass, I suspect that the chief use of the nutriment in the seed is to favour the growth of the young seedling, whilst struggling with other plants growing vigorously all around.

Look at a plant in the midst of its range, why does ◀2 it not double or quadruple its numbers? We know that it can perfectly well withstand a little more heat or cold, dampness or dryness, for elsewhere it ranges

1 Such structures as the seed plumes or flattened hair-fringed legs of the water beetle are adaptive *with respect to* other organisms in their environment, in the same complex web of interaction.

2 This rhetorical question is another way Darwin leads his readers to his central point about struggle; why indeed does a particular species not become far more abundant than it is, given the potent power of intrinsic population increase? The answer is incessant checks.

1 It is extremely important, in Darwin's view, for his readers to appreciate the complexity of the interactions, checks, and balances that control the abundance and distribution of each organism. He seems almost frustrated by the need to convey this difficult concept that is essential to his theory of transmutation by natural selection.

into slightly hotter or colder, damper or drier districts. In this case we can clearly see that if we wished in imagination to give the plant the power of increasing in number, we should have to give it some advantage over its competitors, or over the animals which preyed on it. On the confines of its geographical range, a change of constitution with respect to climate would clearly be an advantage to our plant; but we have reason to believe that only a few plants or animals range so far, that they are destroyed by the rigour of the climate alone. Not until we reach the extreme confines of life, in the arctic regions or on the borders of an utter desert, will competition cease. The land may be extremely cold or dry, yet there will be competition between some few species, or between the individuals of the same species, for the warmest or dampest spots.

Hence, also, we can see that when a plant or animal is placed in a new country amongst new competitors, though the climate may be exactly the same as in its former home, yet the conditions of its life will generally be changed in an essential manner. If we wished to increase its average numbers in its new home, we should have to modify it in a different way to what we should have done in its native country; for we should have to give it some advantage over a different set of competitors or enemies.

1 It is good thus to try in our imagination to give any form some advantage over another. Probably in no single instance should we know what to do, so as to succeed. It will convince us of our ignorance on the mutual relations of all organic beings; a conviction as necessary, as it seems to be difficult to acquire. All that we can do, is to keep steadily in mind that each organic being is striving to increase at a geometrical

ratio; that each at some period of its life, during some season of the year, during each generation or at intervals, has to struggle for life, and to suffer great destruction. When we reflect on this struggle, we may console ourselves with the full belief, that the war of nature is not incessant, that no fear is felt, that death is generally prompt, and that the vigorous, the healthy, and the happy survive and multiply. ◀1

1 The final two sentences of this chapter are remarkable for their content and juxtaposition. In the penultimate sentence Darwin repeats his mantra: if all populations are constantly "striving" to increase (and the reproductive potential is indeed there to do so exponentially), yet numbers always stay about the same, then incessant death and destruction must be occurring on a grand scale.

The final sentence has a completely different, almost consoling tone. Did Darwin, realizing that his argument would horrify his Victorian audience steeped in the Natural Theology worldview, seek to soften the blow? Is he consoling himself, too?

In this chapter Darwin reiterates the two key elements of the natural selection process: the occurrence of abundant and heritable variation and the struggle for existence. He also reveals the depth of his insight into this mechanism: for example, his realization that if successful reproduction is the currency of transmutation, reproductive choices and competition must set into play another kind of selection dynamic, which he termed "sexual selection." Other important revelations include his "divergence of character" model, which produces the spreading branches of the tree of life; the central role that extinction plays as a natural process; and the realization that these processes neatly explain the Linnaean hierarchy of taxonomic groups nested within groups. This magisterial chapter is a nice précis of one of the two main contributions of this book, namely, Darwin's mechanism of transmutation. The subsequent chapters collectively provide the other main contribution: a demonstration that transmutation has occurred. Note that the only figure in the book, a diagram, makes its debut in this chapter. Darwin refers to this versatile diagram to make different but related points in later chapters.

---

1 This first paragraph, which begins with two rhetorical questions, summarizes Darwin's argument in the previous three chapters.

CHAPTER IV.

NATURAL SELECTION.

Natural Selection—its power compared with man's selection—its power on characters of trifling importance—its power at all ages and on both sexes—Sexual Selection—On the generality of intercrosses between individuals of the same species—Circumstances favourable and unfavourable to Natural Selection, namely, intercrossing, isolation, number of individuals—Slow action—Extinction caused by Natural Selection—Divergence of Character, related to the diversity of inhabitants of any small area, and to naturalisation—Action of Natural Selection, through Divergence of Character and Extinction, on the descendants from a common parent—Explains the Grouping of all organic beings.

▶1 How will the struggle for existence, discussed too briefly in the last chapter, act in regard to variation? Can the principle of selection, which we have seen is so potent in the hands of man, apply in nature? I think we shall see that it can act most effectually. Let it be borne in mind in what an endless number of strange peculiarities our domestic productions, and, in a lesser degree, those under nature, vary; and how strong the hereditary tendency is. Under domestication, it may be truly said that the whole organisation becomes in some degree plastic. Let it be borne in mind how infinitely complex and close-fitting are the mutual relations of all organic beings to each other and to their physical conditions of life. Can it, then, be thought improbable, seeing that variations useful to man have undoubtedly occurred, that other variations useful in some way to each being in the great and complex battle of life, should sometimes occur in the course of thousands of generations? If such do occur, can we doubt (remembering that many more individuals are born than can possibly survive) that individuals having any advantage, however slight, over others, would have the best chance of surviving and of procreating their kind? On the other hand, we may feel sure that any variation in the least degree injurious would be rigidly destroyed. This preservation of favourable variations 1 and the rejection of injurious variations, I call Natural Selection. Variations neither useful nor injurious would not be affected by natural selection, and would be left a fluctuating element, as perhaps we see in the species called polymorphic.

We shall best understand the probable course of 2 natural selection by taking the case of a country undergoing some physical change, for instance, of climate. The proportional numbers of its inhabitants would almost immediately undergo a change, and some species might become extinct. We may conclude, from what we have seen of the intimate and complex manner in which the inhabitants of each country are bound together, that any change in the numerical proportions of some of the inhabitants, independently of the change of climate itself, would most seriously affect many of the others. If the country were open on its borders, new forms would certainly immigrate, and this also would seriously disturb the relations of some of the former inhabitants. Let it be remembered how powerful the influence of a single introduced tree or mammal has been shown to be. But in the case of an island, or of 3 a country partly surrounded by barriers, into which new and better adapted forms could not freely enter, we should then have places in the economy of nature which would assuredly be better filled up, if some of the original inhabitants were in some manner modified; for, had the area been open to immigration, these same

1 The opening paragraph concludes with a succinct definition of natural selection and acknowledgment that some traits are selectively neutral, that is, neither favored nor disfavored by selection. The term "natural selection" was criticized for personifying a natural process (see p. 61), and by the third edition (1861) Darwin was defending his choice: "In the literal sense of the word, no doubt, natural selection is a misnomer; but who ever objected to chemists speaking of the elective affinities of the various elements?–and yet an acid cannot strictly be said to elect the base with which it will in preference combine." Darwin continued in the next paragraph: "It has been said that I speak of natural selection as an active power or Deity; but who objects to an author speaking of the attraction of gravity as ruling the movements of the planets? Every one knows what is meant by such metaphorical expressions . . . So again it is difficult to avoid personifying the word Nature; but I mean by Nature, only the aggregate action and product of many natural laws" (Peckham 1959, p. 165). With the appearance of the fifth edition in 1869, he made a minor concession and appended "or Survival of the Fittest" to "Natural Selection" in this closing line.

2 In this first full paragraph Darwin revisits the importance of ecological interconnectedness. Note that he has chosen the example of "a country undergoing some physical change" to illustrate the action of natural selection. The reason for this choice is twofold: besides presenting conditions under which selective pressures are expected to shift, it also demonstrates Darwin's contention that environmental change can engender variations, the fuel of natural selection. This is explicitly stated on the next page.

3 Note the view of island inhabitants revealed here. Isolated forms—literally insular—are protected from the immigration of better-adapted forms. Island biota figured in a complex way in Darwin's thinking, as we will see later in the book (pp. 105–106, 174, 321, 369, 383). The way Darwin looks at such flora and fauna gives us insight into the relative stress he places on isolation versus other factors like competition in driving transmutation. We get the sense here that the island forms are *(continued)*

(continued)
less well adapted than continental forms. This is a view of islands in the mold of "the land time forgot," and there is some truth to this portrayal.

1 Here is a clear statement on the origin of variation: changing environmental conditions are supposed to affect the reproductive system, engendering variability. As these variations are the raw material on which natural selection acts, the potential for evolutionary change is greatest whenever the conditions of life are changing.

2 This is the first of several explicit comparisons between what humans accomplish via artificial selection and what natural selection can accomplish. Note that the word "nature" is personified by capitalization.

3 Organisms are not perfectly adapted, as evidenced by the fact that they are often usurped by invading introduced species. Examples of such usurpation are innumerable; on the previous page Darwin mentions island species, which are especially susceptible to displacement. (This is why Hawaii leads U.S. states in the number of listed endangered species.) But certainly continental species can be outcompeted by exotics, too; examples of highly successful invaders include kudzu, starlings, privet, brown trout, and zebra mussels. Darwin's argument here is that if a native species could have been modified to advantage, enabling it to resist invading species, it could not have been perfectly adapted to begin with. Why stress a lack of perfect adaptation? If there is always room for improvement, there are always opportunities for a struggle for supremacy and for the action of natural selection.

places would have been seized on by intruders. In such case, every slight modification, which in the course of ages chanced to arise, and which in any way favoured the individuals of any of the species, by better adapting them to their altered conditions, would tend to be preserved; and natural selection would thus have free scope for the work of improvement.

1 We have reason to believe, as stated in the first chapter, that a change in the conditions of life, by specially acting on the reproductive system, causes or increases variability; and in the foregoing case the conditions of life are supposed to have undergone a change, and this would manifestly be favourable to natural selection, by giving a better chance of profitable variations occurring; and unless profitable variations do occur, natural selection can do nothing. Not that, as I believe, any extreme amount of variability is necessary; 2 as man can certainly produce great results by adding up in any given direction mere individual differences, so could Nature, but far more easily, from having incomparably longer time at her disposal. Nor do I believe that any great physical change, as of climate, or any unusual degree of isolation to check immigration, is actually necessary to produce new and unoccupied places for natural selection to fill up by modifying and improving some of the varying inhabitants. For as all the inhabitants of each country are struggling together with nicely balanced forces, extremely slight modifications in the structure or habits of one inhabitant would often give it an advantage over others; and still further modifications of the same kind would 3 often still further increase the advantage. No country can be named in which all the native inhabitants are now so perfectly adapted to each other and to the physical conditions under which they live, that none of

them could anyhow be improved; for in all countries, the natives have been so far conquered by naturalised productions, that they have allowed foreigners to take firm possession of the land. And as foreigners have thus everywhere beaten some of the natives, we may safely conclude that the natives might have been modified with advantage, so as to have better resisted such intruders.

As man can produce and certainly has produced a great result by his methodical and unconscious means of selection, what may not nature effect? Man can act only on external and visible characters: nature cares nothing for appearances, except in so far as they may be useful to any being. She can act on every internal organ, on every shade of constitutional difference, on the whole machinery of life. Man selects only for his own good; Nature only for that of the being which she tends. Every selected character is fully exercised by her; and the being is placed under well-suited conditions of life. Man keeps the natives of many climates in the same country; he seldom exercises each selected character in some peculiar and fitting manner; he feeds a long and a short beaked pigeon on the same food; he does not exercise a long-backed or long-legged quadruped in any peculiar manner; he exposes sheep with long and short wool to the same climate. He does not allow the most vigorous males to struggle for the females. He does not rigidly destroy all inferior animals, but protects during each varying season, as far as lies in his power, all his productions. He often begins his selection by some half-monstrous form; or at least by some modification prominent enough to catch his eye, or to be plainly useful to him. Under nature, the slightest difference of structure or constitution may well turn the nicely-balanced scale in the

1 Here, to good effect, Darwin contrasts artificial with natural selection. The intent is to show that while humans have accomplished a great deal through selective breeding, their efforts pale in comparison with natural selection. Natural selection is far more thorough and exacting: "every shade of constitutional difference" may be a point on which selection acts; "the slightest difference in structure" could make the difference between life and death. What's more, natural selection acts incessantly on every point of anatomy and physiology, inside and out. The penultimate sentence of this paragraph, on the next page, rises to a crescendo, invoking finally the vast time that natural selection has been acting—"whole geological periods" in contrast to the paltry thousands of years humans have been around. In the third edition Darwin capitalizes the word "nature" throughout this paragraph, using this as an opportunity to again clarify his point: "Nature (if I may be allowed thus to personify the natural preservation of varying and favoured individuals during the struggle for existence) cares nothing for appearances" (Peckham 1959, p. 167).

1 This culminating passage seems to invoke the divine; that species in nature bear the stamp of "far higher workmanship" is language one might associate with a sermon, yet nature is here personified as the crafter of species.

2 Here Darwin reiterates the incessant action of natural selection. Notice the choice of terms: *daily and hourly* scrutinizing . . . *every* variation . . . working *silently and insensibly whenever and wherever* opportunity for improvement arises. Why is it necessary to underscore this point? Darwin must convince his readers that natural selection is capable of generating the diversity and exquisite adaptations of organisms. He foresees that this will be a stumbling block, more so than acceptance of transmutation itself.

3 The slow, imperceptible action of natural selection brings about changes gradually. We eventually notice that things are different but seem oblivious to the slow transmutation that brought the differences about. This harks back to Darwin's lament on p. 29 that breeders see varietal differences in their animals yet cannot seem to see that the differences represent the sum of slow, insensible changes that the farmers themselves selected over many generations.

4 Protective coloration—camouflage—had been cited as evidence for design by a creator. Darwin here argues that natural selection can produce and maintain such coloration. Individually, characters such as shade of color might seem trivial, but insofar as they affect success in survival, competition, and reproduction, selection can act upon them.

struggle for life, and so be preserved. How fleeting are
the wishes and efforts of man! how short his time! and
consequently how poor will his products be, compared
with those accumulated by nature during whole geolo-
1 gical periods. Can we wonder, then, that nature's productions should be far "truer" in character than man's
productions; that they should be infinitely better
adapted to the most complex conditions of life, and
should plainly bear the stamp of far higher workmanship?

2 It may be said that natural selection is daily and
hourly scrutinising, throughout the world, every variation, even the slightest; rejecting that which is bad,
preserving and adding up all that is good; silently and
insensibly working, whenever and wherever opportunity offers, at the improvement of each organic being
in relation to its organic and inorganic conditions of
3 life. We see nothing of these slow changes in progress,
until the hand of time has marked the long lapse of
ages, and then so imperfect is our view into long past
geological ages, that we only see that the forms of life
are now different from what they formerly were.

Although natural selection can act only through and
for the good of each being, yet characters and structures,
which we are apt to consider as of very trifling import-
4 ance, may thus be acted on. When we see leaf-eating
insects green, and bark-feeders mottled-grey; the
alpine ptarmigan white in winter, the red-grouse the
colour of heather, and the black-grouse that of peaty
earth, we must believe that these tints are of service to
these birds and insects in preserving them from danger.
Grouse, if not destroyed at some period of their lives,
would increase in countless numbers; they are known
to suffer largely from birds of prey; and hawks are
guided by eyesight to their prey,—so much so, that on

parts of the Continent persons are warned not to keep white pigeons, as being the most liable to destruction. Hence I can see no reason to doubt that natural selection might be most effective in giving the proper colour to each kind of grouse, and in keeping that colour, when once acquired, true and constant. Nor ought we to think that the occasional destruction of an animal of any particular colour would produce little effect: we should remember how essential it is in a flock of white sheep to destroy every lamb with the faintest trace of black. In plants the down on the fruit and the colour of the flesh are considered by botanists as characters of the most trifling importance: yet we hear from an excellent horticulturist, Downing, that in the United States smooth-skinned fruits suffer far more from a beetle, a curculio, than those with down; that purple plums suffer far more from a certain disease than yellow plums; whereas another disease attacks yellow-fleshed peaches far more than those with other coloured flesh. If, with all the aids of art, these slight differences make a great difference in cultivating the several varieties, assuredly, in a state of nature, where the trees would have to struggle with other trees and with a host of enemies, such differences would effectually settle which variety, whether a smooth or downy, a yellow or purple fleshed fruit, should succeed.

In looking at many small points of difference between species, which, as far as our ignorance permits us to judge, seem to be quite unimportant, we must not forget that climate, food, &c., probably produce some slight and direct effect. It is, however, far more necessary to bear in mind that there are many unknown laws of correlation of growth, which, when one part of the organisation is modified through variation, and the modifications are accumulated by natural selection for

1 Here is a parallel between the way natural selection acts on coloration and the way humans do so. Just as sheep breeders weed out lambs with the merest suggestion of black in order to develop and maintain purely white-wool flocks, so too does natural selection weed out any shade of difference that could make the organism even slightly more vulnerable. It makes sense for Darwin to use the artificial selection analogy wherever possible, to help put the natural process in human terms.

The fruit example underscores the point that differences in seemingly trivial characters may not be so trivial, affecting the health and welfare of the organism in ways we cannot fathom. The bottom line is that selection is expected to act on such differences.

2 Darwin is citing Andrew Jackson Downing's popular manual *Fruits and Fruit Trees of America* (New York, 1845). "Curculio" is a name for weevils, family Curculionidae.

3 Variation and its source and interrelationships are, again, something of a mystery, with climate, diet, and other factors having some effect. Some traits are correlated, meaning they share some inexplicable developmental or physiological connection.

1 Traits affected at one life stage by selection may affect other traits in other life stages. Darwin asserts that selection can jointly shape these traits; they must be mutually adaptive to persist. Each must be suitable for the needs of the life stage at which it occurs (or at least not be injurious to it).

the good of the being, will cause other modifications, often of the most unexpected nature.

1 As we see that those variations which under domestication appear at any particular period of life, tend to reappear in the offspring at the same period;—for instance, in the seeds of the many varieties of our culinary and agricultural plants; in the caterpillar and cocoon stages of the varieties of the silkworm; in the eggs of poultry, and in the colour of the down of their chickens; in the horns of our sheep and cattle when nearly adult;—so in a state of nature, natural selection will be enabled to act on and modify organic beings at any age, by the accumulation of profitable variations at that age, and by their inheritance at a corresponding age. If it profit a plant to have its seeds more and more widely disseminated by the wind, I can see no greater difficulty in this being effected through natural selection, than in the cotton-planter increasing and improving by selection the down in the pods on his cotton-trees. Natural selection may modify and adapt the larva of an insect to a score of contingencies, wholly different from those which concern the mature insect. These modifications will no doubt affect, through the laws of correlation, the structure of the adult; and probably in the case of those insects which live only for a few hours, and which never feed, a large part of their structure is merely the correlated result of successive changes in the structure of their larvæ. So, conversely, modifications in the adult will probably often affect the structure of the larva; but in all cases natural selection will ensure that modifications consequent on other modifications at a different period of life, shall not be in the least degree injurious: for if they became so, they would cause the extinction of the species.

Natural selection will modify the structure of the

young in relation to the parent, and of the parent in relation to the young. In social animals it will adapt the structure of each individual for the benefit of the community; if each in consequence profits by the selected change. What natural selection cannot do, is to modify the structure of one species, without giving it any advantage, for the good of another species; and though statements to this effect may be found in works of natural history, I cannot find one case which will bear investigation. A structure used only once in an animal's whole life, if of high importance to it, might be modified to any extent by natural selection; for instance, the great jaws possessed by certain insects, and used exclusively for opening the cocoon—or the hard tip to the beak of nestling birds, used for breaking the egg. It has been asserted, that of the best short-beaked tumbler-pigeons more perish in the egg than are able to get out of it; so that fanciers assist in the act of hatching. Now, if nature had to make the beak of a full-grown pigeon very short for the bird's own advantage, the process of modification would be very slow, and there would be simultaneously the most rigorous selection of the young birds within the egg, which had the most powerful and hardest beaks, for all with weak beaks would inevitably perish: or, more delicate and more easily broken shells might be selected, the thickness of the shell being known to vary like every other structure.

Sexual Selection.—Inasmuch as peculiarities often appear under domestication in one sex and become hereditarily attached to that sex, the same fact probably occurs under nature, and if so, natural selection will be able to modify one sex in its functional relations to the other sex, or in relation to wholly different habits of life in the two sexes, as is sometimes the case

1 This is an important insight: natural selection can only favor traits that improve the individual's likelihood of survival and reproductive success. This can result in mutualism but cannot produce a self-sacrificial trait that serves the exclusive good of another species, or even another member of the same species. The operative word is "exclusive" good, and cases of mutualism and symbiosis, or the fascinating brand of cooperation found in certain insect societies, do not involve acting for the *exclusive* good of others.

Accounts of putative altruistic or sacrificial behaviors abounded in the anecdotal natural history literature of the past few centuries, but serious discussions of good-of-the-species behavior could be found even in scientific literature.

2 In the last edition of *Origin* Darwin inserted a paragraph here qualifying the action of natural selection, taking pains to explain that some mortality is random, or stochastic, even at times claiming individuals that might have possessed superior competitive traits. To paraphrase Stephen Jay Gould, imagine that the best-adapted fish of its kind arises in a pond, but then the pond dries up. Gould's point is that stochastic mortality can make no contribution to increasing adaptedness: "It may be well here to remark that with all beings there must be much fortuitous destruction, which can have little or no influence on the course of natural selection. For instance a vast number of eggs or seeds are annually devoured, and these could be modified through natural selection only if they varied in some manner which protected them from their enemies. Yet many of these eggs or seeds would perhaps, if not destroyed, have yielded individuals better adapted to their conditions of life than any of those which happened to survive" (Stauffer 1975, p. 173).

3 Darwin recognized that reproduction is the ultimate evolutionary currency. A special form of selection that he dubbed "sexual selection" can improve the chances of reproductive success. Darwin described the process of sexual selection in his 1842 *Sketch* and 1844 *Essay*, but he does not use the term "sexual selection" until later. It appears in *Natural Selection* (Stauffer 1975) and as an annotation in the D notebook (Barrett et al. 1987, p. 378). That notebook dates to the summer of 1838, but since the term is a later addition it isn't clear exactly when it was coined.

1 This is one of the two forms of sexual selection, now termed "male-male competition." Note the key distinction between this form of selection and natural selection: in this struggle individuals that are not favored may survive but will leave no offspring.

2 The second form of sexual selection, introduced here in the context of birds, is termed "female choice." Darwin's use of language in describing the process of female choice as being "of a more peaceful character" than male-male competition seems to echo cultural biases of warlike males and docile females.

with insects. And this leads me to say a few words on what I call Sexual Selection. This depends, not on a struggle for existence, but on a struggle between the males for possession of the females; the result is not death to the unsuccessful competitor, but few or no offspring. Sexual selection is, therefore, less rigorous than natural selection. Generally, the most vigorous males, those which are best fitted for their places in nature, will leave most progeny. But in many cases, victory will depend not on general vigour, but on having special weapons, confined to the male sex. A hornless stag or spurless cock would have a poor chance of leaving offspring. Sexual selection by always allowing the victor to breed might surely give indomitable courage, length to the spur, and strength to the wing to strike in the spurred leg, as well as the brutal cock-fighter, who knows well that he can improve his breed by careful selection of the best cocks. How low in the scale of nature this law of battle descends, I know not; male alligators have been described as fighting, bellowing, and whirling round, like Indians in a war-dance, for the possession of the females; male salmons have been seen fighting all day long; male stag-beetles often bear wounds from the huge mandibles of other males. The war is, perhaps, severest between the males of polygamous animals, and these seem oftenest provided with special weapons. The males of carnivorous animals are already well armed; though to them and to others, special means of defence may be given through means of sexual selection, as the mane to the lion, the shoulder-pad to the boar, and the hooked jaw to the male salmon; for the shield may be as important for victory, as the sword or spear.

Amongst birds, the contest is often of a more peaceful character. All those who have attended to the subject, believe that there is the severest rivalry between the males of many species to attract by singing the females. The rock-thrush of Guiana, birds of Paradise, and some others, congregate; and successive males display their gorgeous plumage and perform strange antics before the females, which standing by as spectators, at last choose the most attractive partner. Those who have closely attended to birds in confinement well know that they often take individual preferences and dislikes: thus Sir R. Heron has described how one pied peacock was eminently attractive to all his hen birds. It may appear childish to attribute any effect to such apparently weak means: I cannot here enter on the details necessary to support this view; but if man can in a short time give elegant carriage and beauty to his bantams, according to his standard of beauty, I can see no good reason to doubt that female birds, by selecting, during thousands of generations, the most melodious or beautiful males, according to their standard of beauty, might produce a marked effect. I strongly suspect that some well-known laws with respect to the plumage of male and female birds, in comparison with the plumage of the young, can be explained on the view of plumage having been chiefly modified by sexual selection, acting when the birds have come to the breeding age or during the breeding season; the modifications thus produced being inherited at corresponding ages or seasons, either by the males alone, or by the males and females; but I have not space here to enter on this subject.

Thus it is, as I believe, that when the males and females of any animal have the same general habits of life, but differ in structure, colour, or ornament, such differences have been mainly caused by sexual selection; that is, individual males have had, in successive generations, some slight advantage over other

1 Darwin refers here to a paper by Sir Robert Heron read before the Zoological Society of London in 1835. A "pied" color pattern is mottled, blotched, or variegated. The term may be derived from magpies, birds of the crow family (Corvidae) with conspicuous black and white patches. Pied coloration was, and still is, of great interest as a novelty to breeders. Heron's studies with pied peafowl were part of an early nineteenth-century enthusiasm for color variants in domesticated birds. Pied variants per se were first described about that time, but they must have occurred as long as the birds have been domesticated (a considerable time, as they are mentioned in antiquity—in the books of Kings and Chronicles in the Bible, for example). The genetics of pied coloration in peafowl is now well understood (Somes and Burger 1993).

2 This is another instance of Darwin's repeated contrast between artificial and natural (and sexual) selection. The comparison is made for effect, to remind his readers of the power of artificial selection. If by that process people can modify the ornamental feathers of bantam chickens in no time at all, then female birds choosing mates on the basis of song, tail length, or color, can surely produce "a marked effect" over countless generations of selection. Note Darwin's comment that sexual selection is an apparently weak process. He is concerned that readers having trouble accepting natural selection—with the death and destruction the process entails—as a potent causative agent of change will find laughable the idea that mere fancy, as female choice of tail ornament would seem to be, could effect change.

3 Sexual dimorphism is often associated with sexual selection of one type or the other.

1 Note that Darwin's emphasis in this section is on the complex interrelationships of flowering plants and pollinators. The wolf scenario is presented in about a page and a half, while the pollinator example occupies the next three and a half pages. To Darwin plant-pollinator coevolution presents a wonderful example of how a seemingly mundane action like the transfer of pollen by insects can explain not only the evolution of nectar production but also, more subtly, the evolution of separate-sex plants (bearing male-only flowers or female-only flowers). He is fascinated by plant pollination and plant-insect interactions; in fact, his very next book after the *Origin*, appearing in 1862, is a treatise on orchid pollination. Several other remarkable botanical books follow—on insectivorous plants (1875); effects of self-pollination and crossing in plants (1876); plants with differently structured flowers on the same specimen (1877); and climbing and movement in plants (1880). Several of these themes are discussed on the next few pages.

Incidentally, Darwin's orchid book is more than an innocent botanical digression. Orchids presented a powerful demonstration of how pre-existing structures can be modified in various ways, coopted into new functions. Flower structure in orchids is often modified in seemingly bizarre ways, all toward the end of helping the plant achieve cross-pollination. In a telling letter to Asa Gray dated July 23, 1862, shortly after the orchid book was published, Darwin wrote, "Of all the carpenters for knocking the right nail on the head, you are the very best: no one else has perceived that my chief interest in my orchid book, has been that it was a 'flank movement' on the enemy." He goes on to ask Gray what he thinks "about what I say in last [chapter] of Orchid Book on the meaning & cause of the endless diversity of means for same general purpose. It bears on design, that endless question" (*Correspondence* 10: 330). See Dressler's book *The Orchids* (1990) for accounts of orchid structure and pollination.

males, in their weapons, means of defence, or charms; and have transmitted these advantages to their male offspring. Yet, I would not wish to attribute all such sexual differences to this agency: for we see peculiarities arising and becoming attached to the male sex in our domestic animals (as the wattle in male carriers, horn-like protuberances in the cocks of certain fowls, &c.), which we cannot believe to be either useful to the males in battle, or attractive to the females. We see analogous cases under nature, for instance, the tuft of hair on the breast of the turkey-cock, which can hardly be either useful or ornamental to this bird;—indeed, had the tuft appeared under domestication, it would have been called a monstrosity.

▶ *Illustrations of the action of Natural Selection.*—In order to make it clear how, as I believe, natural selection acts, I must beg permission to give one or two imaginary illustrations. Let us take the case of a wolf, which preys on various animals, securing some by craft, some by strength, and some by fleetness; and let us suppose that the fleetest prey, a deer for instance, had from any change in the country increased in numbers, or that other prey had decreased in numbers, during that season of the year when the wolf is hardest pressed for food. I can under such circumstances see no reason to doubt that the swiftest and slimmest wolves would have the best chance of surviving, and so be preserved or selected,—provided always that they retained strength to master their prey at this or at some other period of the year, when they might be compelled to prey on other animals. I can see no more reason to doubt this, than that man can improve the fleetness of his greyhounds by careful and methodical selection, or by that unconscious selection which results from each man trying to keep the best dogs without any thought of modifying the breed.

Even without any change in the proportional numbers of the animals on which our wolf preyed, a cub might be born with an innate tendency to pursue certain kinds of prey. Nor can this be thought very improbable; for we often observe great differences in the natural tendencies of our domestic animals; one cat, for instance, taking to catch rats, another mice; one cat, according to Mr. St. John, bringing home winged game, another hares or rabbits, and another hunting on marshy ground and almost nightly catching woodcocks or snipes. The tendency to catch rats rather than mice is known to be inherited. Now, if any slight innate change of habit or of structure benefited an individual wolf, it would have the best chance of surviving and of leaving offspring. Some of its young would probably inherit the same habits or structure, and by the repetition of this process, a new variety might be formed which would either supplant or coexist with the parent-form of wolf. Or, again, the wolves inhabiting a mountainous district, and those frequenting the lowlands, would naturally be forced to hunt different prey; and from the continued preservation of the individuals best fitted for the two sites, two varieties might slowly be formed. These varieties would cross and blend where they met; but to this subject of intercrossing we shall soon have to return. I may add, that, according to Mr. Pierce, there are two varieties of the wolf inhabiting the Catskill Mountains in the United States, one with a light greyhound-like form, which pursues deer, and the other more bulky, with shorter legs, which more frequently attacks the shepherd's flocks.

Let us now take a more complex case. Certain plants excrete a sweet juice, apparently for the sake of eliminating something injurious from their sap: this is

1 This account of the hunting proclivities of cats came from Charles George William St. John's popular book *Short Sketches of the Wild Sports and Natural History of the Highlands of Scotland* (London, 1846).

2 Darwin is referring to James Pierce's "Memoir on the Catskill Mountains," published in the *American Journal of Science and Arts* in 1823. The existence of these two wolf varieties with different builds, pursuing different kinds of prey, would support his scenario of selection promoting divergence. It seems unlikely that such a striking form as bulky with shorter legs would have evolved in the scant two hundred years or so since sheep were introduced to the New York area; perhaps this putative wolf variety was supposed to have specialized on something else before sheep became available.

3 Here Darwin is referring to *extrafloral nectaries:* plant structures other than the flowers that secrete sugary nectar. Well over 100 plant families have species with extrafloral nectaries. Darwin's explanation of the purpose of these structures—to eliminate some supposed "injurious substance" from the sap—is erroneous; it is now understood that extrafloral nectaries function to attract insects such as ants that, while aggressively attending to and defending the sugar source, effectively defend the plant against herbivores as well (Bentley 1977).

1 Leguminosae is an older botanical name for legumes, plants of the bean family. This is a large and diverse family, called Fabaceae today (after the common bean genus, *Faba,* which also happens to be the Latin word for bean).

2 Darwin's evolutionary scenario for *floral* nectaries is slow, stepwise, and gradual. Modern biologists would agree with him that floral evolution has largely been driven by the benefits of outcrossing, particularly outcrossing mediated by insects. Implied in his scenario is the hypothesis that extrafloral nectaries preceded floral nectaries. This is likely the case, considering that many ferns—nonflowering plants—also bear nectary organs; a good example is the common bracken fern, *Pteridium aquilinum,* which has a pair of nectaries on the stem just below the leaf.

effected by glands at the base of the stipules in some 1 Leguminosæ, and at the back of the leaf of the common laurel. This juice, though small in quantity, is greedily sought by insects. Let us now suppose a little sweet juice or nectar to be excreted by the inner bases of the petals of a flower. In this case insects in seeking the nectar would get dusted with pollen, and would certainly often transport the pollen from one flower to the stigma of another flower. The flowers of two distinct individuals of the same species would thus get crossed; and the act of crossing, we have good reason to believe (as will hereafter be more fully alluded to), would produce very vigorous seedlings, which consequently would have the best chance of flourishing and surviving. Some of these seedlings would probably inherit the nectar-excreting 2 power. Those individual flowers which had the largest glands or nectaries, and which excreted most nectar, would be oftenest visited by insects, and would be oftenest crossed; and so in the long-run would gain the upper hand. Those flowers, also, which had their stamens and pistils placed, in relation to the size and habits of the particular insects which visited them, so as to favour in any degree the transportal of their pollen from flower to flower, would likewise be favoured or selected. We might have taken the case of insects visiting flowers for the sake of collecting pollen instead of nectar; and as pollen is formed for the sole object of fertilisation, its destruction appears a simple loss to the plant; yet if a little pollen were carried, at first occasionally and then habitually, by the pollen-devouring insects from flower to flower, and a cross thus effected, although nine-tenths of the pollen were destroyed, it might still be a great gain to the plant; and those individuals which produced more and more pollen, and had larger and larger anthers, would be selected.

When our plant, by this process of the continued preservation or natural selection of more and more attractive flowers, had been rendered highly attractive to insects, they would, unintentionally on their part, regularly carry pollen from flower to flower; and that they can most effectually do this, I could easily show by many striking instances. I will give only one—not as a very striking case, but as likewise illustrating one step in the separation of the sexes of plants, presently to be alluded to. Some holly-trees bear only male flowers, which have four stamens producing rather a small quantity of pollen, and a rudimentary pistil; other holly-trees bear only female flowers; these have a full-sized pistil, and four stamens with shrivelled anthers, in which not a grain of pollen can be detected. Having found a female tree exactly sixty yards from a male tree, I put the stigmas of twenty flowers, taken from different branches, under the microscope, and on all, without exception, there were pollen-grains, and on some a profusion of pollen. As the wind had set for several days from the female to the male tree, the pollen could not thus have been carried. The weather had been cold and boisterous, and therefore not favourable to bees, nevertheless every female flower which I examined had been effectually fertilised by the bees, accidentally dusted with pollen, having flown from tree to tree in search of nectar. But to return to our imaginary case: as soon as the plant had been rendered so highly attractive to insects that pollen was regularly carried from flower to flower, another process might commence. No naturalist doubts the advantage of what has been called the "physiological division of labour;" hence we may believe that it would be advantageous to a plant to produce stamens alone in one flower or on one whole plant, and pistils alone in

1 Darwin made his holly pollination observations in May 1857. In a letter to Asa Gray the following month, he wrote, "It is really pretty to see how effectual insects are: a short time ago I found a female Holly 60 measured yards from any other Holly & I cut off some twigs & took by chance 20 stigmas, cut off their tops & put them under microscope: there was pollen on every one & in profusion on most! Weather cloudy & stormy & unfavourable, wind in wrong direction to have brought any." Gray replied: "What you say about the pollen carried to a female *Holly* so abundantly, and to such a distance is very remarkable" (*Correspondence* 6: 412, 422).

2 The concept of a "physiological division of labor" refers to tissue and organ specialization. This idea, originating with the French zoologist Henri Milne-Edwards in 1827, holds that overall efficiency is improved by task specialization. Milne-Edwards received his inspiration from the economist Adam Smith. Darwin saw that this theory could be applied to species divergence, as we will see (pp. 115–117).

another flower or on another plant. In plants under culture and placed under new conditions of life, sometimes the male organs and sometimes the female organs become more or less impotent; now if we suppose this to occur in ever so slight a degree under nature, then as pollen is already carried regularly from flower to flower, [1] and as a more complete separation of the sexes of our plant would be advantageous on the principle of the division of labour, individuals with this tendency more and more increased, would be continually favoured or selected, until at last a complete separation of the sexes would be effected.

Let us now turn to the nectar-feeding insects in our imaginary case: we may suppose the plant of which we have been slowly increasing the nectar by continued selection, to be a common plant; and that certain insects depended in main part on its nectar for food. I could give many facts, showing how anxious bees are to save time; for instance, their habit of cutting [2] holes and sucking the nectar at the bases of certain flowers, which they can, with a very little more trouble, enter by the mouth. Bearing such facts in mind, I can see no reason to doubt that an accidental deviation in the size and form of the body, or in the curvature and length of the proboscis, &c., far too slight to be appreciated by us, might profit a bee or other insect, so that an individual so characterised would be able to obtain its food more quickly, and so have a better chance of living and leaving descendants. Its descendants would probably inherit a tendency to a similar slight deviation of structure. The tubes of the corollas of the common [3] red and incarnate clovers (Trifolium pratense and incarnatum) do not on a hasty glance appear to differ in length; yet the hive-bee can easily suck the nectar out of the incarnate clover, but not out of the common red

1 Darwin's model of the evolution of dioecy (separate sexes) in plants rests on the supposition that having two sexes is more pollination-efficient, perhaps because it is less wasteful and reduces the chance of inbreeding. This is consistent with the idea of the "physiological division of labor" as well. When both stamens and pistils are present in the same flower (or male and female flowers are present on the same plant), some of the plant's pollen is wasted on itself, as it is inevitably delivered to its own pistils by insects. Insofar as self-fertilization is undesirable—and that isn't always so—having separate sexes decreases this waste and increases the likelihood of outcrossing, both of which would be favored by selection.

2 This behavior is termed "nectar robbery"; it is commonly found among short-tongued bees like small carpenter bees and bumblebees, which have learned to get at the nectar deep within a flower receptacle by biting through the base of the corolla tube. Blueberries and related plants of the genus *Vaccinium* are commonly "robbed" in this way.

3 *Trifolium* is the clover genus, family Fabaceae, so named for the trio of leaflets common to many species. There are about 300 *Trifolium* species worldwide. *Trifolium pratense* is red clover, while *T. incarnatum* is crimson clover—the two species are similar in flower color but differ in flower length. On reading the passage here, a correspondent named Charles Hardy wrote to inform Darwin that he was mistaken about "hive-bees" (honeybees) never visiting red clover. Hardy reported that when red clover is mown, the flowers grow back smaller and are then accessible to honeybees; he had routinely observed "myriads of bees" visiting this species (*Correspondence* 8: 300). Darwin duly added this information in the third edition in 1861: "I have been informed, that when the red clover has been mown, the flowers of the second crop are somewhat smaller, and that these are abundantly visited by hive-bees" (Peckham 1959, p. 184).

clover, which is visited by humble-bees alone; so that whole fields of the red clover offer in vain an abundant supply of precious nectar to the hive-bee. Thus it might be a great advantage to the hive-bee to have a slightly longer or differently constructed proboscis. On the other hand, I have found by experiment that the fertility of clover greatly depends on bees visiting and moving parts of the corolla, so as to push the pollen on to the stigmatic surface. Hence, again, if humble-bees were to become rare in any country, it might be a great advantage to the red clover to have a shorter or more deeply divided tube to its corolla, so that the hive-bee could visit its flowers. Thus I can understand how a flower and a bee might slowly become, either simultaneously or one after the other, modified and adapted in the most perfect manner to each other, by the continued preservation of individuals presenting mutual and slightly favourable deviations of structure.

I am well aware that this doctrine of natural selection, exemplified in the above imaginary instances, is open to the same objections which were at first urged against Sir Charles Lyell's noble views on "the modern changes of the earth, as illustrative of geology;" but we now very seldom hear the action, for instance, of the coast-waves, called a trifling and insignificant cause, when applied to the excavation of gigantic valleys or to the formation of the longest lines of inland cliffs. Natural selection can act only by the preservation and accumulation of infinitesimally small inherited modifications, each profitable to the preserved being; and as modern geology has almost banished such views as the excavation of a great valley by a single diluvial wave, so will natural selection, if it be a true principle, banish the belief of the continued creation of new organic

1 Darwin envisions coevolution between bees and clovers, with the size and proboscis length of the bees matching the dimensions of the clover flowers. Should bumblebees become rare, he imagines, selection would favor the evolution of smaller flowers that would be more accessible for pollination by honeybees. In the vein of plant-pollinator coevolution, Darwin famously predicted the existence of a large hawkmoth (Sphingidae) with a proboscis long enough to reach deep into the 30 cm (12 inch) corolla tube of the rare comet orchid (*Angraecum sesquipedale*) of Madagascar. Of this species Darwin wrote in his orchid book: "[*A. sesquipedale* has] nectaries 11 and a half inches long, with only the lower inch and a half filled with very sweet nectar . . . [I]t is, however, surprising, that any insect should be able to reach the nectar: our English sphinxes have probosces as long as their bodies; but in Madagascar there must be moths with probosces capable of extension to a length of between ten and eleven inches!" (Darwin 1862, pp. 197–198).

Wallace agreed, and in 1867 he pointed out that a known African species—*Xanthopan morganii*—had a proboscis length of more than seven inches, making it likely that another large moth with a tongue just a few more inches in length must exist. He urged naturalists to search for it, and sixteen years later just such a moth was discovered. It turned out to be a subspecies of the very moth Wallace discussed, and was subsequently named *Xanthopan morgani praedicta* in recognition of Darwin's and Wallace's prediction.

2 Lyell was mentor and source of inspiration to Darwin, who once wrote of the geologist, "I never forget that almost everything which I have done in science I owe to the study of his great works" (F. Darwin 1896, p. 196). One of the most important things that Darwin learned from Lyell was to view biological change in geological terms—change marked by gradualism and uniformity. He envisioned natural selection as a slow, steady process that eventually effected great changes. Here we see Darwin explicitly comparing himself to the master: just as Lyell's "noble views" on changes in the earth initially met with opposition but were now accepted, so too would Darwin's skeptical readers eventually embrace his model of the action of natural selection. Moreover, just as modern geology banished the theory of "single diluvial waves" excavating great valleys, so too would

(continued)

(continued)
acceptance of natural selection—also a non-gradualist, non-uniformitarian view—banish belief in *de novo* special creation of species, or abrupt changes in their structure. Is this a rhetorical ploy whereby Darwin sets himself up for approval by analogy? Note, by the way, that the line "the modern changes of the earth, as illustrative of geology" is a paraphrase of the subtitle of the later editions of Lyell's *Principles:* "Modern Changes of the Earth and Its Inhabitants, Considered as Illustrative of Geology."

1 The effects of intercrossing (what we now call outbreeding, as opposed to inbreeding) are important to Darwin in several ways. Inbreeding, which he refers to as "close intercrossing," was known to have deleterious effects on offspring—something of particular concern to Darwin, who married his first cousin. Also, the positive effects of outbreeding are important to Darwin's ideas on the evolution of the sexes. In this section we see in microcosm Darwin's "strong inference" approach in arguing that organisms must at least occasionally outcross. On the next page he presents "several large classes of facts . . . which on any other view are inexplicable."

2 The horticultural pioneer Thomas Andrew Knight published a paper in 1799 entitled "An account of some experiments on the fecundation of vegetables," in which he wrote of experimentally castrating flowers in order to control their pollination and demonstrated the benefits of outcrossing. "Many experiments, of the same kind, were tried on other plants," Knight noted, "but it is sufficient to say, that all tended to evince, that improved varieties of every fruit and esculent plant may be obtained by this process, and that nature intended that a sexual intercourse should take place between neighboring plants of the same species" (Knight 1799, p. 202). Darwin was impressed. "Nature thus tells us, in the most emphatic manner, that she abhors perpetual self-fertilisation," he concluded in his orchid book (Darwin 1862, p. 359). In *Variation of Animals and Plants under Domestication* (1868b) he declared that "it is a law of nature that organic beings shall not fertilise themselves for perpetuity. This law was first plainly hinted at in 1799, with respect to plants, by Andrew Knight" (*Variation,* vol. 2, p. 175). For a time this principle was known as the Knight-Darwin law.

beings, or of any great and sudden modification in their structure.

1 *On the Intercrossing of Individuals.*—I must here introduce a short digression. In the case of animals and plants with separated sexes, it is of course obvious that two individuals must always unite for each birth; but in the case of hermaphrodites this is far from obvious. Nevertheless I am strongly inclined to believe that with all hermaphrodites two individuals, either occasionally or habitually, concur for the reproduction of their kind. This view, I may add, was first suggested
2 by Andrew Knight. We shall presently see its importance; but I must here treat the subject with extreme brevity, though I have the materials prepared for an ample discussion. All vertebrate animals, all insects, and some other large groups of animals, pair for each birth. Modern research has much diminished the number of supposed hermaphrodites, and of real hermaphrodites a large number pair; that is, two individuals regularly unite for reproduction, which is all that concerns us. But still there are many hermaphrodite animals which certainly do not habitually pair, and a vast majority of plants are hermaphrodites. What reason, it may be asked, is there for supposing in these cases that two individuals ever concur in reproduction? As it is impossible here to enter on details, I must trust to some general considerations alone.

In the first place, I have collected so large a body of facts, showing, in accordance with the almost universal belief of breeders, that with animals and plants a cross between different varieties, or between individuals of the same variety but of another strain, gives vigour and fertility to the offspring; and on the other hand, that *close* interbreeding diminishes vigour and fertility; that these facts alone incline me to believe that it is a general law of nature (utterly ignorant though we be of the meaning of the law) that no organic being self-fertilises itself for an eternity of generations; but that a cross with another individual is occasionally—perhaps at very long intervals—indispensable.

On the belief that this is a law of nature, we can, I think, understand several large classes of facts, such as the following, which on any other view are inexplicable. Every hybridizer knows how unfavourable exposure to wet is to the fertilisation of a flower, yet what a multitude of flowers have their anthers and stigmas fully exposed to the weather! but if an occasional cross be indispensable, the fullest freedom for the entrance of pollen from another individual will explain this state of exposure, more especially as the plant's own anthers and pistil generally stand so close together that self-fertilisation seems almost inevitable. Many flowers, on the other hand, have their organs of fructification closely enclosed, as in the great papilionaceous or pea-family; but in several, perhaps in all, such flowers, there is a very curious adaptation between the structure of the flower and the manner in which bees suck the nectar; for, in doing this, they either push the flower's own pollen on the stigma, or bring pollen from another flower. So necessary are the visits of bees to papilionaceous flowers, that I have found, by experiments published elsewhere, that their fertility is greatly diminished if these visits be prevented. Now, it is scarcely possible that bees should fly from flower to flower, and not carry pollen from one to the other, to the great good, as I believe, of the plant. Bees will act like a camel-hair pencil, and it is quite sufficient just to touch the anthers of one flower and then the stigma of another with the same brush to ensure fertilisation; but it must not be

1 This is the first of Darwin's "large classes of facts": there is a tradeoff in many plant species between full exposure of stamens and pistils to ensure cross-pollination and loss of flower function (presumably owing to loss of pollen) by exposure to rain.

2 The second fact: the intricate structure of the flowers of some species virtually guarantees pollen transfer, and fertility declines if insects are prevented from visiting. Some plants have adaptations that seem to ensure self-fertilization (though the implication is that since this process requires insects, some degree of cross-pollination will occur too), while others have adaptations that prevent the transfer of pollen from the stamens to the stigma of the same flower.

supposed that bees would thus produce a multitude of hybrids between distinct species; for if you bring on the same brush a plant's own pollen and pollen from another species, the former will have such a prepotent effect, that it will invariably and completely destroy, as has been shown by Gärtner, any influence from the foreign pollen.

When the stamens of a flower suddenly spring towards the pistil, or slowly move one after the other towards it, the contrivance seems adapted solely to ensure self-fertilisation; and no doubt it is useful for this end: but, the agency of insects is often required to cause the stamens to spring forward, as Kölreuter has shown to be the case with the barberry; and curiously in this very genus, which seems to have a special contrivance for self-fertilisation, it is well known that if very closely-allied forms or varieties are planted near each other, it is hardly possible to raise pure seedlings, so largely do they naturally cross. In many other cases, far from there being any aids for self-fertilisation, there are special contrivances, as I could show from the writings of C. C. Sprengel and from my own observations, which effectually prevent the stigma receiving pollen from its own flower: for instance, in Lobelia fulgens, there is a really beautiful and elaborate contrivance by which every one of the infinitely numerous pollen-granules are swept out of the conjoined anthers of each flower, before the stigma of that individual flower is ready to receive them; and as this flower is never visited, at least in my garden, by insects, it never sets a seed, though by placing pollen from one flower on the stigma of another, I raised plenty of seedlings; and whilst another species of Lobelia growing close by, which is visited by bees, seeds freely. In very many other cases, though there be no special mechanical contrivance to prevent the stigma of a flower receiving its own pollen, yet, as

1 The spring-action contrivance of barberry flowers—their "irritability"—was first described by the German botanist Joseph Gottlieb Kölreuter. In a paper entitled "Some further observations and experiments on the irritability of the stamens of the Barberry," he related his discovery: "It was about the beginning of May 1772, while busied in attending the singular property of the Barberry, by which the stamens are caused, on the application of a slight external stimulus, to spring forwards towards the pistil, that I discovered, both in respect of the structure and position of the stamens themselves, as well as in relation to that phaenomenon, several circumstances till now entirely unknown, and such I trust as will be thought worthy to be communicated to the lovers of natural history. These discoveries are as follow, and seem to me to throw some light on this species of irritability in vegetables" (reprinted in the *Annals of Botany,* vol. II, pp. 1–10, London, 1806).

2 *Lobelia fulgens* is now synonymous with *Lobelia cardinalis,* the Cardinal-Flower. The spectacular scarlet spike of showy long-petaled blossoms of this Central and North American species attracts hummingbird pollinators. That explains Darwin's aside that this species is never visited by insects in his garden—hummingbirds are strictly New World. This species, by the way, was described in 1809 from specimens collected by Alexander von Humboldt and Aimé Bonpland.

C. C. Sprengel has shown, and as I can confirm, either the anthers burst before the stigma is ready for fertilisation, or the stigma is ready before the pollen of that flower is ready, so that these plants have in fact separated sexes, and must habitually be crossed. How strange are these facts! How strange that the pollen and stigmatic surface of the same flower, though placed so close together, as if for the very purpose of self-fertilisation, should in so many cases be mutually useless to each other! How simply are these facts explained on the view of an occasional cross with a distinct individual being advantageous or indispensable! 1

If several varieties of the cabbage, radish, onion, and of some other plants, be allowed to seed near each other, a large majority, as I have found, of the seedlings thus raised will turn out mongrels: for instance, I raised 233 seedling cabbages from some plants of different varieties growing near each other, and of these only 78 were true to their kind, and some even of these were not perfectly true. Yet the pistil of each cabbage-flower is surrounded not only by its own six stamens, but by those of the many other flowers on the same plant. How, then, comes it that such a vast number of the seedlings are mongrelized? I suspect that it must arise from the pollen of a distinct *variety* having a prepotent effect over a flower's own pollen; and that this is part of the general law of good being derived from the intercrossing of distinct individuals of the same species. When distinct *species* are crossed the case is directly the reverse, for a plant's own pollen is always prepotent over foreign pollen; but to this subject we shall return in a future chapter. 2

In the case of a gigantic tree covered with innumerable flowers, it may be objected that pollen could seldom be carried from tree to tree, and at most only from flower

1 The phenomena that Darwin described on the preceding two pages promote, if only occasionally, the outcrossing of individuals: "Nature abhors perpetual self-fertilization."

2 As we saw in chapter III, Darwin is fond of conducting his own experiments in addition to drawing on the data of others to bolster his arguments. Vegetables are convenient to work with because they are easily propagated and are available in several distinct varieties. Here he found that the varieties readily hybridize: 67 percent of the seedlings in the cabbage study proved to be mixes ("mongrels") of the several varieties he planted in close proximity. Considering that each plant's pistils are literally surrounded by its own stamens, a mechanism clearly exists to promote outcrossing. Darwin explains the phenomenon in terms of pollen from a different variety having a "prepotent effect" over the plant's own pollen; this phenomenon, which we now know is controlled by a simple genetic mechanism, is termed "self-incompatibility" today.

to flower on the same tree, and that flowers on the same tree can be considered as distinct individuals only in a limited sense. I believe this objection to be valid, but that nature has largely provided against it by giving to trees a strong tendency to bear flowers with separated sexes. When the sexes are separated, although the male and female flowers may be produced on the same tree, we can see that pollen must be regularly carried from flower to flower; and this will give a better chance of pollen being occasionally carried from tree to tree. That trees belonging to all Orders have their sexes more often separated than other plants, I find to be the case in this country; and at my request Dr. Hooker tabulated the trees of New Zealand, and Dr. Asa Gray those of the United States, and the result was as I anticipated. On the other hand, Dr. Hooker has recently informed me that he finds that the rule does not hold in Australia; and I have made these few remarks on the sexes of trees simply to call attention to the subject.

Turning for a very brief space to animals: on the land there are some hermaphrodites, as land-mollusca and earth-worms; but these all pair. As yet I have not found a single case of a terrestrial animal which fertilises itself. We can understand this remarkable fact, which offers so strong a contrast with terrestrial plants, on the view of an occasional cross being indispensable, by considering the medium in which terrestrial animals live, and the nature of the fertilising element; for we know of no means, analogous to the action of insects and of the wind in the case of plants, by which an occasional cross could be effected with terrestrial animals without the concurrence of two individuals. Of aquatic animals, there are many self-fertilising hermaphrodites; but here currents in the water offer an obvious means for an occasional cross. And, as in the case of flowers, I have as yet

1 Darwin found it curious that trees seemed to be dioecious (separate-sex individuals) to a greater degree than other plants. He called on his botanical friends to help him assemble some data from different flora around the world to confirm this. Hooker and Gray were perhaps the most eminent botanists of the day, and both became Darwin's friends and confidants (they were among the few people in whom he confided his species theory). The floristic data mentioned here reflect their areas of expertise: Joseph Dalton Hooker was Director of Royal Botanic Gardens at Kew. He botanized extensively in the southern hemisphere as part of the HMS *Erebus* and *Terror* expedition to Antarctica (1839–1843) and later explored the botany of the Himalaya region (1847–1850). He knew the New Zealand plants well, having authored the two-volume manual *Flora Novae-Zelandiae* (London, 1853–1855) and a lengthy "essay" on the flora of New Zealand (London, 1853). Asa Gray was the leading American botanist and Darwin's most ardent supporter in the United States. He was Professor of Natural History at Harvard University, where he founded the great herbarium that now bears his name. He authored many botanical books, including the well-known *Manual of the Botany of the Northern United States* (first published in 1847). In 1856 he published an analysis entitled "Statistics of the flora of the northern United States," the source of the information Darwin cites here.

failed, after consultation with one of the highest authorities, namely, Professor Huxley, to discover a single case of an hermaphrodite animal with the organs of reproduction so perfectly enclosed within the body, that access from without and the occasional influence of a distinct individual can be shown to be physically impossible. Cirripedes long appeared to me to present a case of very great difficulty under this point of view; but I have been enabled, by a fortunate chance, elsewhere to prove that two individuals, though both are self-fertilising hermaphrodites, do sometimes cross.

It must have struck most naturalists as a strange anomaly that, in the case of both animals and plants, species of the same family and even of the same genus, though agreeing closely with each other in almost their whole organisation, yet are not rarely, some of them hermaphrodites, and some of them unisexual. But if, in fact, all hermaphrodites do occasionally intercross with other individuals, the difference between hemaphrodites and unisexual species, as far as function is concerned, becomes very small.

From these several considerations and from the many special facts which I have collected, but which I am not here able to give, I am strongly inclined to suspect that, both in the vegetable and animal kingdoms, an occasional intercross with a distinct individual is a law of nature. I am well aware that there are, on this view, many cases of difficulty, some of which I am trying to investigate. Finally then, we may conclude that in many organic beings, a cross between two individuals is an obvious necessity for each birth; in many others it occurs perhaps only at long intervals; but in none, as I suspect, can self-fertilisation go on for perpetuity.

Circumstances favourable to Natural Selection.—This

1 Even hermaphroditic animals cross at least occasionally, Darwin argues. In his *Natural Selection* manuscript, he wrote, "How it may be asked can [hermaphrodites] occasionally cross? If an organism can from the day of its creation go on most strictly interbreeding, that is self-fertilising itself from the day of its creation to its extinction, one may well doubt all the foregoing facts & put them all down to popular prejudices. I can hardly believe this. The subject has sufficient importance for us, in relation to crossing of *slight* varieties being a powerful means of keeping a breed or species true . . . that I must discuss it at some little length." He penciled in the margin: "Get Huxley to read over for this" (Stauffer 1975, p. 43, emphasis Darwin's), referring to his friend the English zoologist Thomas Henry Huxley. Huxley did review this and related material, and agreed with Darwin that hermaphrodites can and do interbreed.

2 In his work on barnacles (cirripedes) Darwin discovered specialized male forms of hermaphroditic species, which he termed "complemental males." He also discovered what seemed to be a mutant form of the hermaphroditic barnacle *Balanus balanoides.* The barnacle had rudimentary male organs yet was able to reproduce—a "smoking gun" for interbreeding (see Stauffer 1975, pp. 44–45).

3 Self-fertilization and inbreeding do occur in a diversity of species, as does clonal or asexual reproduction. Many flowering plants have high self-pollination rates (Barrett 2003), for example, and according to genetic evidence bdelloid rotifers (microscopic freshwater animals of the phylum Rotifera) may have been reproducing asexually for millions of years (see Birky 2004). Generally, however, modern biologists agree with Darwin that outcrossing is a good thing in the evolutionary long term. We now better understand why this is so: sexual reproduction generates variation both by recombination (chromosomal exchange of DNA during meiosis) and by simply assembling and reassembling different combinations of alleles. The effect of this mixing is a richer pool of genotype variants that may respond to selection. Asexual or inbred species often must rely solely on mutation to generate variation (though some asexual groups like bacteria have their own mechanism of genetic exchange and recombination).

1 Following up on the point made on the previous page, Darwin here reiterates that the availability of a large amount of heritable variation is favorable. Moreover, he points out that population size is important because the greater the number of varying individuals, the greater the chances that favorable variations will arise.

2 Here again is the analogy with artificial selection, but in this case Darwin compares natural selection more explicitly with unconscious artificial selection—for example, people independently selecting their lines of livestock, but in the same "direction" according to a common standard of desired traits. The "common standard" with natural selection would be adaptation to the imperfectly occupied niche.

3 Darwin recognizes a difficulty with his model of how natural selection operates. In an environmentally diverse area, local adaptation of organisms will be undermined by intercrossing . . .

▶1 is an extremely intricate subject. A large amount of inheritable and diversified variability is favourable, but I believe mere individual differences suffice for the work. A large number of individuals, by giving a better chance for the appearance within any given period of profitable variations, will compensate for a lesser amount of variability in each individual, and is, I believe, an extremely important element of success. Though nature grants vast periods of time for the work of natural selection, she does not grant an indefinite period; for as all organic beings are striving, it may be said, to seize on each place in the economy of nature, if any one species does not become modified and improved in a corresponding degree with its competitors, it will soon be exterminated.

▶2 In man's methodical selection, a breeder selects for some definite object, and free intercrossing will wholly stop his work. But when many men, without intending to alter the breed, have a nearly common standard of perfection, and all try to get and breed from the best animals, much improvement and modification surely but slowly follow from this unconscious process of selection, notwithstanding a large amount of crossing with inferior animals. Thus it will be in nature; for within a confined area, with some place in its polity not so perfectly occupied as might be, natural selection will always tend to preserve all the individuals varying in the right direction, though in different degrees, so as better to fill up ▶3 the unoccupied place. But if the area be large, its several districts will almost certainly present different conditions of life; and then if natural selection be modifying and improving a species in the several districts, there will be intercrossing with the other individuals of the same species on the confines of each. And in this case the effects of intercrossing can hardly be counterbalanced by natural selection always tending to modify all the individuals in each district in exactly the same manner to the conditions of each; for in a continuous area, the conditions will generally graduate away insensibly from one district to another. The intercrossing will most affect those animals which unite for each birth, which wander much, and which do not breed at a very quick rate. Hence in animals of this nature, for instance in birds, varieties will generally be confined to separated countries; and this I believe to be the case. In hermaphrodite organisms which cross only occasionally, and likewise in animals which unite for each birth, but which wander little and which can increase at a very rapid rate, a new and improved variety might be quickly formed on any one spot, and might there maintain itself in a body, so that whatever intercrossing took place would be chiefly between the individuals of the same new variety. A local variety when once thus formed might subsequently slowly spread to other districts. On the above principle, nurserymen always prefer getting seed from a large body of plants of the same variety, as the chance of intercrossing with other varieties is thus lessened.

Even in the case of slow-breeding animals, which unite for each birth, we must not overrate the effects of intercrosses in retarding natural selection; for I can bring a considerable catalogue of facts, showing that within the same area, varieties of the same animal can long remain distinct, from haunting different stations, from breeding at slightly different seasons, or from varieties of the same kind preferring to pair together.

Intercrossing plays a very important part in nature in keeping the individuals of the same species, or of the same variety, true and uniform in character. It will obviously thus act far more efficiently with those animals

1 . . . a situation especially problematic for mobile species like birds and mammals that range over a wide area. Selection might be expected to produce a gradient or constellation of adapted forms to match the environmental gradient or mosaic in which they live, but if the different forms freely intercross they become homogenized rather than differentiated. Darwin suggests that in this case the group may still present varieties but at a higher spatial scale, "confined to different countries." Less mobile organisms and hermaphrodites are more likely to become more locally adapted, such that varieties occur at a finer spatial scale.

2 Intercrossing is a double-edged sword. It may, particularly under Darwin's model of inheritance, continually undo what selection has wrought: no sooner do varieties become well adapted to particular conditions than intercrossing mixes them up again. Though Darwin cautions the reader not to "overrate the effects of intercrosses in retarding natural selection," this proved to be a thorny problem, and some of his critics leveled devastating arguments against the efficacy of natural selection on precisely this point. In the sixth edition Darwin added a chapter entitled "Miscellaneous Objections to the Theory of Natural Selection" to tackle such problems. In any case, Darwin was convinced that the benefits to be gained from intercrossing outweighed the problems.

3 Here and continuing onto the next page is a clear statement of the benefits of intercrossing. Lack of intercrossing may promote local adaptation and uniformity of character, but at the risk of being unresponsive to selection should the environment change. Such forms will persist only "as long as their conditions of life remain the same."

1 Physical isolation is an important dimension to consider, since it trumps mobility and prevents intercrossing. Here and on the next page Darwin briefly lays out the key effects of isolation: isolated species can become well adapted to local conditions but, even more important, isolation prevents potentially better-adapted outsiders from immigrating and displacing the local species.

which unite for each birth; but I have already attempted to show that we have reason to believe that occasional intercrosses take place with all animals and with all plants. Even if these take place only at long intervals, I am convinced that the young thus produced will gain so much in vigour and fertility over the offspring from long-continued self-fertilisation, that they will have a better chance of surviving and propagating their kind; and thus, in the long run, the influence of intercrosses, even at rare intervals, will be great. If there exist organic beings which never intercross, uniformity of character can be retained amongst them, as long as their conditions of life remain the same, only through the principle of inheritance, and through natural selection destroying any which depart from the proper type; but if their conditions of life change and they undergo modification, uniformity of character can be given to their modified offspring, solely by natural selection preserving the same favourable variations.

1 Isolation, also, is an important element in the process of natural selection. In a confined or isolated area, if not very large, the organic and inorganic conditions of life will generally be in a great degree uniform; so that natural selection will tend to modify all the individuals of a varying species throughout the area in the same manner in relation to the same conditions. Intercrosses, also, with the individuals of the same species, which otherwise would have inhabited the surrounding and differently circumstanced districts, will be prevented. But isolation probably acts more efficiently in checking the immigration of better adapted organisms, after any physical change, such as of climate or elevation of the land, &c.; and thus new places in the natural economy of the country are left open for the old inhabitants to struggle for, and become adapted to, through modifica-

tions in their structure and constitution. Lastly, isolation, by checking immigration and consequently competition, will give time for any new variety to be slowly improved; and this may sometimes be of importance in the production of new species. If, however, an isolated area be very small, either from being surrounded by barriers, or from having very peculiar physical conditions, the total number of the individuals supported on it will necessarily be very small; and fewness of individuals will greatly retard the production of new species through natural selection, by decreasing the chance of the appearance of favourable variations.

If we turn to nature to test the truth of these remarks, and look at any small isolated area, such as an oceanic island, although the total number of the species inhabiting it, will be found to be small, as we shall see in our chapter on geographical distribution; yet of these species a very large proportion are endemic,—that is, have been produced there, and nowhere else. Hence an oceanic island at first sight seems to have been highly favourable for the production of new species. But we may thus greatly deceive ourselves, for to ascertain whether a small isolated area, or a large open area like a continent, has been most favourable for the production of new organic forms, we ought to make the comparison within equal times; and this we are incapable of doing.

Although I do not doubt that isolation is of considerable importance in the production of new species, on the whole I am inclined to believe that largeness of area is of more importance, more especially in the production of species, which will prove capable of enduring for a long period, and of spreading widely. Throughout a great and open area, not only will there be a better chance of favourable variations arising from the large number of individuals of the same species

1 This final point about isolation touches on endemism and the remarkable patterns of species distribution associated with oceanic islands. Early-arriving varieties become "slowly improved," eventually becoming endemic forms. Both Darwin and Wallace were struck in their respective travels by the patterns of endemism characteristic of remote oceanic islands, patterns that proved important in convincing both men of the reality of transmutation. Darwin's B notebook, written in 1837, is full of insights and speculations concerning islands. One example: "island near continents might have some species same as nearest land, which were late arrivals . . . others old ones . . . might have grown altered . . . Hence type would be of the continent though species all different" (see Barrett et al. 1987, p. 173).

2 Isolation was clearly more important to Darwin's early thinking than to his later thinking. In *Natural Selection* (Stauffer 1975, p. 255) he wrote: "I infer that some degree of isolation would generally be almost indispensable." In the *Origin*, however, Darwin gives the distinct impression that he does not think isolation is the only or even the primary means of species evolution. In fact, he discounts its importance in favor of a selection-driven mechanism. Note that he suggests here that oceanic islands might seem "at first sight" to have been favorable for the formation of new species in comparison to similar continental areas, but that "we may thus greatly deceive ourselves." Darwin suspects that "largeness of area" is more important than isolation; as we shall see later in this chapter, that position is consistent with his model of divergence of character. For now, he sidesteps the question by suggesting that the comparison needs to be standardized with respect to time available for speciation—something that cannot be done, or at least could not be done in Darwin's day. (See the Coda for an account of Darwin's interaction with Moritz Wagner over this issue.) The modern take on speciation, first cogently argued by Mayr (1942), is that isolation nearly always plays a role, and thus oceanic islands are veritable crucibles of evolutionary change.

[1] there supported, but the conditions of life are infinitely complex from the large number of already existing species; and if some of these many species become modified and improved, others will have to be improved in a corresponding degree or they will be exterminated. Each new form, also, as soon as it has been much improved, will be able to spread over the open and continuous area, and will thus come into competition with many others. Hence more new places will be formed, and the competition to fill them will be more severe, on a large than on a small and isolated area. Moreover, [2] great areas, though now continuous, owing to oscillations of level, will often have recently existed in a broken condition, so that the good effects of isolation will generally, to a certain extent, have concurred. Finally, I conclude that, although small isolated areas probably have been in some respects highly favourable for the production of new species, yet that the course of modification will generally have been more rapid on large areas; and what is more important, that the new forms produced on large areas, which already have been victorious over many competitors, will be those that will spread most widely, will give rise to most new varieties and species, and will thus play an important part in the changing history of the organic world.

[3] We can, perhaps, on these views, understand some facts which will be again alluded to in our chapter on geographical distribution; for instance, that the productions of the smaller continent of Australia have formerly yielded, and apparently are now yielding, before those of the larger Europæo-Asiatic area. Thus, also, it is that continental productions have everywhere become so largely naturalised on islands. On a small island, the race for life will have been less severe, and there will have been less modification and less exter-

1 In the remainder of this paragraph Darwin expands on his assertion that "largeness of area"—by which he means continental areas—is more important for the formation of new species than are smaller, isolated areas like islands, for a number of reasons: larger areas support greater populations, which in turn are expected to produce a greater absolute number of variants on which selection may act. The competition in larger areas is greater and stronger as well, owing to their greater population densities. Larger areas also support a greater diversity of habitats or niches, more spaces in the economy of nature to which new varieties may adapt.

2 Darwin would like to have his cake and eat it too: oscillation of sea (or land) level would have the effect of breaking continuous continental areas into archipelagos. That large expanses of continental land were once submerged beneath the sea is beyond doubt—marine sedimentary rock formations and their attendant fossils abound on all continents—and so Darwin suggests that the "good effects" of isolation are felt during those periods, while the even better effects of large continuous land areas are felt when the land is high and dry.

3 Here Darwin makes an observation to back up the argument that largeness of area is more important than isolation: when "continental productions" (i.e., continental species) are introduced into islands or similar isolated areas, they tend to handily outcompete the native species, displacing them and not infrequently even driving them to extinction. Darwin takes this as an indication of the superior competitive ability of continental species, superiority derived from the intense history of competition they experienced.

mination. Hence, perhaps, it comes that the flora of Madeira, according to Oswald Heer, resembles the extinct tertiary flora of Europe. All fresh-water basins, taken together, make a small area compared with that of the sea or of the land; and, consequently, the competition between fresh-water productions will have been less severe than elsewhere; new forms will have been more slowly formed, and old forms more slowly exterminated. And it is in fresh water that we find seven genera of Ganoid fishes, remnants of a once preponderant order: and in fresh water we find some of the most anomalous forms now known in the world, as the Ornithorhynchus and Lepidosiren, which, like fossils, connect to a certain extent orders now widely separated in the natural scale. These anomalous forms may almost be called living fossils; they have endured to the present day, from having inhabited a confined area, and from having thus been exposed to less severe competition.

To sum up the circumstances favourable and unfavourable to natural selection, as far as the extreme intricacy of the subject permits. I conclude, looking to the future, that for terrestrial productions a large continental area, which will probably undergo many oscillations of level, and which consequently will exist for long periods in a broken condition, will be the most favourable for the production of many new forms of life, likely to endure long and to spread widely. For the area will first have existed as a continent, and the inhabitants, at this period numerous in individuals and kinds, will have been subjected to very severe competition. When converted by subsidence into large separate islands, there will still exist many individuals of the same species on each island: intercrossing on the confines of the range of each species will thus be checked: after physical changes of any kind, immigration will be pre-

1 Oswald Heer was a Swiss paleobotanist who worked at the University of Zürich. His specialty was flora of the Tertiary period in Europe. Here Darwin is referring to Heer's paper on the plant fossils of Madeira, *Über die fossilen Pflanzen von St. Jorge in Madeira* (1855).

Here Darwin gives examples to support the assertion that island forms are less well adapted than continental forms. Extant island flora like that of Madeira is supposed to resemble the extinct Tertiary flora of Europe, and archaic animal forms are often found on islands or in island-like habitats (like lakes)—shelters where relictual variants or species hold out. The argument is that such groups are found on islands or island-like environments because they were outcompeted and driven to extinction in continental areas. There is, of course, some truth to both of Darwin's assertions: island species often cannot compete with introduced species, and taxonomic relicts are disproportionately found in island environments (e.g., tuataras of New Zealand, the last remnant of once-flourishing rhynchocephalian reptiles).

Note the origin of the phrase "living fossil," which Darwin coined to describe anomalous or relictual species that have apparently persisted far past their heyday when the groups to which they belong were dominant. This is a slightly different meaning from that often used today, which typically refers to species like horseshoe crabs or gingko trees that have apparently changed little from their distant ancestors known from the fossil record.

The groups Darwin mentions here are all considered representatives of basal lineages. Ganoids are a primitive fish group with thick, bony, diamond-shaped scales. They are best represented in the fossil record (e.g., all Sarcopterygii and some Actinopterygii), but some actinopterygids such as gars (Lepisosteidae) and the bizarre bichirs of Africa (Polypteridae) still exist. *Ornithorhynchus* is the duck-billed platypus of eastern Australia, and *Lepidosiren* is the South American lungfish, represented by the sole species *L. paradoxa* in the family Lepidosirenidae.

2 This paragraph and the following provide an excellent summary of the arguments of the past six pages.

1 Natural selection acts slowly, but "often only at long intervals of time." This statement seemingly contradicts the strict gradualist stance that Darwin typically takes, sounding a bit more punctuationist: selection acts slowly, but at intervals. This theory is echoed in the next sentence, which contains the word "intermittent." In later editions of the *Origin* the wording was changed slightly, however, suggesting a different interpretation. By the fifth edition Darwin wrote: "selection generally acts very slowly in effecting changes, at long intervals of time . . . I further believe that these slow, intermittent results of natural selection accord perfectly with what geology tells us . . ." Darwin could be referring, not to selection acting intermittently, but to the intermittent products ("results") of natural selection over time, as seen in the fossil record. By the next (and last) edition, he reverted to something close to the original wording: "natural selection will generally act very slowly, only at long intervals of time" (see Peckham 1959, p. 202). Although passages like these have been used to suggest that Darwin anticipated the idea of punctuated equilibrium and was not a strict gradualist, the overall tenor of his arguments suggests that slow, steady, gradual change was the way he envisioned the evolutionary process. We will revisit this point in chapter X, on geological succession.

vented, so that new places in the polity of each island will have to be filled up by modifications of the old inhabitants; and time will be allowed for the varieties in each to become well modified and perfected. When, by renewed elevation, the islands shall be re-converted into a continental area, there will again be severe competition: the most favoured or improved varieties will be enabled to spread: there will be much extinction of the less improved forms, and the relative proportional numbers of the various inhabitants of the renewed continent will again be changed; and again there will be a fair field for natural selection to improve still further the inhabitants, and thus produce new species.

That natural selection will always act with extreme slowness, I fully admit. Its action depends on there being places in the polity of nature, which can be better occupied by some of the inhabitants of the country undergoing modification of some kind. The existence of such places will often depend on physical changes, which are generally very slow, and on the immigration of better adapted forms having been checked. But the action of natural selection will probably still oftener depend on some of the inhabitants becoming slowly modified; the mutual relations of many of the other inhabitants being thus disturbed. Nothing can be effected, unless favourable variations occur, and variation itself is apparently always a very slow process. The process will often be greatly retarded by free intercrossing. Many will exclaim that these several causes are amply sufficient wholly to stop the action of natural selection. I do
1 not believe so. On the other hand, I do believe that natural selection will always act very slowly, often only at long intervals of time, and generally on only a very few of the inhabitants of the same region at the same time. I further believe, that this very slow, intermit-

tent action of natural selection accords perfectly well with what geology tells us of the rate and manner at which the inhabitants of this world have changed.

Slow though the process of selection may be, if feeble man can do much by his powers of artificial selection, I can see no limit to the amount of change, to the beauty and infinite complexity of the coadaptations between all organic beings, one with another and with their physical conditions of life, which may be effected in the long course of time by nature's power of selection.

Extinction.—This subject will be more fully discussed in our chapter on Geology; but it must be here alluded to from being intimately connected with natural selection. Natural selection acts solely through the preservation of variations in some way advantageous, which consequently endure. But as from the high geometrical powers of increase of all organic beings, each area is already fully stocked with inhabitants, it follows that as each selected and favoured form increases in number, so will the less favoured forms decrease and become rare. Rarity, as geology tells us, is the precursor to extinction. We can, also, see that any form represented by few individuals will, during fluctuations in the seasons or in the number of its enemies, run a good chance of utter extinction. But we may go further than this; for as new forms are continually and slowly being produced, unless we believe that the number of specific forms goes on perpetually and almost indefinitely increasing, numbers inevitably must become extinct. That the number of specific forms has not indefinitely increased, geology shows us plainly; and indeed we can see reason why they should not have thus increased, for the number of places in the polity of nature is not indefinitely great,—not that we

1 Note again the comparison between natural selection and artificial selection. This sentence is nicely crafted: Darwin sees no limit to the changes that may occur, a magisterial vision of beautiful and complex adaptations shaped and reshaped over time.

2 It is significant that Darwin gives extinction its own section. He sees extinction as a process integral to the operation of natural selection—the death of species stems from a failure or inability to adapt, either to competitors or to changing conditions. This view was embraced early on by Lyell, in contradistinction to the alternative belief that species go extinct as a result of exhaustion of a vital "life force." Ironically, this was Darwin's view—his "generational theory" of extinction—while on the *Beagle* voyage. He eventually came around to Lyell's view. In 1837 Darwin wrote in the B notebook that "death of species is a consequence . . . of non adaptation of circumstances" (Barrett et al. 1987, p. 180).

3 This is another aspect of Lyell's theory that Darwin embraced beginning in 1837. Lyell believed that at any given time the overall number of species remained more or less constant. After reading Malthus, Darwin came around to Lyell's ideas that extinction stemmed from non-adaptation to changing circumstances and that species numbers remained relatively constant. He appended two notes to his tree diagram of the B notebook: "Case must be that one generation then should be as many living as now. To do this & to have many species in same genus (as is) REQUIRES extinction" (Barrett et al. 1987, p. 180). This idea sounds very much like an application of Malthusian struggle to species themselves. Just as displacement and death of individuals are an inevitable consequence of population growth and the attendant struggle for existence, so too species cannot go on increasing in number indefinitely.

4 "for the number of places in the polity of nature is not indefinitely great": recall Darwin's wedges metaphor from p. 67; limiting space, and resources, leads inevitably to a strong and incessant selective pressure.

1 This is another explicit statement about the importance of numbers, or population size, in yielding enough variants for selection to act upon.

2 The rarer species that "stand in closest competition" to the more common and actively adapting species will suffer most—they come into direct conflict with the superior competitor. What's more, Darwin suggests here, the most severe competition is felt by the most closely related forms. The newer, and so presumably better adapted, varieties compete most with their "nearest kindred" because of overlap of food, ecological requirements, etc., and eventually outcompete them. This is a key element of Darwin's divergence of character mechanism, which is introduced on the next page.

have any means of knowing that any one region has as yet got its maximum of species. Probably no region is as yet fully stocked, for at the Cape of Good Hope, where more species of plants are crowded together than in any other quarter of the world, some foreign plants have become naturalised, without causing, as far as we know, the extinction of any natives.

1 Furthermore, the species which are most numerous in individuals will have the best chance of producing within any given period favourable variations. We have evidence of this, in the facts given in the second chapter, showing that it is the common species which afford the greatest number of recorded varieties, or incipient species. Hence, rare species will be less quickly modified or improved within any given period, and they will consequently be beaten in the race for life by the modified descendants of the commoner species.

From these several considerations I think it inevitably follows, that as new species in the course of time are formed through natural selection, others will
2 become rarer and rarer, and finally extinct. The forms which stand in closest competition with those undergoing modification and improvement, will naturally suffer most. And we have seen in the chapter on the Struggle for Existence that it is the most closely-allied forms,—varieties of the same species, and species of the same genus or of related genera,—which, from having nearly the same structure, constitution, and habits, generally come into the severest competition with each other. Consequently, each new variety or species, during the progress of its formation, will generally press hardest on its nearest kindred, and tend to exterminate them. We see the same process of extermination amongst our domesticated productions, through the selection of improved forms by man. Many curious

instances could be given showing how quickly new breeds of cattle, sheep, and other animals, and varieties of flowers, take the place of older and inferior kinds. In Yorkshire, it is historically known that the ancient black cattle were displaced by the long-horns, and that these "were swept away by the short-horns" (I quote the words of an agricultural writer) "as if by some murderous pestilence."

Divergence of Character.—The principle, which I have designated by this term, is of high importance on my theory, and explains, as I believe, several important facts. In the first place, varieties, even strongly-marked ones, though having somewhat of the character of species—as is shown by the hopeless doubts in many cases how to rank them—yet certainly differ from each other far less than do good and distinct species. Nevertheless, according to my view, varieties are species in the process of formation, or are, as I have called them, incipient species. How, then, does the lesser difference between varieties become augmented into the greater difference between species? That this does habitually happen, we must infer from most of the innumerable species throughout nature presenting well-marked differences; whereas varieties, the supposed prototypes and parents of future well-marked species, present slight and ill-defined differences. Mere chance, as we may call it, might cause one variety to differ in some character from its parents, and the offspring of this variety again to differ from its parent in the very same character and in a greater degree; but this alone would never account for so habitual and large an amount of difference as that between varieties of the same species and species of the same genus.

As has always been my practice, let us seek light on

1 Darwin is here quoting William Youatt, who was first mentioned on p. 31; in this analogy, newer domestic cattle varieties have metaphorically outcompeted and displaced their progenitor varieties by appealing to the preferences of the breeders.

2 High importance, indeed. Divergence of character is one of the most significant elements of Darwin's model of transmutation—and one of the most misunderstood of his ideas. In essence divergence of character follows from competition between similar forms, where selection for superior competitive ability enables descendant forms to seize on new niches or habitats as they increase in numbers. These descendant forms tend to be divergent and become successively more dissimilar over time. This mode of change translates into a branching-tree pattern of descendant lineages over many generations. Divergence of character thus generates biodiversity.

The confusion over Darwin's meaning has turned on the action of competition. He has often been interpreted as suggesting that selection reduces competition between similar forms, a result of which is divergence. That is subtly but critically different from suggesting that selection hones competitive ability and leads to displacement of other forms as these forms grow in number and spread. The former view might hold if the descendant groups were moving into *unoccupied* niches, but Darwin says just the opposite in this section. See Tammone (1995) for a discussion of this distinction.

3 Varieties, recall, are incipient species in Darwin's view. How, he asks, "does the lesser difference between varieties become augmented into the greater difference between species?" In other words, how do varieties diverge, such that they become sufficiently distinct (in morphology, physiology, etc.) to be considered different species?

4 Here, by analogy, is an answer to the question of how varieties diverge into new species. Selection favoring distinct and different attributes makes varieties increasingly dissimilar. Intermediates are imagined to be disfavored, and selection against them accentuates, and even accelerates, the divergence of the main selected lines.

this head from our domestic productions. We shall here find something analogous. A fancier is struck by a pigeon having a slightly shorter beak; another fancier is struck by a pigeon having a rather longer beak; and on the acknowledged principle that "fanciers do not and will not admire a medium standard, but like extremes," they both go on (as has actually occurred with tumbler-pigeons) choosing and breeding from birds with longer and longer beaks, or with shorter and shorter beaks. Again, we may suppose that at an early period one man preferred swifter horses; another stronger and more bulky horses. The early differences would be very slight; in the course of time, from the continued selection of swifter horses by some breeders, and of stronger ones by others, the differences would become greater, and would be noted as forming two sub-breeds; finally, after the lapse of centuries, the sub-breeds would become converted into two well-established and distinct breeds. As the differences slowly become greater, the inferior animals with intermediate characters, being neither very swift nor very strong, will have been neglected, and will have tended to disappear.
1 Here, then, we see in man's productions the action of what may be called the principle of divergence, causing differences, at first barely appreciable, steadily to increase, and the breeds to diverge in character both from each other and from their common parent.

2 But how, it may be asked, can any analogous principle apply in nature? I believe it can and does apply most efficiently, from the simple circumstance that the more diversified the descendants from any one species become in structure, constitution, and habits, by so much will they be better enabled to seize on many and widely diversified places in the polity of nature, and so be enabled to increase in numbers.

1 Darwin's "principle of divergence" concept developed over a period of time. Divergence per se came to him early, while his *principle,* or mechanism driving divergence, came later. Browne (1980) suggested that three elements of the principle of divergence (hierarchy of taxonomic relationships, division of labor, and the relationship between size of genus and proportion of varieties it contains) were regarded more or less separately by Darwin until about the summer of 1857, when they were combined to formulate a coherent model of divergence. In his model, selection honed competitive ability such that descendant forms could successfully move into unoccupied or "imperfectly occupied" niches or habitats, in the latter case ousting the current occupants, including their own parents. Selective advantage thereby drives diversification, or divergence: "the varying offspring of each species will try (only a few will succeed) to seize on as many and as diverse places in the economy of nature, as possible. Each new variety or species, when formed will generally take the places of and so exterminate its less well-fitted parent. This I believe to be the origin of the classification or arrangement of all organic beings at all times," Darwin wrote to Asa Gray in September 1857 (*Correspondence* 6: 445). About this time Darwin began referring to a "principle" of divergence, and not just divergence (Browne 1980).

2 A clear statement of the benefits of divergence: the more diversified the descendant forms become, the better they are able to "seize on many and widely diversified places in the polity of nature, and so be enabled to increase in numbers." The phrase "polity of nature," as an aside, may have come from a 1760 dissertation by Linnaeus entitled *Politia Naturae,* which Darwin read in translation in 1841. This work concluded, "Thus we see Nature resemble a well regulated state in which every individual has his proper employment and subsistence, and a proper gradation of offices and officers is appointed to correct and restrain every detrimental excess" (quoted in Stauffer 1960, page 240).

We can clearly see this in the case of animals with simple habits. Take the case of a carnivorous quadruped, of which the number that can be supported in any country has long ago arrived at its full average. If its natural powers of increase be allowed to act, it can succeed in increasing (the country not undergoing any change in its conditions) only by its varying descendants seizing on places at present occupied by other animals: some of them, for instance, being enabled to feed on new kinds of prey, either dead or alive; some inhabiting new stations, climbing trees, frequenting water, and some perhaps becoming less carnivorous. The more diversified in habits and structure the descendants of our carnivorous animal became, the more places they would be enabled to occupy. What applies to one animal will apply throughout all time to all animals—that is, if they vary—for otherwise natural selection can do nothing. So it will be with plants. It has been experimentally ◀1 proved, that if a plot of ground be sown with one species of grass, and a similar plot be sown with several distinct genera of grasses, a greater number of plants and a greater weight of dry herbage can thus be raised. The same has been found to hold good when first one variety and then several mixed varieties of wheat have been sown on equal spaces of ground. Hence, if any one species of grass were to go on varying, and those varieties were continually selected which differed from each other in at all the same manner as distinct species and genera of grasses differ from each other, a greater number of individual plants of this species of grass, including its modified descendants, would succeed in living on the same piece of ground. And we well know that each species and each variety of grass is annually sowing almost countless seeds; and thus, as it may be said, is striving its utmost to increase its numbers. Con-

1 This passage refers to a remarkable experimental garden at Woburn Abbey in Bedfordshire, England, in which grasses and other herbaceous species were planted in various combinations and in different soil types—242 four-foot by four-foot square experimental plots in all—in order to measure performance. The results were first published in the *Hortus Gramineus Woburnensis* in 1816. The third edition in 1824 noted that more diverse plant assemblages were more productive than less diverse ones. The experiment may not stand up to modern standards of statistically rigorous design, but nonetheless it may be the first ecological experiment (Hector and Hooper 2002).

Darwin's idea that taxonomically (or phylogenetically) distant species are more likely not only to coexist stably in communities but also to enjoy greater productivity is largely accepted today. For example, one recent experimental study of mycorrhizal fungal assemblages found that realized species richness (and host plant productivity) was greatest with high phylogenetic diversity of fungi, an outcome the study's authors (Maherali and Klironomos 2007) attributed to reduced competition between the fungal lineages.

sequently, I cannot doubt that in the course of many thousands of generations, the most distinct varieties of any one species of grass would always have the best chance of succeeding and of increasing in numbers, and thus of supplanting the less distinct varieties; and varieties, when rendered very distinct from each other, take the rank of species.

The truth of the principle, that the greatest amount of life can be supported by great diversification of structure, is seen under many natural circumstances. In an extremely small area, especially if freely open to immigration, and where the contest between individual and individual must be severe, we always find great diversity in its inhabitants. For instance, I found that a piece of turf, three feet by four in size, which had been exposed for many years to exactly the same conditions, supported twenty species of plants, and these belonged to eighteen genera and to eight orders, which shows how much these plants differed from each other. So it is with the plants and insects on small and uniform islets; and so in small ponds of fresh water. Farmers find that they can raise most food by a rotation of plants belonging to the most different orders: nature follows what may be called a simultaneous rotation. Most of the animals and plants which live close round any small piece of ground, could live on it (supposing it not to be in any way peculiar in its nature), and may be said to be striving to the utmost to live there; but, it is seen, that where they come into the closest competition with each other, the advantages of diversification of structure, with the accompanying differences of habit and constitution, determine that the inhabitants, which thus jostle each other most closely, shall, as a general rule, belong to what we call different genera and orders.

The same principle is seen in the naturalisation of

1 A common interpretation of this idea is that Darwin is getting at the concept of niche partitioning. In modern ecological terms, species that are too similar in resource needs compete, and inevitably one is beaten out by the other: the principle of competitive exclusion. Selection may therefore favor diversification so as to occupy niches sufficiently dissimilar that competition is reduced. But is selection acting to reduce competition per se, or to increase competitive ability through diversification? There is a difference, perhaps, between the mechanism of divergence *in situ* and the coming together of (already divergent) species differing in resource needs to occupy a given habitat. Darwin is conceptually linking the two here, as is evident in the remainder of this paragraph.

2 This small piece of turf supports a remarkable diversity of plants. This particular study was likely part of a larger survey of plants in the meadows around Down House, which Darwin began in June 1855 together with his children and their governess, Brodie. In 2005 several of Darwin's descendants joined with botanists from the Natural History Museum and English Heritage to repeat the survey.

In his "big species book" Darwin supplements this observation of the three-by-four-foot piece of turf with another from a Mr. C. A. Johns, "who says that he covered with his hat, (I presume broad-brimmed) near to Lands End six species of Trifolium, a Lotus & Anthyllis; & had the brim been a little wider it would have covered another Lotus and Genista; which would have made ten species of Leguminosae, belonging to only four genera!" (Stauffer 1975, p. 230).

plants through man's agency in foreign lands. It might have been expected that the plants which have succeeded in becoming naturalised in any land would generally have been closely allied to the indigenes; for these are commonly looked at as specially created and adapted for their own country. It might, also, perhaps have been expected that naturalised plants would have belonged to a few groups more especially adapted to certain stations in their new homes. But the case is very different; and Alph. De Candolle has well remarked in his great and admirable work, that floras gain by naturalisation, proportionally with the number of the native genera and species, far more in new genera than in new species. To give a single instance: in the last edition of Dr. Asa Gray's 'Manual of the Flora of the Northern United States,' 260 naturalised plants are enumerated, and these belong to 162 genera. We thus see that these naturalised plants are of a highly diversified nature. They differ, moreover, to a large extent from the indigenes, for out of the 162 genera, no less than 100 genera are not there indigenous, and thus a large proportional addition is made to the genera of these States.

By considering the nature of the plants or animals which have struggled successfully with the indigenes of any country, and have there become naturalised, we can gain some crude idea in what manner some of the natives would have had to be modified, in order to have gained an advantage over the other natives; and we may, I think, at least safely infer that diversification of structure, amounting to new generic differences, would have been profitable to them.

The advantage of diversification in the inhabitants of the same region is, in fact, the same as that of the physiological division of labour in the organs of the same individual body—a subject so well elucidated by

1 Plants might be expected to be best adapted to the places in which they were supposedly created, on the supposition that the creator would have matched species perfectly to their native surroundings. In several places in the *Origin* Darwin points out that species are not perfectly well adapted, and the seeming ease with which native plants are outcompeted and displaced by introductions (think any number of European weeds) well illustrates that point. An additional important point here concerns the observation that successfully naturalized introductions tend to be quite different, taxonomically, from the natives. Darwin cites this fact in support of his idea that divergence of character permits coexistence of widely varying species. The "great and admirable work" by the Swiss botanist Alphonse de Candolle is the two-volume *Géographie Botanique Raisonnée,* published in 1855.

2 For these data Darwin consulted the second edition of Asa Gray's famous *Manual,* published in 1856. The *Manual* has never been out of print: the eighth, centennial, edition of 1950 is the one still used today. A reviewer wrote of this volume, "In the United States few botanical books have been expected with more anticipation and interest than has the eighth edition of Gray's Manual of Botany" (Just 1953).

3 The concept of a "physiological division of labor," recall from p. 93, holds that overall efficiency is improved by task specialization—certain organs specialize in digesting, others in sight, blood filtering, movement, etc. The concept was developed by the French zoologist and physiologist Henri Milne-Edwards, who saw it as an analogy to the increased productivity in manufacturing and labor markets posited by the economist Adam Smith in the late eighteenth century. Darwin saw this analogy too: on September 23, 1856, he wrote in his notes: "The advantage in each group becoming as different as possible, may be compared to the fact that by division of labor most people can be supported in each country" (cited in Gould 2002, p. 230).

1 This is a grand vision that brings divergence of character, natural selection, and extinction together into a unitary system. The diagram that follows is the only figure appearing in the entire book. Two related diagrams appear in Darwin's *Natural Selection* manuscript (Stauffer 1975). Darwin's insight into the "divergence of character" principle came to him in a flash, well after his inspiration from reading Malthus in October 1838. As late as 1844, the year he wrote the species *Essay,* Darwin still lacked the fundamental principle that would drive divergence. As he wrote in his autobiography: "at that time I overlooked one problem of great importance; and it is astonishing to me . . . how I could have overlooked it and its solution. This problem is the tendency in organic beings descended from the same stock to diverge in character as they become modified. . . . I can remember the very spot in the road, whilst in my carriage, when to my joy the solution occurred to me . . . The solution, as I believe, is that the modified offspring of all dominant and increasing forms tend to become adapted to many and highly diversified places in the economy of nature" (F. Darwin 1896, vol. I, pp. 68–69).

2 The horizontal (x) axis can be taken to represent "morphological space," meaning the closer that points are on the horizontal, the more similar they are in morphology (and by extension, the more similar their resource use, etc., in the economy of nature). The vertical (y) axis represents time, going forward from bottom to top. Each horizontal line, Darwin says on the next page, represents a thousand generations. There are initially eleven species, A–L. After fourteen thousand generations there are fifteen descendant species, almost all of which are descended from just two initial species, A and I. Eight of the original eleven have no descendants.

Milne Edwards. No physiologist doubts that a stomach by being adapted to digest vegetable matter alone, or flesh alone, draws most nutriment from these substances. So in the general economy of any land, the more widely and perfectly the animals and plants are diversified for different habits of life, so will a greater number of individuals be capable of there supporting themselves. A set of animals, with their organisation but little diversified, could hardly compete with a set more perfectly diversified in structure. It may be doubted, for instance, whether the Australian marsupials, which are divided into groups differing but little from each other, and feebly representing, as Mr. Waterhouse and others have remarked, our carnivorous, ruminant, and rodent mammals, could successfully compete with these well-pronounced orders. In the Australian mammals, we see the process of diversification in an early and incomplete stage of development.

1 After the foregoing discussion, which ought to have been much amplified, we may, I think, assume that the modified descendants of any one species will succeed by so much the better as they become more diversified in structure, and are thus enabled to encroach on places occupied by other beings. Now let us see how this principle of great benefit being derived from divergence of character, combined with the principles of natural selection and of extinction, will tend to act.

2 The accompanying diagram will aid us in understanding this rather perplexing subject.* Let A to L represent the species of a genus large in its own country; these species are supposed to resemble each other in unequal degrees, as is so generally the case in nature, and as is represented in the diagram by the letters standing at unequal distances. I have said a large genus, because we have seen in the second chapter,

*[The diagram referred to follows page 117]

that on an average more of the species of large genera vary than of small genera; and the varying species of the large genera present a greater number of varieties. We have, also, seen that the species, which are the commonest and the most widely-diffused, vary more than rare species with restricted ranges. Let (A) be a common, widely-diffused, and varying species, belonging to a genus large in its own country. The little fan of diverging dotted lines of unequal lengths proceeding from (A), may represent its varying offspring. The variations are supposed to be extremely slight, but of the most diversified nature; they are not supposed all to appear simultaneously, but often after long intervals of time; nor are they all supposed to endure for equal periods. Only those variations which are in some way profitable will be preserved or naturally selected. And here the importance of the principle of benefit being derived from divergence of character comes in; for this will generally lead to the most different or divergent variations (represented by the outer dotted lines) being preserved and accumulated by natural selection. When a dotted line reaches one of the horizontal lines, and is there marked by a small numbered letter, a sufficient amount of variation is supposed to have been accumulated to have formed a fairly well-marked variety, such as would be thought worthy of record in a systematic work. ◀1

The intervals between the horizontal lines in the diagram, may represent each a thousand generations; but it would have been better if each had represented ten thousand generations. After a thousand generations, species (A) is supposed to have produced two fairly well-marked varieties, namely a^1 and m^1. These two varieties will generally continue to be exposed to the same conditions which made their parents variable, ◀2

1 This sentence holds the key to understanding Darwin's diagram. Note that of all the lines in each little fan or spray extending from forms given a letter label, the *outermost* lines tend to persist (that is, continue on to the next level marked by the horizontal lines). This means that the most divergent varieties tend to outcompete and drive to extinction their less divergent sibling varieties, which are indicated by the intermediates in the little sprays. Extending the divergent lines of descent naturally produces a branching tree-like pattern. The most divergent forms are not always the ones to persist, mind you, but they are most of the time in Darwin's thinking. Counting all the instances in which this is the case on the left side of this diagram, descendants of A, we have seventeen out of twenty-nine cases, or 59 percent. More than half of the time, then, Darwin imagines that the most divergent varieties survive. Note that the count for the right side of the diagram, descendants of species I, is also 59 percent. Is that by design?

2 The engine of transmutation and divergence is heritable variation. Here Darwin again invokes environmental causes for variation, imagining that if descendant forms are exposed to essentially the same conditions as their parent forms they will vary to the same degree. He continues on rather shakier ground on the next page, suggesting that "the tendency to variability is itself hereditary." Would we agree with that assertion today?

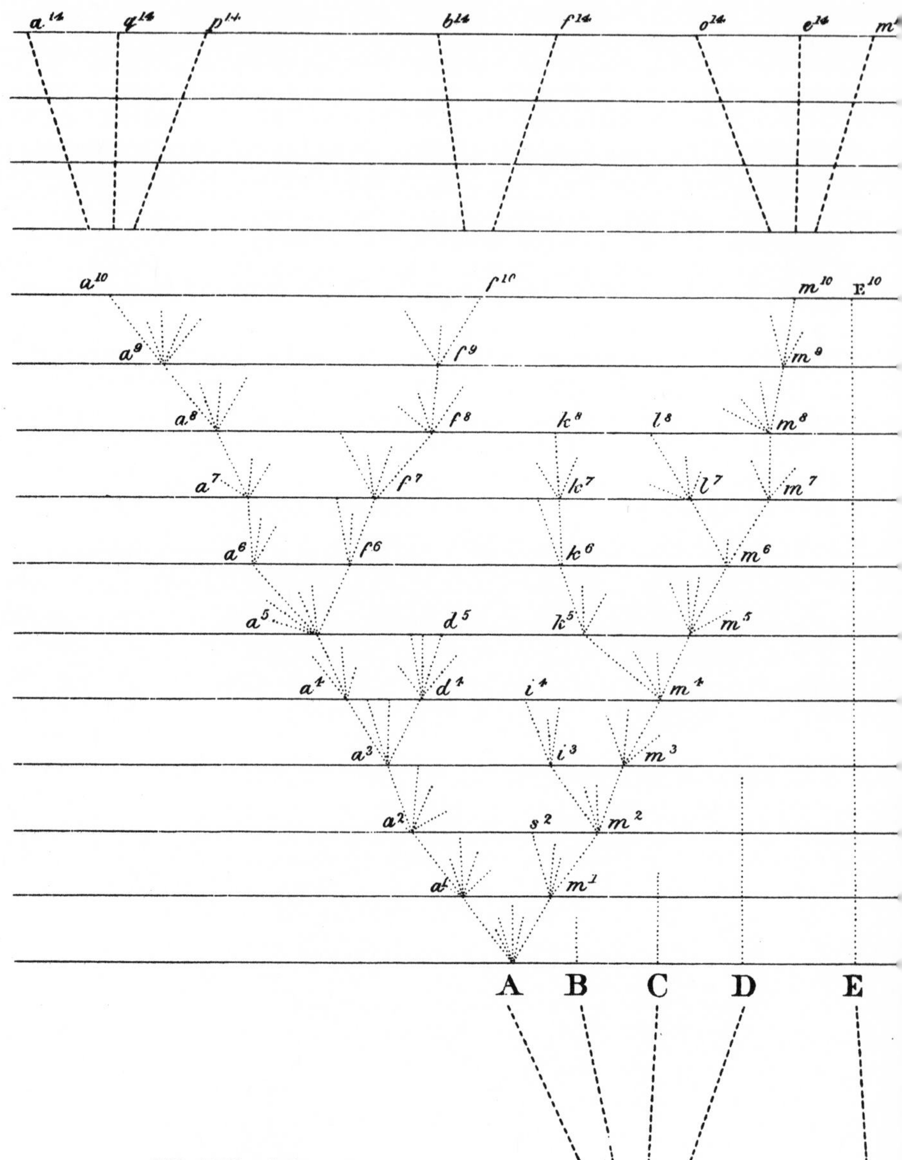

W. West lith. Hatton Garden.

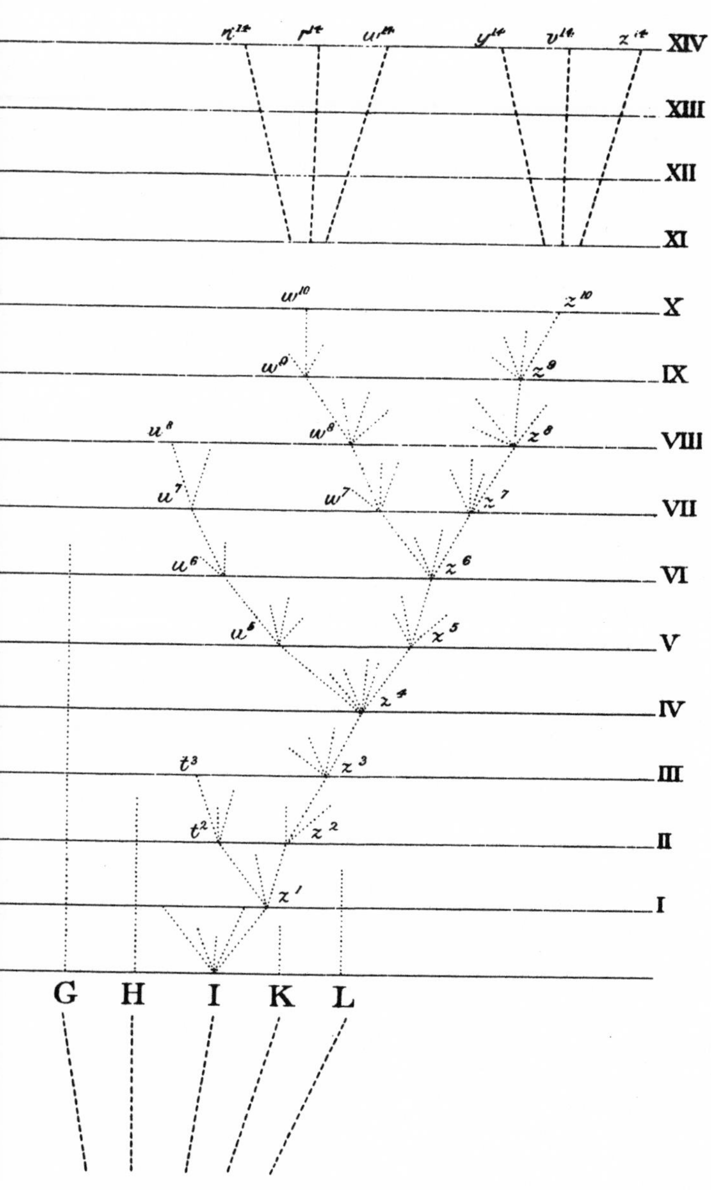
XIV
XIII
XII
XI
X
IX
VIII
VII
VI
V
IV
III
II
I
G
H
I
K
L

1 In general, this is a "success breeds success" model—widespread and varying forms will give rise to widespread and varying forms—that modern biologists would agree with only in the general sense that large, widespread species may exhibit more variation than smaller, more restricted species by virtue of having more individuals that *can* vary.

and the tendency to variability is in itself hereditary, consequently they will tend to vary, and generally to vary in nearly the same manner as their parents varied. Moreover, these two varieties, being only slightly modified forms, will tend to inherit those advantages which made their common parent (A) more numerous than most of the other inhabitants of the same country; they will likewise partake of those more general advantages which made the genus to which the parent-species belonged, a large genus in its own country. And these circumstances we know to be favourable to the production of new varieties.

If, then, these two varieties be variable, the most divergent of their variations will generally be preserved during the next thousand generations. And after this interval, variety a^1 is supposed in the diagram to have produced variety a^2, which will, owing to the principle of divergence, differ more from (A) than did variety a^1. Variety m^1 is supposed to have produced two varieties, namely m^2 and s^2, differing from each other, and more considerably from their common parent (A). We may continue the process by similar steps for any length of time; some of the varieties, after each thousand generations, producing only a single variety, but in a more and more modified condition, some producing two or three varieties, and some failing to produce any. [1] Thus the varieties or modified descendants, proceeding from the common parent (A), will generally go on increasing in number and diverging in character. In the diagram the process is represented up to the ten-thousandth generation, and under a condensed and simplified form up to the fourteen-thousandth generation.

But I must here remark that I do not suppose that the process ever goes on so regularly as is represented in the diagram, though in itself made somewhat irregular.

I am far from thinking that the most divergent varieties will invariably prevail and multiply: a medium form may often long endure, and may or may not produce more than one modified descendant; for natural selection will always act according to the nature of the places which are either unoccupied or not perfectly occupied by other beings; and this will depend on infinitely complex relations. But as a general rule, the more diversified in structure the descendants from any one species can be rendered, the more places they will be enabled to seize on, and the more their modified progeny will be increased. In our diagram the line of succession is broken at regular intervals by small numbered letters marking the successive forms which have become sufficiently distinct to be recorded as varieties. But these breaks are imaginary, and might have been inserted anywhere, after intervals long enough to have allowed the accumulation of a considerable amount of divergent variation.

As all the modified descendants from a common and widely-diffused species, belonging to a large genus, will tend to partake of the same advantages which made their parent successful in life, they will generally go on multiplying in number as well as diverging in character: this is represented in the diagram by the several divergent branches proceeding from (A). The modified offspring from the later and more highly improved branches in the lines of descent, will, it is probable, often take the place of, and so destroy, the earlier and less improved branches: this is represented in the diagram by some of the lower branches not reaching to the upper horizontal lines. In some cases I do not doubt that the process of modification will be confined to a single line of descent, and the number of the descendants will not be increased; although the amount

1 An important point to keep in mind: this is not simply a model in which offspring forms supplant parental forms. If it were, the pattern of transmutation could be linear, with no branching tree necessarily following. Darwin's divergence of character principle does entail branching and spreading. Divergence increases competitive ability against not only parental forms but also other competing forms. Outcompeted forms are supplanted, their place in the economy of nature taken over and occupied.

2 Recall that species A at the bottom left of the diagram is a "common, widely diffused, and varying species, belonging to a genus large in its own country" (p. 117). Implicitly, B, C, and D are not. These are driven to extinction, literally overtopped by the burgeoning tree that is A's descendants. Recalling that horizontal position indicates a certain niche in the economy of nature, we can see that the positions of B–D become co-opted by the group in lineage m; a perpendicular line dropped from m^1–m^6 would intersect the bottom horizontal line more or less at their positions. Extinction, then, is a central feature of Darwin's model.

of divergent modification may have been increased in the
[1] successive generations. This case would be represented in the diagram, if all the lines proceeding from (A) were removed, excepting that from a^1 to a^{10}. In the same way, for instance, the English race-horse and English pointer have apparently both gone on slowly diverging in character from their original stocks, without either having given off any fresh branches or races.

After ten thousand generations, species (A) is supposed to have produced three forms, a^{10}, f^{10}, and m^{10}, which, from having diverged in character during the successive generations, will have come to differ largely, but perhaps unequally, from each other and from their common parent. If we suppose the amount of change between each horizontal line in our diagram to be excessively small, these three forms may still be only well-marked varieties; or they may have arrived at the doubtful
[2] category of sub-species; but we have only to suppose the steps in the process of modification to be more numerous or greater in amount, to convert these three forms into well-defined species: thus the diagram illustrates the steps by which the small differences distinguishing varieties are increased into the larger differences distinguishing species. By continuing the same process for a greater number of generations (as shown in the diagram in a condensed and simplified manner), we get eight species, marked by the letters between a^{14} and m^{14}, all descended from (A). Thus, as I believe, species are multiplied and genera are formed.

In a large genus it is probable that more than one species would vary. In the diagram I have assumed that a second species (I) has produced, by analogous steps, after ten thousand generations, either two well-marked varieties (w^{10} and z^{10}) or two species, according to the amount of change supposed to be represented be-

1 Darwin is here making a distinction between what we now call *anagensis,* or evolution within a lineage over time, and *cladogenesis,* evolution by lineage splitting. The English racehorse or pointer might be examples of anagenetic change, for directional selection has "pushed" these organisms steadily further from the parental or ancestral form. This change is indicated in the diagram by the angle at which the line from A to a^{10} proceeds from the vertical; that is, the lineage is moving away from the form of A. Darwin is careful to build in situations where little or no transmutation is seen in even ancient lineages (presumably because they are so well adapted to their environment). Lineage E is an example; this might represent so-called living fossils like coelacanths or horseshoe crabs.

2 It is worth noting that because transmutation and divergence are a slow and continuous process, the question of whether any two forms descended from a common ancestor are considered well-marked varieties or actual species is something of a judgment call. Competent naturalists can disagree on how much or what kinds of difference would be sufficient to consider a given set of forms species or varieties (see the earlier "doubtful forms" discussion on pp. 16–17). This ambiguity holds true for higher taxonomic levels as well: how much difference is enough to put two species in different genera? Regardless, with continual divergence, groups of descendant species united by common ancestry and distinct from other such groups naturally form—these become arranged into genera, families, orders, classes, etc.: the entire Linnaean hierarchy. This process is explained fully on p. 123.

tween the horizontal lines. After fourteen thousand generations, six new species, marked by the letters n^{14} to z^{14}, are supposed to have been produced. In each genus, the species, which are already extremely different in character, will generally tend to produce the greatest number of modified descendants; for these will have the best chance of filling new and widely different places in the polity of nature: hence in the diagram I have chosen the extreme species (A), and the nearly extreme species (I), as those which have largely varied, and have given rise to new varieties and species. The other nine species (marked by capital letters) of our original genus, may for a long period continue transmitting unaltered descendants; and this is shown in the diagram by the dotted lines not prolonged far upwards from want of space.

But during the process of modification, represented in the diagram, another of our principles, namely that of extinction, will have played an important part. As in each fully stocked country natural selection necessarily acts by the selected form having some advantage in the struggle for life over other forms, there will be a constant tendency in the improved descendants of any one species to supplant and exterminate in each stage of descent their predecessors and their original parent. For it should be remembered that the competition will generally be most severe between those forms which are most nearly related to each other in habits, constitution, and structure. Hence all the intermediate forms between the earlier and later states, that is between the less and more improved state of a species, as well as the original parent-species itself, will generally tend to become extinct. So it probably will be with many whole collateral lines of descent, which will be conquered by later and improved lines of descent. If, however, the

1 Extinction is, again, an integral part of the divergence process. It was not so very long before the *Origin* that the occurrence of extinction was still questioned. Some Enlightenment thinkers held that organisms did not go extinct but rather continued to exist somewhere.

The idea of extinction seemed antithetical to the principle of *plenitude*, which in some forms dates back to Aristotle and was later incorporated into Christian theology. Plenitude holds that all things exist, or that God realizes all possibilities (Lovejoy 1964). Another Enlightenment thinker, the Comte de Buffon (1707–1788), rejected the plenitude concept and embraced extinction as part of his vision of earth's formation and evolution, set forth in his book *Époques de la Nature* (1788). Extinction per se was irrelevant in some biological models, including Lamarck's transmutation theory. Lamarck's brand of evolution held that animals are continually evolving up a supposed scale of organizational complexity, where lower, less complex, levels are constantly replenished by spontaneous generation. Thus all levels of organization are occupied at all times, at least in the contemporary world. Lamarck ran afoul of Georges Cuvier (1769–1832), the leading comparative anatomist of the era, who strongly believed that extinction occurs. Cuvier held that the history of life consisted of episodes of creation and extinction—catastrophes, the most recent of which was the Noachian flood. As late as 1830, in the first edition of *Principles of Geology*, Lyell speculated that plants and animals of past ages may return once suitable environmental conditions cycle back. For this he was lampooned by Henry de la Beche in the famous "Awful Changes" cartoon, in which Professor Ichthyosaurus lectures to young ichthyosauri on the nature of the extinct animal—humans—whose skull with puny jawbone and teeth are before them (see Rudwick 1992, p. 49). By Darwin's day the fact of extinction was firmly established. Helping to crystallize this concept in the public's imagination was the discovery of giant (and apparently quite extinct) reptiles in England in the 1820s—the group Richard Owen later named *Dinosauria.*

1 Tree-pruning: the descendants of A and I are supposed to have "taken the places of, and thus exterminated, not only their parents . . . but likewise some of the original species which were most nearly related to their parents." Note that only F has persisted, the lineage of the species set A–F that differs most from A.

modified offspring of a species get into some distinct country, or become quickly adapted to some quite new station, in which child and parent do not come into competition, both may continue to exist.

If then our diagram be assumed to represent a considerable amount of modification, species (A) and all the earlier varieties will have become extinct, having been replaced by eight new species (a^{14} to m^{14}); and (I) will have been replaced by six (n^{14} to z^{14}) new species.

But we may go further than this. The original species of our genus were supposed to resemble each other in unequal degrees, as is so generally the case in nature; species (A) being more nearly related to B, C, and D, than to the other species; and species (I) more to G, H, K, L, than to the others. These two species (A) and (I), were also supposed to be very common and widely diffused species, so that they must originally have had some advantage over most of the other species of the genus. Their modified descendants, fourteen in number at the fourteen-thousandth generation, will probably have inherited some of the same advantages: they have also been modified and improved in a diversified manner at each stage of descent, so as to have become adapted to many related places in the natural economy of their country. [1] It seems, therefore, to me extremely probable that they will have taken the places of, and thus exterminated, not only their parents (A) and (I), but likewise some of the original species which were most nearly related to their parents. Hence very few of the original species will have transmitted offspring to the fourteen-thousandth generation. We may suppose that only one (F), of the two species which were least closely related to the other nine original species, has transmitted descendants to this late stage of descent.

The new species in our diagram descended from the original eleven species, will now be fifteen in number. Owing to the divergent tendency of natural selection, the extreme amount of difference in character between species a^{14} and z^{14} will be much greater than that between the most different of the original eleven species. The new species, moreover, will be allied to each other ◀1 in a widely different manner. Of the eight descendants from (A) the three marked a^{14}, q^{14}, p^{14}, will be nearly related from having recently branched off from a^{10}; b^{14} and f^{14}, from having diverged at an earlier period from a^{5}, will be in some degree distinct from the three first-named species; and lastly, o^{14}, e^{14}, and m^{14}, will be nearly related one to the other, but from having diverged at the first commencement of the process of modification, will be widely different from the other five species, and may constitute a sub-genus or even a distinct genus.

The six descendants from (I) will form two sub-genera or even genera. But as the original species (I) differed largely from (A), standing nearly at the extreme points of the original genus, the six descendants from (I) will, owing to inheritance, differ considerably from the eight descendants from (A); the two groups, moreover, are supposed to have gone on diverging in different directions. The intermediate species, also (and this is a very important consideration), which connected the original species (A) and (I), have all become, excepting (F), extinct, and have left no descendants. Hence the six new species descended from (I), and the eight descended from (A), will have to be ranked as very distinct genera, or even as distinct sub-families.

Thus it is, as I believe, that two or more genera ◀2 are produced by descent, with modification, from two or more species of the same genus. And the two or

1 The key word here is "allied." Branching and rebranching as lineages diversify produces a nested hierarchy of branches and twigs. This is a nested hierarchy of alliance, or similarity: (a^{14}, q^{14}, p^{14}), (b^{14}, f^{14}), and (o^{14}, e^{14}, m^{14}) are sets of species alliance, but the first two (a^{14}–f^{14}) may represent a genus distinct from genus (o^{14}, e^{14}, m^{14}). Taken together, (a^{14}–m^{14}) could be ranked as one subfamily or family, and the species descending from I (n^{14}–z^{14}) another subfamily or family as described in the following paragraph.

2 In this paragraph it becomes clear that this diagram can be taken as a random slice of time: A–L are descendent forms themselves, as indicated by the dashed lines beneath them. Note that the lines are drawn such that sets of them will converge on a common ancestor. Later in the book Darwin will take this to its logical conclusion: a universal common ancestor for all life forms!

more parent-species are supposed to have descended from some one species of an earlier genus. In our diagram, this is indicated by the broken lines, beneath the capital letters, converging in sub-branches downwards towards a single point; this point representing a single species, the supposed single parent of our several new sub-genera and genera.

It is worth while to reflect for a moment on the character of the new species F[14], which is supposed not to have diverged much in character, but to have retained the form of (F), either unaltered or altered only in a slight degree. In this case, its affinities to the other fourteen new species will be of a curious and circuitous nature. Having descended from a form which stood between the two parent-species (A) and (I), now supposed to be extinct and unknown, it will be in some degree intermediate in character between the two groups descended from these species. 1 But as these two groups have gone on diverging in character from the type of their parents, the new species (F[14]) will not be directly intermediate between them, but rather between types of the two groups; and every naturalist will be able to bring some such case before his mind.

2 In the diagram, each horizontal line has hitherto been supposed to represent a thousand generations, but each may represent a million or hundred million generations, and likewise a section of the successive strata of the earth's crust including extinct remains. We shall, when we come to our chapter on Geology, have to refer again to this subject, and I think we shall then see that the diagram throws light on the affinities of extinct beings, which, though generally belonging to the same orders, or families, or genera, with those now living, yet are often, in some degree, intermediate in character between existing groups; and we can understand this fact, for

1 This point is very important: to some critics, if Darwin's ideas were correct, we would expect to see all manner of intermediate forms linking species—extant, living species. That we do *not* see a continuum of such linking forms could be cited as evidence against descent with modification, in this view. Here, however, Darwin points out that intermediate forms are to be found not directly between extant species but rather between more general groups or types of species. Note the morphological gap between, say, groups (a^{14}, q^{14}, p^{14}) and (b^{14}, f^{14}). The principle of divergence ensures nested groups of related forms, but the pruning effect of extinction ensures that extant groups cannot be arrayed in a perfect continuum of relationship.

2 Here is a glimpse of the heuristic power of the divergence of character diagram. The horizontal lines represent thousands, even millions, of generations in the discussion so far, but they can also represent geological strata. Darwin seems rather understated here: looking at the diagram as a set of strata studded with fossil remains does throw light on the affinities of extinct beings, both with other extinct forms and with extant species. This echoes Alfred Russel Wallace's great insight in his Sarawak Law paper: "Every species has come into existence coincident both in space and time with a pre-existing closely allied species" (Wallace 1855, p. 186). This topic will be discussed in greater detail in chapter X.

the extinct species lived at very ancient epochs when the branching lines of descent had diverged less.

I see no reason to limit the process of modification, as now explained, to the formation of genera alone. If, in our diagram, we suppose the amount of change represented by each successive group of diverging dotted lines to be very great, the forms marked a^{14} to p^{14}, those marked b^{14} and f^{14}, and those marked o^{14} to m^{14}, will form three very distinct genera. We shall also have two very distinct genera descended from (I); and as these latter two genera, both from continued divergence of character and from inheritance from a different parent, will differ widely from the three genera descended from (A), the two little groups of genera will form two distinct families, or even orders, according to the amount of divergent modification supposed to be represented in the diagram. And the two new families, or orders, will have descended from two species of the original genus; and these two species are supposed to have descended from one species of a still more ancient and unknown genus.

We have seen that in each country it is the species of the larger genera which oftenest present varieties or incipient species. This, indeed, might have been expected; for as natural selection acts through one form having some advantage over other forms in the struggle for existence, it will chiefly act on those which already have some advantage; and the largeness of any group ◀1 shows that its species have inherited from a common ancestor some advantage in common. Hence, the struggle for the production of new and modified descendants, will mainly lie between the larger groups, which are all trying to increase in number. One large group will slowly conquer another large group, reduce its numbers, and thus lessen its chance of further variation and improvement. Within the same large

1 This echoes the "success breeds success" idea we met with on p. 118. The very largeness of a group signals that member species have inherited a competitive advantage in common. Note, again, that the chances of "further variation" and therefore improvement by natural selection are linked to population size, with larger sizes having an inherent advantage over smaller sizes.

group, the later and more highly perfected sub-groups, from branching out and seizing on many new places in the polity of Nature, will constantly tend to supplant and destroy the earlier and less improved sub-groups. Small and broken groups and sub-groups will finally [1] tend to disappear. Looking to the future, we can predict that the groups of organic beings which are now large and triumphant, and which are least broken up, that is, which as yet have suffered least extinction, will for a long period continue to increase. But which groups will ultimately prevail, no man can predict; for we well know that many groups, formerly most extensively developed, have now become extinct. Looking still more remotely to the future, we may predict that, owing to the continued and steady increase of the larger groups, a multitude of smaller groups will become utterly extinct, and leave no modified descendants; and consequently that of the species living at any one period, extremely few will transmit descendants to a remote futurity. I shall have to return to this subject in the chapter on Classification, but I may add that on this view of extremely few of the more ancient species having transmitted descendants, and on the view of all the descendants of the same species making a class, [2] we can understand how it is that there exist but very few classes in each main division of the animal and vegetable kingdoms. Although extremely few of the most ancient species may now have living and modified descendants, yet at the most remote geological period, the earth may have been as well peopled with many species of many genera, families, orders, and classes, as at the present day.

[3] *Summary of Chapter.*—If during the long course of ages and under varying conditions of life, organic beings

1 The inherently unpredictable nature of the future macroevolutionary landscape, a seemingly simple concept, is even more difficult to forecast than Darwin and his fellow evolutionists might have imagined. The efforts of Niles Eldredge, Stephen Jay Gould, Elizabeth Vrba, and others in the late twentieth century helped underscore the contingent nature of evolutionary history, where chance events can have profound consequences for evolutionary trajectories. See Vrba and Eldredge's 2005 book *Macroevolution,* a tribute to Gould.

2 Perhaps the overall number of genera, families, etc., has not changed much over time, while the species making up these groups have changed. Whole groups go extinct, but they are replaced with the diversifying descendants of the few remaining groups. It isn't clear if Darwin really means that overall numbers of higher taxa would be found at even the "most remote geological period"; after all, that is incompatible with the existence of a universal common ancestor, and he later speculates that all life has perhaps proceeded from one or a few such common ancestors.

3 It is hard to imagine a more tightly written summary: these three and a half pages manage to capture every nuanced argument of the previous twenty-six pages. I will summarize the key points paragraph by paragraph.

Paragraph 1: Natural selection follows from heritable variation and success in the struggle for existence linked to this variation. Sexual selection is a special form of natural selection.

vary at all in the several parts of their organisation, and I think this cannot be disputed; if there be, owing to the high geometrical powers of increase of each species, at some age, season, or year, a severe struggle for life, and this certainly cannot be disputed; then, considering the infinite complexity of the relations of all organic beings to each other and to their conditions of existence, causing an infinite diversity in structure, constitution, and habits, to be advantageous to them, I think it would be a most extraordinary fact if no variation ever had occurred useful to each being's own welfare, in the same way as so many variations have occurred useful to man. But if variations useful to any organic being do occur, assuredly individuals thus characterised will have the best chance of being preserved in the struggle for life; and from the strong principle of inheritance they will tend to produce offspring similarly characterised. This principle of preservation, I have called, for the sake of brevity, Natural Selection. Natural selection, on the principle of qualities being inherited at corresponding ages, can modify the egg, seed, or young, as easily as the adult. Amongst many animals, sexual selection will give its aid to ordinary selection, by assuring to the most vigorous and best adapted males the greatest number of offspring. Sexual selection will also give characters useful to the males alone, in their struggles with other males.

Whether natural selection has really thus acted in nature, in modifying and adapting the various forms of life to their several conditions and stations, must be judged of by the general tenour and balance of evidence given in the following chapters. But we already see how it entails extinction; and how largely extinction has acted in the world's history, geology plainly declares. Natural selection, also, leads to divergence of 1

1 Paragraph 2: Extinction is an expected outcome of natural selection; a consequence of divergence of character is improved competitive ability, which ultimately permits diverging forms to outcompete and displace intermediates. Divergence permits coexistence of a large number of diverse forms in even a small area. Continuing divergence makes species out of varieties, and diverging sets of species are grouped into genera.

character; for more living beings can be supported on the same area the more they diverge in structure, habits, and constitution, of which we see proof by looking at the inhabitants of any small spot or at naturalised productions. Therefore during the modification of the descendants of any one species, and during the incessant struggle of all species to increase in numbers, the more diversified these descendants become, the better will be their chance of succeeding in the battle of life. Thus the small differences distinguishing varieties of the same species, will steadily tend to increase till they come to equal the greater differences between species of the same genus, or even of distinct genera.

1 We have seen that it is the common, the widely-diffused, and widely-ranging species, belonging to the larger genera, which vary most; and these will tend to transmit to their modified offspring that superiority which now makes them dominant in their own countries. Natural selection, as has just been remarked, leads to divergence of character and to much extinction of the less improved and intermediate forms of life. On these principles, I believe, the nature of the affinities of all organic beings may be explained. It is a truly wonderful fact—the wonder of which we are apt to overlook from familiarity—that all animals and all plants throughout all time and space should be related to each other in group subordinate to group, in the manner which we everywhere behold—namely, varieties of the same species most closely related together, species of the same genus less closely and unequally related together, forming sections and sub-genera, species of distinct genera much less closely related, and genera related in different degrees, forming sub-families, families, orders, sub-classes, and classes. The several subordinate groups in any class cannot be

1 Paragraph 3: Common species are successful species, with large populations that range far and wide, sporting a larger number of variations than more restricted species. These variations ensure continued success by providing the fuel for further adaptation by natural selection. Natural selection thus drives divergence of character, one outcome of which is extinction of intermediate forms. Continual branching and rebranching of some lineages in the divergence process, coupled with extinction of others, naturally produces a nested hierarchy of degrees of relationship—"groups subordinate to groups," as Darwin so eloquently puts it. The divergence model thus explains the Linnaean taxonomic hierarchy: Linnaeus's system works so well because it is in fact genealogical.

ranked in a single file, but seem rather to be clustered round points, and these round other points, and so on in almost endless cycles. On the view that each species has been independently created, I can see no explanation of this great fact in the classification of all organic beings; but, to the best of my judgment, it is explained through inheritance and the complex action of natural selection, entailing extinction and divergence of character, as we have seen illustrated in the diagram.

The affinities of all the beings of the same class have sometimes been represented by a great tree. I believe this simile largely speaks the truth. The green and budding twigs may represent existing species; and those produced during each former year may represent the long succession of extinct species. At each period of growth all the growing twigs have tried to branch out on all sides, and to overtop and kill the surrounding twigs and branches, in the same manner as species and groups of species have tried to overmaster other species in the great battle for life. The limbs divided into great branches, and these into lesser and lesser branches, were themselves once, when the tree was small, budding twigs; and this connexion of the former and present buds by ramifying branches may well represent the classification of all extinct and living species in groups subordinate to groups. Of the many twigs which flourished when the tree was a mere bush, only two or three, now grown into great branches, yet survive and bear all the other branches; so with the species which lived during long-past geological periods, very few now have living and modified descendants. From the first growth of the tree, many a limb and branch has decayed and dropped off; and these lost branches of various sizes may represent those whole orders, families, and genera which have now no living representatives, and

1 Paragraph 4: Here is the evocative "tree of life" metaphor in all its glory, and Darwin's prose glitters with passion and eloquence. Many of the images of this passage appear in modified form in *Natural Selection.* Interestingly, the passage there opens with reference to more than one tree: "The relation of all past & all present beings may be loosely compared with the growth of a few gigantic trees" (Stauffer 1975, p. 248). Did Darwin mean a group of gigantic trees together, as in a forest or wood? The image is alluring: perhaps each tree represented a great taxonomic group, like Kingdoms. In that case, however, they cannot be united in a single, unified system using a multiple-tree metaphor. Perhaps realizing this, Darwin opted to represent the profusion of life with the single-tree metaphor of the *Origin.*

Darwin opens this paragraph by stating that "the affinities of all the beings of the same class have sometimes been represented by a great tree." Tree-like depictions of relationships—even though intended as schematics and not reflective of genealogy—were rare in Darwin's day. In some fields, the preferred way to illustrate relationships was more horizontal, a web of lines connecting points that looked more like a road map than even the bushiest tree; Linnaeus depicted botanical relationships this way. Zoologists, too, favored a reticulate, horizontal depiction, and branching systems were only introduced in the 1820s. For example, in 1828 the zoologist Karl Ernst Von Baer (1792–1876) used a branching pattern (though one that was horizontally arrayed) to depict embryological relationships (Stevens 1984, O'Hara 1991).

which are known to us only from having been found in a fossil state. As we here and there see a thin straggling branch springing from a fork low down in a tree, and which by some chance has been favoured and is still alive on its summit, so we occasionally see an animal like the Ornithorhynchus or Lepidosiren, which in some small degree connects by its affinities two large branches of life, and which has apparently been saved from fatal competition by having inhabited a protected station. As buds give rise by growth to fresh buds, and these, if vigorous, branch out and overtop on all sides many a feebler branch, so by generation I believe it [1] has been with the great Tree of Life, which fills with its dead and broken branches the crust of the earth, and covers the surface with its ever branching and beautiful ramifications.

1 The Rev. John Fleming (1785–1857) was not a transmutationist, yet he may have provided the first explicit tree analogy for taxonomic relationships in his 1829 essay "On systems and methods in natural history." Writing of the dichotomous branching method like that used in modern taxonomic keys, Fleming noted, "The class, in this system, may be compared to the trunk of a tree, the subordinate orders and sections to the branches, and the species to the buds or leaves on the sprays" (Fleming 1829, p. 311). Darwin may have had Fleming in mind when he wrote the opening sentence of the final paragraph in this chapter—note his explicit reference to relationships of beings *of the same class.*

With this essay Fleming provoked the ire of the entomologist William Sharp MacLeay (1792–1865), who published an attack on Fleming's ideas entitled "On the dying struggle of the dichotomous system" (1830). MacLeay had an axe to grind: he originated and staunchly defended the ill-fated quinarian or circular system of classification, which sought to identify patterns of affinity among taxa based on the principle of grouping by fives. This system was abandoned by the mid-1840s.

We might be speaking of a "coral of life" today instead of a tree of life if Darwin had pursued his early musing that branching corals are better representative of evolutionary history. "The tree of life should perhaps be called the coral of life," he wrote in the B notebook, "base of branches dead; so that passages cannot be seen" (Barrett et al. 1987, p. 177). (See Bredekamp 2005, Maderspacher 2006, and Voss 2007.) Ironically, the molecular-driven revolution in microbial ecology may be bringing us full circle. Doolittle and Bapteste (2007) argue that pursuit of a single, universal tree of life is quixotic. Embracing "pattern pluralism"—the idea that different patterns of evolutionary relationship hold for different taxa and at different temporal scales—is the only way to reconcile the hierarchical, branched portion of the tree of life with the web of relationships at its base.

CHAPTER V.

LAWS OF VARIATION.

Effects of external conditions — Use and disuse, combined with natural selection; organs of flight and of vision—Acclimatisation — Correlation of growth — Compensation and economy of growth—False correlations—Multiple, rudimentary, and lowly organised structures variable—Parts developed in an unusual manner are highly variable: specific characters more variable than generic: secondary sexual characters variable—Species of the same genus vary in an analogous manner—Reversions to long lost characters—Summary.

I HAVE hitherto sometimes spoken as if the variations —so common and multiform in organic beings under domestication, and in a lesser degree in those in a state of nature—had been due to chance. This, of course, is a wholly incorrect expression, but it serves to acknowledge plainly our ignorance of the cause of each particular variation. Some authors believe it to be as much the function of the reproductive system to produce individual differences, or very slight deviations of structure, as to make the child like its parents. But the much greater variability, as well as the greater frequency of monstrosities, under domestication or cultivation, than under nature, leads me to believe that deviations of structure are in some way due to the nature of the conditions of life, to which the parents and their more remote ancestors have been exposed during several generations. I have remarked in the first chapter—but a long catalogue of facts which cannot be here given would be necessary to show the truth of the remark—that the reproductive system is eminently susceptible to changes in the conditions of life; and to

Some authors see this chapter as a continuation of Chapter IV, while others view it as the first of a series of chapters in which Darwin explores applications of his ideas. Elements of both views are in evidence. Variation and variability were perplexing in Darwin's day (indeed, until Watson and Crick in 1953). The best Darwin could do was attempt to discern patterns in the expression of variation that might make sense in light of natural selection and descent with modification. In a sense, variability—its causes and the patterns of its expression—presented a difficulty. Darwin thought he glimpsed a telltale signal in the nearly overwhelming noise of variation, and he argues in this chapter that such signals can only be understood in terms of his model. In that way, "Laws of Variation" can thus also be understood as the first of several chapters in which Darwin outlines a problem but then "solves" it by showing that observed patterns resonate with his theory.

---

1 Is this a real or perceived difference? Perhaps points of morphology of domesticated species merely appear to show more variation than corresponding structures in wild species because they are scrutinized more closely. Or perhaps they are actually more variable simply because they can tolerate more variability.

1 This is a reprise of frustrations expressed earlier over the nature or cause of variation.

2 Much of this is qualification. The bottom line is that the environment has a "perturbing" influence on reproductive elements.

this system being functionally disturbed in the parents, I chiefly attribute the varying or plastic condition of the offspring. The male and female sexual elements seem to be affected before that union takes place which is to form a new being. In the case of "sporting" plants, the bud, which in its earliest condition does not apparently differ essentially from an ovule, is alone affected. But why, because the reproductive system is disturbed, this or that part should vary more or less, we are profoundly ignorant. Nevertheless, we can here and there dimly catch a faint ray of light, and we may feel sure that there must be some cause for each deviation of structure, however slight.

How much direct effect difference of climate, food, &c., produces on any being is extremely doubtful. My impression is, that the effect is extremely small in the case of animals, but perhaps rather more in that of plants. We may, at least, safely conclude that such influences cannot have produced the many striking and complex co-adaptations of structure between one organic being and another, which we see everywhere throughout nature. Some little influence may be attributed to climate, food, &c.: thus, E. Forbes speaks confidently that shells at their southern limit, and when living in shallow water, are more brightly coloured than those of the same species further north or from greater depths. Gould believes that birds of the same species are more brightly coloured under a clear atmosphere, than when living on islands or near the coast. So with insects, Wollaston is convinced that residence near the sea affects their colours. Moquin-Tandon gives a list of plants which when growing near the sea-shore have their leaves in some degree fleshy, though not elsewhere fleshy. Several other such cases could be given.

The fact of varieties of one species, when they range

into the zone of habitation of other species, often acquiring in a very slight degree some of the characters of such species, accords with our view that species of all kinds are only well-marked and permanent varieties. Thus the species of shells which are confined to tropical and shallow seas are generally brighter-coloured than those confined to cold and deeper seas. The birds which are confined to continents are, according to Mr. Gould, brighter-coloured than those of islands. The insect-species confined to sea-coasts, as every collector knows, are often brassy or lurid. Plants which live exclusively on the sea-side are very apt to have fleshy leaves. He who believes in the creation of each species, will have to say that this shell, for instance, was created with bright colours for a warm sea; but that this other shell became bright-coloured by variation when it ranged into warmer or shallower waters. ◀1

When a variation is of the slightest use to a being, we cannot tell how much of it to attribute to the accumulative action of natural selection, and how much to the conditions of life. Thus, it is well known to furriers that animals of the same species have thicker and better fur the more severe the climate is under which they have lived; but who can tell how much of this difference may be due to the warmest-clad individuals having been favoured and preserved during many generations, and how much to the direct action of the severe climate? for it would appear that climate has some direct action on the hair of our domestic quadrupeds.

Instances could be given of the same variety being produced under conditions of life as different as can well be conceived; and, on the other hand, of different varieties being produced from the same species under the same conditions. Such facts show how indirectly

1 Darwin points out that striking intraspecies differences exist among populations of widely ranging species. These examples likely differ in root causes: some may reflect physiological adaptation in plastic (inherently variable) species ("brassy or lurid" beetles, fleshy plant leaves), others perhaps more genetic adaptation; for example, are shells of tropical waters more brightly colored to warn off predators? We are in the enviable position of knowing the difference between plasticity and "hard-wired" genetic adaptation, though we may not know which is more important in particular cases. Darwin had to grapple with understanding the basis of variation without this knowledge. Consider, from the perspective of someone in his day, pondering whether phenotypic variations like thicker winter coats of certain mammals, or suntanning of people under conditions of high sun exposure, proceed from the same cause as, say, horticultural sports, or variation in leg length.

1 Effects of use and disuse present a puzzle to Darwin. Is it disuse leading to atrophy that produces change? Or does selection reduce the structure?

2 Richard Owen was a distinguished comparative anatomist and paleontologist, as well as a prolific author of scientific papers and books on fossil and living mammals, birds, and reptiles. From his study of flightless birds, Owen became associated with the extinct giant moas of New Zealand, the existence of which he predicted from the discovery of a single hefty femur. He was proven correct when complete skeletons were found not long afterward. Owen eventually named several moa species. Later he published *Memoirs of the Extinct Wingless Birds of New Zealand* (London, 1879).

The loggerhead is better known as the steamer duck, *Tachyeres brachypterus.* This Falkland Islands endemic is perhaps the largest of ducks, reaching some twenty-two pounds. "Loggerhead" refers to the reputed clumsiness and stupidity of these ducks, which lack an escape response at the approach of ill-intentioned humans—the "tameness" that spelled the end of many an island species. (The extinct dodo's unflattering name has the same origin.) The steamer duck, still extant and thriving, has reduced wings and cannot fly. A close relative on the South American mainland reportedly retains but rarely exercises its flying ability. Darwin's comparison with the Aylesbury duck is apt: this large-bodied, white-feathered domestic duck was bred for meat beginning in the eighteenth century in or near Aylesbury in southern England (and is still that town's emblem).

the conditions of life must act. Again, innumerable instances are known to every naturalist of species keeping true, or not varying at all, although living under the most opposite climates. Such considerations as these incline me to lay very little weight on the direct action of the conditions of life. Indirectly, as already remarked, they seem to play an important part in affecting the reproductive system, and in thus inducing variability; and natural selection will then accumulate all profitable variations, however slight, until they become plainly developed and appreciable by us.

1 *Effects of Use and Disuse.*—From the facts alluded to in the first chapter, I think there can be little doubt that use in our domestic animals strengthens and enlarges certain parts, and disuse diminishes them; and that such modifications are inherited. Under free nature, we can have no standard of comparison, by which to judge of the effects of long-continued use or disuse, for we know not the parent-forms; but many animals have structures which can be explained by the
2 effects of disuse. As Professor Owen has remarked, there is no greater anomaly in nature than a bird that cannot fly; yet there are several in this state. The logger-headed duck of South America can only flap along the surface of the water, and has its wings in nearly the same condition as the domestic Aylesbury duck. As the larger ground-feeding birds seldom take flight except to escape danger, I believe that the nearly wingless condition of several birds, which now inhabit or have lately inhabited several oceanic islands, tenanted by no beast of prey, has been caused by disuse. The ostrich indeed inhabits continents and is exposed to danger from which it cannot escape by flight, but by kicking it can defend itself from enemies, as well as any of the smaller

quadrupeds. We may imagine that the early progenitor of the ostrich had habits like those of a bustard, and that as natural selection increased in successive generations the size and weight of its body, its legs were used more, and its wings less, until they became incapable of flight.

Kirby has remarked (and I have observed the same fact) that the anterior tarsi, or feet, of many male dung-feeding beetles are very often broken off; he examined seventeen specimens in his own collection, and not one had even a relic left. In the Onites apelles the tarsi are so habitually lost, that the insect has been described as not having them. In some other genera they are present, but in a rudimentary condition. In the Ateuchus or sacred beetle of the Egyptians, they are totally deficient. There is not sufficient evidence to induce us to believe that mutilations are ever inherited; and I should prefer explaining the entire absence of the anterior tarsi in Ateuchus, and their rudimentary condition in some other genera, by the long-continued effects of disuse in their progenitors; for as the tarsi are almost always lost in many dung-feeding beetles, they must be lost early in life, and therefore cannot be much used by these insects.

In some cases we might easily put down to disuse modifications of structure which are wholly, or mainly, due to natural selection. Mr. Wollaston has discovered the remarkable fact that 200 beetles, out of the 550 species inhabiting Madeira, are so far deficient in wings that they cannot fly; and that of the twenty-nine endemic genera, no less than twenty-three genera have all their species in this condition! Several facts, namely, that beetles in many parts of the world are very frequently blown to sea and perish; that the beetles in Madeira, as observed by Mr. Wollaston, lie much con-

1 Bustards, family Otididae, are Old World relatives of rails and cranes, widely distributed in southern Europe, Africa, Asia, part of New Guinea, and Australia. Darwin is imagining that an ostrich progenitor might have resembled bustards—both are stout-bodied birds well adapted for running—but he is not suggesting that the two are related. In fact, ostriches belong not only to a different family, Struthionidae, but to a different order. Europe's largest bird is the aptly named great bustard, *Otis tarda*, weighing over thirty pounds and having an impressive eight-foot wingspan. The two have converged on a fast-running lifestyle, but while both have reduced digits for more efficient running, bustards have three toes and ostriches have two.

2 William Kirby coauthored, with William Spence, the celebrated book *Introduction to Entomology*, which came out in several editions between 1815 and 1826. Dung beetles are evidently rough on their tarsi, the distal leg segments, which are often observed to be broken off. Some groups, however, seem to have lost their tarsi altogether—emerging from the pupal stage without them rather than simply losing them in rough-and-tumble activities like courtship, dung-ball rolling, and nest-digging. The fact that these beetles seem to get along just fine without tarsi suggests that the slender segments are expendable. But is it disuse that has led to their apparently permanent loss in species like the Egyptian sacred beetle? Or has selection hastened their elimination because they were somehow a liability? Probably neither is correct from a modern perspective. The most likely explanation is that tarsi, particularly on the front legs, are indeed expendable, so much so that when a developmental mutation eliminated them there was no fitness reduction—the mutation was tolerated, and perhaps even promoted as a result of economizing on resources.

3 Darwin's source here is Thomas Vernon Wollaston's *Insecta Maderensia* (London, 1854).

cealed, until the wind lulls and the sun shines; that the proportion of wingless beetles is larger on the exposed Dezertas than in Madeira itself; and especially the extraordinary fact, so strongly insisted on by Mr. Wollaston, of the almost entire absence of certain large groups of beetles, elsewhere excessively numerous, and which groups have habits of life almost necessitating frequent flight;—these several considerations have made me believe that the wingless condition of so many Madeira beetles is mainly due to the action of natural selection, but combined probably with disuse. For during thousands of successive generations each individual beetle which flew least, either from its wings having been ever so little less perfectly developed or from indolent habit, will have had the best chance of surviving from not being blown out to sea; and, on the other hand, those beetles which most readily took to flight will oftenest have been blown to sea and thus have been destroyed.

The insects in Madeira which are not ground-feeders, and which, as the flower-feeding coleoptera and lepidoptera, must habitually use their wings to gain their subsistence, have, as Mr. Wollaston suspects, their wings not at all reduced, but even enlarged. This is quite compatible with the action of natural selection. For when a new insect first arrived on the island, the tendency of natural selection to enlarge or to reduce the wings, would depend on whether a greater number of individuals were saved by successfully battling with the winds, or by giving up the attempt and rarely or never flying. As with mariners shipwrecked near a coast, it would have been better for the good swimmers if they had been able to swim still further, whereas it would have been better for the bad swimmers if they had not been able to swim at all and had stuck to the wreck.

1 The Dezertas is, in Darwin's words, "a mountainous rock near Madeira, four miles long and about three-quarters in breadth" (*Natural Selection;* Stauffer 1975, p. 292). Note that Darwin makes much of the disproportionate number of flightless forms on this wind-swept rock: "I have ascertained through Mr. Wollaston's kindness, that on the Dezertas . . . there are 54 Beetles; & that of these, 26 are winged & 28 wingless, which is a proportion one-fourth larger, than the Dezertas ought to have had in accordance with the proportions of the winged & wingless coleoptera in the whole archipelago."

Flightlessness in island beetle fauna, particularly common in the ground beetle family Carabidae, has long been of interest from biogeographical and evolutionary perspectives. The noted twentieth-century coleopterist Philip Darlington, Jr., studied these beetles (e.g., Darlington 1943, 1970), and his observations of the carabids of Caribbean islands were important in elucidating the species-area relationship. Darlington did not favor Darwin's explanation that wingless forms are selected because they are less likely to be blown away; he suggested, rather, that characteristics of available habitat area and resources in island and mountain-top environments favored a more sedentary lifestyle.

The eyes of moles and of some burrowing rodents are rudimentary in size, and in some cases are quite covered up by skin and fur. This state of the eyes is probably due to gradual reduction from disuse, but aided perhaps by natural selection. In South America, a burrowing [1] rodent, the tuco-tuco, or Ctenomys, is even more subterranean in its habits than the mole; and I was assured by a Spaniard, who had often caught them, that they were frequently blind; one which I kept alive was certainly in this condition, the cause, as appeared on dissection, having been inflammation of the nictitating membrane. As frequent inflammation of the eyes must [2] be injurious to any animal, and as eyes are certainly not indispensable to animals with subterranean habits, a reduction in their size with the adhesion of the eyelids and growth of fur over them, might in such case be an advantage; and if so, natural selection would constantly aid the effects of disuse.

It is well known that several animals, belonging to the [3] most different classes, which inhabit the caves of Styria and of Kentucky, are blind. In some of the crabs the foot-stalk for the eye remains, though the eye is gone; the stand for the telescope is there, though the telescope with its glasses has been lost. As it is difficult to imagine that eyes, though useless, could be in any way injurious to animals living in darkness, I attribute their loss wholly to disuse. In one of the blind animals, [4] namely, the cave-rat, the eyes are of immense size; and Professor Silliman thought that it regained, after living some days in the light, some slight power of vision. In the same manner as in Madeira the wings of some of the insects have been enlarged, and the wings of others have been reduced by natural selection aided by use and disuse, so in the case of the cave-rat natural selection seems to have struggled with the loss of light and

1 *Ctenomys* are caviomorph rodents of the family Ctenomyidae. Some fifty described species are widely distributed in South America, filling the same ecological niche as the pocket gophers in North America. Darwin describes these curious animals in chapter III of *The Voyage of the Beagle,* referring to them as "tucutuco," an onomatopoeic name: "This animal is universally known by a very peculiar noise which it makes when beneath the ground. A person, the first time he hears it, is much surprised; for it is not easy to tell whence it comes, nor is it possible to guess what kind of creature utters it. The noise consists in a short, but not rough, nasal grunt, which is monotonously repeated about four times in quick succession: the name Tucutuco is given in imitation of the sound. Where this animal is abundant, it may be heard at all times of the day, and sometimes directly beneath one's feet."

2 It may be easy to imagine how selection might reduce or eliminate eyes in fossorial (burrowing subterranean) animals: besides being costly to make in terms of energy, these organs are likely to be prone to irritation or infection in a subterranean environment. It is not disuse but "disneed" that leads to their reduction; unneeded structures become a liability and selection is expected to act to eliminate them.

3 An extensive network of limestone caves occurs in the southern Austrian province of Styria, a vast dolomitic karst region. Kentucky, too, has extensive limestone formations pockmarked with caves, including Mammoth Cave, the world's longest with over 347 miles of passages mapped to date. Both cave systems are home to a complement of unique dark-adapted species, including fish, salamanders, crayfish, arthropods, and other animals. These cavernicoles are typically pigmentless, with reduced visual organs. Darwin's comment about the stand for the telescope remaining even after the telescope is gone is evocative: mentioned only in passing here, such oddities and anomalies are powerful evidence for the action of natural selection. Why should a creator only half-eliminate an unnecessary structure, after all?

(continued)

(*continued*)

4 The initial observation Darwin reports here came from an 1851 article by Benjamin Silliman, Jr., in the *American Journal of Science and Arts.* In late 1860 Silliman wrote to Darwin with further observations, which were duly included in the third edition of *Origin.*

1 In this final long paragraph of the section, Darwin points out that the biogeography of cave forms argues against special creation. Taxonomic affinities of cave organisms fall out along geographic lines, not adaptation to place—the cave forms of Europe are most similar to terrestrial European species, those of the caves of North America are most similar to terrestrial North American species, and so on. A reasonable assumption under special creation would be that cave-adapted species as a group were crafted specially for caves, regardless of where in the world the caves are found. Their affinities should therefore be with each other.

Jørgen Matthias Christian Schiødte is cited twice in this paragraph. He was a zoologist at the Copenhagen Museum who published a paper on cave insects in 1849. An English translation of the paper was read before the Entomological Society of London in 1851. Darwin later wrote to J. O. Westwood for information bearing on cave insects; in his reply of November 23, 1856, Westwood advised him, "I cannot do better than refer you to Dr Wallich's Translation of Schiodtes remarkable paper published in the Trans. Ent. Society New Ser. Vol. 1.—which I think from your letter you cannot have seen" (*Correspondence* 6: 282).

to have increased the size of the eyes; whereas with all the other inhabitants of the caves, disuse by itself seems to have done its work.

▶ It is difficult to imagine conditions of life more similar than deep limestone caverns under a nearly similar climate: so that on the common view of the blind animals having been separately created for the American and European caverns, close similarity in their organisation and affinities might have been expected; but, as Schiödte and others have remarked, this is not the case, and the cave-insects of the two continents are not more closely allied than might have been anticipated from the general resemblance of the other inhabitants of North America and Europe. On my view we must suppose that American animals, having ordinary powers of vision, slowly migrated by successive generations from the outer world into the deeper and deeper recesses of the Kentucky caves, as did European animals into the caves of Europe. We have some evidence of this gradation of habit; for, as Schiödte remarks, "animals not far remote from ordinary forms, prepare the transition from light to darkness. Next follow those that are constructed for twilight; and, last of all, those destined for total darkness." By the time that an animal had reached, after numberless generations, the deepest recesses, disuse will on this view have more or less perfectly obliterated its eyes, and natural selection will often have effected other changes, such as an increase in the length of the antennæ or palpi, as a compensation for blindness. Notwithstanding such modifications, we might expect still to see in the cave-animals of America, affinities to the other inhabitants of that continent, and in those of Europe, to the inhabitants of the European continent. And this is the case with some of the American cave-animals, as I hear from

Professor Dana; and some of the European cave-insects are very closely allied to those of the surrounding country. It would be most difficult to give any rational explanation of the affinities of the blind cave-animals to the other inhabitants of the two continents on the ordinary view of their independent creation. That several of the inhabitants of the caves of the Old and New Worlds should be closely related, we might expect from the well-known relationship of most of their other productions. Far from feeling any surprise that some of the cave-animals should be very anomalous, as Agassiz has remarked in regard to the blind fish, the Amblyopsis, and as is the case with the blind Proteus with reference to the reptiles of Europe, I am only surprised that more wrecks of ancient life have not been preserved, owing to the less severe competition to which the inhabitants of these dark abodes will probably have been exposed.

Acclimatisation.—Habit is hereditary with plants, as in the period of flowering, in the amount of rain requisite for seeds to germinate, in the time of sleep, &c., and this leads me to say a few words on acclimatisation. As it is extremely common for species of the same genus to inhabit very hot and very cold countries, and as I believe that all the species of the same genus have descended from a single parent, if this view be correct, acclimatisation must be readily effected during long-continued descent. It is notorious that each species is adapted to the climate of its own home: species from an arctic or even from a temperate region cannot endure a tropical climate, or conversely. So again, many succulent plants cannot endure a damp climate. But the degree of adaptation of species to the climates under which they live is often overrated.

1 The Yale geologist James Dwight Dana had an extensive correspondence with Darwin. Responding to an inquiry about the fauna of Mammoth Cave, Dana wrote, "Professor Agassiz told me that the family to which the Fishes belong–the Cyprinodonts–was rather strikingly American" (September 8, 1856; *Correspondence* 6: 215). Darwin, in a letter of September 29 thanking Dana for this information and asking for more, briefly mentions his "heterodox" views on species: "You will be rather indignant at hearing that I am becoming, indeed I [should] say have become, sceptical on the permanent immutability of species" (*Correspondence* 6: 235).

2 *Amblyopsis* is a small genus of six cave-dwelling fish species placed in their own family, the Amblyopsidae. In his textbook *Principles of Zoology* (1851), the famed Harvard comparative anatomist and paleontologist Louis Agassiz wrote of "that curious fish *(Amblyopsis spelaeus)*, which lives in the Mammoth Cave, and which appears to want even the orbital cavity." Agassiz published a lengthy treatment of these fish in 1847 and a shorter article in 1851. *Proteus* is not a reptile but a blind cave-dwelling amphibian found in southern Europe; it is also known as the olm. There is only one species of *Proteus, P. anguinus,* family Proteidae. After the *Origin* appeared, correspondents wrote to Darwin with further examples of such "wrecks of ancient life"; several of these are cited in later editions of the book.

3 The ability of species to acclimate to new conditions is important to Darwin in more ways than one. Species must be adaptable, at least slowly and incrementally, given that descendant species are now found in conditions far removed from those of their common ancestor. But remember, too, that Darwin thinks novel variations are generated by exposure to varying environmental conditions. If species were adapted to their climate only, opportunities for creation of new variations would be limited.

1 Darwin cites several botanical authorities for these observations of inherent variability in cold tolerance. Darwin's friend Joseph Dalton Hooker, recall, had direct experience with the flora of the Himalayas. George Henry Kendrick Thwaites was Director of the botanical gardens at Peradeniya, Sri Lanka (Ceylon), and Hewett Cottrell Watson made a special study of the botany of the Azores.

We may infer this from our frequent inability to predict whether or not an imported plant will endure our climate, and from the number of plants and animals brought from warmer countries which here enjoy good health. We have reason to believe that species in a state of nature are limited in their ranges by the competition of other organic beings quite as much as, or more than, by adaptation to particular climates. But whether or not the adaptation be generally very close, we have evidence, in the case of some few plants, of their becoming, to a certain extent, naturally habituated to different temperatures, or becoming acclimatised: thus the pines and rhododendrons, raised from seed collected [1] by Dr. Hooker from trees growing at different heights on the Himalaya, were found in this country to possess different constitutional powers of resisting cold. Mr. Thwaites informs me that he has observed similar facts in Ceylon, and analogous observations have been made by Mr. H. C. Watson on European species of plants brought from the Azores to England. In regard to animals, several authentic cases could be given of species within historical times having largely extended their range from warmer to cooler latitudes, and conversely; but we do not positively know that these animals were strictly adapted to their native climate, but in all ordinary cases we assume such to be the case; nor do we know that they have subsequently become acclimatised to their new homes.

As I believe that our domestic animals were originally chosen by uncivilised man because they were useful and bred readily under confinement, and not because they were subsequently found capable of far-extended transportation, I think the common and extraordinary capacity in our domestic animals of not only withstanding the most different climates but of being perfectly fertile (a far severer test) under them, may be used as an argument that a large proportion of other animals, now in a state of nature, could easily be brought to bear widely different climates. We must not, however, push the foregoing argument too far, on account of the probable origin of some of our domestic animals from several wild stocks: the blood, for instance, of a tropical and arctic wolf or wild dog may perhaps be mingled in our domestic breeds. The rat and mouse cannot be considered as domestic animals, but they have been transported by man to many parts of the world, and now have a far wider range than any other rodent, living free under the cold climate of Faroe in the north and of the Falklands in the south, and on many islands in the torrid zones. Hence I am inclined to look at adaptation to any special climate as a quality readily grafted on an innate wide flexibility of constitution, which is common to most animals. On this view, the capacity of enduring the most different climates by man himself and by his domestic animals, and such facts as that former species of the elephant and rhinoceros were capable of enduring a glacial climate, whereas the living species are now all tropical or sub-tropical in their habits, ought not to be looked at as anomalies, but merely as examples of a very common flexibility of constitution, brought, under peculiar circumstances, into play. [1]

How much of the acclimatisation of species to any peculiar climate is due to mere habit, and how much to the natural selection of varieties having different innate constitutions, and how much to both means combined, is a very obscure question. That habit or custom has some influence I must believe, both from analogy, and from the incessant advice given in agricultural works, even in the ancient Encyclopædias of China, to be very cau- [2]

1 The Faroe Islands are found in northern Europe, between Iceland and Norway. These northerly islands were Norwegian territory until 1814, when they came under Danish authority. The Falkland Islands are a southerly archipelago, in contrast, lying about 300 miles off the coast of Argentina. These islands were claimed by the British in 1833; in 1982 Argentina attempted to reassert authority over the Falklands, precipitating a brief war with Britain. Notably, the 1982 British forces used maps based on the survey work of the *Beagle* when Darwin visited the islands in 1833 and 1834. There is a town named for Darwin on the east Falkland Island.

2 In this discussion, think of "habit" as reflecting plasticity and "adaptation" as stemming from innate differences, owing to the action of natural selection. Darwin's discussion of this topic in his big book, *Natural Selection,* is more thorough, with several additional examples highlighting the capacity of organisms to endure under widely differing conditions: the spread of "escaped" cultivated plants into regions of varying climate, and the stunting of woody plant species in those parts of their range that extend to high latitudes or altitudes (Stauffer 1975, p. 288).

1 Darwin is careful not to dispel the notion that natural selection can play a role in adapting species to novel conditions. Notice he suggests near the end of this paragraph that this point can be demonstrated experimentally, by selecting for cold tolerance in kidney beans.

tious in transposing animals from one district to another; for it is not likely that man should have succeeded in selecting so many breeds and sub-breeds with constitutions specially fitted for their own districts: the result must, I think, be due to habit. On the other hand, I can see no reason to doubt that natural selection will continually tend to preserve those individuals which are born with constitutions best adapted to their native countries. In treatises on many kinds of cultivated plants, certain varieties are said to withstand certain climates better than others: this is very strikingly shown in works on fruit trees published in the United States, in which certain varieties are habitually recommended for the northern, and others for the southern States; and as most of these varieties are of recent origin, they cannot owe their constitutional differences to habit. The case of the Jerusalem artichoke, which is never propagated by seed, and of which consequently new varieties have not been produced, has even been advanced—for it is now as tender as ever it was—as proving that acclimatisation cannot be effected! The case, also, of the kidney-bean has been often cited for a similar purpose, and with much greater weight; but until some one will sow, during a score of generations, his kidney-beans so early that a very large proportion are destroyed by frost, and then collect seed from the few survivors, with care to prevent accidental crosses, and then again get seed from these seedlings, with the same precautions, the experiment cannot be said to have been even tried. Nor let it be supposed that no differences in the constitution of seedling kidney-beans ever appear, for an account has been published how much more hardy some seedlings appeared to be than others.

On the whole, I think we may conclude that habit,

use, and disuse, have, in some cases, played a considerable part in the modification of the constitution, and of the structure of various organs; but that the effects of use and disuse have often been largely combined with, and sometimes overmastered by, the natural selection of innate differences.

Correlation of Growth.—I mean by this expression that the whole organisation is so tied together during its growth and development, that when slight variations in any one part occur, and are accumulated through natural selection, other parts become modified. This is a very important subject, most imperfectly understood. The most obvious case is, that modifications accumulated solely for the good of the young or larva, will, it may safely be concluded, affect the structure of the adult; in the same manner as any malconformation affecting the early embryo, seriously affects the whole organisation of the adult. The several parts of the body which are homologous, and which, at an early embryonic period, are alike, seem liable to vary in an allied manner: we see this in the right and left sides of the body varying in the same manner; in the front and hind legs, and even in the jaws and limbs, varying together, for the lower jaw is believed to be homologous with the limbs. These tendencies, I do not doubt, may be mastered more or less completely by natural selection: thus a family of stags once existed with an antler only on one side; and if this had been of any great use to the breed it might probably have been rendered permanent by natural selection.

Homologous parts, as has been remarked by some authors, tend to cohere; this is often seen in monstrous plants; and nothing is more common than the union of homologous parts in normal structures, as the union of

1 By "correlation of growth" Darwin is referring to the puzzling phenomenon of developmental linkage, where changes in one structure appear to be accompanied by changes in another, seemingly unrelated, structure. He tackles this mystery in his usual way: he searches for patterns.

Over the next few pages Darwin will suggest various causes of real or apparent correlation, with the goal of arguing that in at least some cases correlation follows from the action of natural selection. That is, the correlation is selected for.

2 Here is one cause of correlation: homologous parts or structures change or vary in similar ways, despite being alike or undifferentiated at early embryonic stages. Although unstated here, homology is understood in terms of common ancestry in Darwin's view, so it makes sense that homologous structures should vary in similar ways—we would say their underlying genetics is essentially identical.

the petals of the corolla into a tube. Hard parts seem to affect the form of adjoining soft parts; it is believed by some authors that the diversity in the shape of the pelvis in birds causes the remarkable diversity in the [1] shape of their kidneys. Others believe that the shape of the pelvis in the human mother influences by pressure the shape of the head of the child. In snakes, [2] according to Schlegel, the shape of the body and the manner of swallowing determine the position of several of the most important viscera.

The nature of the bond of correlation is very frequently quite obscure. M. Is. Geoffroy St. Hilaire has forcibly remarked, that certain malconformations very frequently, and that others rarely coexist, without our [3] being able to assign any reason. What can be more singular than the relation between blue eyes and deafness in cats, and the tortoise-shell colour with the female sex; the feathered feet and skin between the outer toes in pigeons, and the presence of more or less down on the young birds when first hatched, with the future colour of their plumage; or, again, the relation between the hair and teeth in the naked Turkish dog, though here probably homology comes into play? With respect to this latter case of correlation, I think it can hardly be accidental, that if we pick out the two orders of mammalia which are most abnormal in their dermal covering, viz. Cetacea (whales) and Edentata (armadilloes, scaly anteaters, &c.), that these are likewise the most abnormal in their teeth.

I know of no case better adapted to show the importance of the laws of correlation in modifying important structures, independently of utility and, therefore, of natural selection, than that of the difference [4] between the outer and inner flowers in some Compositous and Umbelliferous plants. Every one knows the

1 Head shape was imagined to be a significant source of human mental variation in the days of phrenology. That long-discredited "science" held that different intellectual and emotional attributes mapped to particular areas of the brain; that the degree to which those areas were more or less well developed indicated how well (or poorly) developed those mental attributes were; and, finally, that the degree of development of brain areas was manifest in the surface features of the skull. Thus, phrenologists held, one could "read" the little bumps and valleys of the skull to draw inferences on the mental endowments and character of the individual.

2 Darwin consulted an 1843 translation of Hermann Schlegel's authoritative *Essai sur la physiognomie des serpents.*

3 Biologists have learned a lot about some of these curious linkages in the wake of the genetics revolution of the twentieth century. Congenital deafness is prevalent in many mammals with a white hair coat and solid blue eye color, apparently owing to genetic linkage between a gene for a particular coat color and genes associated with the degeneration of the cochlea. The tortoise-shell and calico coat coloration in female cats stems from the location of the pigment genes on the sex chromosomes, combined with the mammalian genetic phenomenon of X chromosome inactivation in females. Which copy is inactivated is random, and this leads to the expression of coat-color alleles in a mosaic pattern. This phenomenon was discovered by the British geneticist Mary F. Lyon in 1961, 102 years after publication of the *Origin.*

4 Compositous and umbelliferous plants, the sunflower and parsley families Asteraceae and Apiaceae, respectively, typically produce clusters of small flowers that collectively appear as a single large flower. Most convincing to pollinators are those cases in which the flowers on the edge of the floral disk have well-developed petals (often just on their outer edge) while those in the center of the disk have no petals at all, as in a daisy. Darwin sees these as examples of modification of correlated traits, for the two flower types also show differences in the shape

(continued)

difference in the ray and central florets of, for instance, the daisy, and this difference is often accompanied with the abortion of parts of the flower. But, in some Compositous plants, the seeds also differ in shape and sculpture; and even the ovary itself, with its accessory parts, differs, as has been described by Cassini. These differences have been attributed by some authors to pressure, and the shape of the seeds in the ray-florets in some Compositæ countenances this idea; but, in the case of the corolla of the Umbelliferæ, it is by no means, as Dr. Hooker informs me, in species with the densest heads that the inner and outer flowers most frequently differ. It might have been thought that the development of the ray-petals by drawing nourishment from certain other parts of the flower had caused their abortion; but in some Compositæ there is a difference in the seeds of the outer and inner florets without any difference in the corolla. Possibly, these several differences may be connected with some difference in the flow of nutriment towards the central and external flowers: we know, at least, that in irregular flowers, those nearest to the axis are oftenest subject to peloria, and become regular. I may add, as an instance of this, and of a striking case of correlation, that I have recently observed in some garden pelargoniums, that the central flower of the truss often loses the patches of darker colour in the two upper petals; and that when this occurs, the adherent nectary is quite aborted; when the colour is absent from only one of the two upper petals, the nectary is only much shortened.

With respect to the difference in the corolla of the central and exterior flowers of a head or umbel, I do not feel at all sure that C. C. Sprengel's idea that the ray-florets serve to attract insects, whose agency is highly advantageous in the fertilisation of plants of ◀1

(continued)
and surface sculpturing of their seeds. He put this a bit more clearly in *Natural Selection:* "As we can hardly suppose that internal and structural differences in the fruit on the same individual plant can be of use to the species, we must attribute the differences in the pericarps,—in their shape, their appendages, & even in the ovary itself with its accessory parts—of the central and marginal florets of many compositae, to some correlation of growth" (Stauffer 1975, p. 300).

1 The morphological difference between disk and ray (central and peripheral) florets could be seen as a division of labor of sorts, toward the end of attracting pollinators with greater efficiency. That interpretation, with which Darwin agrees, was put forth in Christian Konrad Sprengel's 1793 treatise *Das entdeckte Geheimniss der Natur im Bau und in der Befruchtung der Blumen* [The Secret of Nature in the Form and Fertilization of Flowers Discovered]. According to Sprengel, floral scents, shapes, and colors complement one another to attract insects for pollination. "Nature seems to have wished that no flower should be fertilized with its own pollen," Sprengel wrote.

these two orders, is so far-fetched, as it may at first appear: and if it be advantageous, natural selection may have come into play. But in regard to the differences both in the internal and external structure of the seeds, which are not always correlated with any differences in the flowers, it seems impossible that they can be in any way advantageous to the plant: yet in the Umbelliferæ these differences are of such apparent importance—the seeds being in some cases, according to Tausch, orthospermous in the exterior flowers and cœlospermous in the central flowers,—that the elder De Candolle founded his main divisions of the order on analogous differences. Hence we see that modifications of structure, viewed by systematists as of high value, may be wholly due to unknown laws of correlated growth, and without being, as far as we can see, of the slightest service to the species.

We may often falsely attribute to correlation of growth, structures which are common to whole groups of species, and which in truth are simply due to inheritance; for an ancient progenitor may have acquired through natural selection some one modification in structure, and, after thousands of generations, some other and independent modification; and these two modifications, having been transmitted to a whole group of descendants with diverse habits, would naturally be thought to be correlated in some necessary manner. So, again, I do not doubt that some apparent correlations, occurring throughout whole orders, are entirely due to the manner alone in which natural selection can act. For instance, Alph. De Candolle has remarked that winged seeds are never found in fruits which do not open: I should explain the rule by the fact that seeds could not gradually become winged through natural selection, except in fruits which opened; so that the individual plants producing

1 *Orthospermous* is a botanical term meaning bearing straight seeds, as found in the fruits of many species of the parsley family. *Coelospermous* means "hollow-seeded," plants that have the ventral surface of the carpels incurved at the ends; coriander seed is commonly given as an example. In this passage Darwin refers to Ignaz Friedrich Tausch's paper *Classification des Ombellifères,* published in 1835. The elder de Candolle, recall, is Augustin Pyramus de Candolle, a Swiss botanist and plant biogeographer. Darwin is referring to de Candolle's compendium of plant families, *Prodromus systematis naturalis regni vegetabili.*

2 This comment reveals the insights that can be gained by "descent thinking"—what appears to be correlation of characters in species groups may simply reflect inheritance of these characters together from a common ancestor. Ancestrally the character states could even have been picked up independently, separated by a long time period, but they will appear correlated.

3 This example, winged seeds absent from fruits that do not open (indehiscent fruit), is interesting in that, were a case to be found of winged seeds produced by a species with indehiscent fruit, Darwin might cite it as a great example of "historical baggage," an anomaly that reflects an evolutionary past when the seeds were dispersed by the wind.

seeds which were a little better fitted to be wafted further, might get an advantage over those producing seed less fitted for dispersal; and this process could not possibly go on in fruit which did not open.

The elder Geoffroy and Goethe propounded, at about the same period, their law of compensation or balancement of growth; or, as Goethe expressed it, "in order to spend on one side, nature is forced to economise on the other side." I think this holds true to a certain extent with our domestic productions: if nourishment flows to one part or organ in excess, it rarely flows, at least in excess, to another part; thus it is difficult to get a cow to give much milk and to fatten readily. The same varieties of the cabbage do not yield abundant and nutritious foliage and a copious supply of oil-bearing seeds. When the seeds in our fruits become atrophied, the fruit itself gains largely in size and quality. In our poultry, a large tuft of feathers on the head is generally accompanied by a diminished comb, and a large beard by diminished wattles. With species in a state of nature it can hardly be maintained that the law is of universal application; but many good observers, more especially botanists, believe in its truth. I will not, however, here give any instances, for I see hardly any way of distinguishing between the effects, on the one hand, of a part being largely developed through natural selection and another and adjoining part being reduced by this same process or by disuse, and, on the other hand, the actual withdrawal of nutriment from one part owing to the excess of growth in another and adjoining part.

I suspect, also, that some of the cases of compensation which have been advanced, and likewise some other facts, may be merged under a more general principle, namely, that natural selection is continually trying to economise in every part of the organisation. If under

1 "The elder Geoffroy" is Étienne Geoffroy St. Hilaire. He and the German polymath Johann Wolfgang von Goethe held that there is a *loi du balancement,* a "law of compensation" or balancing of growth, such that development of one organ is compensated by underdevelopment of another. The law of compensation was a common nineteenth-century notion of organic development. Think of it as a sort of zero-sum concept of physiology: to increase development of one structure, another must be reduced in compensation, perhaps because there is only so much nutriment to go around. Here is a typical expression of the idea, from Ralph Waldo Emerson's 1841 essay entitled, appropriately enough, *Compensation:*

> Polarity, or action and reaction, we meet in every part of nature . . . For example, in the animal kingdom the physiologist has observed that no creatures are favorites, but a certain compensation balances every gift and every defect. A surplusage given to one part is paid out of a reduction from another part of the same creature. If the head and neck are enlarged, the trunk and extremities are cut short.

Notice that Darwin approaches this idea with suspicion, pointing out that it is not possible to distinguish compensation from either use and disuse or the differential action of selection on certain structures. His bottom line is that selection continually acts "to economise in every part of the organisation" and can reduce or enlarge traits *without* any compensating reduction or enlargement of other traits.

1 Continuing the point of the previous page, Darwin argues that selection will minimize waste: reducing the size of an unnecessary structure will allow the resources that went into its construction to be put toward, say, reproduction. The barnacle *Ibla cumingi,* which Darwin described, provides a fascinating example: while dissecting this southern hemisphere barnacle, he discovered that the males are reduced to tiny parasite-like structures, mere bags of gonads, lodged within the shell of the comparatively gigantic female. Here is an account in a letter to Henslow dated April 1, 1848:

> Talking of Cirripedia, I must tell you a curious case I have just these few last days made out: all the Cirripedia are bisexual, except one genus, & in this the female has the ordinary appearance, whereas the male has no one part of its body like the female & is microscopically minute; but here comes the odd fact, the male or sometimes two males, at the instant they cease being locomotive larvae become parasitic within the sack of the female, & thus fixed & half embedded in the flesh of their wives they pass their whole lives & can never move again. (*Correspondence* 4: 127)

Proteolepas was another puzzling find. Darwin took this to be an aberrant parasitic barnacle, "footless" and with other reduced characters. The scientific name for barnacles is Cirripedia, translating as "feathery-feet" in reference to their modified legs, which function in filter-feeding. Alluding to *Proteolepas* in a letter to Agassiz, he wrote: "some Cirri *pedia* are *apodal,*" and he even named a new order, Apoda, based on this aberrant species. *Proteolepas* turned out to be a parasitic isopod, but Darwin's point remains: selection can greatly reduce some traits while leaving others unaffected.

2 Note that no opportunity is missed to put in a plug for gradualism; *Proteolepas* changed by *slow steps.*

changed conditions of life a structure before useful becomes less useful, any diminution, however slight, in its development, will be seized on by natural selection, for it will profit the individual not to have its nutriment wasted in building up an useless structure. I can thus only understand a fact with which I was much struck when examining cirripedes, and of which many other instances could be given: namely, that when a cirripede is parasitic within another and is thus protected, it loses more or less completely its own shell or carapace. This is the case with the male Ibla, and in a truly extraordinary manner with the Proteolepas: for the carapace in all other cirripedes consists of the three highly-important anterior segments of the head enormously developed, and furnished with great nerves and muscles; but in the parasitic and protected Proteolepas, the whole anterior part of the head is reduced to the merest rudiment attached to the bases of the prehensile antennæ. Now the saving of a large and complex structure, when rendered superfluous by the parasitic habits of the Proteolepas, though effected by slow steps, would be a decided advantage to each successive individual of the species; for in the struggle for life to which every animal is exposed, each individual Proteolepas would have a better chance of supporting itself, by less nutriment being wasted in developing a structure now become useless.

Thus, as I believe, natural selection will always succeed in the long run in reducing and saving every part of the organisation, as soon as it is rendered superfluous, without by any means causing some other part to be largely developed in a corresponding degree. And, conversely, that natural selection may perfectly well succeed in largely developing any organ, without requiring as a necessary compensation the reduction of some adjoining part.

It seems to be a rule, as remarked by Is. Geoffroy St. Hilaire, both in varieties and in species, that when any part or organ is repeated many times in the structure of the same individual (as the vertebræ in snakes, and the stamens in polyandrous flowers) the number is variable; whereas the number of the same part or organ, when it occurs in lesser numbers, is constant. The same author and some botanists have further remarked that multiple parts are also very liable to variation in structure. Inasmuch as this "vegetative repetition," to use Prof. Owen's expression, seems to be a sign of low organisation; the foregoing remark seems connected with the very general opinion of naturalists, that beings low in the scale of nature are more variable than those which are higher. I presume that lowness in this case means that the several parts of the organisation have been but little specialised for particular functions; and as long as the same part has to perform diversified work, we can perhaps see why it should remain variable, that is, why natural selection should have preserved or rejected each little deviation of form less carefully than when the part has to serve for one special purpose alone. In the same way that a knife which has to cut all sorts of things may be of almost any shape; whilst a tool for some particular object had better be of some particular shape. Natural selection, it should never be forgotten, can act on each part of each being, solely through and for its advantage.

Rudimentary parts, it has been stated by some authors, and I believe with truth, are apt to be highly variable. We shall have to recur to the general subject of rudimentary and aborted organs; and I will here only add that their variability seems to be owing to their uselessness, and therefore to natural selection having no power to check deviations in their structure. Thus

1 In the main paragraph on this page Darwin explores another pattern in the occurrence of variation: in cases where a structure or segment is much repeated, the number and form of the units tend to vary more than they do in cases where there are relatively few units. This touches on specialization vs. generalization and "low" vs. "high" position in the supposed scale of nature. The idea, still held today, is that relatively undifferentiated, repeating units (like the segments of millipedes or centipedes) represent a more primitive state than fewer segments variously modified in specialized ways (like the segments of insects). Insofar as the repeated units are more generalized, selection on them is more relaxed and they can be expected to vary more than specialized units. Darwin's knife analogy is apt, and again he puts it even better in *Natural Selection* (Stauffer 1975, p. 299):

> As long as an organ had to act in many ways, its exact form would probably not signify; just as a knife for cutting all sorts of things, may be almost of any shape, but a cutting tool for some particular object had best be of some particular shape; so with an organ as it began to be specialised through natural selection for some particular end, its particular structure would be come more & more important; & this same natural selection would tend to keep the form constant by the rejection of accidental deviations.

2 Today we can understand the high degree of variation found in rudimentary (vestigial) structures in terms of relaxed selection. Once selectively neutral, random fluctuation is expected through genetic drift. This point will be revisited in a page or two.

rudimentary parts are left to the free play of the various laws of growth, to the effects of long-continued disuse, and to the tendency to reversion.

[1] *A part developed in any species in an extraordinary degree or manner, in comparison with the same part in allied species, tends to be highly variable.*—Several years [2] ago I was much struck with a remark, nearly to the above effect, published by Mr. Waterhouse. I infer also from an observation made by Professor Owen, with respect to the length of the arms of the ourang-outang, that he has come to a nearly similar conclusion. It is hopeless to attempt to convince any one of the truth of this proposition without giving the long array of facts which I have collected, and which cannot possibly be here introduced. I can only state my conviction that it is a rule of high generality. I am aware of several causes of error, but I hope that I have made due allowance for them. It should be understood that the rule by no means applies to any part, however unusually developed, unless it be unusually developed in comparison with the same part in closely allied species. Thus, the bat's wing is a most abnormal structure in the class mammalia; but the rule would not here apply, because there is a whole group of bats having wings; it would apply only if some one species of bat had its wings developed in some remarkable manner in comparison with the other species of the same genus. The rule applies very strongly in the case of secondary sexual characters, when [3] displayed in any unusual manner. The term, secondary sexual characters, used by Hunter, applies to characters which are attached to one sex, but are not directly connected with the act of reproduction. The rule applies to males and females; but as females more rarely offer remarkable secondary sexual characters, it applies

1 In this important section, Darwin zeroes in on another pattern he perceives in the occurrence of variation—a pattern of great significance for him in that it seems to signal the action of natural selection. He first apologizes in his usual way for the abbreviated treatment he is about to launch into, and indeed, he did assemble a prodigious range of observations to back up his claim—the corresponding section in *Natural Selection* has subsections devoted to numerous examples from birds, barnacles, and insects.

The pattern of interest here is that highly developed structures will also be highly variable as compared with similar but less well-developed structures in closely related species. The same pattern holds, he argues, for well-developed secondary sexual characteristics, too (again, as compared with similar characters in closely related species that lack the same degree of development). Seemingly simple observations, but ones that make little sense except in terms of evolution by natural selection.

2 Darwin is citing an observation made by Waterhouse in his 1848 book *A Natural History of the Mammalia:* "As a general rule where any species is characterised by a maximum of development of certain parts, those parts are more subject to variation in the different individuals of the species than are parts which approach more nearly to the normal conditions" (p. 452). One of Richard Owen's earliest papers (1830–1831) was "On the Anatomy of the Ourang-outang."

3 Secondary sexual characters make for a good case study of Darwin's assertion that well-developed traits tend to be more variable than the corresponding but less well-developed trait in other species or, in this case, sex. Darwin is referring to the anatomist John Hunter's book *Observations on Certain Parts of the Animal Oeconomy* (London, 1786).

more rarely to them. The rule being so plainly applicable in the case of secondary sexual characters, may be due to the great variability of these characters, whether or not displayed in any unusual manner—of which fact I think there can be little doubt. But that our rule is not confined to secondary sexual characters is clearly shown in the case of hermaphrodite cirripedes; and I may here add, that I particularly attended to Mr. Waterhouse's remark, whilst investigating this Order, and I am fully convinced that the rule almost invariably holds good with cirripedes. I shall, in my future work, give a list of the more remarkable cases; I will here only briefly give one, as it illustrates the rule in its largest application. The opercular valves of sessile cirripedes (rock barnacles) are, in every sense of the word, very important structures, and they differ extremely little even in different genera; but in the several species of one genus, Pyrgoma, these valves 1
present a marvellous amount of diversification: the homologous valves in the different species being sometimes wholly unlike in shape; and the amount of variation in the individuals of several of the species is so great, that it is no exaggeration to state that the varieties differ more from each other in the characters of these important valves than do other species of distinct genera.

As birds within the same country vary in a remarkably small degree, I have particularly attended to them, and the rule seems to me certainly to hold good in this class. I cannot make out that it applies to plants, and this would seriously have shaken my belief in its truth, had not the great variability in plants made it particularly difficult to compare their relative degrees of variability.

When we see any part or organ developed in a 2
remarkable degree or manner in any species, the fair

1 Of some four manuscript pages of barnacle examples in *Natural Selection* Darwin selected just one, *Pyrgoma*, to make his point in the *Origin*. These cirripedes, which are now placed in the genus *Savignium*, are known as the coral-inhabiting barnacles. They are highly specialized for life atop a living coral colony. Darwin notes that the two opercular valves of these barnacles, the little "doors" the cirripede can close to shut itself within the safety of its calcified fortress, are not symmetrical but differ in size and shape even among *Pyrgoma* species. This may reflect their specialization on corals in different genera and even suborders: since barnacles reproduce by internal fertilization, coral specificity would be expected to lead to isolation and, eventually, speciation. This is supported by a phylogenetic study by Mokady et al. (1999), who analyzed several Red Sea *Savignium* species sampled from different coral hosts and found that the phenotypic differences among them correlated with genetic differences, suggesting isolation.

2 The crux of the matter is nicely set up: we might reasonably expect especially well-developed structures to be very important to the organism. Why, then, should such important structures be so liable to vary? Darwin rhetorically asks. This makes no sense under the special creation model, but Darwin suggests that it is precisely what we would expect of structures undergoing selective change—they must have had lots of variation for selection to act upon, and so typically they will still exhibit that variation.

1 The argument is strengthened by looking at comparable patterns in domesticated organisms—recall the profound analogy for common descent by natural selection that the domestication process represents for Darwin. If a similar pattern of variability can be found in structures known to be under rapid change by breeders, it's virtually a smoking gun. Of course, Darwin presents just such a case, pointing to the beak, wattle, and tail feathers of domestic pigeons.

presumption is that it is of high importance to that species; nevertheless the part in this case is eminently liable to variation. Why should this be so? On the view that each species has been independently created, with all its parts as we now see them, I can see no explanation. But on the view that groups of species have descended from other species, and have been modified through natural selection, I think we can obtain some light. In our domestic animals, if any part, or the whole animal, be neglected and no selection be applied, that part (for instance, the comb in the Dorking fowl) or the whole breed will cease to have a nearly uniform character. The breed will then be said to have degenerated. In rudimentary organs, and in those which have been but little specialised for any particular purpose, and perhaps in polymorphic groups, we see a nearly parallel natural case; for in such cases natural selection either has not or cannot come into full play, and thus the organisation is left in a fluctuating condition. [1] But what here more especially concerns us is, that in our domestic animals those points, which at the present time are undergoing rapid change by continued selection, are also eminently liable to variation. Look at the breeds of the pigeon; see what a prodigious amount of difference there is in the beak of the different tumblers, in the beak and wattle of the different carriers, in the carriage and tail of our fantails, &c., these being the points now mainly attended to by English fanciers. Even in the sub-breeds, as in the short-faced tumbler, it is notoriously difficult to breed them nearly to perfection, and frequently individuals are born which depart widely from the standard. There may be truly said to be a constant struggle going on between, on the one hand, the tendency to reversion to a less modified state, as well as an innate tendency to further variability of all kinds, and, on the other hand, the power of steady selection to keep the breed true. In the long run selection gains the day, and we do not expect to fail so far as to breed a bird as coarse as a common tumbler from a good short-faced strain. But as long as selection is rapidly going on, there may always be expected to be much variability in the structure undergoing modification. It further deserves notice that these variable characters, produced by man's selection, sometimes become attached, from causes quite unknown to us, more to one sex than to the other, generally to the male sex, as with the wattle of carriers and the enlarged crop of pouters.

Now let us turn to nature. When a part has been developed in an extraordinary manner in any one species, compared with the other species of the same genus, we may conclude that this part has undergone an extraordinary amount of modification, since the period when the species branched off from the common progenitor of the genus. This period will seldom be remote in any extreme degree, as species very rarely endure for more than one geological period. An extraordinary amount of modification implies an unusually large and long-continued amount of variability, which has continually been accumulated by natural selection for the benefit of the species. But as the variability of the extraordinarily-developed part or organ has been so great and long-continued within a period not excessively remote, we might, as a general rule, expect still to find more variability in such parts than in other parts of the organisation, which have remained for a much longer period nearly constant. And this, I am convinced, is the case. That the struggle between natural selection on the one hand, and the tendency to reversion and variability on the other hand, will in the

1 This paragraph restates the point made in the previous paragraph, but more forcefully. In cases where a structure is "extraordinarily" well-developed in one species compared with closely related species, it is assumed that the structure has undergone great change since the point of common ancestry. To sustain this change, there had to be considerable variability for quite some time, though generally within the timeframe of a geological period. Note the next step in the argument: since this variability has been so great and fairly recent (i.e., a period "not excessively remote"), we should *still* expect to see that variability.

We get a sense here of how Darwin sees variations generated and sustained; recall from chapter I his view that variations can stem from perturbations to the reproductive system, which can result from altered circumstances like a new environment. Although not explicitly stated in this way, this passage suggests Darwin thinks that the rapid and sustained selection the species is experiencing (as evidenced by a structure or organ developing to an extraordinary degree) reflects environmental or other significant change—how else would we see such strong selective pressure?—and this change itself engenders still more variations. In this way the process is self-sustaining, with altered circumstances provoking new variations that are the raw material for selection to act upon. On the next page Darwin calls this process "generative variability." Today we would not agree with this formulation, or the idea that just because a structure varied in the evolutionary past it should still exhibit a high level of variation now.

1 "Generative variability" is variation manifest in structures that have recently experienced rapid and considerable evolutionary change. Darwin envisions this as a dynamic process. Given enough time—after the structure has reached its maximum extent of development—selection weeds out most of the deviations and the trait ends up fixed.

2 Here "specific character" refers to a trait characterizing a species, while "generic character" is a trait characterizing a genus. Almost by definition, if a character is invariant at the genus level it can be used to help define or characterize that genus, but if it is variable among species within the genus it can only be used to help characterize individual species. That is the point of the red and blue flower example. There is more to this than a truism, however: characters used to distinguish species in a group are more variable—*even at the individual level*—than those used to distinguish genera, Darwin argues. He dismisses the suggestion that species-distinguishing characters are more variable because they are more free to vary, as they might be if they were relatively unimportant physiologically. Traits of great physiological importance should vary little, the argument goes, so of course they would end up being generic-level traits.

course of time cease; and that the most abnormally developed organs may be made constant, I can see no reason to doubt. Hence when an organ, however abnormal it may be, has been transmitted in approximately the same condition to many modified descendants, as in the case of the wing of the bat, it must have existed, according to my theory, for an immense period in nearly the same state; and thus it comes to be no more variable than any other structure. It is only in those cases in which the modification has been comparatively recent and extraordinarily great that we ought to find the *generative variability*, as it may be called, still present in a high degree. For in this case the variability will seldom as yet have been fixed by the continued selection of the individuals varying in the required manner and degree, and by the continued rejection of those tending to revert to a former and less modified condition.

The principle included in these remarks may be extended. It is notorious that specific characters are more variable than generic. To explain by a simple example what is meant. If some species in a large genus of plants had blue flowers and some had red, the colour would be only a specific character, and no one would be surprised at one of the blue species varying into red, or conversely; but if all the species had blue flowers, the colour would become a generic character, and its variation would be a more unusual circumstance. I have chosen this example because an explanation is not in this case applicable, which most naturalists would advance, namely, that specific characters are more variable than generic, because they are taken from parts of less physiological importance than those commonly used for classing genera. I believe this explanation is partly, yet only indirectly, true; I shall, however, have to return to this subject in our chapter on Classification. It would be almost superfluous to adduce evidence in support of the above statement, that specific characters are more variable than generic; but I have repeatedly noticed in works on natural history, that when an author has remarked with surprise that some *important* organ or part, which is generally very constant throughout large groups of species, has *differed* considerably in closely-allied species, that it has, also, been *variable* in the individuals of some of the species. And this fact shows that a character, which is generally of generic value, when it sinks in value and becomes only of specific value, often becomes variable, though its physiological importance may remain the same. Something of the same kind applies to monstrosities: at least Is. Geoffroy St. Hilaire seems to entertain no doubt, that the more an organ normally differs in the different species of the same group, the more subject it is to individual anomalies.

On the ordinary view of each species having been independently created, why should that part of the structure, which differs from the same part in other independently-created species of the same genus, be more variable than those parts which are closely alike in the several species? I do not see that any explanation can be given. But on the view of species being only strongly marked and fixed varieties, we might surely expect to find them still often continuing to vary in those parts of their structure which have varied within a moderately recent period, and which have thus come to differ. Or to state the case in another manner:—the points in which all the species of a genus resemble each other, and in which they differ from the species of some other genus, are called generic characters; and these characters in common I attribute to inheritance from a common

1 The example of flower color was chosen, Darwin said on the previous page, to show that the excess of variation in specific traits has nothing to do with degree of physiological importance. That traits varying among species also vary greatly among individuals of those species simply reflects recency in the action of natural selection, particularly if the trait in question differs considerably among species. That point is emphasized here with italics.

2 "Monstrosities"—excessively developed organs or structures —offer a parallel example: structures that vary to a great degree among closely related species are also the most likely structures to become pathological monstrosities. Darwin is here alluding to volume 3 of Isidore Geoffroy St. Hilaire's 1836 work *Histoire des Anomalies.*

3 At last we come to the implications presented by these observations: why, under special creation, should traits that happen to vary at the species level also exhibit disproportionate individual-level variation? If species are independently created, there should be no such pattern, no correspondence among them. But if closely allied species have descended from a common ancestor and were recently but strongly marked varieties, they might be expected still to exhibit the variation that selection acted upon to differentiate them further into species. In other words, the traits that distinguish species might be expected to continue varying, as they must have varied in the recent evolutionary past for selection to act upon them.

These patterns of variation thus present another occasion for Darwin to draw a sharp contrast between pattern and process: he reports an observed pattern and asks which process makes the most sense in explaining it, special creation or descent with modification.

1 Darwin reports a curious observation in these next two long paragraphs, one that would seem to be the icing on the cake of his argument. Secondary sexual traits are not only variable but appear to be more variable than other traits among closely related species. We can see where the argument is going: we can think of secondary sexual traits in the same way as the "extraordinarily modified" traits earlier in this section, but here sexual selection rather than natural selection is acting to shape them.

progenitor, for it can rarely have happened that natural selection will have modified several species, fitted to more or less widely-different habits, in exactly the same manner: and as these so-called generic characters have been inherited from a remote period, since that period when the species first branched off from their common progenitor, and subsequently have not varied or come to differ in any degree, or only in a slight degree, it is not probable that they should vary at the present day. On the other hand, the points in which species differ from other species of the same genus, are called specific characters; and as these specific characters have varied and come to differ within the period of the branching off of the species from a common progenitor, it is probable that they should still often be in some degree variable,—at least more variable than those parts of the organisation which have for a very long period remained constant.

1 In connexion with the present subject, I will make only two other remarks. I think it will be admitted, without my entering on details, that secondary sexual characters are very variable; I think it also will be admitted that species of the same group differ from each other more widely in their secondary sexual characters, than in other parts of their organisation; compare, for instance, the amount of difference between the males of gallinaceous birds, in which secondary sexual characters are strongly displayed, with the amount of difference between their females; and the truth of this proposition will be granted. The cause of the original variability of secondary sexual characters is not manifest; but we can see why these characters should not have been rendered as constant and uniform as other parts of the organisation; for secondary sexual characters have been accumulated by sexual selection, which

is less rigid in its action than ordinary selection, as it does not entail death, but only gives fewer offspring to the less favoured males. Whatever the cause may be of the variability of secondary sexual characters, as they are highly variable, sexual selection will have had a wide scope for action, and may thus readily have succeeded in giving to the species of the same group a greater amount of difference in their sexual characters, than in other parts of their structure.

It is a remarkable fact, that the secondary sexual differences between the two sexes of the same species are generally displayed in the very same parts of the organisation in which the different species of the same genus differ from each other. Of this fact I will give in illustration two instances, the first which happen to stand on my list; and as the differences in these cases are of a very unusual nature, the relation can hardly be accidental. The same number of joints in the tarsi is a character generally common to very large groups of beetles, but in the Engidæ, as Westwood has remarked, the number varies greatly; and the number likewise differs in the two sexes of the same species: again in fossorial hymenoptera, the manner of neuration of the wings is a character of the highest importance, because common to large groups; but in certain genera the neuration differs in the different species, and likewise in the two sexes of the same species. This relation has a clear meaning on my view of the subject: I look at all the species of the same genus as having as certainly descended from the same progenitor, as have the two sexes of any one of the species. Consequently, whatever part of the structure of the common progenitor, or of its early descendants, became variable; variations of this part would, it is highly probable, be taken advantage of by natural and sexual selection, in

1 These may have been the first two examples in Darwin's notebooks, but they are the first and fourth in a list of ten examples drawn from insects, mammals, and birds given in *Natural Selection* (Stauffer 1975, p. 335). Engidae, a fungus beetle family named by Westwood, is now called Erotylidae. In *Natural Selection* Darwin commented on the "numberless differences" these beetles exhibit in tarsal number. The "fossorial Hymenoptera" refers here to *Tiphia* wasps, sleek black wasps about a half-inch long, the females of which make their living by burrowing into the soil to locate subterranean beetle grubs to parasitize. These wasps are widely used for biological control of introduced beetle pests—for example, *T. vernalis*, the spring tiphia, was introduced into the United States to help control Japanese beetles. Here Darwin is commenting on the variable nature of the wing venation found in these wasps.

1 This is Darwin at his synthetic best, tying together the lines of evidence laid out over the past several pages. He builds to the statement, "All are principles closely connected together"—a consilience approach writ small. What is significant here is Darwin's altogether unexpected take on patterns in variation, perceiving signals in the noise of variation that are consistent with his theory.

order to fit the several species to their several places in the economy of nature, and likewise to fit the two sexes of the same species to each other, or to fit the males and females to different habits of life, or the males to struggle with other males for the possession of the females.

1 Finally, then, I conclude that the greater variability of specific characters, or those which distinguish species from species, than of generic characters, or those which the species possess in common;—that the frequent extreme variability of any part which is developed in a species in an extraordinary manner in comparison with the same part in its congeners; and the not great degree of variability in a part, however extraordinarily it may be developed, if it be common to a whole group of species;—that the great variability of secondary sexual characters, and the great amount of difference in these same characters between closely allied species;—that secondary sexual and ordinary specific differences are generally displayed in the same parts of the organisation,—are all principles closely connected together. All being mainly due to the species of the same group having descended from a common progenitor, from whom they have inherited much in common,—to parts which have recently and largely varied being more likely still to go on varying than parts which have long been inherited and have not varied,—to natural selection having more or less completely, according to the lapse of time, overmastered the tendency to reversion and to further variability,—to sexual selection being less rigid than ordinary selection,—and to variations in the same parts having been accumulated by natural and sexual selection, and thus adapted for secondary sexual, and for ordinary specific purposes.

Distinct species present analogous variations; and a variety of one species often assumes some of the characters of an allied species, or reverts to some of the characters of an early progenitor.—These propositions will be most readily understood by looking to our domestic races. The most distinct breeds of pigeons, in countries most widely apart, present sub-varieties with reversed feathers on the head and feathers on the feet,—characters not possessed by the aboriginal rock-pigeon; these then are analogous variations in two or more distinct races. The frequent presence of fourteen or even sixteen tail-feathers in the pouter, may be considered as a variation representing the normal structure of another race, the fantail. I presume that no one will doubt that all such analogous variations are due to the several races of the pigeon having inherited from a common parent the same constitution and tendency to variation, when acted on by similar unknown influences. In the vegetable kingdom we have a case of analogous variation, in the enlarged stems, or roots as commonly called, of the Swedish turnip and Ruta baga, plants which several botanists rank as varieties produced by cultivation from a common parent: if this be not so, the case will then be one of analogous variation in two so-called distinct species; and to these a third may be added, namely, the common turnip. According to the ordinary view of each species having been independently created, we should have to attribute this similarity in the enlarged stems of these three plants, not to the *vera causa* of community of descent, and a consequent tendency to vary in a like manner, but to three separate yet closely related acts of creation. 1

With pigeons, however, we have another case, namely, the occasional appearance in all the breeds, of slaty-blue birds with two black bars on the wings, a white 2

1 The section heading says it all: related species, and varieties of the same species, will often vary in similar ways (i.e., the same parts will vary or will do so in similar ways). Take the cases of the peculiar feather modifications in certain pigeon varieties, like the reversed feathers on the head and feet, or the enlarged corm of the three turnip varieties Darwin mentions. To Darwin, related species or varieties are liable to vary in similar ways; rather than view each case as a separate act of creation, he looks to their common descent to explain their fundamental similarity and tendency to vary in similar ways. Today we understand this in terms of basic similarity in genetic constitution, but that is not to say that variations are not random. They are, but closely related forms are already starting from a point of great similarity.

2 This second paragraph opens with a lengthy and dense discussion of reversion, the tendency for ancestral traits to reappear or even be "reconstituted" by careful crossing. This is an important phenomenon for Darwin, for it is consistent with common descent.

1 This expression is common still, linguistic historical baggage that harks back to the pre-Mendelian days when it was believed that particles specifying all physical attributes circulate in the blood, eventually making their way to the reproductive organs. Under such a model of "blending" inheritance, it was common to speak of proportion of blood ancestry; the phrase "it's in the blood" remains a common expression today. As Darwin appreciated, language bears the stamp of history and often provides intriguing analogies with the biological evolutionary process.

rump, a bar at the end of the tail, with the outer feathers externally edged near their bases with white. As all these marks are characteristic of the parent rock-pigeon, I presume that no one will doubt that this is a case of reversion, and not of a new yet analogous variation appearing in the several breeds. We may I think confidently come to this conclusion, because, as we have seen, these coloured marks are eminently liable to appear in the crossed offspring of two distinct and differently coloured breeds; and in this case there is nothing in the external conditions of life to cause the reappearance of the slaty-blue, with the several marks, beyond the influence of the mere act of crossing on the laws of inheritance.

No doubt it is a very surprising fact that characters should reappear after having been lost for many, perhaps for hundreds of generations. But when a breed has been crossed only once by some other breed, the offspring occasionally show a tendency to revert in character to the foreign breed for many generations—some say, for a dozen or even a score of generations. After twelve generations, the proportion of blood, to use a common expression, of any one ancestor, is only 1 in 2048; and yet, as we see, it is generally believed that a tendency to reversion is retained by this very small proportion of foreign blood. In a breed which has not been crossed, but in which *both* parents have lost some character which their progenitor possessed, the tendency, whether strong or weak, to reproduce the lost character might be, as was formerly remarked, for all that we can see to the contrary, transmitted for almost any number of generations. When a character which has been lost in a breed, reappears after a great number of generations, the most probable hypothesis is, not that the offspring suddenly takes after an ancestor some hundred generations distant, but that in each successive generation there has been a tendency to reproduce the character in question, which at last, under unknown favourable conditions, gains an ascendancy. For instance, it is probable that in each generation of the barb-pigeon, which produces most rarely a blue and black-barred bird, there has been a tendency in each generation in the plumage to assume this colour. This view is hypothetical, but could be supported by some facts; and I can see no more abstract improbability in a tendency to produce any character being inherited for an endless number of generations, than in quite useless or rudimentary organs being, as we all know them to be, thus inherited. Indeed, we may sometimes observe a mere tendency to produce a rudiment inherited: for instance, in the common snapdragon (Antirrhinum) a rudiment of a fifth stamen so often appears, that this plant must have an inherited tendency to produce it. ◀1

As all the species of the same genus are supposed, ◀2 on my theory, to have descended from a common parent, it might be expected that they would occasionally vary in an analogous manner; so that a variety of one species would resemble in some of its characters another species; this other species being on my view only a well-marked and permanent variety. But characters thus gained would probably be of an unimportant nature, for the presence of all important characters will be governed by natural selection, in accordance with the diverse habits of the species, and will not be left to the mutual action of the conditions of life and of a similar inherited constitution. It might further be expected that the species of the same genus would occasionally exhibit reversions to lost ancestral characters. As, however, we never know the exact character of the common ancestor of a group, we could not distinguish these two

1 Darwin would be impressed with how much is now known about this species. The snapdragon *Antirrhinum majus* (Scrophulariaceae) has become an important model organism in the study of floral developmental genetics. The gene responsible for flower development, *cycloidea*, was identified in 1996; this species has five stamen primordia, but only four develop into functional stamens. *Cycloidea* expression is reduced in one, and the fifth stamen remains puny. Darwin is right that *Antirrhinum* has an inherited tendency to express a fifth stamen: all the basic genetic architecture is there (Luo et al. 1996). This is precisely why "reversion" to ancestral characters occurs as well, something mentioned at the bottom of this page.

In 2005 a discovery was made that would have delighted Darwin: *Antirrhinum* and the common mustard *Arabidopsis* inherited, from a distant common ancestor, a duplicated gene involved in the development of flower structure. In each species one of the two copies plays its original or ancestral role, while the other copy has diverged in function. The divergent copy in *Antirrhinum* has apparently retained some of the original functionality, however, as researchers were able to show that either version of the duplicated gene from *Antirrhinum* influences flower development in genetically engineered *Antirrhinum* and *Arabidopsis* (Causier et al. 2005).

2 Closely related species can vary in what Darwin calls an analogous manner for essentially the same reason that the vestigial fifth stamen of *Antirrhinum* occasionally develops fully: the genetic architecture is there and nearly identical, so mutations with similar effects are to be expected. When he says that "important" characters would have stemmed from natural selection, he is referring to what we would term "derived" or "apomorphic" characters that are selected for within each lineage. Darwin is insightful in realizing that character similarity between related species could arise in two ways: reversion in these species to a common ancestral form, or new variations in a similar direction.

1 The discussion of varying characters in this paragraph is connected to an important point made earlier in the book, namely, that "doubtful forms" that are intermediate between species abound. What makes for an intermediate variety or form? Resembling one species in some characteristics and another species in other characteristics; in other words, forms that are mosaics of traits of the species they link. When Darwin says he has collected a long list of such examples, he is not exaggerating: "I will now give in small type such cases as I have collected," he wrote in *Natural Selection.* This statement is followed by no fewer than nine manuscript pages of plant, mammal, insect, bird, reptile, and crustacean examples (Stauffer 1975, pp. 324–328).

cases: if, for instance, we did not know that the rock-pigeon was not feather-footed or turn-crowned, we could not have told, whether these characters in our domestic breeds were reversions or only analogous variations; but we might have inferred that the blueness was a case of reversion, from the number of the markings, which are correlated with the blue tint, and which it does not appear probable would all appear together from simple variation. More especially we might have inferred this, from the blue colour and marks so often appearing when distinct breeds of diverse colours are crossed. Hence, though under nature it must generally be left doubtful, what cases are reversions to an anciently existing character, and what are new but analogous variations, yet we ought, on my theory, sometimes to find the varying offspring of a species assuming characters (either from reversion or from analogous variation) which already occur in some other members of the same group. And this undoubtedly is the case in nature.

A considerable part of the difficulty in recognising a variable species in our systematic works, is due to its varieties mocking, as it were, some of the other species of the same genus. A considerable catalogue, also, 1 could be given of forms intermediate between two other forms, which themselves must be doubtfully ranked as either varieties or species; and this shows, unless all these forms be considered as independently created species, that the one in varying has assumed some of the characters of the other, so as to produce the intermediate form. But the best evidence is afforded by parts or organs of an important and uniform nature occasionally varying so as to acquire, in some degree, the character of the same part or organ in an allied species. I have collected a long list of such cases; but here, as before, I lie under a great disadvantage in not being able to give them. I can only repeat that such cases certainly do occur, and seem to me very remarkable.

I will, however, give one curious and complex case, not indeed as affecting any important character, but from occurring in several species of the same genus, partly under domestication and partly under nature. It is a case apparently of reversion. The ass not rarely has very distinct transverse bars on its legs, like those on the legs of the zebra: it has been asserted that these are plainest in the foal, and from inquiries which I have made, I believe this to be true. It has also been asserted that the stripe on each shoulder is sometimes double. The shoulder-stripe is certainly very variable in length and outline. A white ass, but *not* an albino, has been described without either spinal or shoulder stripe; and these stripes are sometimes very obscure, or actually quite lost, in dark-coloured asses. The koulan of Pallas is said to have been seen with a double shoulder-stripe. The hemionus has no shoulder-stripe; but traces of it, as stated by Mr. Blyth and others, occasionally appear: and I have been informed by Colonel Poole that the foals of this species are generally striped on the legs, and faintly on the shoulder. The quagga, though so plainly barred like a zebra over the body, is without bars on the legs; but Dr. Gray has figured one specimen with very distinct zebra-like bars on the hocks.

With respect to the horse, I have collected cases in England of the spinal stripe in horses of the most distinct breeds, and of *all* colours; transverse bars on the legs are not rare in duns, mouse-duns, and in one instance in a chestnut: a faint shoulder-stripe may sometimes be seen in duns, and I have seen a trace in a

1 The example of the equines is also singled out in *Natural Selection.* The striping patterns of this group, consisting of the horse, donkey, quagga, hemionus, and zebra, make for more than a fascinating study in reversion and intergradation: Darwin sought to document as many cases as he could of usually unstriped members of the group—horses and donkeys—bearing stripes. The zebra is the most strikingly striped member of the group, but all the other species exhibit striping to varying degrees and in different locations on the body. Darwin's masterly presentation culminates on page 166.

2 These authorities all had experience with horses and their relatives. Peter Simon Pallas, who led two expeditions to far eastern Russia, named the hemionus or Onager (*Equus hemionus*), the wild ass of central Asia, and its subspecies the Khulan, the Mongolian wild ass. Edward Blyth, curator of the museum of the Asiatic Society of Bengal in Calcutta, India, and Skeffington Poole, army officer in India, both corresponded with Darwin regarding horses. John Edward Gray was a curator at the British Museum (Natural History); the illustration Darwin is referring to comes from Gray's *Gleanings from the Menagerie and Aviary at Knowsley Hall: Hoofed Quadrupeds* (Liverpool, 1850), a volume on Lord Edward Smith Stanley's celebrated menagerie.

1 Kattywar horses are named for the Indian district in which the breed was developed. W. W. Edwards is cited in both the *Origin* and *Variation of Animals and Plants under Domestication* as providing Darwin with information on racehorse coat-striping patterns.

2 Colonel Charles Hamilton Smith authored a volume on horses in Sir William Jardine's *Naturalist's Library* series (Edinburgh, 1841), the source of Darwin's information here.

bay horse. My son made a careful examination and sketch for me of a dun Belgian cart-horse with a double stripe on each shoulder and with leg-stripes; and a man, whom I can implicitly trust, has examined for me a small dun Welch pony with *three* short parallel stripes on each shoulder.

1 In the north-west part of India the Kattywar breed of horses is so generally striped, that, as I hear from Colonel Poole, who examined the breed for the Indian Government, a horse without stripes is not considered as purely-bred. The spine is always striped; the legs are generally barred; and the shoulder-stripe, which is sometimes double and sometimes treble, is common; the side of the face, moreover, is sometimes striped. The stripes are plainest in the foal; and sometimes quite disappear in old horses. Colonel Poole has seen both gray and bay Kattywar horses striped when first foaled. I have, also, reason to suspect, from information given me by Mr. W. W. Edwards, that with the English race-horse the spinal stripe is much commoner in the foal than in the full-grown animal. Without here entering on further details, I may state that I have collected cases of leg and shoulder stripes in horses of very different breeds, in various countries from Britain to Eastern China; and from Norway in the north to the Malay Archipelago in the south. In all parts of the world these stripes occur far oftenest in duns and mouse-duns; by the term dun a large range of colour is included, from one between brown and black to a close approach to cream-colour.

2 I am aware that Colonel Hamilton Smith, who has written on this subject, believes that the several breeds of the horse have descended from several aboriginal species—one of which, the dun, was striped; and that the above-described appearances are all due to ancient

crosses with the dun stock. But I am not at all satisfied with this theory, and should be loth to apply it to breeds so distinct as the heavy Belgian cart-horse, Welch ponies, cobs, the lanky Kattywar race, &c., inhabiting the most distant parts of the world.

Now let us turn to the effects of crossing the several species of the horse-genus. Rollin asserts, that the common mule from the ass and horse is particularly apt to have bars on its legs. I once saw a mule with its legs so much striped that any one at first would have thought that it must have been the product of a zebra; and Mr. W. C. Martin, in his excellent treatise [1] on the horse, has given a figure of a similar mule. In four coloured drawings, which I have seen, of hybrids between the ass and zebra, the legs were much more plainly barred than the rest of the body; and in one of them there was a double shoulder-stripe. In Lord [2] Moreton's famous hybrid from a chestnut mare and male quagga, the hybrid, and even the pure offspring subsequently produced from the mare by a black Arabian sire, were much more plainly barred across the legs than is even the pure quagga. Lastly, and this is another most remarkable case, a hybrid has been figured by Dr. Gray (and he informs me that he knows of a second case) from the ass and the hemionus; and this hybrid, though the ass seldom has stripes on its legs and the hemionus has none and has not even a shoulder-stripe, nevertheless had all four legs barred, and had three short shoulder-stripes, like those on the dun Welch pony, and even had some zebra-like stripes on the sides of its face. With respect to this last fact, I was so convinced that not even a stripe of colour appears from what would commonly be called an accident, that I was led solely from the occurrence of the face-stripes on this hybrid from the ass and hemionus,

1 William Charles Linnaeus Martin, curator of the museum of the Zoological Society of London, wrote several natural history books, including *The History of the Horse* (London, 1845).

2 George Douglas, the Sixteenth Earl of Morton ("Moreton" is a misspelling) sought to domesticate the quagga, now thought to be a subspecies of the plains zebra *Equus burchelli*, to save it from extinction. He was able to obtain a male but no female. Around 1815 he performed what he thought was the next-best mating: he paired the quagga with a mare. The offspring of the quagga-mare pairing bore quagga-like striping, unsurprisingly. But notice that Darwin says this was true even of "the pure offspring subsequently produced from the mare by a black Arabian sire." You should have done a double-take on reading that passage: how could the quagga continue to influence the offspring of later mates of the mare? This is the most famous example of the nineteenth-century notion of telegony, "offspring at a distance," though Darwin does not use the term here. In the latter part of that century telegony was a major area of research, and to Darwin, who was wedded to a model of blending inheritance, the idea was plausible: could some of the particles of inheritance transmitted with the sperm somehow remain in the female's body? If so, they could continue to influence subsequent offspring.

Ironically, Darwin's initial explanation for the striping pattern was correct: recency of common ancestry, and sharing a basic (genetic) predisposition to vary in similar ways; again, the genetic architecture between close relatives is nearly identical. Over time Darwin increasingly turned to quasi-Lamarckian ideas like telegony as a way to get enough variation for selection to act upon. Telegony was consistent with Darwin's theory of inheritance, pangenesis, but both were abandoned with the rediscovery of Mendel's work at the turn of the twentieth century. See Morton (1821) for the earl's original communication, and Burkhardt (1979) and Gould (1983) for accounts of telegony and the quagga saga in Darwin's thinking.

to ask Colonel Poole whether such face-stripes ever occur in the eminently striped Kattywar breed of horses, and was, as we have seen, answered in the affirmative.

1 What now are we to say to these several facts? We see several very distinct species of the horse-genus becoming, by simple variation, striped on the legs like a zebra, or striped on the shoulders like an ass. In the horse we see this tendency strong whenever a dun tint appears—a tint which approaches to that of the general colouring of the other species of the genus. The appearance of the stripes is not accompanied by any change of form or by any other new character. We see this tendency to become striped most strongly displayed in hybrids from between several of the most distinct species. 2 Now observe the case of the several breeds of pigeons: they are descended from a pigeon (including two or three sub-species or geographical races) of a bluish colour, with certain bars and other marks; and when any breed assumes by simple variation a bluish tint, these bars and other marks invariably reappear; but without any other change of form or character. When the oldest and truest breeds of various colours are crossed, we see a strong tendency for the blue tint and bars and marks to reappear in the mongrels. I have stated that the most probable hypothesis to account for the reappearance of very ancient characters, is—that there is a *tendency* in the young of each successive generation to produce the long-lost character, and that this tendency, from unknown causes, sometimes prevails. And we have just seen that in several species of the horse-genus the stripes are either plainer or appear more commonly in the young than in the old. Call the breeds of pigeons, some of which have bred true for centuries, species; and how exactly parallel is the case with that of the species of the horse-genus!

1 Darwin steps back now and briefly lists the "several facts" concerning striping patterns in equines; how do we make sense of it? Here we see clearly what he is driving at: the subtle and not-so-subtle striping patterns that appear in the equines reveal their close relationship. Recall the similar point made with pigeons on p. 25.

2 Darwin revisits the case of the pigeons here to make a remarkable and insightful point: the way pigeon varieties yield offspring that sometimes resemble the purported common ancestor, the rock pigeon, is exactly analogous to what is seen with the striping of equine crosses—but think of the implications: the pigeons in question are recognized as varieties, while the equines are recognized as good species. This underscores Darwin's point that varieties are essentially no different from species—that species are in fact but "well marked and permanent" varieties. Think of the equines as simply farther along a continuum of divergence than are the pigeons; nonetheless, in both groups the stamp of their shared ancestry is revealed when they are crossed.

For myself, I venture confidently to look back thousands on thousands of generations, and I see an animal striped like a zebra, but perhaps otherwise very differently constructed, the common parent of our domestic horse, whether or not it be descended from one or more wild stocks, of the ass, the hemionus, quagga, and zebra.

He who believes that each equine species was independently created, will, I presume, assert that each species has been created with a tendency to vary, both under nature and under domestication, in this particular manner, so as often to become striped like other species of the genus; and that each has been created with a strong tendency, when crossed with species inhabiting distant quarters of the world, to produce hybrids resembling in their stripes, not their own parents, but other species of the genus. To admit this view is, as it seems to me, to reject a real for an unreal, or at least for an unknown, cause. It makes the works of God a mere mockery and deception; I would almost as soon believe with the old and ignorant cosmogonists, that fossil shells had never lived, but had been created in stone so as to mock the shells now living on the sea-shore. 1

Summary.—Our ignorance of the laws of variation is profound. Not in one case out of a hundred can we pretend to assign any reason why this or that part differs, more or less, from the same part in the parents. But whenever we have the means of instituting a comparison, the same laws appear to have acted in producing the lesser differences between varieties of the same species, and the greater differences between species of the same genus. The external conditions of life, as climate and food, &c., seem to have induced some slight modifications. Habit in producing constitutional dif- 2

1 Why should the ordinarily unstriped members of this group occasionally "revert" to leg or shoulder striping if they are all separately and independently created? Why should hybrids sometimes resemble in this regard other species of the genus more than they do their own parents? To Darwin, the only explanation can be common descent. He seems contemptuous of the notion that pigeon varieties or equine species were simply created to vary in the ways they do, equating that with the old interpretation of fossils as forms specially created to echo the living organisms nearby. He is referring to a non-organic view of marine fossils that prevailed in Medieval times (though was not universally accepted).

The charge of making the works of God "a mere mockery and deception" is serious to Darwin—consider proposals like that of Philip Henry Gosse in his 1857 book *Omphalos,* which held that the creator had deliberately made the world with the false appearance of antiquity and history. The patterns of variation outlined in this chapter spoke clearly to Darwin: they are only explicable on the supposition of common descent. This process in itself inspired wonder in Darwin and some transmutationist colleagues; the deeply religious Asa Gray saw God's handiwork in transmutation. "If [the theist] cannot recognize design in Nature because of evolution," Gray wrote in *Darwiniana* (1876), "he may be ranked with those of whom it was said, 'Except ye see signs and wonders ye will not believe.'"

2 That ignorance of the factors controlling variation was profound in Darwin's day is an understatement; it is hard to imagine the puzzlement and frustration of not understanding reproduction, heredity, and variation. The long road to figuring this out started not with Mendel, whose important work was not "discovered" until the year 1900, but with Friedrich Miescher's 1871 isolation of a mysterious phosphorus-rich substance from cell nuclei: "I cannot help thinking that here lies the most essential physiological role of phosphorus in the organism."

ferences, and use in strengthening, and disuse in weakening and diminishing organs, seem to have been more potent in their effects. Homologous parts tend to vary in the same way, and homologous parts tend to cohere. Modifications in hard parts and in external parts some- [1] times affect softer and internal parts. When one part is largely developed, perhaps it tends to draw nourishment from the adjoining parts; and every part of the structure which can be saved without detriment to the individual, will be saved. Changes of structure at an early age will generally affect parts subsequently developed; and there are very many other correlations of growth, the nature of which we are utterly unable to understand. Multiple parts are variable in number and in structure, perhaps arising from such parts not having been closely specialised to any particular function, so that their modifications have not been closely checked by natural selection. It is probably from this same cause that organic beings low in the scale of nature are more variable than those which have their whole organisation more specialised, and are higher in the scale. Rudimentary organs, from being useless, will be disregarded by natural selection, and hence probably are variable. Specific characters—that is, the characters which have come to differ since the several species of the same genus branched off from a common parent—are more variable than generic characters, or those which have long been inherited, and have not differed [2] within this same period. In these remarks we have referred to special parts or organs being still variable, because they have recently varied and thus come to differ; but we have also seen in the second Chapter that the same principle applies to the whole individual; for in a district where many species of any genus are found—that is, where there has been much former

1 This is a restatement of the "law of compensation."

2 Again, modern biologists would disagree with Darwin's idea that especially well-developed traits vary to a greater degree than expected simply because they have been varying and undergoing selection in the past (as evidenced by their considerable development). The point is asserted again on the next page.

variation and differentiation, or where the manufactory of new specific forms has been actively at work—there, on an average, we now find most varieties or incipient species. Secondary sexual characters are highly variable, and such characters differ much in the species of the same group. Variability in the same parts of the organisation has generally been taken advantage of in giving secondary sexual differences to the sexes of the same species, and specific differences to the several species of the same genus. Any part or organ [1] developed to an extraordinary size or in an extraordinary manner, in comparison with the same part or organ in the allied species, must have gone through an extraordinary amount of modification since the genus arose; and thus we can understand why it should often still be variable in a much higher degree than other parts; for variation is a long-continued and slow process, and natural selection will in such cases not as yet have had time to overcome the tendency to further variability and to reversion to a less modified state. But when a species with any extraordinarily-developed organ has become the parent of many modified descendants —which on my view must be a very slow process, requiring a long lapse of time—in this case, natural selection may readily have succeeded in giving a fixed character to the organ, in however extraordinary a manner it may be developed. Species inheriting nearly the same constitution from a common parent and exposed to similar influences will naturally tend to present analogous variations, and these same species may occasionally revert to some of the characters of their ancient progenitors. Although new and important modifications may not arise from reversion and analogous variation, such modifications will add to the beautiful and harmonious diversity of nature.

1 Is it possible that large and well-developed structures simply *appear* to vary more than their under-endowed counterparts in other species simply because they are larger and more conspicuous?

In any case, real or not, Darwin thought he perceived a significant pattern in the way structures varied, and he held fast to this view, as evidenced by the fact that these passages are little changed in subsequent *Origin* editions. The idea occurred to Darwin early on—in fact, it would seem that it suddenly came to him on his geological excursion to Glen Roy in Scotland in the summer of 1838. Sandwiched between entries of geological observations we suddenly find this question posed in his Glen Roy notebook: "are those animals subject to much variation which have lately acquired their peculiarities?" (Barrett et al. 1987, p. 149). We see Darwin thinking in terms of high degrees of variation as an indicator that selection has been recently acting.

1 Darwin ends his summary with a return to a key point made throughout the book: all change—in this case important modifications of structure—stems from slow, steady action of natural selection. Struggle, variation, selection, and gradualism are all packed into this final sentence.

Whatever the cause may be of each slight difference in the offspring from their parents—and a cause for each must exist—it is the steady accumulation, through natural selection, of such differences, when beneficial to the individual, that gives rise to all the more important modifications of structure, by which the innumerable beings on the face of this earth are enabled to struggle with each other, and the best adapted to survive.

CHAPTER VI.

DIFFICULTIES ON THEORY.

Difficulties on the theory of descent with modification—Transitions—Absence or rarity of transitional varieties—Transitions in habits of life—Diversified habits in the same species—Species with habits widely different from those of their allies—Organs of extreme perfection—Means of transition—Cases of difficulty—Natura non facit saltum—Organs of small importance—Organs not in all cases absolutely perfect—The law of Unity of Type and of the Conditions of Existence embraced by the theory of Natural Selection.

LONG before having arrived at this part of my work, a ◀1 crowd of difficulties will have occurred to the reader. Some of them are so grave that to this day I can never reflect on them without being staggered; but, to the best of my judgment, the greater number are only apparent, and those that are real are not, I think, fatal to my theory.

These difficulties and objections may be classed under the following heads:—Firstly, why, if species have descended from other species by insensibly fine gradations, do we not everywhere see innumerable transitional forms? Why is not all nature in confusion instead of the species being, as we see them, well defined?

Secondly, is it possible that an animal having, for instance, the structure and habits of a bat, could have been formed by the modification of some animal with wholly different habits? Can we believe that natural selection could produce, on the one hand, organs of trifling importance, such as the tail of a giraffe, which serves as a fly-flapper, and, on the other hand, organs of

In this chapter and the next two, Darwin confronts four important sets of objections that his critics are likely to raise. He shrewdly addresses the most serious problems up front, defusing them rather than allowing them to linger and gnaw at the reader. Addressing them here also makes it more likely that a reader entertaining such objections will give the rest of the book a chance and read on. These "difficulties" chapters can be viewed as a transitional section, a pause in which to anticipate objections before forging ahead to explore applications of the theory and supporting evidence. This section thus punctuates the opening chapters, which establish the existence and adequacy of natural selection as a *vera causa,* and the remaining chapters, which explore applicability. Darwin's method in this and the next two chapters is to present problems and then offer responses, attempting to show that the problem is only apparent and not as serious as it first appears; or that the solution or explanation for the problem actually supports his theory. In fact, elements of this "problem–solution/response" pattern can be seen in nearly all the remaining chapters as Darwin explores diverse applications of his theory.

---

1 Darwin tips his hand in this opening sentence: he will attempt to show that the gravest problems merely appear to be problems. He describes them in the remainder of the section; they number four in all: absence of transitional varieties, modification of structures into wholly different uses, evolution of instinct by natural selection, and patterns in hybrid sterility. The first two will be addressed in this chapter, and the last two will each be tackled in chapters of their own. We will see Darwin turning adversity into virtue. Although he certainly appreciates the seriousness of the difficulties, he will strive to show how observed patterns in these areas are actually consistent with—and therefore support—his thesis rather than undermine it.

such wonderful structure, as the eye, of which we hardly as yet fully understand the inimitable perfection?

Thirdly, can instincts be acquired and modified through natural selection? What shall we say to so marvellous an instinct as that which leads the bee to make cells, which have practically anticipated the discoveries of profound mathematicians?

Fourthly, how can we account for species, when crossed, being sterile and producing sterile offspring, whereas, when varieties are crossed, their fertility is unimpaired?

The two first heads shall be here discussed—Instinct and Hybridism in separate chapters.

[1] *On the absence or rarity of transitional varieties.*—
[2] As natural selection acts solely by the preservation of profitable modifications, each new form will tend in a fully-stocked country to take the place of, and finally to exterminate, its own less improved parent or other less-favoured forms with which it comes into competition. Thus extinction and natural selection will, as we have seen, go hand in hand. Hence, if we look at each species as descended from some other unknown form, both the parent and all the transitional varieties will generally have been exterminated by the very process of formation and perfection of the new form.

But, as by this theory innumerable transitional forms must have existed, why do we not find them embedded in countless numbers in the crust of the earth? It will be much more convenient to discuss this question in the chapter on the Imperfection of the geological record; and I will here only state that I believe the answer mainly lies in the record being incomparably less perfect than is generally supposed; the imperfection of the record being chiefly due to organic beings not inhabiting

1 This first difficulty seems grave indeed: absence or rarity of transitional forms is a real problem for a theory that posits continuous transition!

2 There are two contexts for finding transitional forms. The first context is the record of life's history—the fossil record. If, as Darwin suggests, selection and extinction go hand-in-hand, should we not find a multitude of fossil transitional forms? Darwin takes up this topic later, in chapter IX; here he merely hints at his solution—namely, that while such forms must surely have existed, the stringent conditions favoring fossilization, combined with destructive processes like erosion, conspire to give us a woefully incomplete record.

profound depths of the sea, and to their remains being embedded and preserved to a future age only in masses of sediment sufficiently thick and extensive to withstand an enormous amount of future degradation; and such fossiliferous masses can be accumulated only where much sediment is deposited on the shallow bed of the sea, whilst it slowly subsides. These contingencies will concur only rarely, and after enormously long intervals. Whilst the bed of the sea is stationary or is rising, or when very little sediment is being deposited, there will be blanks in our geological history. The crust of the earth is a vast museum; but the natural collections have been made only at intervals of time immensely remote.

But it may be urged that when several closely-allied species inhabit the same territory we surely ought to find at the present time many transitional forms. Let us take a simple case: in travelling from north to south over a continent, we generally meet at successive intervals with closely allied or representative species, evidently filling nearly the same place in the natural economy of the land. These representative species often meet and interlock; and as the one becomes rarer and rarer, the other becomes more and more frequent, till the one replaces the other. But if we compare these species where they intermingle, they are generally as absolutely distinct from each other in every detail of structure as are specimens taken from the metropolis inhabited by each. By my theory these allied species have descended from a common parent; and during the process of modification, each has become adapted to the conditions of life of its own region, and has supplanted and exterminated its original parent and all the transitional varieties between its past and present states. Hence we ought not to expect at the

1 The earth's crust as "vast museum" is nice imagery; if the "collections" were made only haphazardly or at long intervals, the holdings must be incomplete.

2 The second context for finding transitional forms—intermediate or transitional forms among extant species—is considered in the remainder of this section. If the history of life has been a continuous process of slow, steady transition from one form to another, should not the living world consist primarily of forms in transition? Particularly in cases where several closely related species are found in the same area, transitional varieties between them should occur if they all descended from a common ancestor. We tend to see, in contrast, abrupt transitions between such species.

3 Darwin is careful to point out that a multitude of transitional varieties is not expected between living forms because the process of descent with modification involves offspring forms continuously supplanting parental forms. Recall from chapter IV the model of divergence of character. Selection acts to make successive descendant forms increasingly dissimilar, with offspring forms often outcompeting and driving to extinction the parental forms. Transitional varieties or species tend not to survive for this reason, but they might be found in the fossil record.

1 But what of geographically and climatically intermediate regions? Should not such transitional areas harbor transitional species or varieties? No, for two reasons. Darwin first suggests that the current lay of the land may not have any bearing on the former lay of the land, and areas that are now continuous may have been discontinuous in the past. This is a neat sidestep; if a continuous continental area is broken up into islands, the physically intermediate areas cease to exist (or exist only beneath the waves). Insularized, the species undergoes modification in two separate areas. When they come together again later following uplift and emergence of the linking land area from the sea, there is no transitional form linking these now-distinct forms.

2 Here Darwin does not sidestep the issue: he believes that the same abrupt transitions can occur even on land areas that were never insularized. This topic will come up again in connection with the importance of islands for Darwin's thinking. While island endemics provide a compelling demonstration of the effects of isolation and provided Darwin with an important clue that species are not immutable, he later downplayed the necessity of isolation for speciation (and therefore the importance of islands). By what mechanism, then, do species boundaries become sharpened in continuous land areas? Darwin posits several processes in operation, paramount among them competition.

present time to meet with numerous transitional varieties in each region, though they must have existed there, and may be embedded there in a fossil condition. ▶1 But in the intermediate region, having intermediate conditions of life, why do we not now find closely-linking intermediate varieties? This difficulty for a long time quite confounded me. But I think it can be in large part explained.

In the first place we should be extremely cautious in inferring, because an area is now continuous, that it has been continuous during a long period. Geology would lead us to believe that almost every continent has been broken up into islands even during the later tertiary periods; and in such islands distinct species might have been separately formed without the possibility of intermediate varieties existing in the intermediate zones. By changes in the form of the land and of climate, marine areas now continuous must often have existed within recent times in a far less continuous and uniform condition than at present. ▶2 But I will pass over this way of escaping from the difficulty; for I believe that many perfectly defined species have been formed on strictly continuous areas; though I do not doubt that the formerly broken condition of areas now continuous has played an important part in the formation of new species, more especially with freely-crossing and wandering animals.

In looking at species as they are now distributed over a wide area, we generally find them tolerably numerous over a large territory, then becoming somewhat abruptly rarer and rarer on the confines, and finally disappearing. Hence the neutral territory between two representative species is generally narrow in comparison with the territory proper to each. We see the same fact in ascending mountains, and sometimes it is quite remarkable how abruptly, as Alph. De Candolle has observed, a common alpine species disappears. The same fact has been noticed by Forbes in sounding the depths of the sea with the dredge. To those who look at climate and the physical conditions of life as the all-important elements of distribution, these facts ought to cause surprise, as climate and height or depth graduate away insensibly. But when we bear in mind that almost every species, even in its metropolis, would increase immensely in numbers, were it not for other competing species; that nearly all either prey on or serve as prey for others; in short, that each organic being is either directly or indirectly related in the most important manner to other organic beings, we must see that the range of the inhabitants of any country by no means exclusively depends on insensibly changing physical conditions, but in large part on the presence of other species, on which it depends, or by which it is destroyed, or with which it comes into competition; and as these species are already defined objects (however they may have become so), not blending one into another by insensible gradations, the range of any one species, depending as it does on the range of others, will tend to be sharply defined. Moreover, each species on the confines of its range, where it exists in lessened numbers, will, during fluctuations in the number of its enemies or of its prey, or in the seasons, be extremely liable to utter extermination; and thus its geographical range will come to be still more sharply defined.

If I am right in believing that allied or representative species, when inhabiting a continuous area, are generally so distributed that each has a wide range, with a comparatively narrow neutral territory between them, in which they become rather suddenly rarer and rarer; then, as varieties do not essentially differ from species,

1 The late eighteenth and early nineteenth centuries saw a great deal of research aimed at elucidating the relationship between plant distribution and physical factors such as climate. In *Prodromus Systematis Naturalis Regni Vegetabilis* (1824–1870), Alphonse de Candolle and his father, Augustin, discussed both latitudinal and altitudinal zones in great detail. They recognized three zones encountered with increasing elevation: montane (deciduous forest), subalpine (coniferous forest, grading into krummholz), and alpine (treeless). The transition zone to treelessness corresponds fairly closely to the 10° C mean daily isotherm for the warmest month of the year (Löve 1970). Regarding Edward Forbes, Darwin may be referring to dredging work undertaken in connection with Forbes's "azoic theory," which held that marine life is found only to a depth of 300 fathoms.

2 In this insightful passage Darwin argues that biotic processes of interaction and interdependence, for example, competition and predation, will tend to sharpen species boundaries.

3 Is Darwin setting the parameters to be favorable to his thesis? Does intermediate ("neutral") territory between the ranges of species tend to be comparatively more narrow than the main ranges themselves? If so, this observation dovetails nicely with his earlier assertion that range and population size affect variability and the potential for adaptability.

There may be validity to this observation, in that well-marked varieties are often found as geographical isolates peripheral to the main range of a species. Still, notice the caution in Darwin's wording: "practically, as far as I can make out," the rule holds good; "it would appear" from information given by the botanists that intermediate varieties are rare; "if we may trust these facts . . ." He is tentative because his sources are not convinced of the truth of the observation. Information on the relative low population numbers of intermediate varieties gleaned from Thomas Vernon Wollaston's book *On the Varieties of Species, with Especial Reference to the Insects* (London, 1856) was confirmed personally for Darwin. Turning to Hewett Cottrell Watson and Asa Gray, however, he found that while these botanists thought there may be something to the claim, they did not give a ringing endorsement. "Both these botanists concur in this opinion," Dar-

(continued)

the same rule will probably apply to both; and if we in imagination adapt a varying species to a very large area, we shall have to adapt two varieties to two large areas, and a third variety to a narrow intermediate zone. The intermediate variety, consequently, will exist in lesser numbers from inhabiting a narrow and lesser area; and practically, as far as I can make out, this rule holds good with varieties in a state of nature. I have met with striking instances of the rule in the case of varieties intermediate between well-marked varieties in the genus Balanus. And it would appear from information given me by Mr. Watson, Dr. Asa Gray, and Mr. Wollaston, that generally when varieties intermediate between two other forms occur, they are much rarer numerically than the forms which they connect. Now, if we may trust these facts and inferences, and therefore conclude that varieties linking two other varieties together have generally existed in lesser numbers than the forms which they connect, then, I think, we can understand why intermediate varieties should not endure for very long periods;—why as a general rule they should be exterminated and disappear, sooner than the forms which they originally linked together

For any form existing in lesser numbers would, as already remarked, run a greater chance of being exterminated than one existing in large numbers; and in this particular case the intermediate form would be eminently liable to the inroads of closely allied forms existing on both sides of it. But a far more important consideration, as I believe, is that, during the process of further modification, by which two varieties are supposed on my theory to be converted and perfected into two distinct species, the two which exist in larger numbers from inhabiting larger areas, will have a great advantage over the intermediate variety, which exists

(continued)
win wrote in *Natural Selection*, "& Mr. Watson has given me a list of twelve nearly intermediate varieties found in Britain which are rarer than the forms, which they connect. But both these naturalists have insisted strongly on various sources of doubt in forming any decided judgment on this head" (Stauffer 1975, p. 268). So Darwin is circumspect; the rule holds good only "practically," as far as he can make out.

1 Insofar as intermediate territory between species ranges may be smaller in area than the ranges themselves, intermediate varieties adapted to that zone would likely be less abundant than the main populations. Recall from chapter IV how important population size can be: smaller numbers mean fewer novel variations arising for natural selection to act upon, which in turn mean a higher likelihood of being outcompeted and driven to extinction. As the intermediate species or variety is displaced by the dominant flanking species or varieties, the loss of the intermediate makes the transition from one to the other species or variety appear all the more abrupt.

in smaller numbers in a narrow and intermediate zone. For forms existing in larger numbers will always have a better chance, within any given period, of presenting further favourable variations for natural selection to seize on, than will the rarer forms which exist in lesser numbers. Hence, the more common forms, in the race for life, will tend to beat and supplant the less common forms, for these will be more slowly modified and improved. It is the same principle which, as I believe, accounts for the common species in each country, as shown in the second chapter, presenting on an average a greater number of well-marked varieties than do the rarer species. I may illustrate what I mean by supposing three varieties of sheep to be kept, one adapted to an extensive mountainous region; a second to a comparatively narrow, hilly tract; and a third to wide plains at the base; and that the inhabitants are all trying with equal steadiness and skill to improve their stocks by selection; the chances in this case will be strongly in favour of the great holders on the mountains or on the plains improving their breeds more quickly than the small holders on the intermediate narrow, hilly tract; and consequently the improved mountain or plain breed will soon take the place of the less improved hill breed; and thus the two breeds, which originally existed in greater numbers, will come into close contact with each other, without the interposition of the supplanted, intermediate hill-variety. ◀1

To sum up, I believe that species come to be tolerably well-defined objects, and do not at any one period present an inextricable chaos of varying and intermediate links: firstly, because new varieties are very slowly formed, for variation is a very slow process, and natural selection can do nothing until favourable variations chance to occur, and until a place in the natural ◀2

1 The sheep example illustrates Darwin's point, but again one might criticize the example as contrived: must the hilly piedmont always be more limited in area than flanking mountainous and plains areas? What if we imagined the three areas as equal in size, as indeed they roughly are in the southern Appalachian region of North America along a transect from mountains to piedmont to coastal plain?

2 This section ends with four reasons that species are fairly well defined, with marked boundaries, rather than intergraded into one another. Such intergradation would indeed make an "inextricable chaos" of organisms, for naturalists and everyone else. Notice, though, that Darwin must walk a fine line here. Earlier in the book he made much of species grading into one another by "insensible gradations" and of the prevalence of "doubtful forms." Is he being contradictory here in arguing that species are "tolerably well-defined objects?"

Time, and gradualism, may be the key to avoiding contradiction. This first of the four reasons touches on the slow and steady pace of change. If new variants arise slowly, as a function of local environment, population size, etc., and if the action of natural selection on them varies in strength and intensity, then in a given time and place relatively few species are likely to exhibit marked intermediate variants.

1 Here is the geological argument: periodic physical discontinuities in land areas sharpen species boundaries. Species or varieties intermediate in structure are replaced in each isolated subarea, accentuating the differences between species when these land areas are united again.

2 By force of numbers, the species or varieties occupying larger land areas outcompete those occupying smaller land areas, as intermediate zones are imagined to be. Take note of the phrase "accidental extermination" on the next page; by this Darwin means extinction by stochastic processes. Smaller, more geographically restricted populations are more susceptible to catastrophe.

polity of the country can be better filled by some modification of some one or more of its inhabitants. And such new places will depend on slow changes of climate, or on the occasional immigration of new inhabitants, and, probably, in a still more important degree, on some of the old inhabitants becoming slowly modified, with the new forms thus produced and the old ones acting and reacting on each other. So that, in any one region and at any one time, we ought only to see a few species presenting slight modifications of structure in some degree permanent; and this assuredly we do see.

1 Secondly, areas now continuous must often have existed within the recent period in isolated portions, in which many forms, more especially amongst the classes which unite for each birth and wander much, may have separately been rendered sufficiently distinct to rank as representative species. In this case, intermediate varieties between the several representative species and their common parent, must formerly have existed in each broken portion of the land, but these links will have been supplanted and exterminated during the process of natural selection, so that they will no longer exist in a living state.

2 Thirdly, when two or more varieties have been formed in different portions of a strictly continuous area, intermediate varieties will, it is probable, at first have been formed in the intermediate zones, but they will generally have had a short duration. For these intermediate varieties will, from reasons already assigned (namely from what we know of the actual distribution of closely allied or representative species, and likewise of acknowledged varieties), exist in the intermediate zones in lesser numbers than the varieties which they tend to connect. From this cause alone the interme-

diate varieties will be liable to accidental extermination; and during the process of further modification through natural selection, they will almost certainly be beaten and supplanted by the forms which they connect; for these from existing in greater numbers will, in the aggregate, present more variation, and thus be further improved through natural selection and gain further advantages.

Lastly, looking not to any one time, but to all time, if my theory be true, numberless intermediate varieties, linking most closely all the species of the same group together, must assuredly have existed; but the very process of natural selection constantly tends, as has been so often remarked, to exterminate the parent-forms and the intermediate links. Consequently evidence of their former existence could be found only amongst fossil remains, which are preserved, as we shall in a future chapter attempt to show, in an extremely imperfect and intermittent record.

On the origin and transitions of organic beings with peculiar habits and structure.—It has been asked by the opponents of such views as I hold, how, for instance, a land carnivorous animal could have been converted into one with aquatic habits; for how could the animal in its transitional state have subsisted? It would be easy to show that within the same group carnivorous animals exist having every intermediate grade between truly aquatic and strictly terrestrial habits; and as each exists by a struggle for life, it is clear that each is well adapted in its habits to its place in nature. Look at the Mustela vison of North America, which has webbed feet and which resembles an otter in its fur, short legs, and form of tail; during summer this animal dives for and preys on fish, but during the long winter

1 Coming full circle, Darwin returns to the point that nearly infinite intermediates must have existed over time, so we must look for them in the fossil record. A host of problems arise there, too, however, for many processes either limit fossilization or destroy fossils over time.

2 This important section tackles the problem of functionality of transitional forms. A common view in Darwin's day, and our own, is that organisms are so well adapted to their environment that any change would be for the worse. Even more problematically, these adaptations often seem to require full development to be functional. How can transitional forms, with presumably less-than-full development of an organ, be at all functional? What good is half a wing or half an eye? The latter point will be taken up later in this chapter. First, Darwin considers how organisms that seem transitional may have gotten that way. His strategy is to cite examples of single species that seem to bridge a gap, that is, exist in two worlds, then (page 180) to point to groups of species that exemplify a range of morphologies or structures, and finally (page 183) to present cases where individuals of a species seem to exhibit atypical habits that could be a starting point for divergence.

3 Mammals like the American mink *Mustela vison* (family Mustelidae) exemplify the first point: these sleek animals seem equally at home in water and on land. They alternately live the life of an otter and a skunk at different times of year.

it leaves the frozen waters, and preys like other polecats on mice and land animals. If a different case had been taken, and it had been asked how an insectivorous quadruped could possibly have been converted into a flying bat, the question would have been far more difficult, and I could have given no answer. Yet I think such difficulties have very little weight.

Here, as on other occasions, I lie under a heavy disadvantage, for out of the many striking cases which I have collected, I can give only one or two instances of transitional habits and structures in closely allied species of the same genus; and of diversified habits, either constant or occasional, in the same species. And it seems to me that nothing less than a long list of such cases is sufficient to lessen the difficulty in any particular case like that of the bat.

1 Look at the family of squirrels; here we have the finest gradation from animals with their tails only slightly flattened, and from others, as Sir J. Richardson has remarked, with the posterior part of their bodies rather wide and with the skin on their flanks rather full, to the so-called flying squirrels; and flying squirrels have their limbs and even the base of the tail united by a broad expanse of skin, which serves as a parachute and allows them to glide through the air to an astonishing distance from tree to tree. We cannot doubt that each structure is of use to each kind of squirrel in its own country, by enabling it to escape birds or beasts of prey, or to collect food more quickly, or, as there is reason to believe, by lessening the danger from occasional falls. But it does not follow from this fact that the structure of each squirrel is the best that it is possible to conceive under all natural conditions. Let the climate and vegetation change, let other competing rodents or new beasts of prey immigrate, or old ones

1 Squirrels provide an example of a group with a range of adaptations, from running and climbing to gliding. There are over forty species of flying squirrels, which as a group constitute a well-defined tribe of the squirrel family Sciuridae. The parachute-like skin expansion that they use for gliding is called the "patagium."

Sir John Richardson discussed flying squirrels in his treatise on the zoology of northern North America, *Fauna Boreali-Americana* (London, 1829–1837), including a new species from the Rocky Mountains that he named *Pteromys alpinus.* Darwin borrowed Captain Fitzroy's copies of the Richardson volumes for an extended period after returning from the *Beagle* voyage, judging from a letter from Fitzroy dated February 26, 1838. "My dear Darwin," he wrote, "not the slightest inconvenience was caused by your keeping Richardson, I assure you,—Had I wished to look at it—I would have written but it is not in my line" (*Correspondence* 2: 75).

become modified, and all analogy would lead us to believe that some at least of the squirrels would decrease in numbers or become exterminated, unless they also became modified and improved in structure in a corresponding manner. Therefore, I can see no difficulty, more especially under changing conditions of life, in the continued preservation of individuals with fuller and fuller flank-membranes, each modification being useful, each being propagated, until by the accumulated effects of this process of natural selection, a perfect so-called flying squirrel was produced.

Now look at the Galeopithecus or flying lemur, which formerly was falsely ranked amongst bats. It has an extremely wide flank-membrane, stretching from the corners of the jaw to the tail, and including the limbs and the elongated fingers: the flank-membrane is, also, furnished with an extensor muscle. Although no graduated links of structure, fitted for gliding through the air, now connect the Galeopithecus with the other Lemuridæ, yet I can see no difficulty in supposing that such links formerly existed, and that each had been formed by the same steps as in the case of the less perfectly gliding squirrels; and that each grade of structure had been useful to its possessor. Nor can I see any insuperable difficulty in further believing it possible that the membrane-connected fingers and fore-arm of the Galeopithecus might be greatly lengthened by natural selection; and this, as far as the organs of flight are concerned, would convert it into a bat. In bats which have the wing-membrane extended from the top of the shoulder to the tail, including the hind-legs, we perhaps see traces of an apparatus originally constructed for gliding through the air rather than for flight.

If about a dozen genera of birds had become extinct or were unknown, who would have ventured to have

1 Here Darwin asks his readers to imagine selection slowly modifying an ancestral squirrel to produce a flying (actually gliding) squirrel. But read on—notice that these squirrels themselves represent a starting point in an imagined transitional series.

2 The colugo, or flying lemur, is a step removed from the flying squirrels in having the additional adaptation of a specialized muscle for extending or retracting the patagium. These mammals are not primates like true lemurs, but are placed in their own order, the Dermoptera—"skin-winged" animals. They are closely allied with the bats, order Chiroptera, so though Darwin refers to their having been mistakenly classified with the bats, zoologists have come back around to the idea that the colugos are close bat relatives, if not a type of bat themselves.

The bats represent another point further along this series; in this group the arm bones and digits are greatly elongated such that the patagium does not simply link the arm and body but stretches between all digits. Unlike the squirrels and colugos, bats are fully flighted. Notice how Darwin ends this discussion: can we see hints of the ancestral gliding morphology in these flying mammals? Darwin is trying to get his readers to think in terms of transitions. Focused as his Victorian audience was on a highly modified and well-adapted endpoint like the bat, it might be difficult for them to buy the argument that selection could be responsible; but just fill in the gaps with somewhat simpler versions of the morphology and the series becomes apparent. Or so he hopes. Eventual discovery of transitional fossils makes the series more concrete; the oldest bat fossil yet found, for example, the early Eocene *Onychonycteris finneyi* recently discovered in a quarry in Wyoming, is fully flighted but lacks echolocation ability and exhibits limb proportions "intermediate between all other known bats and forelimb-dominated non-volant mammals" (Simmons et al. 2008).

3 Darwin again appeals to imagination here. This brief paragraph makes an interesting point: the diverse modifications and uses of wings in birds illustrate how the same organ can be tweaked more or less slightly to great effect. Had some of these groups gone extinct or been unknown, would people have been

(continued)

(continued)
able to imagine that such uses were possible? Take penguins, for example, a group that uses its wings in fin-like fashion to "fly" through the water with remarkable speed and agility. Yet these wings are useless for flying in air. Darwin seems to be trying to get the reader to push the envelope with imagination—if it might have been difficult to imagine a modification for a bird flying in water, what other modifications are out there that we have not imagined?

The "Micropterus of Eyton" is a reference to the flightless loggerhead, or steamer duck, which we met in the last chapter (p. 134). Thomas Campbell Eyton authored a volume on ducks in 1838: *A Monograph on the Anatidae, or Duck Tribe. Apteryx* is the genus of the flightless kiwi of New Zealand, family Apterygidae, named in 1813 by the Oxford naturalist George Shaw.

1 Following the line of thought of the previous paragraph, Darwin makes the same point with an inverted example: structures initially adapted for swimming later modified for flying. Imagine that flying fish—which are actually gliders—represent an early transitional state of a future lineage of fully flighted fish. Their anatomy would reveal their true taxonomic affinity, but it would be difficult to figure out how or why the group transitioned from an exclusively marine existence, their fins modified from a swimming role to a flying one. Flying fish (Exocoetidae) are primarily tropical, numbering around fifty species in eight genera. These fish can glide for tens to hundreds of meters by means of greatly expanded pectoral fins.

surmised that birds might have existed which used their wings solely as flappers, like the logger-headed duck (Micropterus of Eyton); as fins in the water and front legs on the land, like the penguin; as sails, like the ostrich; and functionally for no purpose, like the Apteryx. Yet the structure of each of these birds is good for it, under the conditions of life to which it is exposed, for each has to live by a struggle; but it is not necessarily the best possible under all possible conditions. It must not be inferred from these remarks that any of the grades of wing-structure here alluded to, which perhaps may all have resulted from disuse, indicate the natural steps by which birds have acquired their perfect power of flight; but they serve, at least, to show what diversified means of transition are possible.

▶1 Seeing that a few members of such water-breathing classes as the Crustacea and Mollusca are adapted to live on the land, and seeing that we have flying birds and mammals, flying insects of the most diversified types, and formerly had flying reptiles, it is conceivable that flying-fish, which now glide far through the air, slightly rising and turning by the aid of their fluttering fins, might have been modified into perfectly winged animals. If this had been effected, who would have ever imagined that in an early transitional state they had been inhabitants of the open ocean, and had used their incipient organs of flight exclusively, as far as we know, to escape being devoured by other fish?

When we see any structure highly perfected for any particular habit, as the wings of a bird for flight, we should bear in mind that animals displaying early transitional grades of the structure will seldom continue to exist to the present day, for they will have been supplanted by the very process of perfection through natural selection. Furthermore, we may conclude that transitional grades between structures fitted for very different habits of life will rarely have been developed at an early period in great numbers and under many subordinate forms. Thus, to return to our imaginary illustration of the flying-fish, it does not seem probable that fishes capable of true flight would have been developed under many subordinate forms, for taking prey of many kinds in many ways, on the land and in the water, until their organs of flight had come to a high stage of perfection, so as to have given them a decided advantage over other animals in the battle for life. Hence the chance of discovering species with transitional grades of structure in a fossil condition will always be less, from their having existed in lesser numbers, than in the case of species with fully developed structures.

I will now give two or three instances of diversified and of changed habits in the individuals of the same species. When either case occurs, it would be easy for natural selection to fit the animal, by some modification of its structure, for its changed habits, or exclusively for one of its several different habits. But it is difficult to tell, and immaterial for us, whether habits generally change first and structure afterwards; or whether slight modifications of structure lead to changed habits; both probably often change almost simultaneously. Of cases of changed habits it will suffice merely to allude to that of the many British insects which now feed on exotic plants, or exclusively on artificial substances. Of diversified habits innumerable instances could be given: I have often watched a tyrant flycatcher (Saurophagus sulphuratus) in South America, hovering over one spot and then proceeding to another, like a kestrel, and at other times standing stationary on the margin of water, and then dashing like a kingfisher at a fish. In our own country the larger titmouse (Parus major) may be

1 It is especially important for Darwin to examine individual-level diversity in habits, since it is at this level that selection must initially act. Divergence at the species level starts with individual variation. The bird examples may not seem especially inspired, but the bear example on the next page became rather famous (or infamous).

2 Which changes first, structure or habits? Darwin asks. He concludes that both probably change almost simultaneously. He treated this topic in a bit more detail in *Natural Selection*, with a point worth emphasizing: "In the case of no organic being can we pretend to conjecture through what exact lines of life its progenitors have passed. We may use our knowledge of the habits of existing animals as a guide to conjecture . . . in as much as it is probable that amongst the many living & greatly diversified descendents of some ancient & extinct form, some would retain the habits not greatly modified of their several progenitors at different stages of descent" (Stauffer 1975, p. 346).

Darwin sees that under a genealogical model of descent, while truly transitional forms lived long ago and might be sought in the fossil record, some aspects of the ancestral species should be manifest in descendant species.

3 The tyrant flycatchers, family Tyrannidae, constitute the largest bird family. Exclusively found in the New World, these insectivorous birds often hunt in a characteristic way, sitting watchfully on an exposed perch from which they suddenly sally forth to catch insects on the wing. The account Darwin gives of *Saurophagus sulphuratus* stems from direct experience. He reported this bird's un-tyrannid-like behavior in chapter III of *Voyage.*

seen climbing branches, almost like a creeper; it often, like a shrike, kills small birds by blows on the head; and I have many times seen and heard it hammering the seeds of the yew on a branch, and thus breaking [1] them like a nuthatch. In North America the black bear was seen by Hearne swimming for hours with widely open mouth, thus catching, like a whale, insects in the water. Even in so extreme a case as this, if the supply of insects were constant, and if better adapted competitors did not already exist in the country, I can see no difficulty in a race of bears being rendered, by natural selection, more and more aquatic in their structure and habits, with larger and larger mouths, till a creature was produced as monstrous as a whale.

As we sometimes see individuals of a species following habits widely different from those both of their own species and of the other species of the same genus, we might expect, on my theory, that such individuals would occasionally have given rise to new species, having anomalous habits, and with their structure either slightly or considerably modified from that of their proper type. And such instances do occur in nature. Can a more striking instance of adaptation be given than that of a woodpecker for climbing trees and for seizing insects in the chinks of the bark? Yet in North America there are woodpeckers which feed largely on fruit, and others with elongated wings which chase insects on the wing; [2] and on the plains of La Plata, where not a tree grows, there is a woodpecker, which in every essential part of its organisation, even in its colouring, in the harsh tone of its voice, and undulatory flight, told me plainly of its close blood-relationship to our common species; yet it is a woodpecker which never climbs a tree!

Petrels are the most aërial and oceanic of birds, yet in the quiet Sounds of Tierra del Fuego, the Puffinuria

1 The explorer Samuel Hearne compared swimming black bears to whales when he encountered them in lakes north of Churchill, Canada: "This was in the month of June, long before any fruit was ripe, for the want of which [the bears] fed entirely on water insects, which in some of the lakes we crossed that day were in astonishing multitudes. The method by which the bears catch those insects is by swimming with their mouths open, in the same manner as the whales do" (Hearne 1795, p. 370).

The bear and whale comparison became a sore point. When Darwin told Richard Owen that he dropped the example for the next *Origin* edition, Owen replied, "Oh have you, well I was more struck with this than any other passage; you little know of the remarkable & essential relationship between bears & whales." Darwin swallowed the line and restored the passage; far from believing in any such relationship, however, Owen scathingly wrote in his review of the *Origin:* "We look . . . in vain for any instance of hypothetical transmutation in Lamarck so gross as the one above cited" (Owen 1860, p. 518). In the remaining editions of the *Origin,* Darwin simply inserted a qualifier: ". . . catching, *almost* like a whale, insects in the water" (emphasis mine).

2 Darwin's example of the plains woodpeckers of South America also got him in trouble. The nearly treeless grassland through which the Río de la Plata flows is home to the pampas woodpecker, *Colaptes campestris.* Like the North American flicker, the pampas woodpecker is a ground dweller that uses its long beak to probe for worms and grubs. It can also peck and probe trees in usual woodpecker fashion, though, and Darwin was stung by naturalist William Henry Hudson's suggestion that he deliberately omitted this fact in the *Origin:* "I trust that Mr. Hudson is mistaken when he says that any one acquainted with the habits of this bird might be induced to believe that I 'had purposely wrested the truth in order to prove' my theory," Darwin wrote in an 1870 paper. He clarified the point in later editions of the *Origin.*

berardi, in its general habits, in its astonishing power of diving, its manner of swimming, and of flying when unwillingly it takes flight, would be mistaken by any one for an auk or grebe; nevertheless, it is essentially a petrel, but with many parts of its organisation profoundly modified. On the other hand, the acutest observer by examining the dead body of the water-ouzel would never have suspected its sub-aquatic habits; yet this anomalous member of the strictly terrestrial thrush family wholly subsists by diving,—grasping the stones with its feet and using its wings under water.

He who believes that each being has been created as we now see it, must occasionally have felt surprise when he has met with an animal having habits and structure not at all in agreement. What can be plainer than that the webbed feet of ducks and geese are formed for swimming? yet there are upland geese with webbed feet which rarely or never go near the water; and no one except Audubon has seen the frigate-bird, which has all its four toes webbed, alight on the surface of the sea. On the other hand, grebes and coots are eminently aquatic, although their toes are only bordered by membrane. What seems plainer than that the long toes of grallatores are formed for walking over swamps and floating plants, yet the water-hen is nearly as aquatic as the coot; and the landrail nearly as terrestrial as the quail or partridge. In such cases, and many others could be given, habits have changed without a corresponding change of structure. The webbed feet of the upland goose may be said to have become rudimentary in function, though not in structure. In the frigate-bird, the deeply-scooped membrane between the toes shows that structure has begun to change.

He who believes in separate and innumerable acts of creation will say, that in these cases it has pleased the

1 The water-ouzel, better known as the water-thrush or American dipper (*Cinclus mexicanus*, Cinclidae), is a songbird found along fast-flowing streams in western North America and Central America. These curious birds walk along stream beds under water by holding on to rocks, probing for insects and other tidbits; as perching birds with a penchant for the aquatic life, they presented a fascinating example of transitional habits to Darwin.

2 Having "habits and structure not at all in agreement" begs an explanation. The geese that never go near water yet sport webbed feet, like the upland or Magellanic goose (*Chloephaga picta*) of South America, provide a case in point. So do frigatebirds (*Fregata magnificens*), which Audubon painted in *Birds of America*. Audubon included an inset painting of the birds' feet, showing that they are half-webbed—perhaps underscoring that they are half-marine, or transitional (Weissmann 1998), roosting in trees and only rarely alighting on the ocean.

Anatomical anomalies arise when habits have changed but structure has remained essentially the same. But why? In the B notebook Darwin wrote: "There certainly appears attempt in each dominant structure to accommodate itself to as many situations as possible.—Why should we have in open country a ground parrot.—woodpecker—a desert. Kingfisher.—mountain tringas.—Upland goose.—water chionis water rat with land structures . . .—This is but . . . attempt at adaptation of each element" (Barrett et al. 1987, p. 185).

3 This paragraph echoes the appeal made on p. 167; asserting that the disconnect between structure and habit can be explained by the creator's pleasing to do so sidesteps and does not address the question. Why do upland geese have wings they don't use? "Because they were made that way" is no answer. By contrast, the observation is entirely consistent with descent with modification, driven by natural selection.

1 This section is important to Darwin's argument. Here he anticipates and confronts head-on one of the most difficult arguments against transmutation: the argument from design, based on structures of great complexity. Indeed, notice that Darwin immediately takes on the epitome of organic complexity, the eye. The theologian William Paley's (see p. 201) teleological argument for the existence of a creator is well known: he opens *Natural Theology* (1802) with the famous case of a watch found on a heath. The intricacy and complexity of the watch imply a watchmaker. He then turned to the eye as a case of great organic complexity: "Every observation which was made in our first chapter concerning the watch may be repeated with strict propriety concerning the eye," he wrote. Darwin is taking the bull by the horns in making a case for an evolutionary origin for the eye.

Creator to cause a being of one type to take the place of one of another type; but this seems to me only restating the fact in dignified language. He who believes in the struggle for existence and in the principle of natural selection, will acknowledge that every organic being is constantly endeavouring to increase in numbers; and that if any one being vary ever so little, either in habits or structure, and thus gain an advantage over some other inhabitant of the country, it will seize on the place of that inhabitant, however different it may be from its own place. Hence it will cause him no surprise that there should be geese and frigate-birds with webbed feet, either living on the dry land or most rarely alighting on the water; that there should be long-toed corncrakes living in meadows instead of in swamps; that there should be woodpeckers where not a tree grows; that there should be diving thrushes, and petrels with the habits of auks.

1 *Organs of extreme perfection and complication.*—To suppose that the eye, with all its inimitable contrivances for adjusting the focus to different distances, for admitting different amounts of light, and for the correction of spherical and chromatic aberration, could have been formed by natural selection, seems, I freely confess, absurd in the highest possible degree. Yet reason tells me, that if numerous gradations from a perfect and complex eye to one very imperfect and simple, each grade being useful to its possessor, can be shown to exist; if further, the eye does vary ever so slightly, and the variations be inherited, which is certainly the case; and if any variation or modification in the organ be ever useful to an animal under changing conditions of life, then the difficulty of believing that a perfect and complex eye could be formed by natural

selection, though insuperable by our imagination, can hardly be considered real. How a nerve comes to be sensitive to light, hardly concerns us more than how life itself first originated; but I may remark that several facts make me suspect that any sensitive nerve may be rendered sensitive to light, and likewise to those coarser vibrations of the air which produce sound.

In looking for the gradations by which an organ in any species has been perfected, we ought to look exclusively to its lineal ancestors; but this is scarcely ever possible, and we are forced in each case to look to species of the same group, that is to the collateral descendants from the same original parent-form, in order to see what gradations are possible, and for the chance of some gradations having been transmitted from the earlier stages of descent, in an unaltered or little altered condition. Amongst existing Vertebrata, we find but a small amount of gradation in the structure of the eye, and from fossil species we can learn nothing on this head. In this great class we should probably have to descend far beneath the lowest known fossiliferous stratum to discover the earlier stages, by which the eye has been perfected.

In the Articulata we can commence a series with an optic nerve merely coated with pigment, and without any other mechanism; and from this low stage, numerous gradations of structure, branching off in two fundamentally different lines, can be shown to exist, until we reach a moderately high stage of perfection. In certain crustaceans, for instance, there is a double cornea, the inner one divided into facets, within each of which there is a lens-shaped swelling. In other crustaceans the transparent cones which are coated by pigment, and which properly act only by excluding lateral pencils of light, are convex at their upper ends

1 This is an important point. Recall the divergence of character diagram from chapter IV, showing ancestral forms successively replaced by descendant forms as lineages diverge. By and large, structurally intermediate forms are ancestral, not extant, and so can be found only in the fossil record. Thus Darwin here says that to see gradations we should look to the lineal ancestors. Since the record of those ancestors is spotty, the next best thing is to look at close relatives all descended from the same common ancestor.

This insight underlies modern phylogenetic reconstruction, which recognizes that organisms are mosaics, in a sense, of traits that are derived relative to a common ancestor (apomorphies) and traits that were inherited from that ancestor (plesiomorphies). In modern cladistic analysis, identifying groups of derived traits shared with different sets of related species is central to reconstructing the branching order of the lineages making up that species group. It is all about comparing traits among "collateral relatives."

2 Applying this idea, Darwin points out that we cannot learn much from the vertebrates—all have pretty much identical eye structure. The arthropods, by contrast (Darwin's Articulata), are an ideal group in which to seek gradations in eye structure. Arthropods comprise far more species than do vertebrates. For example, Wilson and Perlman (2000) give an estimated figure of 46,500 vertebrate species worldwide, and a conservative 945,000 species for the dominant arthropod groups—insects, arachnids, and crustaceans. Moreover, arthropods have existed far longer on earth, are more widely distributed, and are typically many times more numerically abundant than vertebrates. Put all these together and the result is far better representation in the fossil record.

and must act by convergence; and at their lower ends there seems to be an imperfect vitreous substance. With these facts, here far too briefly and imperfectly given, which show that there is much graduated diversity in the eyes of living crustaceans, and bearing in mind how small the number of living animals is in proportion to those which have become extinct, I can see no very [1] great difficulty (not more than in the case of many other structures) in believing that natural selection has converted the simple apparatus of an optic nerve merely coated with pigment and invested by transparent membrane, into an optical instrument as perfect as is possessed by any member of the great Articulate class.

[2] He who will go thus far, if he find on finishing this treatise that large bodies of facts, otherwise inexplicable, can be explained by the theory of descent, ought not to hesitate to go further, and to admit that a structure even as perfect as the eye of an eagle might be formed by natural selection, although in this case he does not know any of the transitional grades. His reason ought to conquer his imagination; though I have felt the difficulty far too keenly to be surprised at any degree of hesitation in extending the principle of natural selection to such startling lengths.

[3] It is scarcely possible to avoid comparing the eye to a telescope. We know that this instrument has been perfected by the long-continued efforts of the highest human intellects; and we naturally infer that the eye has been formed by a somewhat analogous process. But may not this inference be presumptuous? Have we any right to assume that the Creator works by intellectual powers like those of man? If we must compare the eye to an optical instrument, we ought in imagination to take a thick layer of transparent tissue, with a nerve sensitive to light beneath, and then suppose every

1 Darwin's scenario in general is accepted today: looking at the constellation of photoreception systems and organs, a range of complexity is apparent. From simple photoreceptor molecules to eye spots, to pit or cup eyes, to camera-type eyes, eyes vary in complexity in terms of epithelial folding, focusing mechanisms, and cellular structures housing the photo-pigments. And yet, even vertebrate and invertebrate eyes appear to have a common genetic underpinning.

In recent decades, biologists studying eyes and eye evolution have been debating whether eyes have a single evolutionary origin or several. Salvini-Plawen and Mayr (1977) showed that eye spots alone have arisen independently between forty and sixty-five times. The compound eye type found in insects and certain crustaceans may also have arisen independently. This eye type either arose twice or was lost in one lineage but, intriguingly, its genetic architecture somehow remained and was later re-expressed (Oakley and Cunningham 2002, Oakley 2003). The loss and reappearance of expression underscore the important point that evolutionary independence is relative. Common underlying genetic architecture—which has been demonstrated for mammals, arthropods, and cephalopods (Gehring 1998, Tomarev et al. 1997)—can lead to *parallelism,* separate but similar evolutionary pathways stemming from underlying homologies. Eyes have thus arisen independently in various lineages, but not from scratch, thanks to their having inherited the same basic genetic architecture. See Land and Fernald (1992) and Nilsson (2005) for reviews of eyes and eye evolution.

2 In these paragraphs Darwin seems to be making a plea, knowing how difficult it will be for readers to accept a naturalistic explanation for such complex structures. Urging that "his reason ought to conquer his imagination" may seem odd for a scientific treatise; evolution by natural selection boggles the mind, and one must suspend the natural tendency toward disbelief.

3 Perhaps this is a psychological explanation for the tendency to assume the action of a creator when presented with examples of organic complexity. We are drawn to analogy, in this case with an instrument like a telescope. That instrument was made by design and effort so we naturally assume the same must be true of its organic counterpart.

part of this layer to be continually changing slowly in density, so as to separate into layers of different densities and thicknesses, placed at different distances from each other, and with the surfaces of each layer slowly changing in form. Further we must suppose that there is a power always intently watching each slight accidental alteration in the transparent layers; and carefully selecting each alteration which, under varied circumstances, may in any way, or in any degree, tend to produce a distincter image. We must suppose each new state of the instrument to be multiplied by the million; and each to be preserved till a better be produced, and then the old ones to be destroyed. In living bodies, variation will cause the slight alterations, generation will multiply them almost infinitely, and natural selection will pick out with unerring skill each improvement. Let this process go on for millions on millions of years; and during each year on millions of individuals of many kinds; and may we not believe that a living optical instrument might thus be formed as superior to one of glass, as the works of the Creator are to those of man?

If it could be demonstrated that any complex organ existed, which could not possibly have been formed by numerous, successive, slight modifications, my theory would absolutely break down. But I can find out no such case. No doubt many organs exist of which we do not know the transitional grades, more especially if we look to much-isolated species, round which, according to my theory, there has been much extinction. Or again, if we look to an organ common to all the members of a large class, for in this latter case the organ must have been first formed at an extremely remote period, since which all the many members of the class have been developed; and in order to discover the early transitional grades through which the organ has

1 Here is a grand vision (so to speak) for the evolution of the eye: the slow action of variation, struggle, and selection over the course of millions of generations. The last sentence of the paragraph is worth noting: implicit here is that natural selection is playing the role of the Creator.

2 This comment strikes at the heart of the modern "intelligent design" (ID) movement. The old saying goes that just one flying pig is proof that pigs can fly, and so too with Darwin's grand vision for descent with modification by natural selection: just one example of a complex organ that cannot be explained by gradual evolution undermines that explanation for all organs. ID proponents have revived the argument from design, maintaining that enzymes and their pathways are the molecular equivalent of the eye in putatively being "irreducibly complex" and cannot be explained by gradualism. This is Paley in modern guise, but so-called irreducible complexity is assertion rather than empirical conclusion. One strong indication that enzymes and enzymatic pathways evolve is the great variability in which the same reactions are catalyzed in different taxa; there is considerable convergence in basic enzyme structure and function, but the different starting points for such analogous enzymes are clear in their structure and genetics (Galperin et al. 1998).

passed, we should have to look to very ancient ancestral forms, long since become extinct.

1 We should be extremely cautious in concluding that an organ could not have been formed by transitional gradations of some kind. Numerous cases could be given amongst the lower animals of the same organ performing at the same time wholly distinct functions; thus the alimentary canal respires, digests, and excretes in the larva of the dragon-fly and in the fish Cobites. In the Hydra, the animal may be turned inside out, and the exterior surface will then digest and the stomach respire. In such cases natural selection might easily specialise, if any advantage were thus gained, a part or organ, which had performed two functions, for one function alone, and thus wholly change its nature by insensible steps. Two distinct organs sometimes perform simultaneously the same function in the same individual; to give one instance, there are fish with gills or branchiæ that breathe the air dissolved in the water, at the same time that they breathe free air in their swimbladders, this latter organ having a ductus pneumaticus for its supply, and being divided by highly vascular partitions. In these cases, one of the two organs might with ease be modified and perfected so as to perform all the work by itself, being aided during the process of modification by the other organ; and then this other organ might be modified for some other and quite distinct purpose, or be quite obliterated.

2 The illustration of the swimbladder in fishes is a good one, because it shows us clearly the highly important fact that an organ originally constructed for one purpose, namely flotation, may be converted into one for a wholly different purpose, namely respiration. The swimbladder has, also, been worked in as an accessory to the auditory organs of certain fish, or, for I do not know which

1 Don't be too quick to dismiss the notion that even complex organs could have evolved gradually, Darwin cautions. This important paragraph introduces two interrelated ideas about transitional stages. In the first, Darwin suggests that ancestral organisms may have been more generalized, with multiple functions performed by the same structure or organ. Natural selection may have acted over time to specialize function, even giving new functions, transitioning the structure or organ to new roles. The second idea is *co-optation,* in which a structure that evolved for one function is then modified to serve another function. Darwin realizes that co-optation (a modern term, not used by him) offers not only an explanation for how complex organs might arise (transitioning from other uses), but also potent evidence for evolution in underscoring the odd and often jury-rigged nature of complex organs. The contrived ways in which some structures are built argue against design, yet this is precisely what is expected if the structures arose in stepwise fashion through a process that can only modify or build on what came before.

2 The swimbladder-lung relationship is not as clear as Darwin believed. In his day, and for many years afterward, it was assumed that lungs evolved in vertebrates in the transition from a marine to a terrestrial environment, and the hydrostatic swimbladders of fish, being air-filled pouches, were therefore the likely ancestral structures that became modified into lungs. But lung evolution is more complex than this; ironically, some evidence suggests that lungs may be ancestral to swimbladders rather than the other way around. Other studies suggest that both swimbladders and lungs evolved from common gill pouch structures: respiratory swimbladders evolved from the dorsal portion of the pouch, and lungs (twice) evolved from the ventral portion (see Perry et al. 2001).

view is now generally held, a part of the auditory apparatus has been worked in as a complement to the swimbladder. All physiologists admit that the swimbladder is homologous, or "ideally similar," in position and structure with the lungs of the higher vertebrate animals: hence there seems to me to be no great difficulty in believing that natural selection has actually converted a swimbladder into a lung, or organ used exclusively for respiration.

I can, indeed, hardly doubt that all vertebrate animals having true lungs have descended by ordinary generation from an ancient prototype, of which we know nothing, furnished with a floating apparatus or swimbladder. We can thus, as I infer from Professor Owen's interesting description of these parts, understand the strange fact that every particle of food and drink which we swallow has to pass over the orifice of the trachea, with some risk of falling into the lungs, notwithstanding the beautiful contrivance by which the glottis is closed. In the higher Vertebrata the branchiæ have wholly disappeared—the slits on the sides of the neck and the loop-like course of the arteries still marking in the embryo their former position. But it is conceivable that the now utterly lost branchiæ might have been gradually worked in by natural selection for some quite distinct purpose: in the same manner as, on the view entertained by some naturalists that the branchiæ and dorsal scales of Annelids are homologous with the wings and wing-covers of insects, it is probable that organs which at a very ancient period served for respiration have been actually converted into organs of flight.

In considering transitions of organs, it is so important to bear in mind the probability of conversion from one function to another, that I will give one more instance. Pedunculated cirripedes have two minute folds of skin,

1 The swimbladder and lung may not be strictly homologous, but Darwin is correct that there is a close relationship between the two. Regardless of the relative order of evolution and of the precise ancestral relationship of these organs, Darwin's point remains valid: he maintains that selection can slowly and gradually modify structures that have a certain function into other structures with a rather different function.

Note, by the way, the curious phrase "ideally similar." This is a now-archaic use of the word "ideal," with roots in Plato's doctrine of ideas, or *eidos.*

2 Strange fact, indeed. What Darwin is getting at here is evidence for *poor* design. Good design is compatible with the idea of special creation as well as descent with modification, but poor design would seem to make sense only in terms of the latter. This "designed" feature of our bodies—using the same passageway for both breathing and conducting chewed food—seems like a bad idea given that it sets up the possibility of death by choking. Why wouldn't a creator simply make two completely separate passages and avoid that horrible possibility? But we might understand this in terms of transmutation and cooptation: this less-than-ideal design stems from evolution's tinkering, building structure and function on structure and function. Our anatomy bears the stamp of history, a message reiterated in Darwin's second example: the persistent slits of ancestral gill branchiae still evident in the embryos of "higher" vertebrates. To Darwin this is evolutionary baggage. To Owen and others wed to the archetype idea, such features simply represented baggage of a different sort: evidence that species are created as variations on an underlying archetypal theme.

3 Cirripedes (barnacles) were Darwin's group of choice for detailed study. Between 1851 and 1854 he published four volumes on fossil and living barnacles, two on the pedunculated (stalked) gooseneck barnacles, Lepadidae. The latter are mostly open-ocean barnacles that anchor themselves to flotsam by means of a thick flexible stalk that resembles a goose's head and neck. There is more to the common name than that, however: an odd notion developed in Medieval times held that these barnacles gave rise to barnacle geese (*Branta* spp.), supposedly owing to

(continued)

(continued)
similarities in coloration and the gooseneck appearance of the barnacles. The specific name of the Brent goose, *Branta bernicla,* refers to this barnacle origin. The name was conferred by Linnaeus in 1758—acknowledging the myth, not embracing it.

1 Here Darwin gives an interesting example of significant functional differences between homologous structures in a group of organisms he knew very well. The sessile barnacles (Balanidae, the common encrusting barnacles) respire by means of structures called branchiae. In the gooseneck barnacles these serve not for respiration but as egg-retaining devices; Darwin himself first described this device in one of his monographs, calling them *ovigerous frenae,* a term still in use today. Note that he believes the gooseneck barnacles are more primitive and the sessile barnacles more derived (evolutionarily advanced), an inference upheld by modern analysis based on DNA sequences (e.g., Harris et al. 2000). His point in the final sentence is worth noting: if we lacked gooseneck barnacles altogether and had only sessile species to study, it would have required a leap of imagination to see that their branchiae could have served another, quite different, function—namely, egg retention. The ovigerous frenae were co-opted into a respiratory role.

2 Imagine the amazement and wonder that the discovery of electric organs excited, the astonishing ability "to wield at will the artillery of the skies," as Sir Richard Owen so poetically put it. Here is Owen in his lecture on electric eels: "Extraordinary as are the modifications and appendages of the peripheral extremities of the nerves of smell, sight, and hearing, other nerves in fishes are subject to still stranger combinations, and constitute organs quite unknown in any other class of Vertebrate Animals; those, viz., which endow a fish with the wonderful property of accumulating, concentrating, and applying in its own behoof an imponderable agent of a purely physical nature, which gives it the power to communicate electric shocks" (Owen 1846, p. 212).

called by me the ovigerous frena, which serve, through the means of a sticky secretion, to retain the eggs until they are hatched within the sack. These cirripedes have no branchiæ, the whole surface of the body and sack, including the small frena, serving for respiration. ▶1 The Balanidæ or sessile cirripedes, on the other hand, have no ovigerous frena, the eggs lying loose at the bottom of the sack, in the well-enclosed shell; but they have large folded branchiæ. Now I think no one will dispute that the ovigerous frena in the one family are strictly homologous with the branchiæ of the other family; indeed, they graduate into each other. Therefore I do not doubt that little folds of skin, which originally served as ovigerous frena, but which, likewise, very slightly aided the act of respiration, have been gradually converted by natural selection into branchiæ, simply through an increase in their size and the obliteration of their adhesive glands. If all pedunculated cirripedes had become extinct, and they have already suffered far more extinction than have sessile cirripedes, who would ever have imagined that the branchiæ in this latter family had originally existed as organs for preventing the ova from being washed out of the sack?

Although we must be extremely cautious in concluding that any organ could not possibly have been produced by successive transitional gradations, yet, undoubtedly, grave cases of difficulty occur, some of which will be discussed in my future work.

One of the gravest is that of neuter insects, which are often very differently constructed from either the males or fertile females; but this case will be treated ▶2 of in the next chapter. The electric organs of fishes offer another case of special difficulty; it is impossible to conceive by what steps these wondrous organs have been produced; but, as Owen and others have remarked, their intimate structure closely resembles that of common muscle; and as it has lately been shown that Rays have an organ closely analogous to the electric apparatus, and yet do not, as Matteuchi asserts, discharge any electricity, we must own that we are far too ignorant to argue that no transition of any kind is possible.

The electric organs offer another and even more [1] serious difficulty; for they occur in only about a dozen fishes, of which several are widely remote in their affinities. Generally when the same organ appears in several members of the same class, especially if in members having very different habits of life, we may attribute its presence to inheritance from a common ancestor; and its absence in some of the members to its loss through disuse or natural selection. But if the electric organs had been inherited from one ancient progenitor thus provided, we might have expected that all electric fishes would have been specially related to each other. Nor does geology at all lead to the belief that formerly most fishes had electric organs, which most of their modified descendants have lost. The presence of luminous organs in a few insects, belong- [2] ing to different families and orders, offers a parallel case of difficulty. Other cases could be given; for instance in plants, the very curious contrivance of a mass of pollen-grains, borne on a foot-stalk with a sticky gland at the end, is the same in Orchis and Asclepias,—genera almost as remote as possible amongst flowering plants. In all these cases of two very distinct species furnished with apparently the same anomalous organ, it should be observed that, although the general appearance and function of the organ may be the same, yet some fundamental difference can generally be detected. I am inclined to believe that in nearly the same [3] way as two men have sometimes independently hit on

1 Darwin highlights in this paragraph the serious problem of complex organs occurring in different taxonomic groups. He puts his finger on the difference between sharing a structure or organ as a result of common ancestry and sharing through convergence. Notice that Darwin is implicitly using a parsimony-based method for distinguishing between these possibilities: if the electric organs of all fish species bearing such organs had been inherited from a common ancestor, all electric fish must be related. If the electric organ-bearing fish are distant taxonomic relatives, the common ancestor must have occurred more anciently. But if this is the case, we must suppose that the organ has been lost by many descendant groups, leaving just a few rays, eels, and others with these organs. It is more parsimonious to maintain that these organs have independently arisen in a few scattered groups than to assume they were lost independently in many groups. In a word, *convergence.*

2 So, too, with the other examples: a few flies, beetles, and other groups have organs of bioluminescence, and it is more parsimonious to conclude that this phenomenon arose a few times independently than that it was lost many times independently. The pollen packets Darwin mentions here are termed "pollinia." Pollinia are found in orchids (*Orchis,* family Orchidaceae) and milkweeds (*Asclepias,* family Asclepiaceae), such distantly related groups that this mode of pollen packaging must have arisen convergently. Modern ultrastructural and genetic analysis upholds the convergent nature of electric and bioluminescent organs and pollinia. Keeping in mind that structure bears the stamp of history, however similar these things appear superficially, they are often radically different at the molecular and genetic level, and this underscores their independent origin.

3 The analogy in the final sentence of this paragraph hits the nail on the head: just as two people might come up with the same invention independently, so too can natural selection act similarly on structures in unrelated species to effect the same result. The different starting points, however, reveal their convergent nature.

1 *Natura non facit saltum:* Nature makes no leap. This phrase, which appears seven times in the *Origin* (this is the first of them), would seem to resonate with Darwin's gradualist vision. "Nature makes no leap" is a reference to linking forms, smooth transitions between taxa. The expression dates to ancient Greece and Rome. References can be found in seventeenth-century physics and law, but the first clear application to a biological or taxonomic context appears to come from Linnaeus, who applied it in the context of taxonomic continuity in section 77 of *Philosophia Botanica* (1761): "Groups are to be diligently sought out. This is first and foremost what is required in Botany. Nature does not make leaps. All plants show an affinity with those around them, according to their geographical location."

Milne-Edwards's phrase comes from his 1851 *Introduction à la Zoologie Générale* (Paris), which Darwin cited in *Natural Selection.* Yet a third authority, Augustin François St. Hilaire, is invoked there too: "in the same spirit, a great Botanist says 'Nature, as we have seen a thousand times always proceeds by transitions'" (Stauffer 1975, p. 354).

2 By now we have seen Darwin pose this question many times. Why indeed should such patterns exist on the hypothesis of special creation? It is not that they are altogether inconsistent with creation, but more importantly that they represent yet another body of observation, very different from the others he has covered thus far, consistent with descent with modification. *Natura non facit saltum* is precisely what evolution by natural selection should yield, and this harmonizes with Darwin's earlier argument about "doubtful forms" that seem to blur species boundaries.

3 Seemingly trivial structures pose another kind of problem: their triviality would suggest that natural selection does not act on them, yet how, then, did they come to be? If selected for, such traits must be adaptive, a proposition that sometimes seems laughable. Is the horse's mane that important? The particular shape of an oak leaf, or a maple leaf? The greater petiole length in the latter compared with the former? This is a more serious problem than many contemporary evolutionists realize, since it can lead to the slippery slope of invoking an adaptive explanation for virtually anything.

the very same invention, so natural selection, working for the good of each being and taking advantage of analogous variations, has sometimes modified in very nearly the same manner two parts in two organic beings, which owe but little of their structure in common to inheritance from the same ancestor.

Although in many cases it is most difficult to conjecture by what transitions an organ could have arrived at its present state; yet, considering that the proportion of living and known forms to the extinct and unknown is very small, I have been astonished how rarely an organ can be named, towards which no transitional grade is known to lead. The truth of this remark is indeed shown by that old canon in natural history of ▶1 "Natura non facit saltum." We meet with this admission in the writings of almost every experienced naturalist; or, as Milne Edwards has well expressed it, nature is prodigal in variety, but niggard in innovation. ▶2 Why, on the theory of Creation, should this be so? Why should all the parts and organs of many independent beings, each supposed to have been separately created for its proper place in nature, be so invariably linked together by graduated steps? Why should not Nature have taken a leap from structure to structure? On the theory of natural selection, we can clearly understand why she should not; for natural selection can act only by taking advantage of slight successive variations; she can never take a leap, but must advance by the shortest and slowest steps.

▶3 *Organs of little apparent importance.*—As natural selection acts by life and death,—by the preservation of individuals with any favourable variation, and by the destruction of those with any unfavourable deviation of structure,—I have sometimes felt much difficulty in understanding the origin of simple parts, of which the importance does not seem sufficient to cause the preservation of successively varying individuals. I have sometimes felt as much difficulty, though of a very different kind, on this head, as in the case of an organ as perfect and complex as the eye.

In the first place, we are much too ignorant in regard to the whole economy of any one organic being, to say what slight modifications would be of importance or not. In a former chapter I have given instances of most trifling characters, such as the down on fruit and the colour of the flesh, which, from determining the attacks of insects or from being correlated with constitutional differences, might assuredly be acted on by natural selection. The tail of the giraffe looks like an artificially constructed fly-flapper; and it seems at first incredible that this could have been adapted for its present purpose by successive slight modifications, each better and better, for so trifling an object as driving away flies; yet we should pause before being too positive even in this case, for we know that the distribution and existence of cattle and other animals in South America absolutely depends on their power of resisting the attacks of insects: so that individuals which could by any means defend themselves from these small enemies, would be able to range into new pastures and thus gain a great advantage. It is not that the larger quadrupeds are actually destroyed (except in some rare cases) by the flies, but they are incessantly harassed and their strength reduced, so that they are more subject to disease, or not so well enabled in a coming dearth to search for food, or to escape from beasts of prey.

Organs now of trifling importance have probably in some cases been of high importance to an early progenitor, and, after having been slowly perfected at a

1 Still, as Darwin points out here, we often do not know how or why a trait may be important to the survival or reproductive success of the individual—something as silly-looking as the giraffe's tail could be highly adaptive in reducing the likelihood of the animal's being bitten by disease-carrying flies. Consider, too, that a trait could be used just once and fleetingly in an organism's life yet still be of great importance; for example, the egg tooth that hatchling birds use to break out of the shell.

2 Or consider that traits which are now rather unimportant were of much greater importance ancestrally.

former period, have been transmitted in nearly the same state, although now become of very slight use; and any actually injurious deviations in their structure will always have been checked by natural selection. Seeing how important an organ of locomotion the tail is in most aquatic animals, its general presence and use for many purposes in so many land animals, which in their lungs or modified swimbladders betray their aquatic origin, may perhaps be thus accounted for. A well-developed tail having been formed in an aquatic animal, it might subsequently come to be worked in for all sorts of purposes, as a fly-flapper, an organ of prehension, or as an aid in turning, as with the dog, though the aid must be slight, for the hare, with hardly any tail, can double quickly enough.

In the second place, we may sometimes attribute importance to characters which are really of very little importance, and which have originated from quite secondary causes, independently of natural selection. We should remember that climate, food, &c., probably have some little direct influence on the organisation; that characters reappear from the law of reversion; that correlation of growth will have had a most important influence in modifying various structures; and finally, that sexual selection will often have largely modified the external characters of animals having a will, to give one male an advantage in fighting with another or in charming the females. Moreover when a modification of structure has primarily arisen from the above or other unknown causes, it may at first have been of no advantage to the species, but may subsequently have been taken advantage of by the descendants of the species under new conditions of life and with newly acquired habits.

To give a few instances to illustrate these latter

1 Ancestral traits like tails or dentition can be modified in many different ways in descendent species, and in some of those cases the trait could be reduced to the point of seeming non-utility.

2 By "secondary causes" Darwin is referring to a trait that is influenced by other traits or by the environment and not directly shaped by natural selection. To complicate matters further, even traits that sometimes arise in this incidental manner may then be seized upon by selection and modified for various uses in descendant species. Note the difficulty in ascribing adaptedness and functionality to traits.

remarks. If green woodpeckers alone had existed, and we did not know that there were many black and pied kinds, I dare say that we should have thought that the green colour was a beautiful adaptation to hide this tree-frequenting bird from its enemies; and consequently that it was a character of importance and might have been acquired through natural selection; as it is, I have no doubt that the colour is due to some quite distinct cause, probably to sexual selection. A trailing bamboo in the Malay Archipelego climbs the loftiest trees by the aid of exquisitely constructed hooks clustered around the ends of the branches, and this contrivance, no doubt, is of the highest service to the plant; but as we see nearly similar hooks on many trees which are not climbers, the hooks on the bamboo may have arisen from unknown laws of growth, and have been subsequently taken advantage of by the plant undergoing further modification and becoming a climber. The naked skin on the head of a vulture is generally looked at as a direct adaptation for wallowing in putridity; and so it may be, or it may possibly be due to the direct action of putrid matter; but we should be very cautious in drawing any such inference, when we see that the skin on the head of the clean-feeding male turkey is likewise naked. The sutures in the skulls of young mammals have been advanced as a beautiful adaptation for aiding parturition, and no doubt they facilitate, or may be indispensable for this act; but as sutures occur in the skulls of young birds and reptiles, which have only to escape from a broken egg, we may infer that this structure has arisen from the laws of growth, and has been taken advantage of in the parturition of the higher animals.

We are profoundly ignorant of the causes producing slight and unimportant variations; and we are immedi-

1 The green woodpecker *Picus viridis* is widely distributed in Europe and western Asia. As the largest of the three woodpecker species that breed in Britain, it would have been well known to Darwin and many of his readers.

Darwin observed the trailing plant in the "great hot house" at Kew Gardens; in December 1860 Hooker wrote to tell him that he thought the plant was a palm, not a bamboo. Darwin duly made the correction in the third *Origin* edition (*Correspondence* 8: 522).

Vultures are familiar scavenging birds. The Old World species belong to the family Accipitridae, which also includes hawks and eagles, while those of the New World are placed in their own family, the Cathartidae. Many vultures of both families have featherless heads, perhaps an adaptive feature given that the way these birds feed would leave their head feathers matted and putrid. Some have speculated that carrion gore itself may make the feathers fall out, though this is unlikely since captive birds fed in a more tidy manner also lose their head feathers. Darwin's point here is caution in too quickly ascribing an adaptive function; he underscores our ignorance by pointing out that male turkeys also have naked heads, and yet they do not feed in the same manner.

Perhaps the most interesting example Darwin provides is that of vertebrate skull sutures. This is an excellent example of a trait that can become highly adaptive, though it likely arose for another reason entirely, or was simply incidental to development. In mammals with a narrow birth canal the sutures are critical in allowing the head to be squeezed through. Much later, after birth, the sutures close through ossification. This trait is highly adaptive, but Darwin points out that even birds and reptiles have skull sutures, yet because these organisms hatch from eggs, the sutures are irrelevant to the birth process. They are a consequence of the way the vertebrate skeleton develops, but in one lineage (the mammals) this trait becomes advantageous.

1 Continuing his argument that even seemingly trivial traits can be shaped by natural selection, Darwin here points out that color can be affected by selection through correlation with other traits. He does not mention camouflage or mimicry coloration here—clearly in those cases selection acts on coloration directly. Darwin does not identify the "good observer" mentioned in this passage, but in *Variation of Animals and Plants under Domestication,* he cited Johann Matthias Bechstein's 1801 *Naturgeschichte Deutschlands* as reporting that light-colored cattle are not favored in Thuringia because they are attacked by flies to a greater degree than darker-colored cattle; Darwin recorded reading this work in 1842 (Vorzimmer 1977). Later research on parasitic tsetse flies (*Glossina* spp.) in Africa showed that darker, not lighter, individuals are preferentially attacked—in fact, traps for these scourges are colored black or blue for this reason (see Green 1990, Kappmeier and Nevill 1999).

ately made conscious of this by reflecting on the differences in the breeds of our domesticated animals in different countries,—more especially in the less civilized countries where there has been but little artificial selection. Careful observers are convinced that a damp climate affects the growth of the hair, and that with the hair the horns are correlated. Mountain breeds always differ from lowland breeds; and a mountainous country would probably affect the hind limbs from exercising them more, and possibly even the form of the pelvis; and then by the law of homologous variation, the front limbs and even the head would probably be affected. The shape, also, of the pelvis might affect by pressure the shape of the head of the young in the womb. The laborious breathing necessary in high regions would, we have some reason to believe, increase the size of the chest; and again correlation would come into play. Animals kept by savages in different countries often have to struggle for their own subsistence, and would be exposed to a certain extent to natural selection, and individuals with slightly different constitutions would succeed best under different climates; and there is reason to believe that constitution and colour are correlated. ▶1 A good observer, also, states that in cattle susceptibility to the attacks of flies is correlated with colour, as is the liability to be poisoned by certain plants; so that colour would be thus subjected to the action of natural selection. But we are far too ignorant to speculate on the relative importance of the several known and unknown laws of variation; and I have here alluded to them only to show that, if we are unable to account for the characteristic differences of our domestic breeds, which nevertheless we generally admit to have arisen through ordinary generation, we ought not to lay too much stress on our ignorance of the precise cause of the slight analogous differences between species. I might have adduced for this same purpose the differences between the races of man, which are so strongly marked; I may add that some little light can apparently be thrown on the origin of these differences, chiefly through sexual selection of a particular kind, but without here entering on copious details my reasoning would appear frivolous.

The foregoing remarks lead me to say a few words on the protest lately made by some naturalists, against the utilitarian doctrine that every detail of structure has been produced for the good of its possessor. They believe that very many structures have been created for beauty in the eyes of man, or for mere variety. This doctrine, if true, would be absolutely fatal to my theory. Yet I fully admit that many structures are of no direct use to their possessors. Physical conditions probably have had some little effect on structure, quite independently of any good thus gained. Correlation of growth has no doubt played a most important part, and a useful modification of one part will often have entailed on other parts diversified changes of no direct use. So again characters which formerly were useful, or which formerly had arisen from correlation of growth, or from other unknown cause, may reappear from the law of reversion, though now of no direct use. The effects of sexual selection, when displayed in beauty to charm the females, can be called useful only in rather a forced sense. But by far the most important consideration is that the chief part of the organisation of every being is simply due to inheritance; and consequently, though each being assuredly is well fitted for its place in nature, many structures now have no direct relation to the habits of life of each species. Thus, we can hardly believe that the webbed feet of the upland

1 This is one of just two references to human origins in the *Origin*, a subject that Darwin avoided here for fear that his general argument would be ignored in the furor that any overt treatment of human evolution was likely to provoke. Notice the tentative wording: some *little* light can *apparently* be thrown on this subject. He is referring not to human origins per se but to the origin of human races.

2 There are two separate issues here, and both are commonly held prejudices that Darwin works to dispel. The first, addressed earlier in this chapter, is that each structure and detail exists for the good of its possessor. Recall that Darwin urged caution in ascribing a functional role for every trait: traits can arise or be influenced by correlation, or their importance now may differ from their ancestral importance. The other issue is more pernicious; Darwin must argue against the belief that beauty and variety in nature have been created for the pleasure of humans. In the days of natural theology it was a given that all things were created for a purpose relating to humans, some for sheer utility, others to offer moral lessons, and still others to please the senses. Darwin correctly notes, rather pointedly, that this idea is completely antithetical to his theory. Beginning with the fourth edition, Darwin expanded the discussion of beauty, arguing for its adaptive value to the species possessing the characteristic, not to human aesthetics.

3 Here is a point well worth repeating: inheritance, he argues, is the primary factor behind the "organization" or basic structure of organisms. This is a simple yet profoundly important idea. Common ancestry provides the basic template of body plan and physiology, a template that is modified in various ways but continues to bear the unmistakable signs of its ancestry. Why else, "though each being assuredly is well fitted for its place in nature," do we often find traits that seem to have no bearing at all on the habits, home, and behavior of organisms, like the still-webbed feet of upland geese?

1 The frigatebird (genus *Fregata,* family Fregatidae), too, has partially webbed feet, betraying its relationship to seagoing birds. Frigatebirds are oceanic, but unlike their close relatives the pelicans, they do not alight on water or swim. They have one of the largest wingspan-to-body-weight ratios of any bird, making them supreme aerialists capable of staying aloft for more than a week. The large gular, or throat pouch, of frigatebirds also reveals an affinity with pelicans, but in the case of the frigatebird the bright red pouch of the males is used in courtship signaling.

2 In the remainder of this section Darwin offers several lucid statements on what natural selection can and cannot do:

Natural selection cannot produce a trait that exists solely for the benefit of another species. This paragraph is a clear rejection of the idea that any structures exist to benefit other species, the key word being "solely," since mutualists can benefit one another. In the third sentence Darwin uses the word "annihilate" to describe the effect on his theory should any such structures be found to exist—strong language. The happy and harmonious natural theology view of nature might hold that a rattlesnake's rattle exists to give fair warning to prey, but this is antithetical to the way natural selection works. Under natural selection each individual is striving to maximize its reproductive output, and this means outcompeting other individuals to secure resources and mates. A hypothetical trait that induced its possessor to help another species at a cost to its own likelihood of survival or reproduction would quickly be selected against as other individuals not so behaving ended up with greater reproductive success. Darwin did consider how altruism could be favored; see chapter VII.

1 goose or of the frigate-bird are of special use to these birds; we cannot believe that the same bones in the arm of the monkey, in the fore leg of the horse, in the wing of the bat, and in the flipper of the seal, are of special use to these animals. We may safely attribute these structures to inheritance. But to the progenitor of the upland goose and of the frigate-bird, webbed feet no doubt were as useful as they now are to the most aquatic of existing birds. So we may believe that the progenitor of the seal had not a flipper, but a foot with five toes fitted for walking or grasping; and we may further venture to believe that the several bones in the limbs of the monkey, horse, and bat, which have been inherited from a common progenitor, were formerly of more special use to that progenitor, or its progenitors, than they now are to these animals having such widely diversified habits. Therefore we may infer that these several bones might have been acquired through natural selection, subjected formerly, as now, to the several laws of inheritance, reversion, correlation of growth, &c. Hence every detail of structure in every living creature (making some little allowance for the direct action of physical conditions) may be viewed, either as having been of special use to some ancestral form, or as being now of special use to the descendants of this form—either directly, or indirectly through the complex laws of growth.

2 Natural selection cannot possibly produce any modification in any one species exclusively for the good of another species; though throughout nature one species incessantly takes advantage of, and profits by, the structure of another. But natural selection can and does often produce structures for the direct injury of other species, as we see in the fang of the adder, and in the ovipositor of the ichneumon, by which its eggs are depo-

sited in the living bodies of other insects. If it could be proved that any part of the structure of any one species had been formed for the exclusive good of another species, it would annihilate my theory, for such could not have been produced through natural selection. Although many statements may be found in works on natural history to this effect, I cannot find even one which seems to me of any weight. It is admitted that the rattlesnake has a poison-fang for its own defence and for the destruction of its prey; but some authors suppose that at the same time this snake is furnished with a rattle for its own injury, namely, to warn its prey to escape. I would almost as soon believe that the cat curls the end of its tail when preparing to spring, in order to warn the doomed mouse. But I have not space here to enter on this and other such cases.

Natural selection will never produce in a being any- [1] thing injurious to itself, for natural selection acts solely by and for the good of each. No organ will be formed, as Paley has remarked, for the purpose of causing pain or for doing an injury to its possessor. If a fair balance be struck between the good and evil caused by each part, each will be found on the whole advantageous. After the lapse of time, under changing conditions of life, if any part comes to be injurious, it will be modified; or if it be not so, the being will become extinct, as myriads have become extinct.

Natural selection tends only to make each organic [2] being as perfect as, or slightly more perfect than, the other inhabitants of the same country with which it has to struggle for existence. And we see that this is the degree of perfection attained under nature. The endemic productions of New Zealand, for instance, are perfect one compared with another; but they are now rapidly yielding before the advancing legions of plants

1 *Natural selection will never produce a trait injurious to its possessor—it acts solely for the benefit of individuals.* Traits that do compromise fitness in some ways are balanced by fitness gains in other ways. A good example is the colorful plumage in males of many bird species. Their conspicuous coloration might put them at greater risk for predation, a fitness cost, but for this trait to be selected, the cost must be more than compensated by fitness payoffs in the form of access to mates. In this case natural and sexual selection balance each other.

As we have seen, Rev. William Paley (1743–1805) was an influential English theologian. His *View of the Evidences of Christianity* (1794) was required reading at Cambridge, and *Natural Theology* (originally published in 1802), in which the famous formulation of the teleological argument for the existence of God was developed, became a classic in its own time. It is here that we find the remark to which Darwin alludes: "We never discover a train of contrivance to bring about an evil purpose. No anatomist ever discovered a system of organization calculated to produce pain and disease; or, in explaining the parts of the human body, ever said, this is to irritate, this to inflame; this duct is to convey the gravel to the kidneys; this gland to secrete the humour which forms the gout; if by chance he come at a part of which he knows not the use, the most he can say is, that it is useless; no one ever suspects that it is put there to incommode, to annoy, or to torment" (Paley 1860, pp. 258–259).

2 *Adaptedness is relative to other organisms in the same environment.* Natural selection acts on all organisms in their common environment, which means that degree of adaptedness of any one is relative to the time, place, and field of players (other individuals of the same and different species). This echoes a point made earlier in the book, that adaptation is relative, which is why exquisitely adapted island endemics nonetheless are readily outcompeted and displaced by invading species from continental areas.

1 *Natural selection cannot produce absolute perfection—indeed, even examples of exquisitely adapted structures have their flaws.* Even that icon of organic complexity, the eye, has its problems. Its optical properties are not perfect. Darwin mentions the bee's sting with its backward-directed barbs as a less-than-perfect design in that the act of stinging eviscerates the insect, killing it. This is not a good example for a couple of reasons, one of which Darwin understood and one of which he did not. He understood, as he touches on in the following paragraph and treats more fully in the next chapter, that selection might act at the level of the family to favor the evolution of traits injurious to the individual but highly beneficial to the individual's kin. So the eviscerated bee is giving its life in the defense of the family-colony. What Darwin did not understand about this process is why evisceration might be beneficial. In this case, the sting is left behind with the poison sac (with autonomous muscle that continues to contract and pump venom into the victim) as well as a gland that produces a chemical marker that other defender bees can zero in on to continue the attack. This is how agitated bees are able to chase their fleeing victims —not so much by sight, but by following the scent. One might argue that Darwin is still correct, that surely there is a better way to impart pain and a chemical tag short of suicide.

2 The sting of bees, wasps, and ants is thought to be a modified egg-laying structure, or ovipositor. This is why only females are sting-bearing. As an aside from the annals of natural history, Aristotle in his *Historia Animalia* argued that stinging bees must be male, since nature would not provide weaponry to females of any species—ironic in that *only* the females bear stings!

3 Still on the topic of a want of perfection, Darwin leaves the reader with a profound observation here. Nature has much imperfection, and if we admire the perfection we must at least admit or acknowledge the imperfection. The seeming waste and inefficiency of some natural processes are staggering to Darwin: the production in bee societies of thousands of near-useless drones, virtually all of which will end up slaughtered, or the considerable inefficiency of wind pollination, with untold billions of pollen grains produced so that only the tiniest, slender-

(continued)

1 and animals introduced from Europe. Natural selection will not produce absolute perfection, nor do we always meet, as far as we can judge, with this high standard under nature. The correction for the aberration of light is said, on high authority, not to be perfect even in that most perfect organ, the eye. If our reason leads us to admire with enthusiasm a multitude of inimitable contrivances in nature, this same reason tells us, though we may easily err on both sides, that some other contrivances are less perfect. Can we consider the sting of the wasp or of the bee as perfect, which, when used against many attacking animals, cannot be withdrawn, owing to the backward serratures, and so inevitably causes the death of the insect by tearing out its viscera?

2 If we look at the sting of the bee, as having originally existed in a remote progenitor as a boring and serrated instrument, like that in so many members of the same great order, and which has been modified but not perfected for its present purpose, with the poison originally adapted to cause galls subsequently intensified, we can perhaps understand how it is that the use of the sting should so often cause the insect's own death: for if on the whole the power of stinging be useful to the community, it will fulfil all the requirements of natural selection, though it may cause the
3 death of some few members. If we admire the truly wonderful power of scent by which the males of many insects find their females, can we admire the production for this single purpose of thousands of drones, which are utterly useless to the community for any other end, and which are ultimately slaughtered by their industrious and sterile sisters? It may be difficult, but we ought to admire the savage instinctive hatred of the queen-bee, which urges her instantly to destroy the

young queens her daughters as soon as born, or to perish herself in the combat; for undoubtedly this is for the good of the community; and maternal love or maternal hatred, though the latter fortunately is most rare, is all the same to the inexorable principle of natural selection. If we admire the several ingenious contrivances, by which the flowers of the orchis and of many other plants are fertilised through insect agency, can we consider as equally perfect the elaboration by our fir-trees of dense clouds of pollen, in order that a few granules may be wafted by a chance breeze on to the ovules?

Summary of Chapter.—We have in this chapter discussed some of the difficulties and objections which may be urged against my theory. Many of them are very grave; but I think that in the discussion light has been thrown on several facts, which on the theory of independent acts of creation are utterly obscure. We have seen that species at any one period are not indefinitely variable, and are not linked together by a multitude of intermediate gradations, partly because the process of natural selection will always be very slow, and will act, at any one time, only on a very few forms; and partly because the very process of natural selection almost implies the continual supplanting and extinction of preceding and intermediate gradations. Closely allied species, now living on a continuous area, must often have been formed when the area was not continuous, and when the conditions of life did not insensibly graduate away from one part to another. When two varieties are formed in two districts of a continuous area, an intermediate variety will often be formed, fitted for an intermediate zone; but from reasons assigned, the intermediate variety will usually exist in lesser numbers than

(continued)

est number ever achieve pollination, and then only by chance breezes. There are many other examples, and these undermine the idea of divine beneficence—for instance, those bird species in which the first-hatched nestling kills its siblings by pushing them out of the nest. Limiting resources may not permit rearing more than one chick, but surely a beneficent creator could have come up with a less overtly cruel means of family planning in these species. To Darwin, natural selection is blind to cruelty; it is simply a natural process unfolding.

1 This summary is characteristically thorough and lucid, making commentaries nearly superfluous. Note his bold opening sentences.

the two forms which it connects; consequently the two latter, during the course of further modification, from existing in greater numbers, will have a great advantage over the less numerous intermediate variety, and will thus generally succeed in supplanting and exterminating it.

We have seen in this chapter how cautious we should be in concluding that the most different habits of life could not graduate into each other; that a bat, for instance, could not have been formed by natural selection from an animal which at first could only glide through the air.

We have seen that a species may under new conditions of life change its habits, or have diversified habits, with some habits very unlike those of its nearest congeners. Hence we can understand, bearing in mind that each organic being is trying to live wherever it can live, how it has arisen that there are upland geese with webbed feet, ground woodpeckers, diving thrushes, and petrels with the habits of auks.

Although the belief that an organ so perfect as the eye could have been formed by natural selection, is more than enough to stagger any one; yet in the case of any organ, if we know of a long series of gradations in complexity, each good for its possessor, then, under changing conditions of life, there is no logical impossibility in the acquirement of any conceivable degree of perfection through natural selection. In the cases in which we know of no intermediate or transitional states, we should be very cautious in concluding that none could have existed, for the homologies of many organs and their intermediate states show that wonderful metamorphoses in function are at least possible. For instance, a swim-bladder has apparently been converted into an air-breathing lung. The same organ having performed simultaneously very different functions, and then having been specialised for one function; and two very distinct organs having performed at the same time the same function, the one having been perfected whilst aided by the other, must often have largely facilitated transitions.

We are far too ignorant, in almost every case, to be enabled to assert that any part or organ is so unimportant for the welfare of a species, that modifications in its structure could not have been slowly accumulated by means of natural selection. But we may confidently believe that many modifications, wholly due to the laws of growth, and at first in no way advantageous to a species, have been subsequently taken advantage of by the still further modified descendants of this species. We may, also, believe that a part formerly of high importance has often been retained (as the tail of an aquatic animal by its terrestrial descendants), though it has become of such small importance that it could not, in its present state, have been acquired by natural selection,—a power which acts solely by the preservation of profitable variations in the struggle for life.

Natural selection will produce nothing in one species for the exclusive good or injury of another; though it [1] may well produce parts, organs, and excretions highly useful or even indispensable, or highly injurious to another species, but in all cases at the same time useful to the owner. Natural selection in each well-stocked country, must act chiefly through the competition of the inhabitants one with another, and consequently will produce perfection, or strength in the battle for life, only according to the standard of that country. Hence the inhabitants of one country, generally the smaller one, will often yield, as we see they do yield, to the inhabitants of another and generally larger country. For in

1 The "highly useful" excretions mentioned here likely refer to the honeydew of aphids and related insects like scales and treehoppers. Excess water secreted by these sap-sucking insects nearly always contains a bit of sugar from the plant's phloem, and this commodity forms the basis of an often-elaborate mutualism with ants and sometimes wasps. They harvest the honeydew for food and in exchange fiercely protect the aphids. Indeed, I suspect that many partners in this mutualism find the honeydew "indispensable."

There are many examples of excretions that are injurious to other species: a tremendous pharmacopoeia of defensive chemicals has been identified, some of the most toxic from the world of arthropods. The most deadly are produced by caterpillars, such as the Neotropical silkmoth caterpillar *Lonomia obliqua.* Its sharp spines can deliver a toxin that even in tiny doses induces severe hemorrhaging, hemolysis, and kidney failure. The rate of human fatality is comparable to that of rattlesnake bites, yet with a minute fraction of the venom typically injected in a snake bite. See Eisner (2003) for accounts of diverse forms of chemical defense in arthropods.

the larger country there will have existed more individuals, and more diversified forms, and the competition will have been severer, and thus the standard of perfection will have been rendered higher. Natural selection will not necessarily produce absolute perfection; nor, as far as we can judge by our limited faculties, can absolute perfection be everywhere found.

On the theory of natural selection we can clearly understand the full meaning of that old canon in natural history, "Natura non facit saltum." This canon, if we look only to the present inhabitants of the world, is not strictly correct, but if we include all those of past times, it must by my theory be strictly true.

1 It is generally acknowledged that all organic beings have been formed on two great laws—Unity of Type, and the Conditions of Existence. By unity of type is meant that fundamental agreement in structure, which we see in organic beings of the same class, and which is quite independent of their habits of life. On my theory, unity of type is explained by unity of descent. The expression of conditions of existence, so often insisted on by the illustrious Cuvier, is fully embraced by the principle of natural selection. For natural selection acts by either now adapting the varying parts of each being to its organic and inorganic conditions of life; or by having adapted them during long-past periods of time: the adaptations being aided in some cases by use and disuse, being slightly affected by the direct action of the external conditions of life, and being in all cases subjected to the several laws of growth. Hence, in fact, the law of the Conditions of Existence is the higher law; as it includes, through the inheritance of former adaptations, that of Unity of Type.

1 These "two great laws" refer to a debate between the two leading French anatomists who championed very different ways of interpreting animal structure. Their differences turned on whether structure stems from functional needs or developmental laws, and whether all animals are related in a series or are divided into fundamentally different groups with no relation among them. Georges Cuvier looked at structure and function in adaptive terms: structures like limbs are created for adaptation to "conditions of existence." Each animal is adapted to its environment, and its needs determine its structure. He also divided animals into four body plans, *embranchements,* with no relationship among them. In contrast, Étienne Geoffroy St. Hilaire's central principle was "unity of type" (what we call "homologies"). He held that all animals could be united in one great series because all were created on the same basic archetype.

Cuvier and Geoffroy St. Hilaire were initially friends and colleagues at the *Museum d'Histoire Naturelle* in Paris, but they later became bitter rivals over this issue. Their differences grew over time until the gauntlet was thrown at a meeting of the *Académie des Sciences* in 1830, when two other young biologists presented a paper asserting homologies between cephalopods and vertebrates. Over the course of the next several meetings of the *Académie* the two masters squared off in what has become known as the great Cuvier-Geoffroy Debate (Appel 1987).

Darwin's theory put the opposing ideas of these savants into a very different context, and at once explained both principles. Unity of type—indeed, universal relationship—stems from common descent. And the importance of "conditions of existence" is clear: organisms become well adapted to their environment through natural selection. Darwin concludes that the "law" of conditions of existence is higher than that of unity of type, since regardless of the basic structure inherited from distant ancestors, selection acts to adapt organisms according to the current demands of their environment (though the stamp of history is always present).

CHAPTER VII.

Instinct.

Instincts comparable with habits, but different in their origin—Instincts graduated — Aphides and ants — Instincts variable—Domestic instincts, their origin—Natural instincts of the cuckoo, ostrich, and parasitic bees — Slave-making ants — Hive-bee, its cell-making instinct—Difficulties on the theory of the Natural Selection of instincts—Neuter or sterile insects—Summary.

The subject of instinct might have been worked into the previous chapters; but I have thought that it would be more convenient to treat the subject separately, especially as so wonderful an instinct as that of the hive-bee making its cells will probably have occurred to many readers, as a difficulty sufficient to overthrow my whole theory. I must premise, that I have nothing to do with the origin of the primary mental powers, any more than I have with that of life itself. We are concerned only with the diversities of instinct and of the other mental qualities of animals within the same class.

I will not attempt any definition of instinct. It would be easy to show that several distinct mental actions are commonly embraced by this term; but every one understands what is meant, when it is said that instinct impels the cuckoo to migrate and to lay her eggs in other birds' nests. An action, which we ourselves should require experience to enable us to perform, when performed by an animal, more especially by a very young one, without any experience, and when performed by many individuals in the same way, without their knowing for what purpose it is performed, is usually said to be instinctive.

In this chapter Darwin attempts to show that instincts, however attained, are modified through natural selection. In presenting his argument, he offers some profound insights into the nature of behavior. Like morphology, instinctive behaviors vary considerably within and between species. And as with morphology, such traits can often be arrayed along a continuum, with graduated differences evident among related species. Behavioral patterns found in domesticated animals, too, are as instructive as the morphological traits of these groups and argue for the power of accumulative selection. Darwin also addresses behavioral versions of exquisitely complex structures —instincts like hive-building in bees that at first appear far too complex to have evolved, but which on closer inspection can be seen in the context of descent with modification. Along with the exquisite are also macabre instincts, cases that underscore Darwin's earlier argument about waste and cruelty: poor design, or adaptations so horrific that he could not accept that a benevolent creator would fashion them. The bottom line is that selection is simply a natural, morally neutral process; just as morphological traits evolve by natural selection, so too do instinctive behaviors.

---

1 If Darwin only knew just how wonderful honeybees are! The waggle-dance of these remarkable insects—a symbolic system of behavioral communication that conveys information on direction and relative distance of flower patches from the hive, and is perhaps the most complex form of communication known outside the world of humans—would not be discovered for another hundred years. The German ethologist Karl von Frisch shared the Nobel Prize in 1973 for this work (see von Frisch 1967, Riley et al. 2005).

2 Despite declaring that he will not attempt any definition of instinct, two sentences later Darwin gives us this rather concise definition. It is reasonably close to modern definitions, which typically conceive instinct as an innate motivation or impulse that does not stem from learning. We think of instinctive behaviors as genetically "hard-wired" or pre-programmed.

But I could show that none of these characters of [1] instinct are universal. A little dose, as Pierre Huber expresses it, of judgment or reason, often comes into play, even in animals very low in the scale of nature.

Frederick Cuvier and several of the older metaphysicians have compared instinct with habit. This comparison gives, I think, a remarkably accurate notion of the frame of mind under which an instinctive action is performed, but not of its origin. How unconsciously many habitual actions are performed, indeed not rarely in direct opposition to our conscious will! yet they may be modified by the will or reason. Habits easily become associated with other habits, and with certain periods of time and states of the body. When once acquired, they often remain constant throughout life. Several other points of resemblance between instincts and habits could be pointed out. As in repeating a well-known song, so in instincts, one action follows another by a sort of rhythm; if a person be interrupted in a song, or in repeating anything by rote, he is generally forced to go back to recover the habitual train of thought: so P. [2] Huber found it was with a caterpillar, which makes a very complicated hammock; for if he took a caterpillar which had completed its hammock up to, say, the sixth stage of construction, and put it into a hammock completed up only to the third stage, the caterpillar simply re-performed the fourth, fifth, and sixth stages of construction. If, however, a caterpillar were taken out of a hammock made up, for instance, to the third stage, and were put into one finished up to the sixth stage, so that much of its work was already done for it, far from feeling the benefit of this, it was much embarrassed, and, in order to complete its hammock, seemed forced to start from the third stage, where it had left off, and thus tried to complete the already finished work.

1 Jean Pierre Huber was a Swiss entomologist known primarily for his work on Hymenoptera (ants, bees, wasps, and relatives). His work on insect instincts is less well known. Darwin agreed with Huber's assessment that insects have "une petite dose de raison," and he argues in this chapter that even highly complex instincts can evolve by natural selection. Some behaviorists of Darwin's day disagreed strongly with this theory, most notably the French naturalist Jean Henri Fabre. In an 1880 letter to Fabre concerning some observations of a predacious wasp made by his grandfather Erasmus, Darwin restated his agreement with Huber: "I must believe, with Pierre Huber, that insects have 'une petite dose de raison'" (F. Darwin 1896, p. 397). The quotation comes from the final sentence of Huber's 1836 paper on the hammock caterpillar, which cites their "judgment," not reason: "Tout ceci confirme ce que j'ai toujours observé chez les insectes, c'est qu'à côté de l'instinct la nature leur a accordé à tous une petite dose de jugement." ("All this confirms what I have always observed among the insects, which is that, along with instinct, nature has given to them a small measure of judgment.")

2 The "hammock caterpillar" was misidentified as *Tinea harrisella* by Huber in the 1836 paper Darwin consulted. It is in fact the apple leaf miner *Lyonetia clerkella,* a widely distributed member of the micro-lepidopteran family Lyonetiidae. This species was originally named by Linnaeus as *Phalaena clerkella* in the tenth edition of *Systema Naturae* (1758). Like many members of this and related micro-lepidopterans, tiny *L. clerkella* caterpillars are miners, small enough to eat their way through the parenchyma layer inside apple leaves. Once ready to metamorphose, however, they emerge and construct a distinctive cocoon slung with silk lines like a hammock between the gently upcurved edges of the leaf. Huber's manipulative experiments with *L. clerkella* cocoon-building behavior, which he originally conducted as a doctoral student in 1812, may well be the earliest documented behavioral experiments—as opposed to observational studies, which were common—with any insect.

If we suppose any habitual action to become inherited—and I think it can be shown that this does sometimes happen—then the resemblance between what originally was a habit and an instinct becomes so close as not to be distinguished. If Mozart, instead of playing the pianoforte at three years old with wonderfully little practice, had played a tune with no practice at all, he might truly be said to have done so instinctively. But it would be the most serious error to suppose that the greater number of instincts have been acquired by habit in one generation, and then transmitted by inheritance to succeeding generations. It can be clearly shown that the most wonderful instincts with which we are acquainted, namely, those of the hive-bee and of many ants, could not possibly have been thus acquired.

It will be universally admitted that instincts are as important as corporeal structure for the welfare of each species, under its present conditions of life. Under changed conditions of life, it is at least possible that slight modifications of instinct might be profitable to a species; and if it can be shown that instincts do vary ever so little, then I can see no difficulty in natural selection preserving and continually accumulating variations of instinct to any extent that may be profitable. It is thus, as I believe, that all the most complex and wonderful instincts have originated. As modifications of corporeal structure arise from, and are increased by, use or habit, and are diminished or lost by disuse, so I do not doubt it has been with instincts. But I believe that the effects of habit are of quite subordinate importance to the effects of the natural selection of what may be called accidental variations of instincts;—that is of variations produced by the same unknown causes which produce slight deviations of bodily structure.

No complex instinct can possibly be produced through

1 Darwin is careful to argue that learned traits do not become inherited, just as acquired changes in structure or physiology cannot be inherited. So even prodigies like Mozart had to learn—albeit with an astonishing rate and facility. Truly instinctive behaviors would not require any learning at all. Note, however, Darwin's statement that acquired habits can *sometimes* become heritable. There is some truth to this: the phenomenon is now called genetic assimilation, or the Baldwin effect after the American biologist James Mark Baldwin, who first proposed it in 1896. It is noteworthy that genetic assimilation was nearly simultaneously proposed by Baldwin and other workers at a time when neo-Lamarckism was gaining prominence and before Mendel's work was discovered. In a 1953 review of the subject, George Gaylord Simpson wrote, "The Baldwin effect ostensibly provides a reconciliation between neo-Darwinism and neo-Lamarckism. To the extent that it may really occur, it provides a mechanism that is capable of making acquired characters hereditary—or of seeming to do so" (Simpson 1953, p. 110). In the Baldwin effect, beneficial phenotypic variants, whether morphological or behavioral, initially arise by plasticity. Being adaptive, any mutations that arise to produce the trait and render it heritable will be favored and spread through the population. In that way, an initially nonhereditary learned or plastic response can become hard-wired and hereditary over time.

2 This paragraph makes explicit the parallel between corporeal structure and incorporeal behavior as traits that can evolve. The parallel is complete: behaviors vary; even slight variations in behavior that are profitable to an animal will be preserved by natural selection; accumulative selection can give rise to even the most "complex and wonderful" behaviors; and while habit, use, and disuse can affect the expression of behaviors, they are secondary to natural selection.

natural selection, except by the slow and gradual accumulation of numerous, slight, yet profitable, variations. [1] Hence, as in the case of corporeal structures, we ought to find in nature, not the actual transitional gradations by which each complex instinct has been acquired—for these could be found only in the lineal ancestors of each species—but we ought to find in the collateral lines of descent some evidence of such gradations; or we ought at least to be able to show that gradations of some kind are possible; and this we certainly can do. I have been surprised to find, making allowance for the instincts of animals having been but little observed except in Europe and North America, and for no instinct being known amongst extinct species, how very generally gradations, leading to the most complex instincts, can be discovered. The canon of "Natura non facit saltum" applies with almost equal force to instincts as to bodily organs. Changes of instinct may sometimes be facilitated by the same species having different instincts at different periods of life, or at different seasons of the year, or when placed under different circumstances, &c.; in which case either one or the other instinct might be preserved by natural selection. And such instances of diversity of instinct in the same species can be shown to occur in nature.

Again as in the case of corporeal structure, and conformably with my theory, the instinct of each species is good for itself, but has never, as far as we can judge, [2] been produced for the exclusive good of others. One of the strongest instances of an animal apparently performing an action for the sole good of another, with which I am acquainted, is that of aphides voluntarily yielding their sweet excretion to ants: that they do so voluntarily, the following facts show. I removed all the ants from a group of about a dozen aphides on a dock-

1 The comparison continues: as with structure, behavioral intermediates between forms that descended from a common ancestor are to be found in the lineal ancestors. Behavior, however, does not fossilize well, a problem often also encountered with structures. What to do? Look to *collateral lines of descent*—to relatives—for insight into earlier expressions of the trait. In that way gradations of behavior can be reconstructed in much the same way as gradations of structure.

2 Ant-aphid interactions have been observed for centuries, but the nature of this relationship was probed relatively recently. Pierre Huber, whom we met on p. 208, was one of the first to report (in 1810) that extrusion of honeydew droplets by aphids is influenced by the presence of ants. Honeydew is a sugary aqueous solution excreted by the aphids as they imbibe sap from the plant on which they are feeding. When feeding, they seek to eliminate excess water and concentrate the sugar, but in this inefficient process some sugar is unavoidably lost. This becomes a boon for ants and other insects, which have evolved often elaborate tending behaviors to obtain the sugar; in many cases, the aphids can actually release droplets on demand when they are stroked by the ants—hence Darwin's comment that they "voluntarily [yield] their sweet excretion."

Darwin rightly sees this as mutualism, but for incorrect reasons. He suggests that the ants are doing the aphids a favor by removing the honeydew: it is their "convenience . . . to have it removed." Today we know that the ants provide defense for the aphids, and it is very much in the aphids' interest to pay off these myrmecine bodyguards with sugar water. The aphid species Darwin observed is not clear, but it could have been *Aphis rumicis* L., which specializes on dock plants *(Rumex).*

plant, and prevented their attendance during several hours. After this interval, I felt sure that the aphides would want to excrete. I watched them for some time through a lens, but not one excreted; I then tickled and stroked them with a hair in the same manner, as well as I could, as the ants do with their antennæ; but not one excreted. Afterwards I allowed an ant to visit them, and it immediately seemed, by its eager way of running about, to be well aware what a rich flock it had discovered; it then began to play with its antennæ on the abdomen first of one aphis and then of another; and each aphis, as soon as it felt the antennæ, immediately lifted up its abdomen and excreted a limpid drop of sweet juice, which was eagerly devoured by the ant. Even the quite young aphides behaved in this manner, showing that the action was instinctive, and not the result of experience. But as the excretion is extremely viscid, it is probably a convenience to the aphides to have it removed; and therefore probably the aphides do not instinctively excrete for the sole good of the ants. Although I do not believe that any animal in the world performs an action for the exclusive good of another of a distinct species, yet each species tries to take advantage of the instincts of others, as each takes advantage of the weaker bodily structure of others. So again, in some few cases, certain instincts cannot be considered as absolutely perfect; but as details on this and other such points are not indispensable, they may be here passed over.

As some degree of variation in instincts under a state of nature, and the inheritance of such variations, are indispensable for the action of natural selection, as many instances as possible ought to have been here given; but want of space prevents me. I can only assert, that instincts certainly do vary—for instance,

1 In *Natural Selection,* Darwin argued explicitly against the idea that species can possess traits for the exclusive good of other species: "As in nature selection can act only through the good of the individual, including both sexes, the young, & in social animals the community, no modification can be effected in it for the advantage of another species; & if in any organism structure formed exclusively to profit other species could be shown to exist, it would be fatal to our theory." He goes on to give examples, among them the case of the ants and aphids: "The aphis excretes a sweet fluid, highly useful to ants, & necessary, I presume, to those species which keep the root-feeding aphides in their subterranean nests; but must we infer from this, that aphides were created for the sake of the Ants?" (Stauffer 1975, p. 382).

1 Darwin opened his discussion of variable nesting habits on p. 69 of his *Natural Selection* manuscript with the statement, "We will now consider the variability of the nesting instinct." He then launched into ten manuscript pages of examples! His concluding sentence is characteristically understated: "I think sufficient facts have now been given to show that the nests of birds do sometimes vary" (Stauffer 1975, pp. 502–506).

2 It is a fascinating attribute of island animals that they have often lost defenses; in the case of birds, this includes a flight response at the approach of animals (like humans) larger than themselves. Darwin writes from experience; in chapter XVII of the *Voyage of the Beagle* he offers this account of the birds of the Galapagos:

> I will conclude my description of the natural history of these islands, by giving an account of the extreme tameness of the birds . . . This disposition is common to all the terrestrial species; namely, to the mocking-thrushes, the finches, wrens, tyrant-flycatchers, the dove, and carrion-buzzard. All of them are often approached sufficiently near to be killed with a switch, and sometimes, as I myself tried, with a cap or hat. A gun is here almost superfluous; for with the muzzle I pushed a hawk off the branch of a tree. One day, whilst lying down, a mocking-thrush alighted on the edge of a pitcher, made of the shell of a tortoise, which I held in my hand, and began very quietly to sip the water; it allowed me to lift it from the ground whilst seated on the vessel: I often tried, and very nearly succeeded, in catching these birds by their legs.

3 The point here is that behaviors and ineffable traits like "disposition" exhibit individual variation; such variation, particularly heritable variation, is necessary if these traits evolve by natural selection.

the migratory instinct, both in extent and direction, and in its total loss. [1] So it is with the nests of birds, which vary partly in dependence on the situations chosen, and on the nature and temperature of the country inhabited, but often from causes wholly unknown to us: Audubon has given several remarkable cases of differences in nests of the same species in the northern and southern United States. Fear of any particular enemy is certainly an instinctive quality, as may be seen in nestling birds, though it is strengthened by experience, and by the sight of fear of the same enemy in other animals. [2] But fear of man is slowly acquired, as I have elsewhere shown, by various animals inhabiting desert islands; and we may see an instance of this, even in England, in the greater wildness of all our large birds than of our small birds; for the large birds have been most persecuted by man. We may safely attribute the greater wildness of our large birds to this cause; for in uninhabited islands large birds are not more fearful than small; and the magpie, so wary in England, is tame in Norway, as is the hooded crow in Egypt.

[3] That the general disposition of individuals of the same species, born in a state of nature, is extremely diversified, can be shown by a multitude of facts. Several cases also, could be given, of occasional and strange habits in certain species, which might, if advantageous to the species, give rise, through natural selection, to quite new instincts. But I am well aware that these general statements, without facts given in detail, can produce but a feeble effect on the reader's mind. I can only repeat my assurance, that I do not speak without good evidence.

The possibility, or even probability, of inherited variations of instinct in a state of nature will be strengthened by briefly considering a few cases under domestication. We shall thus also be enabled to see the respective parts which habit and the selection of so-called accidental variations have played in modifying the mental qualities of our domestic animals. A number of curious and authentic instances could be given of the inheritance of all shades of disposition and tastes, and likewise of the oddest tricks, associated with certain frames of mind or periods of time. But let us look to the familiar case of the several breeds of dogs: it cannot be doubted that young pointers (I have myself seen a striking instance) will sometimes point and even back other dogs the very first time that they are taken out; retrieving is certainly in some degree inherited by retrievers; and a tendency to run round, instead of at, a flock of sheep, by shepherd-dogs. I cannot see that these actions, performed without experience by the young, and in nearly the same manner by each individual, performed with eager delight by each breed, and without the end being known,—for the young pointer can no more know that he points to aid his master, than the white butterfly knows why she lays her eggs on the leaf of the cabbage,—I cannot see that these actions differ essentially from true instincts. If we were to see one kind of wolf, when young and without any training, as soon as it scented its prey, stand motionless like a statue, and then slowly crawl forward with a peculiar gait; and another kind of wolf rushing round, instead of at, a herd of deer, and driving them to a distant point, we should assuredly call these actions instinctive. Domestic instincts, as they may be called, are certainly far less fixed or invariable than natural instincts; but they have been acted on by far less rigorous selection, and have been transmitted for an incomparably shorter period, under less fixed conditions of life.

How strongly these domestic instincts, habits, and dis-

1 What better example of a domesticated animal than dogs to illustrate behavioral variation in the service of humans? Darwin had an intimate knowledge of dogs, having grown up with hunting and herding dogs. Even while a student at Cambridge he kept a dog named Sappho that accompanied him on his beetling and shooting expeditions.

2 Darwin likely thinks that instincts, like structures, vary more in domesticated than in wild organisms (recall his argument from the opening paragraph of chapter I, that individuals of domesticated varieties vary more than do those of natural varieties). This could be why he comments that instincts of domesticated animals "are certainly far less fixed or invariable than natural instincts."

positions are inherited, and how curiously they become mingled, is well shown when different breeds of dogs are crossed. Thus it is known that a cross with a bull-dog has affected for many generations the courage and obstinacy of greyhounds; and a cross with a greyhound has given to a whole family of shepherd-dogs a tendency to hunt hares. These domestic instincts, when thus tested by crossing, resemble natural instincts, which in a like manner become curiously blended together, and for a long period exhibit traces of the instincts of either [1] parent: for example, Le Roy describes a dog, whose great-grandfather was a wolf, and this dog showed a trace of its wild parentage only in one way, by not coming in a straight line to his master when called.

Domestic instincts are sometimes spoken of as actions which have become inherited solely from long-continued and compulsory habit, but this, I think, is not true. [2] No one would ever have thought of teaching, or probably could have taught, the tumbler-pigeon to tumble,—an action which, as I have witnessed, is performed by young birds, that have never seen a pigeon tumble. We may believe that some one pigeon showed a slight tendency to this strange habit, and that the long-continued selection of the best individuals in successive generations made tumblers what they now are; and near Glasgow there are house-tumblers, as I hear from Mr. Brent, which cannot fly eighteen inches high without going head over heels. It may be doubted whether any one would have thought of training a dog to point, had not some one dog naturally shown a tendency in this line; and this is known occasionally to happen, as I once saw in a pure terrier. When the first tendency was once displayed, methodical selection and the inherited effects of compulsory training in each successive generation would soon complete the work; and unconscious

1 Darwin's source here is an edition of Charles Georges LeRoy's *Lettres Philosophiques sur l'intelligence et la perfectibilité des animaux, avec quelques letters sur l'homme,* originally published in 1764 and translated into English in 1870.

2 Tumblers are a most unusual breed of pigeon, characterized by a periodic tendency to lose orientation or equilibrium while flying, resulting in the birds' falling in a somersaulting or tumbling manner until they stabilize. The breed has evidently been around for several centuries at least. Genetic study suggests that a single autosomal gene combined with the action of one or more genetic modifiers is responsible for tumbling behavior (Entrikin and Erway 1972).

The "house tumbler" represents an extreme product of breeding, being scarcely capable of getting off the ground for tumbling: "To these extreme varieties," wrote William Tegetmeier in his 1869 treatise on pigeons, "belong the breeds known as House Tumblers—so named because they tumble in the house." By "house" he means pigeon-house or dovecote. Bernard Brent, who bred house tumblers, probably first met Darwin at a meeting of the Columbarian Society, a London pigeon club. Here is Darwin in a letter to his son William in 1855 about the eccentrics of the pigeon clubs: "I think I shall belong to [the Columbarian Society] where, I fancy, I shall meet a strange set of odd men.—Mr Brent was a very queer little fish; but I suppose Mamma told you about him; after dinner he handed me a clay pipe, saying 'here is your pipe' as if it was a matter of course that I [should] smoke.—Another odd little man (N.B. all Pigeons Fanciers are little men, I begin to think)" (*Correspondence* 5: 508).

selection is still at work, as each man tries to procure, without intending to improve the breed, dogs which will stand and hunt best. On the other hand, habit alone in some cases has sufficed; no animal is more difficult to tame than the young of the wild rabbit; scarcely any animal is tamer than the young of the tame rabbit; but I do not suppose that domestic rabbits have ever been selected for tameness; and I presume that we must attribute the whole of the inherited change from extreme wildness to extreme tameness, simply to habit and long-continued close confinement.

Natural instincts are lost under domestication: a remarkable instance of this is seen in those breeds of fowls which very rarely or never become "broody," that is, never wish to sit on their eggs. Familiarity alone prevents our seeing how universally and largely the minds of our domestic animals have been modified by domestication. It is scarcely possible to doubt that the love of man has become instinctive in the dog. All wolves, foxes, jackals, and species of the cat genus, when kept tame, are most eager to attack poultry, sheep, and pigs; and this tendency has been found incurable in dogs which have been brought home as puppies from countries, such as Tierra del Fuego and Australia, where the savages do not keep these domestic animals. How rarely, on the other hand, do our civilised dogs, even when quite young, require to be taught not to attack poultry, sheep, and pigs! No doubt they occasionally do make an attack, and are then beaten; and if not cured, they are destroyed; so that habit, with some degree of selection, has probably concurred in civilising by inheritance our dogs. On the other hand, young chickens have lost, wholly by habit, that fear of the dog and cat which no doubt was originally instinctive in them, in the same way as it is so plainly instinctive in

1 Darwin's point here is well taken: domesticated animals have been drastically altered not only morphologically but mentally. Tameness is perhaps the most obvious manifestation of this fact.

young pheasants, though reared under a hen. It is not that chickens have lost all fear, but fear only of dogs and cats, for if the hen gives the danger-chuckle, they will run (more especially young turkeys) from under her, and conceal themselves in the surrounding grass or thickets; and this is evidently done for the instinctive purpose of allowing, as we see in wild ground-birds, their mother to fly away. But this instinct retained by our chickens has become useless under domestication, for the mother-hen has almost lost by disuse the power of flight.

Hence, we may conclude, that domestic instincts have been acquired and natural instincts have been lost partly by habit, and partly by man selecting and accumulating during successive generations, peculiar mental habits and actions, which at first appeared from what we must in our ignorance call an accident. In some cases compulsory habit alone has sufficed to produce such inherited mental changes; in other cases compulsory habit has done nothing, and all has been the result of selection, pursued both methodically and unconsciously; but in most cases, probably, habit and selection have acted together.

1 We shall, perhaps, best understand how instincts in a state of nature have become modified by selection, by considering a few cases. I will select only three, out of the several which I shall have to discuss in my future work,—namely, the instinct which leads the cuckoo to lay her eggs in other birds' nests; the slave-making instinct of certain ants; and the comb-making power of the hive-bee: these two latter instincts have generally, and most justly, been ranked by naturalists as the most wonderful of all known instincts.

It is now commonly admitted that the more immediate and final cause of the cuckoo's instinct is, that

1 This marks the beginning of an important section. The object of the next eighteen pages is to make a case for the evolution of complex behaviors. Darwin's approach is to demonstrate variation in the expression of these behaviors, and (using his own suggestion of looking to "collateral relatives") to identify behavioral series among related species—a series that suggests an actual transition. The examples Darwin has chosen are striking, representing behavioral equivalents of highly complex structures (like the eye, discussed in the last chapter). Comb-building in honeybees in particular might have struck readers as a unique behavior, too complex to represent a point in a transitional series. The other two behaviors he discusses (cuckoo parasitism and so-called slave-making in ants) might have been less familiar to some of Darwin's readers, but they are no less wonderful in their complexity.

2 Note the use of the term "final cause" here. Different forms of "causality" were a hallmark of Aristotle's natural philosophy explaining how the world works (set forth in his *Physics, Metaphysics,* and *Posterior Analytics*). There are several levels of causality, but the final cause in Aristotelian terms is the ultimate purpose or reason behind a given phenomenon. The more modern take on final cause is that it is the ultimate mechanism behind the phenomenon. The Greek term for final cause was *telos,* root of the word "teleology." Causality language lasted well into the nineteenth century, and offshoot terms like *vera causa*—"true cause"—were its philosophical cousins.

she lays her eggs, not daily, but at intervals of two or three days; so that, if she were to make her own nest and sit on her own eggs, those first laid would have to be left for some time unincubated, or there would be eggs and young birds of different ages in the same nest. If this were the case, the process of laying and hatching might be inconveniently long, more especially as she has to migrate at a very early period; and the first hatched young would probably have to be fed by the male alone. But the American cuckoo is in this predicament; for she makes her own nest and has eggs and young successively hatched, all at the same time. It has been asserted that the American cuckoo occasionally lays her eggs in other birds' nests; but I hear on the high authority of Dr. Brewer, that this is a mistake. Nevertheless, I could give several instances of various birds which have been known occasionally to lay their eggs in other birds' nests. Now let us suppose that the ancient progenitor of our European cuckoo had the habits of the American cuckoo; but that occasionally she laid an egg in another bird's nest. If the old bird profited by this occasional habit, or if the young were made more vigorous by advantage having been taken of the mistaken maternal instinct of another bird, than by their own mother's care, encumbered as she can hardly fail to be by having eggs and young of different ages at the same time; then the old birds or the fostered young would gain an advantage. And analogy would lead me to believe, that the young thus reared would be apt to follow by inheritance the occasional and aberrant habit of their mother, and in their turn would be apt to lay their eggs in other birds' nests, and thus be successful in rearing their young. By a continued process of this nature, I believe that the strange instinct of our cuckoo could be, and has been,

1 The cuckoos, Cuculidae, consist of about five subfamilies. Most Old World species, including the common European cuckoo (*Culculus canorus*), are brood parasites that lay their eggs in the nests of other birds. The cuckoo young eventually kill the young of the host species by monopolizing the hapless parents' rearing efforts. This insidious lifestyle seemed to Darwin to argue against a beneficent creator, as we shall see, and has, incidentally, provided a potent metaphor in literature and poetry. In Act 5, Scene I, of Shakespeare's *Henry IV,* Part I, the Earl of Worcester says to Hal:

> And being fed by us you used us so
> As that ungentle hull, the cuckoo's bird,
> Useth the sparrow; did oppress our nest;
> Grew by our feeding to so great a bulk
> That even our love durst not come near your sight

A few New World species are parasites. Darwin refers to "the" American cuckoo here, but there are two common North American species: the yellow-billed cuckoo, *Coccyzus americanus,* and the black-billed, *C. erythropthalmus.* Of the two, only *C. erythropthalmus* occasionally parasitizes.

Note the gradualistic nature of Darwin's argument: the "occasional and aberrant habit" of a cuckoo progenitor depositing its eggs in the nests of other species could have conferred an advantage. If natural selection thus favored the behavior it would transition from occasional to common.

In early 1858 Darwin wrote to A. A. Gould (Louis Agassiz's coauthor on *Principles of Zoology*) for information about the egg-laying habits of North American cuckoos. Gould forwarded the query to the ornithologist Thomas M. Brewer, who replied (in error) that neither cuckoo parasitizes other birds: "[I know] of few birds more faithful apparently to their own young than both species of our cuckoos" (*Correspondence* 7:40). In this same letter Brewer noted that he once found a chipping sparrow nest with the egg of a black-billed cuckoo, but he concluded—mistakenly—that a person had placed it there.

1 generated. I may add that, according to Dr. Gray and to some other observers, the European cuckoo has not utterly lost all maternal love and care for her own offspring.

2 The occasional habit of birds laying their eggs in other birds' nests, either of the same or of a distinct species, is not very uncommon with the Gallinaceæ; and this perhaps explains the origin of a singular instinct in the allied group of ostriches. For several hen ostriches, at least in the case of the American species, unite and lay first a few eggs in one nest and then in another; and these are hatched by the males. This instinct may probably be accounted for by the fact of the hens laying a large number of eggs; but, as in the case of the cuckoo, at intervals of two or three days.
3 This instinct, however, of the American ostrich has not as yet been perfected; for a surprising number of eggs lie strewed over the plains, so that in one day's hunting I picked up no less than twenty lost and wasted eggs.

Many bees are parasitic, and always lay their eggs in the nests of bees of other kinds. This case is more re-
4 markable than that of the cuckoo; for these bees have not only their instincts but their structure modified in accordance with their parasitic habits; for they do not possess the pollen-collecting apparatus which would be necessary if they had to store food for their own young.
5 Some species, likewise, of Sphegidæ (wasp-like insects) are parasitic on other species; and M. Fabre has lately shown good reason for believing that although the Tachytes nigra generally makes its own burrow and stores it with paralysed prey for its own larvæ to feed on, yet that when this insect finds a burrow already made and stored by another sphex, it takes advantage of the prize, and becomes for the occasion parasitic. In this case, as with the supposed case of the cuckoo, I can

1 The Dr. Gray referred to here is not Asa but John Edward Gray, keeper of Zoology at the British Museum (Natural History).

2 Intra- as well as interspecific brood parasitism is actually rather widespread in birds. See MacWhirter (1989) for a brief review of the topic.

3 In chapter V of *Voyage of the Beagle,* Darwin wrote: "When we were at Bahia Blanca in the months of September and October, [ostrich] eggs, in extraordinary numbers, were found all over the country. They lie either scattered and single, in which case they are never hatched, and are called by the Spaniards huachos; or they are collected together into a shallow excavation, which forms the nest."

4 Darwin is quite right here: brood parasitism in ants, bees, and wasps is very common. Appropriately enough, one group of widespread brood-parasitizing wasps are known as the cuckoo wasps, family Chrysididae. These small, metallic blue-green wasps have an especially thick cuticle dotted all over with small concave depressions—both features apparently help these insects resist the stings of the bees and wasps they attempt to parasitize.

5 *Tachytes nigra* is an example of a predatory solitary wasp, which makes its living by excavating or building a nest, into which it places paralyzed prey. It lays an egg on the prey item, and when the young wasp larva hatches, it has fresh meat, so to speak. The specialized venom of these wasps paralyzes rather than kills the prey outright to keep the food from decaying. Darwin's point here is that this species has been known to opportunistically parasitize other wasps should it come upon a nest—as with the cuckoos, occasional opportunistic parasitism can set the stage for selection to favor obligate parasitism. Parasitism of the yellow-winged sphex by *T. nigra* was reported by the French naturalist Jean Henri Fabre in an 1856 paper, which Darwin read; Fabre later detailed these observations in one of his *Souvenirs Entomologiques* volumes, translated and published in *The Hunting Wasps* (New York, 1915).

see no difficulty in natural selection making an occasional habit permanent, if of advantage to the species, and if the insect whose nest and stored food are thus feloniously appropriated, be not thus exterminated.

Slave-making instinct.—This remarkable instinct was first discovered in the Formica (Polyerges) rufescens by Pierre Huber, a better observer even than his celebrated father. This ant is absolutely dependent on its slaves; without their aid, the species would certainly become extinct in a single year. The males and fertile females do no work. The workers or sterile females, though most energetic and courageous in capturing slaves, do no other work. They are incapable of making their own nests, or of feeding their own larvæ. When the old nest is found inconvenient, and they have to migrate, it is the slaves which determine the migration, and actually carry their masters in their jaws. So utterly helpless are the masters, that when Huber shut up thirty of them without a slave, but with plenty of the food which they like best, and with their larvæ and pupæ to stimulate them to work, they did nothing; they could not even feed themselves, and many perished of hunger. Huber then introduced a single slave (F. fusca), and she instantly set to work, fed and saved the survivors; made some cells and tended the larvæ, and put all to rights. What can be more extraordinary than these well-ascertained facts? If we had not known of any other slave-making ant, it would have been hopeless to have speculated how so wonderful an instinct could have been perfected.

Formica sanguinea was likewise first discovered by P. Huber to be a slave-making ant. This species is found in the southern parts of England, and its habits have been attended to by Mr. F. Smith, of the British

1 Slave-making in ants, known more technically as *dulosis*, is a widespread phenomenon. The so-called slave-makers typically raid the colonies of their victims, making off with the larvae and pupae, which they raise to maturity. Reared in the nest of the foreign ants, the "slaves" then treat it as their colony-family, performing some or all of the work of the colony, depending on slave species. Slave-making behavior was apparently first noted by Pierre Huber in 1804, and reported in his *Recherches sur les Moeurs des Fourmis Indigenes* (Paris, 1810). It was later reported in several British species by Frederick Smith. Although the ant enslavement concept derives from the human practice of slavery, it is clearly different in that it is an interspecific, not intraspecific, interaction. Herbers (2007) suggested abandoning the term "slavery" in ants because of its association with the human practice.

2 Frederick Smith, an entomologist at the British Museum (Natural History), provided Darwin with a great deal of information on slave-making ants. In 1854 Smith authored an "Essay on the genera and species of British Formicidae," and it is likely from this paper that Darwin was first made aware of the habits of the slave-making ant *Formica sanguinea.* He probably read Smith's essay in late 1857 or early 1858; in February 1858 he was asking Smith for advice on when to observe *F. sanguinea* raids.

Museum, to whom I am much indebted for information on this and other subjects. Although fully trusting to the statements of Huber and Mr. Smith, I tried to approach the subject in a sceptical frame of mind, as any one may well be excused for doubting the truth of so extraordinary and odious an instinct as that of making slaves. Hence I will give the observations which I have myself made, in some little detail. I opened fourteen nests of F. sanguinea, and found a few slaves in all. Males and fertile females of the slave-species are found only in their own proper communities, and have never been observed in the nests of F. sanguinea. The slaves are black and not above half the size of their red masters, so that the contrast in their appearance is very great. When the nest is slightly disturbed, the slaves occasionally come out, and like their masters are much agitated and defend the nest: when the nest is much disturbed and the larvæ and pupæ are exposed, the slaves work energetically with their masters in carrying them away to a place of safety. Hence, it is clear, that the slaves feel quite at home. During the months of June and July, on three successive years, I have watched for many hours several nests in Surrey and Sussex, and never saw a slave either leave or enter a nest. As, during these months, the slaves are very few in number, I thought that they might behave differently when more numerous; but Mr. Smith informs me that he has watched the nests at various hours during May, June and August, both in Surrey and Hampshire, and has never seen the slaves, though present in large numbers in August, either leave or enter the nest. Hence he considers them as strictly household slaves. The masters, on the other hand, may be constantly seen bringing in materials for the nest, and food of all kinds. During the present year, however, in the month

1 Darwin was fascinated by these ants. He wrote to Smith to confirm identification of *F. sanguinea* and to ask for pointers on observing a raid. Smith readily obliged, sharing his findings and suggestions for when to look for raids in a letter dated April 30, 1859 (*Correspondence* 7: 287).

2 Some slave-making species are completely dependent on their victims for foraging, nest-building, and defense. In other species like *F. sanguinea*, the slaves do not appear to work outside the colony. Smith wrote to Darwin, "I have been induced from my own observations to consider them—Household-Slaves—which perform some drudgery in the nest" (*Correspondence* 7: 287). This species is now known to be a facultative slave-maker.

of July, I came across a community with an unusually large stock of slaves, and I observed a few slaves mingled with their masters leaving the nest, and marching along the same road to a tall Scotch-fir-tree, twenty-five yards distant, which they ascended together, probably in search of aphides or cocci. According to Huber, who had ample opportunities for observation, in Switzerland the slaves habitually work with their masters in making the nest, and they alone open and close the doors in the morning and evening; and, as Huber expressly states, their principal office is to search for aphides. This difference in the usual habits of the masters and slaves in the two countries, probably depends merely on the slaves being captured in greater numbers in Switzerland than in England.

One day I fortunately chanced to witness a migration from one nest to another, and it was a most interesting spectacle to behold the masters carefully carrying, as Huber has described, their slaves in their jaws. Another day my attention was struck by about a score of the slave-makers haunting the same spot, and evidently not in search of food; they approached and were vigorously repulsed by an independent community of the slave species (F. fusca); sometimes as many as three of these ants clinging to the legs of the slave-making F. sanguinea. The latter ruthlessly killed their small opponents, and carried their dead bodies as food to their nest, twenty-nine yards distant; but they were prevented from getting any pupæ to rear as slaves. I then dug up a small parcel of the pupæ of F. fusca from another nest, and put them down on a bare spot near the place of combat; they were eagerly seized, and carried off by the tyrants, who perhaps fancied that, after all, they had been victorious in their late combat.

1 This scene is reminiscent of Henry David Thoreau's famous account of the "battle of the ants" in chapter 12 of *Walden*. Thoreau noted that the "battles of ants have long been celebrated," and that according to Kirby and Spence, authors of the volume *An Introduction to Entomology*, "Huber is the only modern author who appears to have witnessed them." Thoreau was thrilled to witness them too, and also followed Huber's lead in comparing the ants' conflict with human warfare: "The Herculean ant yielded to numbers; it either perished, the victim of its temerity, or was conducted a prisoner to the enemy's camp" (Huber 1810, quoted in Woodson 1975, p. 554).

At the same time I laid on the same place a small parcel of the pupæ of another species, F. flava, with a few of these little yellow ants still clinging to the fragments of the nest. This species is sometimes, though rarely, made into slaves, as has been described by Mr. Smith. Although so small a species, it is very courageous, and I have seen it ferociously attack other ants. In one instance I found to my surprise an independent community of F. flava under a stone beneath a nest of the slave-making F. sanguinea; and when I had accidentally disturbed both nests, the little ants attacked their big neighbours with surprising courage. Now I was curious to ascertain whether F. sanguinea could distinguish the pupæ of F. fusca, which they habitually make into slaves, from those of the little and furious F. flava, which they rarely capture, and it was evident that they did at once distinguish them: for we have seen that they eagerly and instantly seized the pupæ of F. fusca, whereas they were much terrified when they came across the pupæ, or even the earth from the nest of F. flava, and quickly ran away; but in about a quarter of an hour, shortly after all the little yellow ants had crawled away, they took heart and carried off the pupæ.

One evening I visited another community of F. sanguinea, and found a number of these ants entering their nest, carrying the dead bodies of F. fusca (showing that it was not a migration) and numerous pupæ. I traced the returning file burthened with booty, for about forty yards, to a very thick clump of heath, whence I saw the last individual of F. sanguinea emerge, carrying a pupa; but I was not able to find the desolated nest in the thick heath. The nest, however, must have been close at hand, for two or three individuals of F. fusca were rushing about in the greatest agitation, and one was perched motionless with its own pupa in its mouth on the top of a spray of heath over its ravaged home.

Such are the facts, though they did not need confirmation by me, in regard to the wonderful instinct of making slaves. Let it be observed what a contrast the instinctive habits of F. sanguinea present with those of the F. rufescens. The latter does not build its own nest, does not determine its own migrations, does not collect food for itself or its young, and cannot even feed itself: it is absolutely dependent on its numerous slaves. Formica sanguinea, on the other hand, possesses much fewer slaves, and in the early part of the summer extremely few. The masters determine when and where a new nest shall be formed, and when they migrate, the masters carry the slaves. Both in Switzerland and England the slaves seem to have the exclusive care of the larvæ, and the masters alone go on slave-making expeditions. In Switzerland the slaves and masters work together, making and bringing materials for the nest: both, but chiefly the slaves, tend, and milk as it may be called, their aphides; and thus both collect food for the community. In England the masters alone usually leave the nest to collect building materials and food for themselves, their slaves and larvæ. So that the masters in this country receive much less service from their slaves than they do in Switzerland.

By what steps the instinct of F. sanguinea originated I will not pretend to conjecture. But as ants, which are not slave-makers, will, as I have seen, carry off pupæ of other species, if scattered near their nests, it is possible that pupæ originally stored as food might become developed; and the ants thus unintentionally reared would then follow their proper instincts, and do what work they could. If their presence proved useful to the species which had seized them—if it were more advan-

1 Even such complex and remarkable behaviors as slave-making vary among species, Darwin is pointing out. I mentioned this on p. 220, but here Darwin gives an example of a species, *Formica* (now *Polyergus*) *rufescens*, that is utterly dependent on its slaves for food, nest-building, and defense. Known by the common name of Amazon ants, the genus *Polyergus* consists of about five species worldwide, all obligate slave-makers that specialize on *Formica* species.

2 This is what Darwin has been driving at: just as ancestral cuckoos are envisioned to have occasionally or accidentally deposited eggs in the nests of other species, so too are ancestral ants conjectured to have occasionally collected larvae or (more likely) pupae of other species, bringing them back to the nest (perhaps for food). Note that Darwin cites intraspecific variation too: the English and Swiss populations of *F. sanguinea* differ in important ways. The main idea is that a starting point for this complex behavior is hypothetically possible, and from there a series is envisioned, from opportunistic to facultative to obligate parasitism.

tageous to this species to capture workers than to procreate them—the habit of collecting pupæ originally for food might by natural selection be strengthened and rendered permanent for the very different purpose of raising slaves. When the instinct was once acquired, if carried out to a much less extent even than in our British F. sanguinea, which, as we have seen, is less aided by its slaves than the same species in Switzerland, I can see no difficulty in natural selection increasing and modifying the instinct—always supposing each modification to be of use to the species—until an ant was formed as abjectly dependent on its slaves as is the Formica rufescens.

Cell-making instinct of the Hive-Bee.—I will not here enter on minute details on this subject, but will merely give an outline of the conclusions at which I have arrived. ▶1 He must be a dull man who can examine the exquisite structure of a comb, so beautifully adapted to its end, without enthusiastic admiration. We hear from mathematicians that bees have practically solved a recondite problem, and have made their cells of the proper shape to hold the greatest possible amount of honey, with the least possible consumption of precious wax in their construction. It has been remarked that a skilful workman, with fitting tools and measures, would find it very difficult to make cells of wax of the true form, though this is perfectly effected by a crowd of bees working in a dark hive. Grant whatever instincts you please, and it seems at first quite inconceivable how they can make all the necessary angles and planes, or even perceive when they are correctly made. But the difficulty is not nearly so great as it at first appears: all this beautiful work can be shown, I think, to follow from a few very simple instincts.

1 Only "a dull man" indeed would not appreciate the extraordinary complexity of honeybee comb, with its orderly bilayer of hexagonal cells fashioned of beeswax—all the more remarkable for being fashioned in the dark of the hive. And you, dear reader, should by now appreciate Darwin's method: note that he opens with a statement of the gravity of the problem, acknowledging that it seems "quite inconceivable" how the bees can execute such precise work. In *Natural Selection* he is even more forceful on this point. After describing the geometrical construction of the cells, Darwin concludes that "our admiration is silenced into bewilderment. Here, then, it might be thought we assuredly have a case of a specially endowed instinct, which could not possibly have been arrived at by slow & successive modifications, such as are indispensably necessary on our theory" (Stauffer 1975, pp. 513–514).

Think of honeybee comb as the behavioral equivalent of the eye for Darwin. Indeed, like the eye, honeybee comb has long been a favorite example of design in nature. Henry Lord Brougham said so explicitly in his illustrated edition of Paley's *Natural Theology* in 1839. In the main section, "On Instinct," Lord Brougham cited honeybee comb among other examples specifically arguing against Lamarck's hypothesis that instincts arise from acquired habits, asserting instead that the complexity of the behavior, and its expression without any prior experience, showed that God provided the necessary "knowledge and design" (see Richards 1981). This is the crux of Darwin's difficulty: critics find him caught between Paley and Lamarck.

Darwin ends this paragraph by sweeping aside concerns: "all this beautiful work can be shown . . . to follow from a few very simple instincts." If this is true, it should be possible to look to related species for simpler versions of the behavior that give insight into an evolutionary transition; this is where the argument is going. If, as with the eye, Darwin can make a case for a gradualist transmutation scenario for comb-building, he realizes he will make great strides toward convincing skeptical readers of the truth of his theory.

I was led to investigate this subject by Mr. Waterhouse, who has shown that the form of the cell stands in close relation to the presence of adjoining cells; and the following view may, perhaps, be considered only as a modification of his theory. Let us look to the great principle of gradation, and see whether Nature does not reveal to us her method of work. At one end of a short series we have humble-bees, which use their old cocoons to hold honey, sometimes adding to them short tubes of wax, and likewise making separate and very irregular rounded cells of wax. At the other end of the series we have the cells of the hive-bee, placed in a double layer: each cell, as is well known, is an hexagonal prism, with the basal edges of its six sides bevelled so as to join on to a pyramid, formed of three rhombs. These rhombs have certain angles, and the three which form the pyramidal base of a single cell on one side of the comb, enter into the composition of the bases of three adjoining cells on the opposite side. In the series between the extreme perfection of the cells of the hive-bee and the simplicity of those of the humble-bee, we have the cells of the Mexican Melipona domestica, carefully described and figured by Pierre Huber. The Melipona itself is intermediate in structure between the hive and humble bee, but more nearly related to the latter: it forms a nearly regular waxen comb of cylindrical cells, in which the young are hatched, and, in addition, some large cells of wax for holding honey. These latter cells are nearly spherical and of nearly equal sizes, and are aggregated into an irregular mass. But the important point to notice, is that these cells are always made at that degree of nearness to each other, that they would have intersected or broken into each other, if the spheres had been completed; but this is never permitted, the bees building perfectly flat walls of wax between the spheres

1 Zoologist George Robert Waterhouse evidently told Darwin that he had some ideas about hive-bee cell construction. In an April 1857 letter, he mentioned that "at the last meeting of the Entomological Society there was some discussion upon the subject of the Hive-bee's cell—introduced by Mr. Westwood from whose brief observations on the subject I conclude that his views are the same as mine" (*Correspondence* 6: 377).

2 Darwin's point is clear: bringing the great principle of gradation to bear should yield insights that would otherwise be lost. And note the explicit mention of a transitional *series.*

3 In the short series we have at one end the bumblebees *(Bombus),* which recycle their cocoons as honeypots, and at the other, the hexagonal-comb-building honeybees. Between them is the stingless bee *Melipona domestica*—called the Mexican bee by Darwin and his contemporaries—which makes a cluster of round (cylindrical) cells. Stingless bees are a large group of primarily tropical bees, in the same family as honey- and bumblebees but in their own taxonomic tribe, the Meliponini.

4 *Melipona domestica* appears not to have been a formal taxonomic appellation; this bee is almost certainly *Melipona beecheii,* the celebrated *Xunan-kab* of the Maya, or Royal Mayan bee. These bees have been kept domestically in Mexico for thousands of years. Colonies naturally nest in hollow logs; collectors cut these to a manageable size and plug the open ends with wood or pottery. The rustic "apiary" is then hung from trees or the eaves of dwellings. Darwin's information came from Pierre Huber (1839), who in turn attributed his account to the Scottish naval officer and explorer Basil Hall (1788–1844), famous for his accounts of travels to the far east and the New World. Hall's bee observations were likely first reported in his memoir *Extracts from a Journal Written on the Coasts of Chili, Peru, and Mexico, in the years 1820, 1821, and 1822* (Edinburgh, 1826).

which thus tend to intersect. Hence each cell consists of an outer spherical portion and of two, three, or more perfectly flat surfaces, according as the cell adjoins two, three, or more other cells. When one cell comes into contact with three other cells, which, from the spheres being nearly of the same size, is very frequently and necessarily the case, the three flat surfaces are united into a pyramid; and this pyramid, as Huber has remarked, is manifestly a gross imitation of the three-sided pyramidal basis of the cell of the hive-bee. As in the cells of the hive-bee, so here, the three plane surfaces in any one cell necessarily enter into the construction of three adjoining cells. It is obvious that the Melipona saves wax by this manner of building; for the flat walls between the adjoining cells are not double, but are of the same thickness as the outer spherical portions, and yet each flat portion forms a part of two cells.

Reflecting on this case, it occurred to me that if the Melipona had made its spheres at some given distance from each other, and had made them of equal sizes and had arranged them symmetrically in a double layer, the resulting structure would probably have been as perfect as the comb of the hive-bee. Accordingly I wrote to Professor Miller, of Cambridge, and this geometer has kindly read over the following statement, drawn up from his information, and tells me that it is strictly correct:—

If a number of equal spheres be described with their centres placed in two parallel layers; with the centre of each sphere at the distance of radius $\times \sqrt{2}$, or radius $\times$ 1·41421 (or at some lesser distance), from the centres of the six surrounding spheres in the same layer; and at the same distance from the centres of the adjoining spheres in the other and parallel layer; then, if planes of intersection between the several spheres in

1 Darwin met with William Hallowes Miller, Professor of Mineralogy at Cambridge, in April 1858 "to get wisdom on the geometry of Bees cells," as he put it in a letter to Hooker (*Correspondence* 7: 65). Miller designed a model for the cells and sent Darwin instructions in May 1858. Darwin thought the model had a design flaw, and he asked his son William Erasmus to construct it to identify the problem. Apparently a subtle hint to Miller sufficed: "I have just received your nice note & the [model] for which very many thanks," Darwin wrote to William, "but I hope & think I shall not have to use it, as I had intended, which was delicately to hint to one of the greatest mathematicians that he had made a blunder in his geometry, & sure enough there came a letter yesterday wholly altering what he had previously told me, & which makes my Bees cell go all the better" (*Correspondence* 7: 87). See D'Arcy Thompson's thorough essay on the geometry and construction of honeybee cells, "Of the Bees Cell" (Thompson 1959, pp. 525–544). He points out that honeybees are *not* particularly economical with wax, and writes dismissively of Lord Brougham, whose analysis of the geometry of the cells contains "as striking examples of bad reasoning as are often to be met with in writings relating to mathematical subjects."

both layers be formed, there will result a double layer of hexagonal prisms united together by pyramidal bases formed of three rhombs; and the rhombs and the sides of the hexagonal prisms will have every angle identically the same with the best measurements which have been made of the cells of the hive-bee.

Hence we may safely conclude that if we could slightly modify the instincts already possessed by the Melipona, and in themselves not very wonderful, this bee would make a structure as wonderfully perfect as that of the hive-bee. We must suppose the Melipona to make her cells truly spherical, and of equal sizes; and this would not be very surprising, seeing that she already does so to a certain extent, and seeing what perfectly cylindrical burrows in wood many insects can make, apparently by turning round on a fixed point. We must suppose the Melipona to arrange her cells in level layers, as she already does her cylindrical cells; and we must further suppose, and this is the greatest difficulty, that she can somehow judge accurately at what distance to stand from her fellow-labourers when several are making their spheres; but she is already so far enabled to judge of distance, that she always describes her spheres so as to intersect largely; and then she unites the points of intersection by perfectly flat surfaces. We have further to suppose, but this is no difficulty, that after hexagonal prisms have been formed by the intersection of adjoining spheres in the same layer, she can prolong the hexagon to any length requisite to hold the stock of honey; in the same way as the rude humble-bee adds cylinders of wax to the circular mouths of her old cocoons. By such modifications of instincts in themselves not very wonderful,—hardly more wonderful than those which guide a bird to make its nest,—I believe that the hive-bee

1 In this paragraph we see Darwin arguing that *Melipona* is mere steps away from the complexity of honey (hive) bees, and indeed, presented in this way the two do not seem very far apart. Note his approach of breaking down the whole into component parts (spherical cells, spaced just so in a layered arrangement, extend the length as needed, etc.). Insofar as each part already has expression to some degree, the steps from *Melipona*-style cell construction to that of the honeybee seem few indeed. George Waterhouse, by the way, undertook studies in 1864 showing that individual, isolated cell excavations of honeybees were hemispherical, becoming faceted only when cells were being built side-by-side (*Transactions of the Entomological Society of London* 2: 115). Darwin was excited by this result, since it supported his scenario for a spherical cell precursor in a less highly social bee.

1 William Bernhard Tegetmeier was best known as a pigeon fancier and poultry expert, but he also had a passion for bee-keeping. In this and the following few paragraphs we see Darwin the experimentalist again. Note that the newly excavated cells initially appear rounded, like a sphere.

has acquired, through natural selection, her inimitable architectural powers.

But this theory can be tested by experiment. Following the example of Mr. Tegetmeier, I separated two combs, and put between them a long, thick, square strip of wax: the bees instantly began to excavate minute circular pits in it; and as they deepened these little pits, they made them wider and wider until they were converted into shallow basins, appearing to the eye perfectly true or parts of a sphere, and of about the diameter of a cell. It was most interesting to me to observe that wherever several bees had begun to excavate these basins near together, they had begun their work at such a distance from each other, that by the time the basins had acquired the above stated width (*i. e.* about the width of an ordinary cell), and were in depth about one sixth of the diameter of the sphere of which they formed a part, the rims of the basins intersected or broke into each other. As soon as this occurred, the bees ceased to excavate, and began to build up flat walls of wax on the lines of intersection between the basins, so that each hexagonal prism was built upon the festooned edge of a smooth basin, instead of on the straight edges of a three-sided pyramid as in the case of ordinary cells.

I then put into the hive, instead of a thick, square piece of wax, a thin and narrow, knife-edged ridge, coloured with vermilion. The bees instantly began on both sides to excavate little basins near to each other, in the same way as before; but the ridge of wax was so thin, that the bottoms of the basins, if they had been excavated to the same depth as in the former experiment, would have broken into each other from the opposite sides. The bees, however, did not suffer this to happen, and they stopped their excavations in due time; so that the basins, as soon as they had been a little deepened, came to have flat bottoms; and these flat bottoms, formed by thin little plates of the vermilion wax having been left ungnawed, were situated, as far as the eye could judge, exactly along the planes of imaginary intersection between the basins on the opposite sides of the ridge of wax. In parts, only little bits, in other parts, large portions of a rhombic plate had been left between the opposed basins, but the work, from the unnatural state of things, had not been neatly performed. The bees must have worked at very nearly the same rate on the opposite sides of the ridge of vermilion wax, as they circularly gnawed away and deepened the basins on both sides, in order to have succeeded in thus leaving flat plates between the basins, by stopping work along the intermediate planes or planes of intersection.

Considering how flexible thin wax is, I do not see that there is any difficulty in the bees, whilst at work on the two sides of a strip of wax, perceiving when they have gnawed the wax away to the proper thinness, and then stopping their work. In ordinary combs it has appeared to me that the bees do not always succeed in working at exactly the same rate from the opposite sides; for I have noticed half-completed rhombs at the base of a just-commenced cell, which were slightly concave on one side, where I suppose that the bees had excavated too quickly, and convex on the opposed side, where the bees had worked less quickly. In one well-marked instance, I put the comb back into the hive, and allowed the bees to go on working for a short time, and again examined the cell, and I found that the rhombic plate had been completed, and had become *perfectly flat*: it was absolutely impossible, from the extreme thinness of the little rhombic plate, that they could have effected this by gnawing away the convex side; and I suspect that the bees in such cases stand in the opposed cells and push and bend the ductile and warm wax (which as I have tried is easily done) into its proper intermediate plane, and thus flatten it.

From the experiment of the ridge of vermilion wax, we can clearly see that if the bees were to build for themselves a thin wall of wax, they could make their cells of the proper shape, by standing at the proper distance from each other, by excavating at the same rate, and by endeavouring to make equal spherical hollows, but never allowing the spheres to break into each other. Now bees, as may be clearly seen by examining the edge of a growing comb, do make a rough, circumferential wall or rim all round the comb; and they gnaw into this from the opposite sides, always working circularly as they deepen each cell. They do not make the whole three-sided pyramidal base of any one cell at the same time, but only the one rhombic plate which stands on the extreme growing margin, or the two plates, as the case may be; and they never complete the upper edges of the rhombic plates, until the hexagonal walls are commenced. Some of these statements differ from those made by the justly celebrated elder Huber, but I am convinced of their accuracy; and if I had space, I could show that they are conformable with my theory.

Huber's statement that the very first cell is excavated out of a little parallel-sided wall of wax, is not, as far as I have seen, strictly correct; the first commencement having always been a little hood of wax; but I will not here enter on these details. We see how important a part excavation plays in the construction of the cells; but it would be a great error to suppose that the bees cannot build up a rough wall of wax in the proper

1 The elder Huber is François Huber (1750–1830), the father of Jean Pierre. François was a Swiss apiarist honored as the father of modern beekeeping. Darwin had the second edition of Huber's important book *Nouvelles Observations sur les Abeilles* [New Observations of Bees] (Paris, 1814).

position—that is, along the plane of intersection between two adjoining spheres. I have several specimens showing clearly that they can do this. Even in the rude circumferential rim or wall of wax round a growing comb, flexures may sometimes be observed, corresponding in position to the planes of the rhombic basal plates of future cells. But the rough wall of wax has in every case to be finished off, by being largely gnawed away on both sides. The manner in which the bees build is curious; they always make the first rough wall from ten to twenty times thicker than the excessively thin finished wall of the cell, which will ultimately be left. We shall [1] understand how they work, by supposing masons first to pile up a broad ridge of cement, and then to begin cutting it away equally on both sides near the ground, till a smooth, very thin wall is left in the middle; the masons always piling up the cut-away cement, and adding fresh cement, on the summit of the ridge. We shall thus have a thin wall steadily growing upward; but always crowned by a gigantic coping. From all the cells, both those just commenced and those completed, being thus crowned by a strong coping of wax, the bees can cluster and crawl over the comb without injuring the delicate hexagonal walls, which are only about one four-hundredth of an inch in thickness; the plates of the pyramidal basis being about twice as thick. By this singular manner of building, strength is continually given to the comb, with the utmost ultimate economy of wax.

It seems at first to add to the difficulty of understanding how the cells are made, that a multitude of bees all work together; one bee after working a short time at one cell going to another, so that, as Huber has stated, a score of individuals work even at the commencement of the first cell. I was able practically to show this fact, by covering the edges of the hexagonal walls

1 Darwin turns to analogy to help get his point across, something we see him do to good effect throughout the book.

of a single cell, or the extreme margin of the circumferential rim of a growing comb, with an extremely thin layer of melted vermilion wax; and I invariably found that the colour was most delicately diffused by the bees —as delicately as a painter could have done with his brush—by atoms of the coloured wax having been taken from the spot on which it had been placed, and worked into the growing edges of the cells all round. The work of construction seems to be a sort of balance struck between many bees, all instinctively standing at the same relative distance from each other, all trying to sweep equal spheres, and then building up, or leaving ungnawed, the planes of intersection between these spheres. It was really curious to note in cases of difficulty, as when two pieces of comb met at an angle, how often the bees would entirely pull down and rebuild in different ways the same cell, sometimes recurring to a shape which they had at first rejected.

When bees have a place on which they can stand in their proper positions for working,—for instance, on a slip of wood, placed directly under the middle of a comb growing downwards so that the comb has to be built over one face of the slip—in this case the bees can lay the foundations of one wall of a new hexagon, in its strictly proper place, projecting beyond the other completed cells. It suffices that the bees should be enabled to stand at their proper relative distances from each other and from the walls of the last completed cells, and then, by striking imaginary spheres, they can build up a wall intermediate between two adjoining spheres; but, as far as I have seen, they never gnaw away and finish off the angles of a cell till a large part both of that cell and of the adjoining cells has been built. This capacity in bees of laying down under certain circumstances a rough wall in its proper place between two just-commenced cells, is important, as it bears on a fact, which seems at first quite subversive of the foregoing theory; namely, that the cells on the extreme margin of wasp-combs are sometimes strictly hexagonal; but I have not space here to enter on this subject. Nor does there seem to me any great difficulty in a single insect (as in the case of a queen-wasp) making hexagonal cells, if she work alternately on the inside and outside of two or three cells commenced at the same time, always standing at the proper relative distance from the parts of the cells just begun, sweeping spheres or cylinders, and building up intermediate planes. It is even conceivable that an insect might, by fixing on a point at which to commence a cell, and then moving outside, first to one point, and then to five other points, at the proper relative distances from the central point and from each other, strike the planes of intersection, and so make an isolated hexagon: but I am not aware that any such case has been observed; nor would any good be derived from a single hexagon being built, as in its construction more materials would be required than for a cylinder.

As natural selection acts only by the accumulation of slight modifications of structure or instinct, each profitable to the individual under its conditions of life, it may reasonably be asked, how a long and graduated succession of modified architectural instincts, all tending towards the present perfect plan of construction, could have profited the progenitors of the hive-bee? I think the answer is not difficult: it is known that bees are often hard pressed to get sufficient nectar; and I am informed by Mr. Tegetmeier that it has been experimentally found that no less than from twelve to fifteen pounds of dry sugar are consumed by a hive of bees for the secretion of each pound of wax; so that a prodigious quantity of fluid nectar must be collected and consumed by the bees in a hive for ◀1

1 Here and through the next page Darwin suggests that the gradual, stepwise evolution of hive-bee cell-building is resource-driven. The need for efficient storage of large quantities of honey is likely to have led to the evolution of hexagonal cells. Recall Darwin's comment on p. 224 that "we hear from mathematicians that bees have practically solved a recondite problem." That problem is how to pack the maximum number of equal-sized cells into an area of given size. The cells could be cylindrical, triangular, quadrangular, pentangular, etc. It turns out that hexagons afford maximum cell-packing, and that is just what honeybees build. This is more than assertion: the "Honeycomb Conjecture" in mathematics posits that a hexagonal form represents the best way to divide a surface into cells or compartments of equal area with the smallest total perimeter. Although this conjecture was known to the ancient Greeks, a general proof was only recently developed (see Hales 2001).

1 Darwin argues that economy of wax and the need to maximize honey storage provided the selective pressure for hexagonal cell evolution. This passage underscores a potential problem for Darwin, though there is no indication here that he is aware of it. The problem stems from using species like bumblebees as a hypothetical ancestral form: "it would be an advantage to our humble-bee, if a slight modification of her instinct led her to make her waxen cells near together," Darwin writes. And again: "it would continually be more and more advantageous to our humble-bee, if she were to make her cells more and more regular." You see the problem: the idea of transmutation through steps like that of bumblebee and *Melipona* stages makes sense, but if the hexagon strategy is so advantageous, why do we still have bumblebees making their honeypots the old and inefficient way? We might explain away *Melipona*'s less efficient cell-building on the basis of environment, but the bumblebee is subject to the same harsh winters as is the honeybee. There is a danger of being teleological—are bumblebees evolving toward the honeybee solution to the honeycomb conjecture? Of course not.

the secretion of the wax necessary for the construction of their combs. Moreover, many bees have to remain idle for many days during the process of secretion. A large store of honey is indispensable to support a large stock of bees during the winter; and the security of the hive is known mainly to depend on a large number of bees being supported. Hence the saving of wax by largely saving honey must be a most important element of success in any family of bees. Of course the success of any species of bee may be dependent on the number of its parasites or other enemies, or on quite distinct causes, and so be altogether independent of the quantity of honey which the bees could collect. But let us suppose that this latter circumstance determined, as it probably often does determine, the numbers of a humble-bee which could exist in a country; and let us further suppose that the community lived throughout the winter, and consequently required a store of honey: there can in this case be no doubt that it would be an advantage to our humble-bee, if a slight modification of her instinct led her to make her waxen cells near together, so as to intersect a little; for a wall in common even to two adjoining cells, would save some little wax. Hence it would continually be more and more advantageous to our humble-bee, if she were to make her cells more and more regular, nearer together, and aggregated into a mass, like the cells of the Melipona; for in this case a large part of the bounding surface of each cell would serve to bound other cells, and much wax would be saved. Again, from the same cause, it would be advantageous to the Melipona, if she were to make her cells closer together, and more regular in every way than at present; for then, as we have seen, the spherical surfaces would wholly disappear, and would all be replaced by plane surfaces; and the Melipona would make a comb as perfect as that of the hive-bee. Beyond this stage of perfection in architecture, natural selection could not lead; for the comb of the hive-bee, as far as we can see, is absolutely perfect in economising wax.

Thus, as I believe, the most wonderful of all known instincts, that of the hive-bee, can be explained by natural selection having taken advantage of numerous, successive, slight modifications of simpler instincts; natural selection having by slow degrees, more and more perfectly, led the bees to sweep equal spheres at a given distance from each other in a double layer, and to build up and excavate the wax along the planes of intersection. The bees, of course, no more knowing that they swept their spheres at one particular distance from each other, than they know what are the several angles of the hexagonal prisms and of the basal rhombic plates. The motive power of the process of natural selection having been economy of wax; that individual swarm which wasted least honey in the secretion of wax, having succeeded best, and having transmitted by inheritance its newly acquired economical instinct to new swarms, which in their turn will have had the best chance of succeeding in the struggle for existence. ◀1

No doubt many instincts of very difficult explanation could be opposed to the theory of natural selection,—cases, in which we cannot see how an instinct could possibly have originated; cases, in which no intermediate gradations are known to exist; cases of instinct of apparently such trifling importance, that they could hardly have been acted on by natural selection; cases of instincts almost identically the same in animals so remote in the scale of nature, that we cannot account ◀2

1 Darwin's voluminous notes, correspondence, and experiments concerning bees all point to the singular importance that these organisms held for him. Their behavioral complexity seems even more important than the eye example in morphology. Indeed, Darwin's critics seemed to think so too—some of his opponents attacked this discussion of bee behavior, and perhaps the most vehement of these was the Rev. Haughton. "Have you seen Haughton's coarsely-abusive article of me in Dublin Mag. of Nat. History," Darwin wrote to Hooker in June 1860. "I never knew anything so unfair as in discussing cells of Bees, his ignoring the case of Melipona which builds combs almost exactly intermediate between Hive & Humble-bee.—What has Haughton done that he feels so immeasurably superior to all us wretched naturalists?" (*Correspondence* 8: 237). Haughton followed with an article in the *Annals and Magazine of Natural History* in 1863 that even provoked the ire of Wallace, who published a spirited rebuttal that same year (Wallace 1863).

2 In this final section of the chapter Darwin tackles some puzzling behavioral phenomena that might seem to undermine his theory. The "instincts of very difficult explanation" in this paragraph are all thorny problems, but he selects the thorniest of all for full discussion: the nonreproductive castes of social insects. In the most complex insect societies—ants, termites, and many bees and wasps—there occur specialized castes that never reproduce. That in itself would seem to be a serious problem (for how can selection act on them if they do not pass on their traits?), but added to it is the complication that those neuters are often strikingly differentiated into morphological castes. Some castes differ in size alone, but in others there are remarkable differences between castes specialized for different tasks, such as soldier forms and worker forms. And in general, nonreproductive castes tend to differ dramatically from the reproductives. This is discussed on the next page, but here Darwin identifies three related aspects to the problem: (1) why sterility exists, (2) how natural selection can favor sterility, and (3) how natural selection can yield morphological divergence in sterile individuals.

for their similarity by inheritance from a common parent, and must therefore believe that they have been acquired by independent acts of natural selection. I will not here enter on these several cases, but will confine myself to one special difficulty, which at first appeared to me insuperable, and actually fatal to my whole theory. I allude to the neuters or sterile females in insect-communities: for these neuters often differ widely in instinct and in structure from both the males and fertile females, and yet, from being sterile, they cannot propagate their kind.

The subject well deserves to be discussed at great length, but I will here take only a single case, that of working or sterile ants. How the workers have been rendered sterile is a difficulty; but not much greater than that of any other striking modification of structure; for it can be shown that some insects and other articulate animals in a state of nature occasionally become sterile; and if such insects had been social, and it had been profitable to the community that a number should have been annually born capable of work, but incapable of procreation, I can see no very great difficulty in this being effected by natural selection. But I must pass over this preliminary difficulty. The great difficulty lies in the working ants differing widely from both the males and the fertile females in structure, as in the shape of the thorax and in being destitute of wings and sometimes of eyes, and in instinct. As far as instinct alone is concerned, the prodigious difference in this respect between the workers and the perfect females, would have been far better exemplified by the hive-bee. If a working ant or other neuter insect had been an animal in the ordinary state, I should have unhesitatingly assumed that all its characters had been slowly acquired through natural selection; namely, by an individual

1 Of the first two difficulties—sterility and morphological differentiation of sterile individuals—Darwin is less concerned with the first. He considers this "preliminary difficulty" no more problematic than explaining highly specialized, complex organs. Sterility is known from the odd taxon here and there, after all. Note, though, that he touches on a point that becomes highly significant in terms of the modern understanding of the evolution of these colonies: "if . . . it had been *profitable to the community* that *a number* should have been annually born capable of work, but incapable of procreation, I can see no great difficulty in this being effected by natural selection" (italics mine). Darwin seems to appreciate that selection might favor the specialization of some individuals (not everyone in the colony) even to the point of sterility if that would profit the colony as a whole. This is revisited on the next page. The great difficulty, Darwin says, focusing on ants, is the great divergence in morphology between workers and reproductive individuals.

2 Part of the problem is that the remarkable divergence in morphology occurs within the context of a single family; the offspring differ from their parents yet are themselves sterile. How can their unique structure evolve when they do not reproduce and so cannot pass on the trait?

having been born with some slight profitable modification of structure, this being inherited by its offspring, which again varied and were again selected, and so onwards. But with the working ant we have an insect differing greatly from its parents, yet absolutely sterile; so that it could never have transmitted successively acquired modifications of structure or instinct to its progeny. It may well be asked how is it possible to reconcile this case with the theory of natural selection?

First, let it be remembered that we have innumerable instances, both in our domestic productions and in those in a state of nature, of all sorts of differences of structure which have become correlated to certain ages, and to either sex. We have differences correlated not only to one sex, but to that short period alone when the reproductive system is active, as in the nuptial plumage of many birds, and in the hooked jaws of the male salmon. We have even slight differences in the horns of different breeds of cattle in relation to an artificially imperfect state of the male sex; for oxen of certain breeds have longer horns than in other breeds, in comparison with the horns of the bulls or cows of these same breeds. Hence I can see no real difficulty in any character having become correlated with the sterile condition of certain members of insect-communities: the difficulty lies in understanding how such correlated modifications of structure could have been slowly accumulated by natural selection. ◀1

This difficulty, though appearing insuperable, is lessened, or, as I believe, disappears, when it is remembered that selection may be applied to the family, as well as to the individual, and may thus gain the desired end. Thus, a well-flavoured vegetable is cooked, and the individual is destroyed; but the horticulturist sows seeds of the same stock, and confidently expects to ◀2

1 Darwin invokes correlation to answer this question, as well as "delayed expression" (i.e., instances in which a trait is expressed at a particular life stage, or by only one sex or the other). He points to cases of domesticated animals' exhibiting correlated traits and suggests that these insect castes do the same.

2 Here is a fuller expression of the insight given on p. 236: selection can operate at the level of the family as well as that of the individual! This idea is more fully expressed in *Natural Selection* (see Stauffer 1975, p. 366): "it seems not improbable . . . that a certain number of females, working like the others, but without any waste of time or vital force from breeding, might be of immense service to the community."

Thus does Darwin anticipate elements of what is now called "kin selection." This form of selection operates through inclusive fitness: the transmission of genes directly, through personal reproduction, and indirectly, though the reproduction of relatives sharing the same genes. In modern parlance, having little or no direct reproduction can be more than compensated by enough indirect reproduction. That is what occurs in the most complex social insects: sterile insects have no direct reproduction, but insofar as their efforts assist in the production of more siblings than might otherwise occur, the genes they carry are passed on indirectly. This idea was most fully developed by the English evolutionary biologist William D. Hamilton (1936–2000) in a pair of papers published in 1964.

With his usual knack for metaphor and analogy, Darwin compares this example to domesticated vegetables bred for flavor: the harvested and cooked individuals do not reproduce, but the variety is propagated through other individuals. This leads him, on p. 238, to claim that a specialized breed could be progressively developed by careful selection of the parents, even if the breed itself is never bred. This seems to run counter to the steady directional-selection scenario he gives earlier for domesticated organisms. This may be analogous to the practice of generating varieties by careful hybridization, yielding offspring that have unique traits yet are sterile.

get nearly the same variety; breeders of cattle wish the flesh and fat to be well marbled together; the animal has been slaughtered, but the breeder goes with confidence to the same family. I have such faith in the powers of selection, that I do not doubt that a breed of cattle, always yielding oxen with extraordinarily long horns, could be slowly formed by carefully watching which individual bulls and cows, when matched, produced oxen with the longest horns; and yet no one ox could ever have propagated its kind. Thus I believe it has been with social insects: a slight modification of structure, or instinct, correlated with the sterile condition of certain members of the community, has been advantageous to the community: consequently the fertile males and females of the same community flourished, and transmitted to their fertile offspring a tendency to produce sterile members having the same modification. And I believe that this process has been repeated, until that prodigious amount of difference between the fertile and sterile females of the same species has been produced, which we see in many social insects.

▶1 But we have not as yet touched on the climax of the difficulty; namely, the fact that the neuters of several ants differ, not only from the fertile females and males, but from each other, sometimes to an almost incredible degree, and are thus divided into two or even three castes. The castes, moreover, do not generally graduate into each other, but are perfectly well defined; being as distinct from each other, as are any two species of the same genus, or rather as any two genera of the

▶2 same family. Thus in Eciton, there are working and soldier neuters, with jaws and instincts extraordinarily different: in Cryptocerus, the workers of one caste alone carry a wonderful sort of shield on their heads, the use of which is quite unknown: in the Mexican Myrme-

1 As considerable a difficulty as sterile castes may be, Darwin thinks an even bigger problem is morphological differentiation *among* sterile castes within a given species. The problem here is compounded by the fact that these distinctive subcastes are often well defined, with no intermediate linking forms.

2 The ants cited here are wonderful indeed. *Eciton* are the army ants of the American tropics, a large group of over 200 species known for their formidable foraging raids, in which marauding columns sweep through the forest, overwhelming virtually all invertebrates (and not a few hapless vertebrates) that fail to escape their path. *Myrmecocystus* are "honeypot" ants that feature a specialized sedentary caste which serves as a food repository, or larder, for the colony. These ants are engorged by the workers to produce enormously distended abdomens. The food is later extracted to help the colony get through lean times. A few different ant groups have independently evolved a honeypot caste, the best known of which are *Myrmecocystus* spp. (which occur in western North America) and *Camponotus inflatus* (an Australian species).

The *Cryptocerus* species Darwin mentions was described by Frederick Smith in 1853. In Smith's words, the margin of the head "[curves] upwards, forming the exact model of a dish or bowl, which has a few large scattered punctures within" (Smith 1853, p. 222). In 1857 Henry Walter Bates sent specimens of a smaller caste lacking this remarkable head, and Smith in turn told Darwin about them: "Some time ago you asked me to furnish you with remarkable instances of disparity in form &c in workers of Insects living in community . . . I send you [an example] that is a truly remarkable instance" (*Correspondence* 6: 481).

cocystus, the workers of one caste never leave the nest; they are fed by the workers of another caste, and they have an enormously developed abdomen which secretes a sort of honey, supplying the place of that excreted by the aphides, or the domestic cattle as they may be called, which our European ants guard or imprison.

It will indeed be thought that I have an overweening confidence in the principle of natural selection, when I do not admit that such wonderful and well-established facts at once annihilate my theory. In the simpler case of neuter insects all of one caste or of the same kind, which have been rendered by natural selection, as I believe to be quite possible, different from the fertile males and females,—in this case, we may safely conclude from the analogy of ordinary variations, that each successive, slight, profitable modification did not probably at first appear in all the individual neuters in the same nest, but in a few alone; and that by the long-continued selection of the fertile parents which produced most neuters with the profitable modification, all the neuters ultimately came to have the desired character. On this view we ought occasionally to find neuter-insects of the same species, in the same nest, presenting gradations of structure; and this we do find, even often, considering how few neuter-insects out of Europe have been carefully examined. Mr. F. Smith has shown how surprisingly the neuters of several British ants differ from each other in size and sometimes in colour; and that the extreme forms can sometimes be perfectly linked together by individuals taken out of the same nest: I have myself compared perfect gradations of this kind. It often happens that the larger or the smaller sized workers are the most numerous; or that both large and small are numerous, with those of an intermediate size scanty in numbers. Formica flava has larger and

1 Here we see Darwin positing a gradualistic transmutation scenario for worker castes in general, and the highly specialized castes like honeypots in particular. A couple of aspects of the argument are worthy of notice. First, Darwin reiterates his belief that specialized castes arise through "successive, slight, profitable" modifications. Second, such modifications are not manifest in all individuals of the proto-worker caste at once, but initially appear in a few individuals. Selection may then favor producing all workers with these modifications over time. Note his argument that graduated series of workers within the same colony supports this hypothesis—presumably these colonies would be intermediate between those with a single undifferentiated worker caste and those with differentiated castes that include highly specialized forms like honeypots.

2 *Formica flava* is now called *Lasius flavus*, the common yellow ant or meadow ant of the northern hemisphere. One aspect of *L. flavus*'s biology that Darwin would have found fascinating is its interactions with subterranean root aphids. The ants harvest honeydew from the older aphids and prey on the young ones; one study (Pontin 1978) found the populations of aphids associated with *L. flavus* nests to be so high that the ants could sustain themselves on the aphids alone and forgo other prey! John Lubbock attributed the pale color of this ant to its subterranean habitat: "One is tempted to suggest that the brown species which live so much in open air . . . have retained and even deepened their dark colour; while others, such as *Lasius flavus*, the yellow meadow ant, which lives almost entirely below ground, has become much paler" (Lubbock 1884, pp. 68–69).

smaller workers, with some of intermediate size; and, in this species, as Mr. F. Smith has observed, the larger workers have simple eyes (ocelli), which though small can be plainly distinguished, whereas the smaller workers have their ocelli rudimentary. Having carefully dissected several specimens of these workers, I can affirm that the eyes are far more rudimentary in the smaller workers than can be accounted for merely by their proportionally lesser size; and I fully believe, though I dare not assert so positively, that the workers of intermediate size have their ocelli in an exactly intermediate condition. So that we here have two bodies of sterile workers in the same nest, differing not only in size, but in their organs of vision, yet connected by some few members in an intermediate condition. I may digress by adding, that if the smaller workers had been the most useful to the community, and those males and females had been continually selected, which produced more and more of the smaller workers, until all the workers had come to be in this condition; we should then have had a species of ant with neuters very nearly in the same condition with those of Myrmica. [1] For the workers of Myrmica have not even rudiments of ocelli, though the male and female ants of this genus have well-developed ocelli.

I may give one other case: so confidently did I expect to find gradations in important points of structure between the different castes of neuters in the same species, that I gladly availed myself of Mr. F. Smith's offer of numerous specimens from the same nest of the driver [2] ant (Anomma) of West Africa. The reader will perhaps best appreciate the amount of difference in these workers, by my giving not the actual measurements, but a strictly accurate illustration: the difference was the same as if we were to see a set of workmen building

1 *Myrmica* ants are among the most common ants in the world; the genus was described in 1804 by the French entomologist Pierre André Latreille.

2 The tropical African driver-ants, *Anomma,* behave much like the New World army ants of the genus *Eciton*—formidable nomadic predators have long, curved mandibles. Until recently it was believed that the Old and New World "army ant syndrome" had arisen convergently, but molecular phylogenetic analysis led to the startling conclusion that the two groups are sister-lineages that shared a common ancestor in mid-Cretaceous Gondwanaland about 105 million years ago (Brady 2003).

Anomma workers are often produced in a long, graduated series; in fact, they are among the most polymorphic of ant species. The largest worker morph of one species, *A. nigricans,* has an estimated seventy times the volume of the smallest morph (Hollingsworth 1960)! These ants later became important subjects in the study of ant allometry (the proportional relationships of morphological traits among individuals of different sizes).

a house of whom many were five feet four inches high, and many sixteen feet high; but we must suppose that the larger workmen had heads four instead of three times as big as those of the smaller men, and jaws nearly five times as big. The jaws, moreover, of the working ants of the several sizes differed wonderfully in shape, and in the form and number of the teeth. But the important fact for us is, that though the workers can be grouped into castes of different sizes, yet they graduate insensibly into each other, as does the widely-different structure of their jaws. I speak confidently on this latter point, as Mr. Lubbock made drawings for me with the camera lucida of the jaws which I had dissected from the workers of the several sizes.

With these facts before me, I believe that natural selection, by acting on the fertile parents, could form a species which should regularly produce neuters, either all of large size with one form of jaw, or all of small size with jaws having a widely different structure; or lastly, and this is our climax of difficulty, one set of workers of one size and structure, and simultaneously another set of workers of a different size and structure; —a graduated series having been first formed, as in the case of the driver ant, and then the extreme forms, from being the most useful to the community, having been produced in greater and greater numbers through the natural selection of the parents which generated them; until none with an intermediate structure were produced. ◀1

Thus, as I believe, the wonderful fact of two distinctly defined castes of sterile workers existing in the same nest, both widely different from each other and from their parents, has originated. We can see how useful their production may have been to a social community of insects, on the same principle that the division of ◀2

1 This is a clearer statement of Darwin's scenario for the evolution of castes: a single morph followed by a graduated morph series followed by more abruptly delineated morphs with highly specialized castes.

2 It would have been surprising if Darwin had *not* made this point. We have already seen the division of labor discussed in the context of niche specialization (see pp. 93, 115–117). Caste specialization is a form of division of labor and an even closer analogy with human society. Note the added point on the next page that sterility is a useful feature in the insect societies as a means of preventing intercrossing. In *Natural Selection* Darwin noted that "a division of labour is possible in communities of man, through his intellect, his traditions & artificial instruments; but in communities of insects, which have . . . only their jaws & limbs, a division of labour seems to be possible only by the production of sterile individuals; for had the different workmen been capable of breeding together, the several castes would have been blended together and so lost" (Stauffer 1975, p. 374). This line of reasoning is a bit surprising in that Darwin knew that the workers of these insect societies were all female (see p. 236, line 7) and so could not interbreed (at least not among one another) even had they been fertile.

labour is useful to civilised man. As ants work by inherited instincts and by inherited tools or weapons, and not by acquired knowledge and manufactured instruments, a perfect division of labour could be effected with them only by the workers being sterile; for had they been fertile, they would have intercrossed, and their instincts and structure would have become blended. And nature has, as I believe, effected this admirable division of labour in the communities of ants, by the means of natural selection. But I am bound to confess, that, with all my faith in this principle, I should never have anticipated that natural selection could have been efficient in so high a degree, had not the case of these neuter insects convinced me of the fact. I have, therefore, discussed this case, at some little but wholly insufficient length, in order to show the power of natural selection, and likewise because this is by far the most serious special difficulty, which my theory has encountered. [1] The case, also, is very interesting, as it proves that with animals, as with plants, any amount of modification in structure can be effected by the accumulation of numerous, slight, and as we must call them accidental, variations, which are in any manner profitable, without exercise or habit having come into play. For no amount of exercise, or habit, or volition, in the utterly sterile members of a community could possibly have affected the structure or instincts of the fertile members, which alone leave descendants. I am surprised that no one has advanced this demonstrative case of neuter insects, against the well-known doctrine of Lamarck. [2]

Summary.—I have endeavoured briefly in this chapter to show that the mental qualities of our domestic animals vary, and that the variations are inherited. Still more briefly I have attempted to show that in-

1 These final few sentences are very interesting. First, note Darwin's assertion that "*any* amount of modification in structure" (emphasis mine) can be effected by natural selection. More interesting, however, is his realization that sterile social insect castes provide an eloquent argument against Lamarck's model of transmutation by inheritance of acquired traits. Jean Baptiste de Lamarck developed a bold transmutation theory positing that organisms evolved over time as a result of an inner driving force. There were two components to this evolution: a primary process of progressive change toward greater complexity, and a secondary process of adaptation, generally mediated by behavior, in which organisms adjusted to the demands of their environment. The secondary process, which invoked heritable acquired traits, drew the most criticism (Bowler 2003). Darwin allowed for some transmutation through the action of the environment (i.e., acquired traits), but to him this was a limited and constrained form of evolutionary change. His comment here points out that sterile individuals can neither pass on their traits, however acquired, nor influence the fertile members of the colony. Thus the sterile worker castes of social insects are incompatible with a Lamarckian process of change.

The German biologist August Weismann (1834–1914), originator of the germline theory positing a separation of germ and somatic cells, criticized Lamarck using precisely the same argument. This was published in an 1883 essay entitled "On Inheritance," the same paper in which he proposed his germline theory.

2 Lamarck was ridiculed by Cuvier, Lyell, and others, including Darwin. Yet Darwin acknowledged his debt to Lamarck in his introduction to later editions of *Origin:* "[Lamarck] first did the eminent service of arousing attention to the probability of all changes in the organic, as well as in the inorganic world, being the result of law, and not of miraculous interposition."

stincts vary slightly in a state of nature. No one will dispute that instincts are of the highest importance to each animal. Therefore I can see no difficulty, under changing conditions of life, in natural selection accumulating slight modifications of instinct to any extent, in any useful direction. In some cases habit or use and disuse have probably come into play. I do not pretend that the facts given in this chapter strengthen in any great degree my theory; but none of the cases of difficulty, to the best of my judgment, annihilate it. On the other hand, the fact that instincts are not always absolutely perfect and are liable to mistakes;—that no instinct has been produced for the exclusive good of other animals, but that each animal takes advantage of the instincts of others;—that the canon in natural history, of "natura non facit saltum" is applicable to instincts as well as to corporeal structure, and is plainly explicable on the foregoing views, but is otherwise inexplicable,—all tend to corroborate the theory of natural selection.

This theory is, also, strengthened by some few other facts in regard to instincts; as by that common case of closely allied, but certainly distinct, species, when inhabiting distant parts of the world and living under considerably different conditions of life, yet often retaining nearly the same instincts. For instance, we can understand on the principle of inheritance, how it is that the thrush of South America lines its nest with mud, in the same peculiar manner as does our British thrush: how it is that the male wrens (Troglodytes) of North America, build "cock-nests," to roost in, like the males of our distinct Kitty-wrens,—a habit wholly unlike that of any other known bird. Finally, it may not be a logical deduction, but to my imagination it is far more satisfactory to look at such instincts as the young

1 This may be an understatement; it is true enough that Darwin set out to show that the difficulties posed by instinct and behavior are non-issues, but we can also look at this chapter in the context of consilience: it *does* strengthen his argument, by showing how the theory is applicable to yet another (and quite different) sphere of the biological world.

2 Behaviors, then, can be shared by common ancestry in exactly the same way as structures can. These trans-Atlantic thrushes and wrens are bound by inheritance; thus confamilials on each side of the ocean behave in fundamentally the same manner. A "cock's nest," by the way, is a nest built by a male bird as part of the courtship process. Male wrens often build several nests, and the female then chooses the one she likes best.

3 To me, this chapter ends on a poignant note. Putting aside the reason and logic inherent in his foregoing arguments, here Darwin seems to be offering his emotional response to many of the instincts of animals. His letters and notebooks suggest that he is repulsed at the cruel and macabre behaviors out there, and prefers to think that a benign and benevolent deity would not create such horrors.

[1] cuckoo ejecting its foster-brothers,—ants making slaves, —the larvæ of ichneumonidæ feeding within the live bodies of caterpillars,—not as specially endowed or created instincts, but as small consequences of one general law, leading to the advancement of all organic beings, namely, multiply, vary, let the strongest live and the weakest die.

1 Cuckoos, we have seen, either forcibly eject from the nest or cause the starvation of the host bird's young. "Slave-making" in ants is a horror, and maybe all the more so for its superficial similarity to the hated human institution in the mind of the anti-slavery Darwin. Wasps like the Ichneumonidae present a repulsive lifestyle all their own: these wasps can only survive by laying their eggs within the body of another insect, often a caterpillar. The wasp grub slowly feeds on its prey from within, eventually killing it. Behaviors such as these argued strongly against design to Darwin. "With respect to the theological view of the question; this is always painful to me," he wrote to Asa Gray. "I had no intention to write atheistically. But I own that I cannot see, as plainly as others do, & as I [should] wish to do, evidence of design & beneficence on all sides of us. There seems to me too much misery in the world. I cannot persuade myself that a beneficent & omnipotent God would have designedly created the Ichneumonidae with the express intention of their feeding within the living bodies of caterpillars, or that a cat should play with mice. Not believing this, I see no necessity in the belief that the eye was expressly designed" (*Correspondence* 8: 223).

It was one thing to create beings that had to kill to survive, but quite another to create beings that made their living by inflicting unending pain and malady, or that killed in what seemed an unnecessarily horrific manner. Darwin preferred to think of the biological world as governed by laws, these being far more preferable to a creator with a penchant for cruelty or indifference to suffering. The final part of this remarkable paragraph proclaims his vision: the general law of natural selection leads to the "advancement of all organic beings"; through the grand sweep of time they vary, multiply, and survive or not as best they can.

CHAPTER VIII.

HYBRIDISM.

Distinction between the sterility of first crosses and of hybrids — Sterility various in degree, not universal, affected by close interbreeding, removed by domestication—Laws governing the sterility of hybrids — Sterility not a special endowment, but incidental on other differences — Causes of the sterility of first crosses and of hybrids — Parallelism between the effects of changed conditions of life and crossing — Fertility of varieties when crossed and of their mongrel offspring not universal — Hybrids and mongrels compared independently of their fertility — Summary.

THE view generally entertained by naturalists is that species, when intercrossed, have been specially endowed with the quality of sterility, in order to prevent the confusion of all organic forms. This view certainly seems at first probable, for species within the same country could hardly have kept distinct had they been capable of crossing freely. The importance of the fact that hybrids are very generally sterile, has, I think, been much underrated by some late writers. On the theory of natural selection the case is especially important, inasmuch as the sterility of hybrids could not possibly be of any advantage to them, and therefore could not have been acquired by the continued preservation of successive profitable degrees of sterility. I hope, however, to be able to show that sterility is not a specially acquired or endowed quality, but is incidental on other acquired differences.

In treating this subject, two classes of facts, to a large extent fundamentally different, have generally been confounded together; namely, the sterility of two

In this chapter Darwin attempts to explain patterns of hybridization—the facility of making different crosses, and the products of those crosses—in terms of his model of descent with modification. The overarching point will be familiar by now: species and varieties are essentially equivalent. Sterility and unequal outcomes of reciprocal crosses were quite mysterious in Darwin's time. The prevailing view held that the inability of distinct species to cross, or the sterility of their offspring if they did cross, was a divine endowment to preserve the purity of species. Showing that species interfertility and sterility actually vary, Darwin argues instead that sterility and fertility in different crosses arise by natural law. Though he lacks any real understanding of heredity and therefore the causes of these patterns, the observations reported here resonate with his argument that species and varieties do not differ in any meaningful way, but represent degrees of divergence along a continuum. This continuum suggests transition and thus transmutation.

---

1 The idea that hybrid sterility exists "in order to prevent the confusion of all organic forms" was taken very seriously in Darwin's time. The pioneer Enlightenment plant breeder Joseph Gottlieb Kölreuter undertook his influential experiments in hybridization explicitly to refute Linnaeus's speculation that new species could result from hybridization (Müller-Wille and Orel 2007). Kölreuter discovered hybrid sterility in tobacco plants in 1761 and declared it "one of the most wonderful of all events that have ever occurred upon the wide field of nature." He referred to his hybrids as "the first botanical mule which has been produced by art" (Quoted in Hankins 1985, p. 137). Kölreuter's sterile hybrids reinforced the notion that species are immutable and remain as created.

2 This is a characteristically clever point: how could hybrid sterility be advantageous to the hybrid plant or animal? It cannot result from selection favoring succeeding degrees of sterility, because the more sterile the offspring become, the less able they are to pass on the trait! Darwin is suggesting that rather than being "specially endowed," sterility arises secondarily, as a consequence of other factors.

species when first crossed, and the sterility of the hybrids produced from them.

Pure species have of course their organs of reproduction in a perfect condition, yet when intercrossed they produce either few or no offspring. Hybrids, on the other hand, have their reproductive organs functionally impotent, as may be clearly seen in the state of the male element in both plants and animals; though the organs themselves are perfect in structure, as far as the microscope reveals. In the first case the two sexual elements which go to form the embryo are perfect; in the second case they are either not at all developed, or
1 are imperfectly developed. This distinction is important, when the cause of the sterility, which is common to the two cases, has to be considered. The distinction has probably been slurred over, owing to the sterility in both cases being looked on as a special endowment, beyond the province of our reasoning powers.

The fertility of varieties, that is of the forms known or believed to have descended from common parents, when intercrossed, and likewise the fertility of their mongrel offspring, is, on my theory, of equal importance with the sterility of species; for it seems to make a broad and clear distinction between varieties and species.

First, for the sterility of species when crossed and of
2 their hybrid offspring. It is impossible to study the several memoirs and works of those two conscientious and admirable observers, Kölreuter and Gärtner, who almost devoted their lives to this subject, without being deeply impressed with the high generality of
3 some degree of sterility. Kölreuter makes the rule universal; but then he cuts the knot, for in ten cases in which he found two forms, considered by most authors as distinct species, quite fertile together, he

1 More than one mechanism for sterility further argues against sterility as some special condition divinely endowed.

2 The work of "those two conscientious and admirable observers," Joseph Gottlieb Kölreuter and Karl Friedrich Gärtner, is discussed extensively in this chapter. These German botanists were the leaders in experimental plant hybridization in Darwin's time.

3 Here and over the next few pages Darwin explores sterility and interfertility, making the argument that there is no hard-and-fast rule for which species or varieties will intercross, or how fertile any resulting offspring will be. Kölreuter uses interfertility as a criterion for species diagnosis, in principle an approach we embrace today in our concept of biological species. He applies this so consistently that when he comes across forms commonly regarded as distinct species that nonetheless will cross, he "demotes" then to the status of varieties—defining the problem away. "Cutting the knot" is a reference, of course, to the legendary Gordian Knot cut by Alexander the Great: this is a dig by Darwin, suggesting the problem is not so much solved as eliminated by a bold stroke. Gärtner, too, applied the criterion of interfertility for species diagnosis but came to different conclusions regarding the species Kölreuter studied. We have seen Darwin make use of disagreements among naturalists before.

unhesitatingly ranks them as varieties. Gärtner, also, makes the rule equally universal; and he disputes the entire fertility of Kölreuter's ten cases. But in these and in many other cases, Gärtner is obliged carefully to count the seeds, in order to show that there is any degree of sterility. He always compares the maximum number of seeds produced by two species when crossed and by their hybrid offspring, with the average number produced by both pure parent-species in a state of nature. But a serious cause of error seems to me to be here introduced: a plant to be hybridised must be castrated, and, what is often more important, must be secluded in order to prevent pollen being brought to it by insects from other plants. Nearly all the plants experimentised on by Gärtner were potted, and apparently were kept in a chamber in his house. That these processes are often injurious to the fertility of a plant cannot be doubted; for Gärtner gives in his table about a score of cases of plants which he castrated, and artificially fertilised with their own pollen, and (excluding all cases such as the Leguminosæ, in which there is an acknowledged difficulty in the manipulation) half of these twenty plants had their fertility in some degree impaired. Moreover, as Gärtner during several years repeatedly crossed the primrose and cowslip, which we have such good reason to believe to be varieties, and only once or twice succeeded in getting fertile seed; as he found the common red and blue pimpernels (Anagallis arvensis and cœrulea), which the best botanists rank as varieties, absolutely sterile together; and as he came to the same concluson in several other analogous cases; it seems to me that we may well be permitted to doubt whether many other species are really so sterile, when intercrossed, as Gärtner believes.

1 The table Darwin refers to comes from Gärtner's 1849 treatise *Versuche und Beobachtungen über die Bastarderzeugung im Pflanzenreich* (Experiments and Observations on the Production of Hybrids in the Plant Kingdom). Though Darwin is critical here, he was a great admirer of Gärtner's work; in 1864 he asked Hooker to propose that the Ray Society publish an English translation of this treatise: "I believe I have read with attention everything that has been published on hybridisation & worked a little practically on the subject, & I do not hesitate to affirm that there is more useful & trust worthy matter in Gärtners work than in all others combined" (*Correspondence* 12: 336).

2 Darwin is adept at identifying the problem cases. It may not be clear what he is getting at here until the final sentence: his main objective is to make the point that sterility and interfertility are not absolute. The question of intercrossing was important to Darwin as one way in which to generate intermediate forms. This is reflected in an entry from his Questions and Experiments notebook: in one lengthy list of plant experiments, he wrote, "Experimentise on Primrose seeds—it really is an important case—cross with cowslip pollen.—as these are wild varieties. Is any intermediate form found wild[?]" (Barrett et al. 1987, p. 495).

1 Here Darwin stresses the insensible gradation of sterility/interfertility. His point is that a continuum in facility of intercrossing is consistent with a continuum in divergence.

2 Just as Darwin uses the botanists' disagreement over species and varieties ("doubtful forms") to support his argument that there is no real distinction between the two, so too does he exploit the differing conclusions of Gärtner and Kölreuter.

3 It is important to know not just the immediate results of hybridization, but whether subsequent generations will continue to show the same degree of interfertility. It appears to decline, which Darwin attributes to inbreeding. Though lacking a modern theory of heredity, Darwin did appreciate the deleterious effects of inbreeding.

[1] It is certain, on the one hand, that the sterility of various species when crossed is so different in degree and graduates away so insensibly, and, on the other hand, that the fertility of pure species is so easily affected by various circumstances, that for all practical purposes it is most difficult to say where perfect fertility ends and sterility begins. [2] I think no better evidence of this can be required than that the two most experienced observers who have ever lived, namely, Kölreuter and Gärtner, should have arrived at diametrically opposite conclusions in regard to the very same species. It is also most instructive to compare—but I have not space here to enter on details—the evidence advanced by our best botanists on the question whether certain doubtful forms should be ranked as species or varieties, with the evidence from fertility adduced by different hybridisers, or by the same author, from experiments made during different years. It can thus be shown that neither sterility nor fertility affords any clear distinction between species and varieties; but that the evidence from this source graduates away, and is doubtful in the same degree as is the evidence derived from other constitutional and structural differences.

[3] In regard to the sterility of hybrids in successive generations; though Gärtner was enabled to rear some hybrids, carefully guarding them from a cross with either pure parent, for six or seven, and in one case for ten generations, yet he asserts positively that their fertility never increased, but generally greatly decreased. I do not doubt that this is usually the case, and that the fertility often suddenly decreases in the first few generations. Nevertheless I believe that in all these experiments the fertility has been diminished by an independent cause, namely, from close interbreeding. I have collected so large a body of facts, showing

that close interbreeding lessens fertility, and, on the other hand, that an occasional cross with a distinct individual or variety increases fertility, that I cannot doubt the correctness of this almost universal belief amongst breeders. Hybrids are seldom raised by experimentalists in great numbers; and as the parent-species, or other allied hybrids, generally grow in the same garden, the visits of insects must be carefully prevented during the flowering season: hence hybrids will generally be fertilised during each generation by their own individual pollen; and I am conviced that this would be injurious to their fertility, already lessened by their hybrid origin. I am strengthened in this conviction by a remarkable statement repeatedly made by Gärtner, namely, that if even the less fertile hybrids be artificially fertilised with hybrid pollen of the same kind, their fertility, notwithstanding the frequent ill effects of manipulation, sometimes decidedly increases, and goes on increasing. Now, in artificial fertilisation pollen is as often taken by chance (as I know from my own experience) from the anthers of another flower, as from the anthers of the flower itself which is to be fertilised; so that a cross between two flowers, though probably on the same plant, would be thus effected. Moreover, whenever complicated experiments are in progress, so careful an observer as Gärtner would have castrated his hybrids, and this would have insured in each generation a cross with the pollen from a distinct flower, either from the same plant or from another plant of the same hybrid nature. And thus, the strange fact of the increase of fertility in the successive generations of *artificially fertilised* hybrids may, I believe, be accounted for by close interbreeding having been avoided.

Now let us turn to the results arrived at by the third most experienced hybridiser, namely, the Hon. and

1 Darwin pursues the line of thought introduced near the bottom of the previous page: inbreeding plays a role in reduced fertility.

2 Here is an interesting observation: artificial fertilization (by hand-pollination) leads to increased fertility, while natural fertilization does not. Darwin points out that hand-pollination introduces pollen from other individuals; in other words, fertility increases as a result of outcrossing.

1 Rev. W. Herbert. He is as emphatic in his conclusion that some hybrids are perfectly fertile—as fertile as the pure parent-species—as are Kölreuter and Gärtner that some degree of sterility between distinct species is a universal law of nature. He experimentised on some of the very same species as did Gärtner. The difference in their results may, I think, be in part accounted for by Herbert's great horticultural skill, and by his having hothouses at his command. Of his many important statements I will here give only a single one as an example, namely, that "every ovule in a pod of Crinum capense fertilised by C. revolutum produced a plant, which (he says) I never saw to occur in a case of its natural fecundation." So that we here have perfect, or even more than commonly perfect, fertility in a first cross between two distinct species.

2 This case of the Crinum leads me to refer to a most singular fact, namely, that there are individual plants, as with certain species of Lobelia, and with all the species of the genus Hippeastrum, which can be far more easily fertilised by the pollen of another and distinct species, than by their own pollen. For these plants have been found to yield seed to the pollen of a distinct species, though quite sterile with their own pollen, notwithstanding that their own pollen was found to be perfectly good, for it fertilised distinct species. So that certain individual plants and all the individuals of certain species can actually be hybridised much more readily than they can be self-fertilised! For instance, a bulb of Hippeastrum aulicum produced four flowers; three were fertilised by Herbert with their own pollen, and the fourth was subsequently fertilised by the pollen of a compound hybrid descended from three other and distinct species: the result was that "the ovaries of the three first flowers soon ceased to grow, and after a

1 Darwin brings in the Rev. Herbert's results as a foil to those of Gärtner and Kölreuter. In the opening sentence of this paragraph he calls Herbert the "third most experienced hybridiser," but he is even more laudatory in *Natural Selection,* where he calls him "the third greatest Hybridiser who ever lived" (Stauffer 1975, p. 398). Whereas Gärtner and Kölreuter see some degree of sterility between species as a "universal law of nature," Herbert asserts that some hybrids between species—including some of the very same species the German botanists studied—are completely interfertile. In his treatise *Amaryllidaceae* of 1837, Herbert wrote, "In fact, there is no real or natural line of difference between species and permanent or descendible variety" (Herbert 1837, p. 341). It is interesting to note that Darwin realized the significance of Herbert's work very early on. He wrote to him in 1839 via John Stevens Henslow with a list of ten detailed questions about crossing and interfertility (see *Correspondence* 2: 179), and more questions followed Herbert's response. Darwin cites Herbert, of course, to reinforce the point that there is no hard-and-fast rule concerning sterility and fertility.

As the title of his 1837 work suggests, Herbert made a special study of lilies. *Crinum* is a lily genus named by Linnaeus, who was probably inspired to give this name by the long, trailing white petals of the species *C. americanum* (the name is derived from the Greek *Krinos,* meaning "trailing hair" or the "tail of a comet"). *Crinum capense* and *C. revolutum,* names applied by Herbert to South African species, are now known as *C. bulbispermum* and *C. lineare,* respectively.

2 Even more remarkably—and supportive of Darwin's argument—is the finding that variation also exists *within* species in facility of intercrossing, yielding the decidedly odd situation where certain individual plants are more easily crossed with an entirely different species than fertilized with their own pollen. *Lobelia* is a genus in the bluebell family; *Hippeastrum* is a South American lily genus.

few days perished entirely, whereas the pod impregnated by the pollen of the hybrid made vigorous growth and rapid progress to maturity, and bore good seed, which vegetated freely." In a letter to me, in 1839, Mr. Herbert told me that he had then tried the experiment during five years, and he continued to try it during several subsequent years, and always with the same result. This result has, also, been confirmed by other observers in the case of Hippeastrum with its sub-genera, and in the case of some other genera, as Lobelia, Passiflora and Verbascum. Although the plants in these experiments appeared perfectly healthy, and although both the ovules and pollen of the same flower were perfectly good with respect to other species, yet as they were functionally imperfect in their mutual self-action, we must infer that the plants were in an unnatural state. Nevertheless these facts show on what slight and mysterious causes the lesser or greater fertility of species when crossed, in comparison with the same species when self-fertilised, sometimes depends.

The practical experiments of horticulturists, though not made with scientific precision, deserve some notice. It is notorious in how complicated a manner the species of Pelargonium, Fuchsia, Calceolaria, Petunia, Rhododendron, &c., have been crossed, yet many of these hybrids seed freely. For instance, Herbert asserts that a hybrid from Calceolaria integrifolia and plantaginea, species most widely dissimilar in general habit, "reproduced itself as perfectly as if it had been a natural species from the mountains of Chile." I have taken some pains to ascertain the degree of fertility of some of the complex crosses of Rhododendrons, and I am assured that many of them are perfectly fertile. Mr. C. Noble, for instance, informs me that he raises stocks for grafting from a hybrid

1 This information was included in the letter Herbert wrote on April 5, 1839, answering the questions Darwin sent via Henslow. The precise passage reads:

> The experience of 4 seasons has now shewn that it is certain that if, taking two hybrid Hippeastra wh. have (say) each a 4-flowered stem, 3 flowers on each are set with the dust of the plant itself & one on each with the dust of the other plant, that one on each of them will take the lead & ripen abundant seed, & the other 3 either fail or proceed more tardily & produce an inferior capsule of seed. I am at this moment trying the experiment in a bulb of that genus fresh from the Organ Mountains in Brazil to see whether its own natural pollen or that of a mule will take the lead. (*Correspondence* 2: 182)

2 Calceolarias, slipper-flowers, are South American plants related to snapdragons. Darwin is quoting a paper by Herbert published in the second volume of the *Journal of the Horticultural Society of London* in 1847 (p. 86).

between Rhod. Ponticum and Catawbiense, and that this hybrid "seeds as freely as it is possible to imagine." Had hybrids, when fairly treated, gone on decreasing in fertility in each successive generation, as Gärtner believes to be the case, the fact would have been notorious to nurserymen. Horticulturists raise large beds of the same hybrids, and such alone are fairly treated, for by insect agency the several individuals of the same hybrid variety are allowed to freely cross with each other, and the injurious influence of close interbreeding is thus prevented. Any one may readily convince himself of the efficiency of insect-agency by examining the flowers of the more sterile kinds of hybrid rhododendrons, which produce no pollen, for he will find on their stigmas plenty of pollen brought from other flowers.

In regard to animals, much fewer experiments have been carefully tried than with plants. If our systematic arrangements can be trusted, that is if the genera of animals are as distinct from each other, as are the genera of plants, then we may infer that animals more widely separated in the scale of nature can be more easily crossed than in the case of plants; but the hybrids themselves are, I think, more sterile. I doubt whether any case of a perfectly fertile hybrid animal can be considered as thoroughly well authenticated. It should, however, be borne in mind that, owing to few animals breeding freely under confinement, few experiments have been fairly tried: for instance, the canary-bird has been crossed with nine other finches, but as not one of these nine species breeds freely in confinement, we have no right to expect that the first crosses between them and the canary, or that their hybrids, should be perfectly fertile. Again, with respect to the fertility in successive generations of the more fertile

1 This is further argument against the view of Gärtner and others that hybrids are always sterile. These rhododendrons (*R. ponticum* is a Eurasian species; *R. catawbiense* is native to southeastern North America) readily hybridize, and the hybrids freely continue to intercross and produce fertile seeds. Note the clever observation establishing the efficiency of insect pollination: stigmas in flowers with sterile anthers are still pollen-covered. *Rhododendron* flowers are pollinated primarily by bees, by the way, and honey produced by some species is toxic owing to the presence of grayanotoxin, a terpenoid hydrocarbon in the nectar used to make the honey. This is the basis of "honey intoxication" or "mad honey" poisoning.

hybrid animals, I hardly know of an instance in which two families of the same hybrid have been raised at the same time from different parents, so as to avoid the ill effects of close interbreeding. On the contrary, brothers and sisters have usually been crossed in each successive generation, in opposition to the constantly repeated admonition of every breeder. And in this case, it is not at all surprising that the inherent sterility in the hybrids should have gone on increasing. If we were to act thus, and pair brothers and sisters in the case of any pure animal, which from any cause had the least tendency to sterility, the breed would assuredly be lost in a very few generations.

Although I do not know of any thoroughly well-authenticated cases of perfectly fertile hybrid animals, I have some reason to believe that the hybrids from Cervulus vaginalis and Reevesii, and from Phasianus colchicus with P. torquatus and with P. versicolor are perfectly fertile. The hybrids from the common and Chinese geese (A. cygnoides), species which are so different that they are generally ranked in distinct genera, have often bred in this country with either pure parent, and in one single instance they have bred *inter se*. This was effected by Mr. Eyton, who raised two hybrids from the same parents but from different hatches; and from these two birds he raised no less than eight hybrids (grandchildren of the pure geese) from one nest. In India, however, these cross-bred geese must be far more fertile; for I am assured by two eminently capable judges, namely Mr. Blyth and Capt. Hutton, that whole flocks of these crossed geese are kept in various parts of the country; and as they are kept for profit, where neither pure parent-species exists, they must certainly be highly fertile.

A doctrine which originated with Pallas, has been

1 The animals Darwin calls "Cervulus" here are muntjacs, Asian deer now known by the genus *Muntiacus. Muntiacus reevesi* is the endemic dwarf muntjac of Taiwan, while the animal Darwin calls *C. vaginalis* is now recognized as a subspecies of the Indian species *M. muntjac. Phasianus* is a pheasant genus: *P. colchicus* and *P. torquatus,* ring-necked pheasants, are now regarded as subspecies of the same species. *P. versicolor* is the green or Taiwanese pheasant.

2 Eyton, Blyth, and Hutton are just a few of the naturalists Darwin consulted for information on animal hybridization. Many more can be found in his discussion of hybridism in *Natural Selection* (Stauffer 1975, pp. 426–440).

3 In the first volume of *Variation,* p. 16, Darwin noted that Peter Simon Pallas originated this idea in a 1780 paper published in the "Act. Acad. St. Petersburgh," an anglicization of the academy's journal *Nova Acta Academiae Scientiarum Imperialis Petropolitanae.*

1 Domestic dogs likely arose from gray wolves in East Asia. In one sense this was a single origin in that dogs are monophyletic and are clearly sister taxon to wolves (some authors even consider dogs a wolf subspecies, *Canis lupus familiaris*). On the other hand, genetic evidence suggests that there was a period of repeated introgression of wolf genetic material into dogs, which could mean multiple domestication events or, more likely, periodic crossing or hybridizing with wolves. People may have bred dogs in this way purposefully, with the effect of injecting fresh variation to act upon by artificial selection. Timing estimates for dog domestication vary from as recently as 15,000 years ago to as long as 100,000 years (Vilà et al. 1997, Savolainen et al. 2002, and Lindblad-Toh et al. 2005). Cattle, in contrast, were likely domesticated in at least two distinct events. Indian humped cattle (zebu, *Bos indicus*) and humpless European, north Asian, and African cattle *(Bos taurus)* represent two events, though genetic evidence suggests that some humpless breeds of African cattle may derive from a third event, with a wild ox progenitor different from that of the remaining humpless clade (Hanotte et al. 2002).

Darwin's suggestion that interfertility increases under domestication likely has merit, since the process often involves some degree of repeated backcrossing and hybridizing with parental (progenitor) species. This would have the effect of introgressing genes of progenitor species into various descendant varieties, promoting genetic compatibility. Ability to hybridize, too, was also likely selected for, as new breeds can be developed through this process.

2 In modern parlance, "first crosses" are in essence parental crosses yielding an F_1, or what Darwin terms "hybrids." A "hybrid cross" is then a cross between F_1s, yielding F_2s.

1 largely accepted by modern naturalists; namely, that most of our domestic animals have descended from two or more aboriginal species, since commingled by intercrossing. On this view, the aboriginal species must either at first have produced quite fertile hybrids, or the hybrids must have become in subsequent generations quite fertile under domestication. This latter alternative seems to me the most probable, and I am inclined to believe in its truth, although it rests on no direct evidence. I believe, for instance, that our dogs have descended from several wild stocks; yet, with perhaps the exception of certain indigenous domestic dogs of South America, all are quite fertile together; and analogy makes me greatly doubt, whether the several aboriginal species would at first have freely bred together and have produced quite fertile hybrids. So again there is reason to believe that our European and the humped Indian cattle are quite fertile together; but from facts communicated to me by Mr. Blyth, I think they must be considered as distinct species. On this view of the origin of many of our domestic animals, we must either give up the belief of the almost universal sterility of distinct species of animals when crossed; or we must look at sterility, not as an indelible characteristic, but as one capable of being removed by domestication.

Finally, looking to all the ascertained facts on the intercrossing of plants and animals, it may be concluded that some degree of sterility, both in first crosses and in hybrids, is an extremely general result; but that it cannot, under our present state of knowledge, be considered as absolutely universal.

2 *Laws governing the Sterility of first Crosses and of Hybrids.*—We will now consider a little more in detail the

circumstances and rules governing the sterility of first crosses and of hybrids. Our chief object will be to see whether or not the rules indicate that species have specially been endowed with this quality, in order to prevent their crossing and blending together in utter confusion. The following rules and conclusions are chiefly drawn up from Gärtner's admirable work on the hybridisation of plants. I have taken much pains to ascertain how far the rules apply to animals, and considering how scanty our knowledge is in regard to hybrid animals, I have been surprised to find how generally the same rules apply to both kingdoms.

It has been already remarked, that the degree of fertility, both of first crosses and of hybrids, graduates from zero to perfect fertility. It is surprising in how many curious ways this gradation can be shown to exist; but only the barest outline of the facts can here be given. When pollen from a plant of one family is placed on the stigma of a plant of a distinct family, it exerts no more influence than so much inorganic dust. From this absolute zero of fertility, the pollen of different species of the same genus applied to the stigma of some one species, yields a perfect gradation in the number of seeds produced, up to nearly complete or even quite complete fertility; and, as we have seen, in certain abnormal cases, even to an excess of fertility, beyond that which the plant's own pollen will produce. So in hybrids themselves, there are some which never have produced, and probably never would produce, even with the pollen of either pure parent, a single fertile seed: but in some of these cases a first trace of fertility may be detected, by the pollen of one of the pure parent-species causing the flower of the hybrid to wither earlier than it otherwise would have done; and the early withering of the flower is well known to be a sign

1 This is another potshot at the idea that hybrid sterility is a special endowment with the purpose of preventing "utter confusion" of species. The admirable work referenced here is Gärtner's plant hybridization treatise *Versuche und Beobachtungen über die Bastarderzeugung im Pflanzenreich* (Stuttgart, 1849).

2 A continuum. Here Darwin is positing a broad gray area of incremental interfertility between the two extremes of no interfertility and complete interfertility. The impossibility of crossing plants of two families is a good reference point for zero interfertility. The likelihood of being able to cross increases with what Darwin calls "systematic affinity" on the next page.

of incipient fertilisation. From this extreme degree of sterility we have self-fertilised hybrids producing a greater and greater number of seeds up to perfect fertility.

1 Hybrids from two species which are very difficult to cross, and which rarely produce any offspring, are generally very sterile; but the parallelism between the difficulty of making a first cross, and the sterility of the hybrids thus produced—two classes of facts which are generally confounded together—is by no means strict. There are many cases, in which two pure species can be united with unusual facility, and produce numerous hybrid-offspring, yet these hybrids are remarkably sterile. On the other hand, there are species which can be crossed very rarely, or with extreme difficulty, but the hybrids, when at last produced, are very fertile. Even within the limits of the same genus, for instance in Dianthus, these two opposite cases occur.

The fertility, both of first crosses and of hybrids, is more easily affected by unfavourable conditions, than is the fertility of pure species. But the degree of fertility is likewise innately variable; for it is not always the same when the same two species are crossed under the same circumstances, but depends in part upon the constitution of the individuals which happen to have been chosen for the experiment. So it is with hybrids, for their degree of fertility is often found to differ greatly in the several individuals raised from seed out of the same capsule and exposed to exactly the same conditions.

2 By the term systematic affinity is meant, the resemblance between species in structure and in constitution, more especially in the structure of parts which are of high physiological importance and which differ little in the allied species. Now the fertility of first crosses

1 A diversity of genetic mechanisms underlies the basic patterns Darwin reports here—that presumably explains why variation in things like facility of crossing, non-equivalence of reciprocal crosses, relative degree of fertility of hybrids, etc., would have seemed so mysterious. Chromosome number can play a significant role, not only in simple genetic divergence, but also in determining whether a cross will produce viable or fertile progeny. Consider this early study of the primroses *Primula verticillata* (found in the Himalayas) and *P. floribunda* (found in the Arabian peninsula). Being geographically isolated, these species would ordinarily never have an opportunity to cross, but when they were brought together in the Royal Botanical Garden at Kew, the opportunity presented itself. Only sterile hybrids are typically produced—botanical equivalents of mules—owing to differences in chromosome number between the parent species. However, on rare occasions, the hybrid experiences spontaneous chromosome doubling, restoring a balanced number (thirty-six) that permits fertility. This is in effect a new species, christened *Primula kewensis* in honor of its "birthplace." See J. B. S. Haldane's book *Causes of Evolution* (1990) for a discussion of this and other examples of chromosome-mediated sterility and interfertility.

2 This makes sense intuitively: the facility of crossing is related to taxonomic relationship, or systematic affinity.

between species, and of the hybrids produced from them, is largely governed by their systematic affinity. This is clearly shown by hybrids never having been raised between species ranked by systematists in distinct families; and on the other hand, by very closely allied species generally uniting with facility. But the correspondence between systematic affinity and the facility of crossing is by no means strict. A multitude of cases could be given of very closely allied species which will not unite, or only with extreme difficulty; and on the other hand of very distinct species which unite with the utmost facility. In the ◀1 same family there may be a genus, as Dianthus, in which very many species can most readily be crossed; and another genus, as Silene, in which the most persevering efforts have failed to produce between extremely close species a single hybrid. Even within the limits of the same genus, we meet with this same difference; for instance, the many species of Nicotiana have been more largely crossed than the species of almost any other genus; but Gärtner found that N. acuminata, which is not a particularly distinct species, obstinately failed to fertilise, or to be fertilised by, no less than eight other species of Nicotiana. Very many analogous facts could be given.

No one has been able to point out what kind, or what amount, of difference in any recognisable character is sufficient to prevent two species crossing. It can be shown that plants most widely different in habit and general appearance, and having strongly marked differences in every part of the flower, even in the pollen, in the fruit, and in the cotyledons, can be crossed. Annual and perennial plants, deciduous and evergreen trees, plants inhabiting different stations and fitted for extremely different climates, can often be crossed with ease.

1 *Dianthus* and *Silene* are members of the pink family, Caryophyllaceae, which includes such familiar garden flowers as phlox, sweet william, and campion. The earliest use of the name "pink" for these flowers is 1566 (*Oxford English Dictionary*), perhaps referring to the common flower color of species in this group. *Nicotiana* is the tobacco genus, in the nightshade family (Solanaceae); nicotine takes its name from the genus.

By a reciprocal cross between two species, I mean the case, for instance, of a stallion-horse being first crossed with a female-ass, and then a male-ass with a mare: these two species may then be said to have been [1] reciprocally crossed. There is often the widest possible difference in the facility of making reciprocal crosses. Such cases are highly important, for they prove that the capacity in any two species to cross is often completely independent of their systematic affinity, or of any recognisable difference in their whole organisation. On the other hand, these cases clearly show that the capacity for crossing is connected with constitutional differences imperceptible by us, and confined to the reproductive system. This difference in the result of reciprocal crosses between the same two species was long ago observed by Kölreuter. To give an instance: [2] Mirabilis jalappa can easily be fertilised by the pollen of M. longiflora, and the hybrids thus produced are sufficiently fertile; but Kölreuter tried more than two hundred times, during eight following years, to fertilise reciprocally M. longiflora with the pollen of M. jalappa, and utterly failed. Several other equally striking cases [3] could be given. Thuret has observed the same fact with certain sea-weeds or Fuci. Gärtner, moreover, found that this difference of facility in making reciprocal crosses is extremely common in a lesser degree. He has observed it even between forms so closely related [4] (as Matthiola annua and glabra) that many botanists rank them only as varieties. It is also a remarkable fact, that hybrids raised from reciprocal crosses, though of course compounded of the very same two species, the one species having first been used as the father and then as the mother, generally differ in fertility in a small, and occasionally in a high degree.

Several other singular rules could be given from

1 How puzzling such observations must have been! Why should a cross work well one way and not the other? Darwin was right when he noted that it clearly had something to do with "constitutional differences" of the reproductive system, "imperceptible" to biologists of the age. We now know that one reason for unequal outcomes in reciprocal crossing stems from genetic differences of the sex chromosomes. Generally among offspring produced in a cross the heterogametic sex—the sex that bears dissimilar sex chromosomes, such as XY males in mammals—is sterile or absent altogether. This is known as *Haldane's Rule* (see Jones and Barton 1985). Another reason is *genomic imprinting*, whereby parent-of-origin of genes influences their expression (Wilkins and Haig 2003). Imprinting effects on developmental genes are responsible for asymmetric sterility or inviability in some reciprocal crosses.

2 *Mirabilis* is the four o'clock flower genus, family Nyctaginaceae. *M. jalapa* (as it is spelled now), sometimes called the "Marvel of Peru" (hence "mirabilis"), is the most widely planted ornamental four o'clock species. These plants were later used in important studies showing that some genetic material is inherited in a non-Mendelian manner, through the chloroplasts.

3 Darwin was familiar with Gustave Adolphe Thuret's 1853 and 1855 works on reproduction in the brown algae, family Fucaceae, which are referred to here as Fuci, as well as other publications by Thuret. In an 1857 letter, Darwin wrote to a correspondent, "No doubt you are familiar with Thurets admirable papers on the Sexes of the Algae, in various Vols. of An. des Sc. Nat. If not, you should look to them" (*Correspondence* 6: 316).

4 *Matthiola* is a genus of flowering plant in the mustard family, Brassicaceae, common in Europe and Asia.

Gärtner: for instance, some species have a remarkable power of crossing with other species; other species of the same genus have a remarkable power of impressing their likeness on their hybrid offspring; but these two powers do not at all necessarily go together. There are certain hybrids which instead of having, as is usual, an intermediate character between their two parents, always closely resemble one of them; and such hybrids, though externally so like one of their pure parent-species, are with rare exceptions extremely sterile. So again amongst hybrids which are usually intermediate in structure between their parents, exceptional and abnormal individuals sometimes are born, which closely resemble one of their pure parents; and these hybrids are almost always utterly sterile, even when the other hybrids raised from seed from the same capsule have a considerable degree of fertility. These facts show how completely fertility in the hybrid is independent of its external resemblance to either pure parent.

Considering the several rules now given, which govern the fertility of first crosses and of hybrids, we see that when forms, which must be considered as good and distinct species, are united, their fertility graduates from zero to perfect fertility, or even to fertility under certain conditions in excess. That their fertility, besides being eminently susceptible to favourable and unfavourable conditions, is innately variable. That it is by no means always the same in degree in the first cross and in the hybrids produced from this cross. That the fertility of hybrids is not related to the degree in which they resemble in external appearance either parent. And lastly, that the facility of making a first cross between any two species is not always governed by their systematic affinity or ◀1

1 This concise summary of the foregoing observations sets the stage for the logical question: why?

1 This is a rare show of passion in an otherwise dense chapter. It is in keeping with Darwin's rhetorical strategy: identify patterns, and then ask which process most likely explains them. It is clear that the "strange arrangements" we see in the peculiar reproductive abilities and inabilities of so many species (some cross, some cannot; some produce sterile hybrids, some fertile ones) cannot have been specially created—they are too messy, too inconsistent.

degree of resemblance to each other. This latter statement is clearly proved by reciprocal crosses between the same two species, for according as the one species or the other is used as the father or the mother, there is generally some difference, and occasionally the widest possible difference, in the facility of effecting an union. The hybrids, moreover, produced from reciprocal crosses often differ in fertility.

Now do these complex and singular rules indicate that species have been endowed with sterility simply to prevent their becoming confounded in nature? I think not. For why should the sterility be so extremely different in degree, when various species are crossed, all of which we must suppose it would be equally important to keep from blending together? Why should the degree of sterility be innately variable in the individuals of the same species? Why should some species cross with facility, and yet produce very sterile hybrids; and other species cross with extreme difficulty, and yet produce fairly fertile hybrids? Why should there often be so great a difference in the result of a reciprocal cross between the same two species? Why, it may even be asked, has the production of hybrids been permitted? to grant to species the special power of producing hybrids, and then to stop their further propagation by different degrees of sterility, not strictly related to the facility of the first union between their parents, seems to be a strange arrangement.

The foregoing rules and facts, on the other hand, appear to me clearly to indicate that the sterility both of first crosses and of hybrids is simply incidental or dependent on unknown differences, chiefly in the reproductive systems, of the species which are crossed. The differences being of so peculiar and limited a nature, that, in reciprocal crosses between two species the male sexual element of the one will often freely act on the female sexual element of the other, but not in a reversed direction. It will be advisable to explain a little more fully by an example what I mean by sterility being incidental on other differences, and not a specially endowed quality. As the capacity of one plant to be grafted or budded on another is so entirely unimportant for its welfare in a state of nature, I presume that no one will suppose that this capacity is a *specially* endowed quality, but will admit that it is incidental on differences in the laws of growth of the two plants. We can sometimes see the reason why one tree will not take on another, from differences in their rate of growth, in the hardness of their wood, in the period of the flow or nature of their sap, &c.; but in a multitude of cases we can assign no reason whatever. Great diversity in the size of two plants, one being woody and the other herbaceous, one being evergreen and the other deciduous, and adaptation to widely different climates, does not always prevent the two grafting together. As in hybridisation, so with grafting, the capacity is limited by systematic affinity, for no one has been able to graft trees together belonging to quite distinct families; and, on the other hand, closely allied species, and varieties of the same species, can usually, but not invariably, be grafted with ease. But this capacity, as in hybridisation, is by no means absolutely governed by systematic affinity. Although many distinct genera within the same family have been grafted together, in other cases species of the same genus will not take on each other. The pear can be grafted far more readily on the quince, which is ranked as a distinct genus, than on the apple, which is a member of the same genus. Even different varieties of the pear take

1 The remainder of this section, beginning here and continuing on for the next page and a half, offers a wonderful example of Darwin's ingenuity. He realized that the facility of grafting offers an analog for hybridizing. The likelihood that a graft "takes" is analogous to the production of fertile hybrids—and more than analogous, since the same general pattern of systematic affinity as indicator of the likelihood of a graft (or cross) holds here. And yet grafting is wholly artificial, a human invention. The patterns of facility cannot be due to any "specially endowed quality," Darwin points out, but flow from degree of taxonomic relationship, just as in the case of facility of hybridization. Darwin even sees a parallel in the idiosyncrasies of hybridizing and grafting; in *Natural Selection* he wrote, "[The French horticulturist] Sageret, moreover, gives reasons for believing that *individual* stocks have a repugnance to receive the grafts of certain varieties, in the same way as we have seen the *individual* plants resist being hybridized" (Stauffer 1975, p. 420; emphasis in original).

with different degrees of facility on the quince; so do different varieties of the apricot and peach on certain varieties of the plum.

As Gärtner found that there was sometimes an innate difference in different *individuals* of the same two species in crossing; so Sagaret believes this to be the case with different individuals of the same two species in being grafted together. As in reciprocal crosses, the facility of effecting an union is often very far from equal, so it sometimes is in grafting; the common gooseberry, for instance, cannot be grafted on the currant, whereas the currant will take, though with difficulty, on the gooseberry.

We have seen that the sterility of hybrids, which have their reproductive organs in an imperfect condition, is a very different case from the difficulty of uniting two pure species, which have their reproductive organs perfect; yet these two distinct cases run to a certain extent parallel. Something analogous occurs in grafting; for Thouin found that three species of Robinia, which seeded freely on their own roots, and which could be grafted with no great difficulty on another species, when thus grafted were rendered barren. On the other hand, certain species of Sorbus, when grafted on other species, yielded twice as much fruit as when on their own roots. We are reminded by this latter fact of the extraordinary case of Hippeastrum, Lobelia, &c., which seeded much more freely when fertilised with the pollen of distinct species, than when self-fertilised with their own pollen.

We thus see, that although there is a clear and fundamental difference between the mere adhesion of grafted stocks, and the union of the male and female elements in the act of reproduction, yet that there is a rude degree of parallelism in the results of grafting and

1 Augustin Sageret (incorrectly spelled Sagaret by Darwin) published *Pomologie physiologique* (Paris, 1830), on the biology of fruit trees; Darwin cited this work several times in *Natural Selection* and *Variation,* and probably got the information mentioned here from this work.

2 The plants mentioned in this paragraph in connection with grafting experiments include locust *(Robinia),* a legume, and rowans, or mountain-ashes *(Sorbus),* found in the rose family. Note the curious observation Darwin cites here: certain *Sorbus* bore twice as much fruit when grafted onto other species as when grafted onto rootstock of their own species. This is a remarkable phenomenon that Darwin thinks is parallel to the increased productivity that results from outcrossed pollination. The observation came from an 1810 paper by the French horticulturist André Thouin, author of treatises on arboriculture and grafting techniques.

of crossing distinct species. And as we must look at the curious and complex laws governing the facility with which trees can be grafted on each other as incidental on unknown differences in their vegetative systems, so I believe that the still more complex laws governing the facility of first crosses, are incidental on unknown differences, chiefly in their reproductive systems. These differences, in both cases, follow to a certain extent, as might have been expected, systematic affinity, by which every kind of resemblance and dissimilarity between organic beings is attempted to be expressed. The facts by no means seem to me to indicate that the greater or lesser difficulty of either grafting or crossing together various species has been a special endowment; although in the case of crossing, the difficulty is as important for the endurance and stability of specific forms, as in the case of grafting it is unimportant for their welfare.

Causes of the Sterility of first Crosses and of Hybrids.— We may now look a little closer at the probable causes of the sterility of first crosses and of hybrids. These two cases are fundamentally different, for, as just remarked, in the union of two pure species the male and female sexual elements are perfect, whereas in hybrids they are imperfect. Even in first crosses, the greater or lesser difficulty in effecting a union apparently depends on several distinct causes. There must sometimes be a physical impossibility in the male element reaching the ovule, as would be the case with a plant having a pistil too long for the pollen-tubes to reach the ovarium. It has also been observed that when pollen of one species is placed on the stigma of a distantly allied species, though the pollen-tubes protrude, they do not penetrate the stigmatic surface. Again, the

1 Keep in mind the terminology Darwin uses: "first crosses" are matings of distinct "parental" forms, producing hybrids. "Hybrid crosses" are subsequent matings of hybrid progeny from the first crosses. Darwin is saying here that the causes of sterility in the cases of first crosses and of hybrid crosses are different.

2 There are three stages at which things can go wrong in first crosses: 1) There may be a physical or mechanical problem with sperm (or pollen) reaching the egg (or ovule); 2) Sperm or pollen may reach the egg or ovule, but for some reason be unable to achieve fertilization; 3) Fertilization may occur, but the embryo fails to develop, or develops to some degree and then aborts.

Noteworthy here is the fact that Darwin is unpacking a black box, namely, "inability to cross," and showing, not only that there is more to such observations than casual consideration would suggest, but that there are explicable causes behind them. He does not begin to understand the causes, but mapping out the various subtleties of the observed phenomenon is a critical first step.

male element may reach the female element, but be incapable of causing an embryo to be developed, as seems to have been the case with some of Thuret's experiments on Fuci. No explanation can be given of these facts, any more than why certain trees cannot be grafted on others. Lastly, an embryo may be developed, and then perish at an early period. This latter alternative has not been sufficiently attended to; but I believe, from observations communicated to me by [1] Mr. Hewitt, who has had great experience in hybridising gallinaceous birds, that the early death of the embryo is a very frequent cause of sterility in first crosses. I was at first very unwilling to believe in this view; as hybrids, when once born, are generally healthy and long-lived, as we see in the case of the common mule. Hybrids, however, are differently circumstanced before and after birth: when born and living in a country where their two parents can live, they are generally placed under suitable conditions of life. But a hybrid partakes of only half of the nature and constitution of its mother, and therefore before birth, as long as it is nourished within its mother's womb or within the egg or seed produced by the mother, it may be exposed to conditions in some degree unsuitable, and consequently be liable to perish at an early period; more especially as all very young beings seem eminently sensitive to injurious or unnatural conditions of life.

[2] In regard to the sterility of hybrids, in which the sexual elements are imperfectly developed, the case is very different. I have more than once alluded to a large body of facts, which I have collected, showing that when animals and plants are removed from their natural conditions, they are extremely liable to have their reproductive systems seriously affected. This, in fact, is

1 Mr. Hewitt did indeed have great experience hybridizing birds. This is Edward Hewitt, poultry show judge and avid poultry breeder. In a letter responding to a query by Darwin (December 1857), Hewitt wrote that his long interest in hybridizing varieties of pheasant and other fowl "was alike my 'hobby' and 'my folly'" (*Correspondence* 6: 509).

2 Here Darwin is building a curious argument. He maintains that hybrid sterility is related to the supposed disturbance of the reproductive system caused by exposing organisms to altered environmental conditions. Recall from chapters I and II that this "disturbance" was considered an important source of variation that could then be acted upon in artificial selection.

the great bar to the domestication of animals. Between the sterility thus superinduced and that of hybrids, there are many points of similarity. In both cases the sterility is independent of general health, and is often accompanied by excess of size or great luxuriance. In both cases, the sterility occurs in various degrees; in both, the male element is the most liable to be affected; but sometimes the female more than the male. In both, the tendency goes to a certain extent with systematic affinity, for whole groups of animals and plants are rendered impotent by the same unnatural conditions; and whole groups of species tend to produce sterile hybrids. On the other hand, one species in a group will sometimes resist great changes of conditions with unimpaired fertility; and certain species in a group will produce unusually fertile hybrids. No one can tell, till he tries, whether any particular animal will breed under confinement or any plant seed freely under culture; nor can he tell, till he tries, whether any two species of a genus will produce more or less sterile hybrids. Lastly, when organic beings are placed during several generations under conditions not natural to them, they are extremely liable to vary, which is due, as I believe, to their reproductive systems having been specially affected, though in a lesser degree than when sterility ensues. So it is with hybrids, for hybrids in successive generations are eminently liable to vary, as every experimentalist has observed.

Thus we see that when organic beings are placed under new and unnatural conditions, and when hybrids are produced by the unnatural crossing of two species, the reproductive system, independently of the general state of health, is affected by sterility in a very similar manner. In the one case, the conditions of life have been disturbed, though often in so slight a degree as to [1]

1 Darwin is suggesting that the reproductive system is similarly disturbed by placing organisms of the same species in an unnatural environment or by hybridizing organisms of different species. The outer environment creates the problem in the first case, while in the second the hybrid offspring's "inner environment" is disturbed by blending its parents' constitutions.

1 Vague to say the least; making sense of sterility and interfertility patterns was a difficult (and, one imagines, frustrating) proposition in Darwin's time. One of Darwin's strengths is his ability to synthesize, to bring together data and observations in a cogent fashion in an attempt to explain a phenomenon. His efforts with sterility and interfertility are valiant, but, linked to a limited understanding of heredity and causes of variation, his model is itself flawed—and he himself acknowledges the difficulty and does not claim to have solved the problem. In the final two paragraphs of this section we see Darwin reach out for yet one more pattern he thinks he glimpses, while acknowledging that it may be nothing more than his fancy. This is described on the next page.

be inappreciable by us; in the other case, or that of hybrids, the external conditions have remained the same, but the organisation has been disturbed by two different structures and constitutions having been blended into one. For it is scarcely possible that two organisations should be compounded into one, without some disturbance occurring in the development, or periodical action, or mutual relation of the different parts and organs one to another, or to the conditions of life. When hybrids are able to breed *inter se*, they transmit to their offspring from generation to generation the same compounded organisation, and hence we need not be surprised that their sterility, though in some degree variable, rarely diminishes.

1 It must, however, be confessed that we cannot understand, excepting on vague hypotheses, several facts with respect to the sterility of hybrids; for instance, the unequal fertility of hybrids produced from reciprocal crosses; or the increased sterility in those hybrids which occasionally and exceptionally resemble closely either pure parent. Nor do I pretend that the foregoing remarks go to the root of the matter: no explanation is offered why an organism, when placed under unnatural conditions, is rendered sterile. All that I have attempted to show, is that in two cases, in some respects allied, sterility is the common result,—in the one case from the conditions of life having been disturbed, in the other case from the organisation having been disturbed by two organisations having been compounded into one.

It may seem fanciful, but I suspect that a similar parallelism extends to an allied yet very different class of facts. It is an old and almost universal belief, founded, I think, on a considerable body of evidence, that slight changes in the conditions of life are beneficial to all living things. We see this acted on by farmers and gardeners in their frequent exchanges of seed, tubers, &c., from one soil or climate to another, and back again. During the convalescence of animals, we plainly see that great benefit is derived from almost any change in the habits of life. Again, both with plants and animals, there is abundant evidence, that a cross between very distinct individuals of the same species, that is between members of different strains or sub-breeds, gives vigour and fertility to the offspring. I believe, indeed, from the facts alluded to in our fourth chapter, that a certain amount of crossing is indispensable even with hermaphrodites; and that close interbreeding continued during several generations between the nearest relations, especially if these be kept under the same conditions of life, always induces weakness and sterility in the progeny.

Hence it seems that, on the one hand, slight changes in the conditions of life benefit all organic beings, and on the other hand, that slight crosses, that is crosses between the males and females of the same species which have varied and become slightly different, give vigour and fertility to the offspring. But we have seen that greater changes, or changes of a particular nature, often render organic beings in some degree sterile; and that greater crosses, that is crosses between males and females which have become widely or specifically different, produce hybrids which are generally sterile in some degree. I cannot persuade myself that this parallelism is an accident or an illusion. Both series of facts seem to be connected together by some common but unknown bond, which is essentially related to the principle of life. ◀1

Fertility of Varieties when crossed, and of their Mongrel offspring.—It may be urged, as a most forcible argu- ◀2

1 Having argued on the previous page that altered or unnatural conditions, internal or external, can lead to problems, Darwin must now reconcile this idea with his even earlier assertion that outbreeding is beneficial whereas inbreeding is not. For what is hybridization but outbreeding, albeit a rather extreme form? Indeed, it is the "extreme" quality that is the problem, according to Darwin. Here he argues that slight environmental changes and what we may think of as "slight outcrossing" (crosses between varieties of the same species) are both beneficial, while more dramatic environmental changes and more extreme outcrossing can lead to such problems as sterility. He sees a parallel here, a "common but unknown bond" that lies at the heart of the essence of life itself, but he cannot see any further.

In broad terms some of these observations are not in dispute: close inbreeding *can* compromise health and fertility, and hybrids generated from crosses between distinct species often *can* be sterile or subfertile. Health and vigor lie between the two: crosses between more closely related forms, like varieties of the same species, tend to be quite robust. The environment's supposed effect is the weak link in this parallelism. A change of conditions can be healthy or not, after all, depending on myriad circumstances. And environment has no perturbing effect on the reproductive system, except under certain circumstances (mutagenic chemicals or irradiation).

2 A note on Darwin's terminology: here "mongrel" refers to the offspring of crossed varieties. So in this section Darwin is considering the fertility of varieties themselves, as compared with the fertility of the mongrels produced. The word "mongrel," by the way, appears to be derived from Old and Middle English words meaning "mixture" or "mingling" *(gemang, mong)*. In English the suffix "-rel" tends to be pejorative (e.g., scoundrel, wastrel), so we get a sense that "mongrels"—mixed breeds—are undesirable (though many dog lovers will point out that mixed breeds tend to have the best dispositions).

1 This is a reiteration of Darwin's earlier argument that there is no essential difference between species and varieties, and that the "rule of thumb" criterion of the interfertility of varieties but not of species does not hold up.

2 Sterility and interfertility vary even with domesticated varieties. Most interbreed readily, but some do not.

1 ment, that there must be some essential distinction between species and varieties, and that there must be some error in all the foregoing remarks, inasmuch as varieties, however much they may differ from each other in external appearance, cross with perfect facility, and yield perfectly fertile offspring. I fully admit that this is almost invariably the case. But if we look to varieties produced under nature, we are immediately involved in hopeless difficulties; for if two hitherto reputed varieties be found in any degree sterile together, they are at once ranked by most naturalists as species. For instance, the blue and red pimpernel, the primrose and cowslip, which are considered by many of our best botanists as varieties, are said by Gärtner not to be quite fertile when crossed, and he consequently ranks them as undoubted species. If we thus argue in a circle, the fertility of all varieties produced under nature will assuredly have to be granted.

2 If we turn to varieties, produced, or supposed to have been produced, under domestication, we are still involved in doubt. For when it is stated, for instance, that the German Spitz dog unites more easily than other dogs with foxes, or that certain South American indigenous domestic dogs do not readily cross with European dogs, the explanation which will occur to every one, and probably the true one, is that these dogs have descended from several aboriginally distinct species. Nevertheless the perfect fertility of so many domestic varieties, differing widely from each other in appearance, for instance of the pigeon or of the cabbage, is a remarkable fact; more especially when we reflect how many species there are, which, though resembling each other most closely, are utterly sterile when intercrossed. Several considerations, however, render the fertility of domestic varieties less remarkable than

at first appears. It can, in the first place, be clearly shown that mere external dissimilarity between two species does not determine their greater or lesser degree of sterility when crossed; and we may apply the same rule to domestic varieties. In the second place, some eminent naturalists believe that a long course of domestication tends to eliminate sterility in the successive generations of hybrids, which were at first only slightly sterile; and if this be so, we surely ought not to expect to find sterility both appearing and disappearing under nearly the same conditions of life. Lastly, and this seems to me by far the most important consideration, new races of animals and plants are produced under domestication by man's methodical and unconscious power of selection, for his own use and pleasure: he neither wishes to select, nor could select, slight differences in the reproductive system, or other constitutional differences correlated with the reproductive system. He supplies his several varieties with the same food; treats them in nearly the same manner, and does not wish to alter their general habits of life. Nature acts uniformly and slowly during vast periods of time on the whole organisation, in any way which may be for each creature's own good; and thus she may, either directly, or more probably indirectly, through correlation, modify the reproductive system in the several descendants from any one species. Seeing this difference in the process of selection, as carried on by man and nature, we need not be surprised at some difference in the result.

I have as yet spoken as if the varieties of the same species were invariably fertile when intercrossed. But it seems to me impossible to resist the evidence of the existence of a certain amount of sterility in the few following cases, which I will briefly abstract. The evidence is at least as good as that from which we believe

1 This "most important consideration" for why we should expect domestic varieties to be interfertile is interesting. Darwin suggests that in domestication people artificially select traits of importance to them, but do not select for differences in the reproductive system. Recalling his belief that conditions of life can affect the reproductive system, it makes sense that Darwin would then point out that, insofar as domestic varieties are provided with the same food and are kept under the same conditions, their reproductive system would remain unaltered.

Other considerations suggest that people can alter the reproductive systems of their domesticated plants and animals, though Darwin seems to discount these. Can increased interfertility be selected for? On this page and back in chapter I, p. 26, Darwin mentions that some authors believe that long-continued domestication can reduce sterility. He allows this but does not endorse the idea because it undermines his argument that varieties are developed by selection and not merely by crossing aboriginal species. If hybrid sterility could be easily reduced over time, the door is open to a hybrid origin for all domestic varieties. On this page the idea is not endorsed because, he says, if hybrid sterility could be reduced over time we should not expect to find sterility appearing and disappearing under the same physical conditions, and yet we do. I am not sure we should agree with this assertion; after all, other traits of domesticated organisms are altered by long-running artificial selection, all the while maintaining these organisms under more or less the same physical conditions.

2 In *Natural Selection,* Darwin noted, "On the view . . . of sterility being commonly lost between species when under domestication, it would be most strange if sterility were, also, to supervene between varieties under raised domestication" (Stauffer 1975, p. 444). That sentence did not make it into the *Origin,* yet it would have been a fitting opener to this section, where Darwin finds it "impossible to resist" the evidence for various degrees of sterility in varieties of domesticated species. Note, too, that this evidence comes from "hostile witnesses"—those esteemed plant breeders Gärtner and Kölreuter, who regarded sterility or interfertility as diagnostic tests for species status. It is worth noting that Darwin devotes perhaps two pages to this evidence here, but in *Natural Selection* had amassed fourteen pages of examples from plants and animals.

1 In *Variation* Darwin discussed the work of the agronomist and plant physiologist Louis François Charles Girou de Buzareingues, who experimented with crossing three French gourd varieties—*Barbarines, Pastissons,* and *Giraumous*—and reported the results in a paper published in 1833. Their ability to hybridize was not surprising because, as Augustin Sageret argued in *Memoire sur les Cucurbitaceae* of 1826, they are varieties of the same species, after all. But why should the facility of hybridizing these varieties decline as the differences between them grow greater?

2 *Verbascum* is the mullein genus, a large group of several hundred species belonging to the snapdragon family (Scrophulariaceae). Many mulleins are popular ornamental plants and were a favorite experimental plant with breeders like Gärtner. In *Variation,* Darwin gave a bit more information about Gärtner's *Verbascum* studies: he crossed an astounding 1,085 flowers and counted the seed yield over an eighteen-year period! (*Variation,* v. II, p. 83). Note the key observation here: crosses *within* varieties are more productive than crosses *between* varieties. Darwin concludes that "in the genus Verbascum the similarly and dissimilarly-coloured varieties of the same species behave, when crossed, like closely allied but distinct species" (*Variation,* vol. II, p. 85). By now the significance of varieties behaving like species should be clear.

in the sterility of a multitude of species. The evidence is, also, derived from hostile witnesses, who in all other cases consider fertility and sterility as safe criterions of specific distinction. Gärtner kept during several years a dwarf kind of maize with yellow seeds, and a tall variety with red seeds, growing near each other in his garden; and although these plants have separated sexes, they never naturally crossed. He then fertilised thirteen flowers of the one with the pollen of the other; but only a single head produced any seed, and this one head produced only five grains. Manipulation in this case could not have been injurious, as the plants have separated sexes. No one, I believe, has suspected that these varieties of maize are distinct species; and it is important to notice that the hybrid plants thus raised were themselves *perfectly* fertile; so that even Gärtner did not venture to consider the two varieties as specifically distinct.

1 Girou de Buzareingues crossed three varieties of gourd, which like the maize has separated sexes, and he asserts that their mutual fertilisation is by so much the less easy as their differences are greater. How far these experiments may be trusted, I know not; but the forms experimentised on, are ranked by Sagaret, who mainly founds his classification by the test of infertility, as varieties.

2 The following case is far more remarkable, and seems at first quite incredible; but it is the result of an astonishing number of experiments made during many years on nine species of Verbascum, by so good an observer and so hostile a witness, as Gärtner: namely, that yellow and white varieties of the same species of Verbascum when intercrossed produce less seed, than do either coloured varieties when fertilised with pollen from their own coloured flowers. Moreover, he asserts that when

yellow and white varieties of one species are crossed with yellow and white varieties of a *distinct* species, more seed is produced by the crosses between the same coloured flowers, than between those which are differently coloured. Yet these varieties of Verbascum present no other difference besides the mere colour of the flower; and one variety can sometimes be raised from the seed of the other.

From observations which I have made on certain varieties of hollyhock, I am inclined to suspect that they present analogous facts.

Kölreuter, whose accuracy has been confirmed by every subsequent observer, has proved the remarkable fact, that one variety of the common tobacco is more fertile, when crossed with a widely distinct species, than are the other varieties. He experimentised on five forms, which are commonly reputed to be varieties, and which he tested by the severest trial, namely, by reciprocal crosses, and he found their mongrel offspring perfectly fertile. But one of these five varieties, when used either as father or mother, and crossed with the Nicotiana glutinosa, always yielded hybrids not so sterile as those which were produced from the four other varieties when crossed with N. glutinosa. Hence the reproductive system of this one variety must have been in some manner and in some degree modified.

From these facts; from the great difficulty of ascertaining the infertility of varieties in a state of nature, for a supposed variety if infertile in any degree would generally be ranked as species; from man selecting only external characters in the production of the most distinct domestic varieties, and from not wishing or being able to produce recondite and functional differences in the reproductive system; from these several considerations and facts, I do not think that the very general

1 Tobacco, genus *Nicotiana,* was a favorite experimental plant of Kölreuter's. Recall that he undertook his studies in the early 1760s in order to refute Linnaeus's suggestion that new forms could arise by hybridization. Early on he distinguished between "hybrid varieties," which produced fertile offspring, and "hybrid species," which did not. In the experiments Darwin describes here, the so-called mongrel offspring are interfertile *hybrid varieties* of common tobacco, *Nicotiana tabacum.* When these varieties are crossed with the species *N. glutinosa*—*hybrid species* crosses—the results are predictable sterility with the exception of one of the crosses. This Darwin cites as having an altered reproductive system, allowing it to be interfertile with another species. Ironically, given Kölreuter's opposition to the idea of hybrid origins, modern genetic analysis supports a hybrid origin for the polyploid *N. tabacum* (Ren and Timko 2001).

1 If the fertility of varieties is not universal, the door is opened to the possibility of degrees of fertility, which itself opens the door to the notion of transition stemming from slowly accumulated modifications.

2 Gärtner did regard species and varieties as fundamentally different, even if they were externally similar. He developed a classification of hybrids based on whether they were produced by species or variety crosses, and which supplied the pollen and which the ovule (Müller-Wille and Orel 2007). Darwin's point here is that offspring produced by species crosses really do not differ much from those produced by variety crosses, yet there are some puzzling distinctions.

▶1 fertility of varieties can be proved to be of universal occurrence, or to form a fundamental distinction between varieties and species. The general fertility of varieties does not seem to me sufficient to overthrow the view which I have taken with respect to the very general, but not invariable, sterility of first crosses and of hybrids, namely, that it is not a special endowment, but is incidental on slowly acquired modifications, more especially in the reproductive systems of the forms which are crossed.

Hybrids and Mongrels compared, independently of their fertility.—Independently of the question of fertility, the offspring of species when crossed and of varieties when crossed may be compared in several other respects. ▶2 Gärtner, whose strong wish was to draw a marked line of distinction between species and varieties, could find very few and, as it seems to me, quite unimportant differences between the so-called hybrid offspring of species, and the so-called mongrel offspring of varieties. And, on the other hand, they agree most closely in very many important respects.

I shall here discuss this subject with extreme brevity. The most important distinction is, that in the first generation mongrels are more variable than hybrids; but Gärtner admits that hybrids from species which have long been cultivated are often variable in the first generation; and I have myself seen striking instances of this fact. Gärtner further admits that hybrids between very closely allied species are more variable than those from very distinct species; and this shows that the difference in the degree of variability graduates away. When mongrels and the more fertile hybrids are propagated for several generations an extreme amount of variability in their offspring is notorious; but some few cases both of hybrids and mongrels long retaining uniformity of character could be given. The variability, however, in the successive generations of mongrels is, perhaps, greater than in hybrids.

This greater variability of mongrels than of hybrids does not seem to me at all surprising. For the parents of mongrels are varieties, and mostly domestic varieties (very few experiments having been tried on natural varieties), and this implies in most cases that there has been recent variability; and therefore we might expect that such variability would often continue and be superadded to that arising from the mere act of crossing. The slight degree of variability in hybrids from the first cross or in the first generation, in contrast with their extreme variability in the succeeding generations, is a curious fact and deserves attention. For it bears on and corroborates the view which I have taken on the cause of ordinary variability; namely, that it is due to the reproductive system being eminently sensitive to any change in the conditions of life, being thus often rendered either impotent or at least incapable of its proper function of producing offspring identical with the parent-form. Now hybrids in the first generation are descended from species (excluding those long cultivated) which have not had their reproductive systems in any way affected, and they are not variable; but hybrids themselves have their reproductive systems seriously affected, and their descendants are highly variable.

But to return to our comparison of mongrels and hybrids: Gärtner states that mongrels are more liable than hybrids to revert to either parent-form; but this, if it be true, is certainly only a difference in degree. Gärtner further insists that when any two species, although most closely allied to each other, are

1 One distinguishing feature of hybrids (produced by species crosses) and mongrels (produced by variety crosses) is their relative variability: mongrels tend to be more variable than hybrids. Darwin attributes this fact to perturbing the reproductive system by the act of crossing, arguing that hybrids, the first generation produced by crossing, should not be notably variable since they were produced by crossing parents that had stable and unaffected reproductive systems. The *reproductive system* of these hybrids, however, should be affected, so that when these reproduce, their offspring are variable. The modern take, of course, is that first-generation progeny from a species cross will be almost uniformly heterozygous, and that uniformity translates into low phenotypic variability. The second-generation progeny, in contrast, have a diversity of genotypes and so are highly variable.

crossed with a third species, the hybrids are widely different from each other; whereas if two very distinct varieties of one species are crossed with another species, the hybrids do not differ much. But this conclusion, as far as I can make out, is founded on a single experiment; and seems directly opposed to the results of several experiments made by Kölreuter.

These alone are the unimportant differences, which Gärtner is able to point out, between hybrid and mongrel plants. On the other hand, the resemblance in mongrels and in hybrids to their respective parents, more especially in hybrids produced from nearly related species, follows according to Gärtner the same laws. When two species are crossed, one has sometimes a prepotent power of impressing its likeness on the hybrid; and so I believe it to be with varieties of plants. With animals one variety certainly often has this prepotent power over another variety. Hybrid plants produced from a reciprocal cross, generally resemble each other closely; and so it is with mongrels from a reciprocal cross. Both hybrids and mongrels can be reduced to either pure parent-form, by repeated crosses in successive generations with either parent.

These several remarks are apparently applicable to animals; but the subject is here excessively complicated, partly owing to the existence of secondary sexual characters; but more especially owing to prepotency in transmitting likeness running more strongly in one sex than in the other, both when one species is crossed with another, and when one variety is crossed with another variety. For instance, I think those authors are right, who maintain that the ass has a prepotent power over the horse, so that both the mule and the hinny more resemble the ass than the horse; but that the prepotency runs more strongly in the male-ass than in

1 "Prepotency" is a distinctively nineteenth-century term derived from the combination of "pre-" as intensifier and "potency" in the sense of transmission of hereditary properties. Prepotency thus refers to disproportionate transmission of hereditary characteristics by one parent vs. the other parent. We now understand this tendency of offspring to more closely resemble one or the other parent in various traits in terms of chromosomal recombination and dominance interactions.

2 Kölreuter and Gärtner undertook what they termed "transformation experiments" to show that new species could not arise from hybridization. These experiments consisted of repeatedly backcrossing hybrids with one of the parents, which would result in reversion to one of the parental forms. To Kölreuter this demonstrated that the parental elements were mixed or blended in the offspring, contrary to Linnaeus's assertion that traits of both parents were individually represented. By Darwin's time the so-called reversion to parental form through backcrossing, or "reversion to type," as it was often called, was well known and cited by some as evidence against species mutability.

the female, so that the mule, which is the offspring of the male-ass and mare, is more like an ass, than is the hinny, which is the offspring of the female-ass and stallion.

Much stress has been laid by some authors on the supposed fact, that mongrel animals alone are born closely like one of their parents; but it can be shown that this does sometimes occur with hybrids; yet I grant much less frequently with hybrids than with mongrels. Looking to the cases which I have collected of cross-bred animals closely resembling one parent, the resemblances seem chiefly confined to characters almost monstrous in their nature, and which have suddenly appeared—such as albinism, melanism, deficiency of tail or horns, or additional fingers and toes; and do not relate to characters which have been slowly acquired by selection. Consequently, sudden reversions to the perfect character of either parent would be more likely to occur with mongrels, which are descended from varieties often suddenly produced and semi-monstrous in character, than with hybrids, which are descended from species slowly and naturally produced. On the whole I entirely agree with Dr. Prosper Lucas, who, after arranging an enormous body of facts with respect to animals, comes to the conclusion, that the laws of resemblance of the child to its parents are the same, whether the two parents differ much or little from each other, namely in the union of individuals of the same variety, or of different varieties, or of distinct species. ◀1

Laying aside the question of fertility and sterility, in all other respects there seems to be a general and close similarity in the offspring of crossed species, and of crossed varieties. If we look at species as having been specially created, and at varieties as having been produced by secondary laws, this similarity would be an ◀2

1 The observation that mongrels seem to closely resemble their parents while hybrids do not can be seen as another argument for reversion and species immutability. Darwin dismisses this notion, suggesting that resemblance to parents is limited to extraordinary characters and is not indicative of the kind of evolutionary change he posits. Note the final sentence of this paragraph, in which he drives home an old (but crucial) point: there is no essential difference between species and varieties. This statement is amplified in the final paragraph of this section.

2 Through all the preceding discussion of hybrid crosses and sterility and interfertility, the point that Darwin has been driving at is stated here in three sentences. Again, he seeks to disabuse his readers of the notion that fundamental differences separate species and varieties as underscored by the essential similarity in patterns of sterility and interfertility seen in species crosses and variety crosses. Note, however, the brilliant and insightful culminating point: supposing species to have been specially created and varieties to be human creations (via domestication), why should expressions of sterility and interfertility be the same in the two cases?

1 The summary reiterates the key points. First, in species crosses and hybrid crosses, sterility varies and is seen in graduated degrees.

2 Second, the grafting parallel is illuminating. Crossing capacity and grafting capacity vary in the same way; given how recent and artificial grafting is, this parallel would seem to torpedo (to use a modern metaphor) the idea that sterility was created to prevent intercrossing. Both, rather, flow from compatibility or incompatibility of basic constitutional elements and so vary with overall systematic affinity, or taxonomic closeness.

3 The causes of sterility are several, but all are related to unknown aspects of the reproductive system. Darwin sees a fundamental connection among the ease with which a cross can be made initially, the relative fertility of the hybrids produced from that cross, and the facility of grafting with these species and varieties.

astonishing fact. But it harmonises perfectly with the view that there is no essential distinction between species and varieties.

1 *Summary of Chapter.*—First crosses between forms sufficiently distinct to be ranked as species, and their hybrids, are very generally, but not universally, sterile. The sterility is of all degrees, and is often so slight that the two most careful experimentalists who have ever lived, have come to diametrically opposite conclusions in ranking forms by this test. The sterility is innately variable in individuals of the same species, and is eminently susceptible of favourable and unfavourable conditions. The degree of sterility does not strictly follow systematic affinity, but is governed by several curious and complex laws. It is generally different, and sometimes widely different, in reciprocal crosses between the same two species. It is not always equal in degree in a first cross and in the hybrid produced from this cross.

2 In the same manner as in grafting trees, the capacity of one species or variety to take on another, is incidental on generally unknown differences in their vegetative systems, so in crossing, the greater or less facility of one species to unite with another, is incidental on unknown differences in their reproductive systems. There is no more reason to think that species have been specially endowed with various degrees of sterility to prevent them crossing and blending in nature, than to think that trees have been specially endowed with various and somewhat analogous degrees of difficulty in being grafted together in order to prevent them becoming inarched in our forests.

3 The sterility of first crosses between pure species, which have their reproductive systems perfect, seems

to depend on several circumstances; in some cases largely on the early death of the embryo. The sterility of hybrids, which have their reproductive systems imperfect, and which have had this system and their whole organisation disturbed by being compounded of two distinct species, seems closely allied to that sterility which so frequently affects pure species, when their natural conditions of life have been disturbed. This view is supported by a parallelism of another kind;—namely, that the crossing of forms only slightly different is favourable to the vigour and fertility of their offspring; and that slight changes in the conditions of life are apparently favourable to the vigour and fertility of all organic beings. It is not surprising that the degree of difficulty in uniting two species, and the degree of sterility of their hybrid-offspring should generally correspond, though due to distinct causes; for both depend on the amount of difference of some kind between the species which are crossed. Nor is it surprising that the facility of effecting a first cross, the fertility of the hybrids produced, and the capacity of being grafted together—though this latter capacity evidently depends on widely different circumstances—should all run, to a certain extent, parallel with the systematic affinity of the forms which are subjected to experiment; for systematic affinity attempts to express all kinds of resemblance between all species.

First crosses between forms known to be varieties, or sufficiently alike to be considered as varieties, and their mongrel offspring, are very generally, but not quite universally, fertile. Nor is this nearly general and perfect fertility surprising, when we remember how liable we are to argue in a circle with respect to varieties in a state of nature; and when we remember that the greater number of varieties have been produced under domesti- [1]

1 There is no real difference, finally, between hybrids and mongrels (the offspring of crossed varieties) with respect to their reproductive constitution.

cation by the selection of mere external differences, and not of differences in the reproductive system. In all other respects, excluding fertility, there is a close general resemblance between hybrids and mongrels. Finally, ▶ then, the facts briefly given in this chapter do not seem to me opposed to, but even rather to support the view, that there is no fundamental distinction between species and varieties.

1 Here Darwin makes a virtue out of adversity: the mysteries and complexities of hybrid sterility and interfertility not only fail to undermine the assertion that species and varieties are fundamentally similar, but they even support that view. In later *Origin* editions Darwin modified this sentence slightly, acknowledging ignorance of the causes of sterility while asserting that the patterns reported in this chapter nonetheless support his thesis.

Darwin recognized that hybridism was a critically important area for his theory. It was a sticking point even among supporters like Huxley, who felt that until domesticated varieties could be shown to be sterile when crossed, the jury was out on the efficacy of natural selection. Darwin disagreed; he felt that the variation in sterility and interfertility was significant. On this issue he chided Huxley in uncharacteristically harsh terms for a misstatement on hybrid sterility made in a public lecture: "it is quite certain that . . . hybrids are often absolutely infertile with one another . . . Can we find any approximation to this [hybrid sterility] in the different races known to be produced by selective breeding from a common stock? Up to the present time the answer . . . is absolutely a negative one" (Huxley 1970, p. 424).

Darwin wrote him after reading a copy of the lecture:

> You say the answer to varieties when crossed being at all sterile is "absolutely a negative." Do you mean to say that Gärtner lied, after experiments by the hundred (and he a hostile witness), when he showed that this was the case with *Verbascum* and with maize (and here you have selected races): does Kölreuter lie when he speaks about the varieties of tobacco? My God, is not the case difficult enough, without its being, as I must think, falsely made more difficult? I believe it is my own fault-my d-d candour: I ought to have made ten times more fuss about these most careful experiments. I did put it stronger in the third edition of the *Origin*. (*Correspondence* 10: 611)

CHAPTER IX.

ON THE IMPERFECTION OF THE GEOLOGICAL RECORD.

On the absence of intermediate varieties at the present day — On the nature of extinct intermediate varieties; on their number — On the vast lapse of time, as inferred from the rate of deposition and of denudation — On the poorness of our palæontological collections — On the intermittence of geological formations — On the absence of intermediate varieties in any one formation — On the sudden appearance of groups of species — On their sudden appearance in the lowest known fossiliferous strata.

IN the sixth chapter I enumerated the chief objections which might be justly urged against the views maintained in this volume. Most of them have now been discussed. One, namely the distinctness of specific forms, and their not being blended together by innumerable transitional links, is a very obvious difficulty. I assigned reasons why such links do not commonly occur at the present day, under the circumstances apparently most favourable for their presence, namely on an extensive and continuous area with graduated physical conditions. I endeavoured to show, that the life of each species depends in a more important manner on the presence of other already defined organic forms, than on climate; and, therefore, that the really governing conditions of life do not graduate away quite insensibly like heat or moisture. I endeavoured, also, to show that intermediate varieties, from existing in lesser numbers than the forms which they connect, will generally be beaten out and exterminated during the course of further modification and improvement. The main cause, however, of innumerable intermediate links not now occurring everywhere throughout nature de-

This first of the two geological chapters in the *Origin* follows the precedent set with chapter VI, namely, to address difficulties before advancing to supporting evidence. The problems Darwin addresses here—apparent gaps, sudden appearance, and lack of intermediate forms in the fossil record—are serious and perhaps constitute his single greatest concern. His worry was borne out, for the harshest critics of the *Origin* in Darwin's day cited the fossil record as not merely unsupportive of the theory of descent with modification but antithetical to it. These included some of the leading paleontological lights, such as Richard Owen and Louis Agassiz. Darwin anticipated this criticism, and here he attempts to show that the vagaries of the fossilization process and the time scales involved conspire to give the *impression* of gaps and abrupt appearance of forms. He argues that it makes sense that relatively few species in general—let alone short-lived intermediates—will be preserved in the fossil record. In these ways Darwin tries to make a virtue of adversity, aiming to convince his readers that the "poorness of our paleontological collections," as he puts it, is to be expected. He is explaining how to "see" in the modern sense—a great triumph of reconceptualization.

1 This is an important point. Truly intermediate evolutionary forms must be sought in evolutionary lineages. They are to be sought between, say, an extant species and an ancient progenitor of that species. Extant (still-living) forms in some way "in between" two other extant forms may seem like intermediates in a sense, but they are not lineal intermediates.

pends on the very process of natural selection, through which new varieties continually take the places of and exterminate their parent-forms. But just in proportion as this process of extermination has acted on an enormous scale, so must the number of intermediate varieties, which have formerly existed on the earth, be truly enormous. Why then is not every geological formation and every stratum full of such intermediate links? Geology assuredly does not reveal any such finely graduated organic chain; and this, perhaps, is the most obvious and gravest objection which can be urged against my theory. The explanation lies, as I believe, in the extreme imperfection of the geological record.

In the first place it should always be borne in mind what sort of intermediate forms must, on my theory, have formerly existed. [1] I have found it difficult, when looking at any two species, to avoid picturing to myself, forms *directly* intermediate between them. But this is a wholly false view; we should always look for forms intermediate between each species and a common but unknown progenitor; and the progenitor will generally have differed in some respects from all its modified descendants. To give a simple illustration: the fantail and pouter pigeons have both descended from the rock-pigeon; if we possessed all the intermediate varieties which have ever existed, we should have an extremely close series between both and the rock-pigeon; but we should have no varieties directly intermediate between the fantail and pouter; none, for instance, combining a tail somewhat expanded with a crop somewhat enlarged, the characteristic features of these two breeds. These two breeds, moreover, have become so much modified, that if we had no historical or indirect evidence regarding their origin, it would not have been possible to have determined from a mere comparison of their structure with that of the rock-pigeon, whether they had descended from this species or from some other allied species, such as C. oenas. [1]

So with natural species, if we look to forms very [2] distinct, for instance to the horse and tapir, we have no reason to suppose that links ever existed directly intermediate between them, but between each and an unknown common parent. The common parent will have had in its whole organisation much general resemblance to the tapir and to the horse; but in some points of structure may have differed considerably from both, even perhaps more than they differ from each other. Hence in all such cases, we should be unable to recognise the parent-form of any two or more species, even if we closely compared the structure of the parent with that of its modified descendants, unless at the same time we had a nearly perfect chain of the intermediate links.

It is just possible by my theory, that one of two living [3] forms might have descended from the other; for instance, a horse from a tapir; and in this case *direct* intermediate links will have existed between them. But such a case would imply that one form had remained for a very long period unaltered, whilst its descendants had undergone a vast amount of change; and the principle of competition between organism and organism, between child and parent, will render this a very rare event; for in all cases the new and improved forms of life will tend to supplant the old and unimproved forms.

By the theory of natural selection all living species [4] have been connected with the parent-species of each genus, by differences not greater than we see between the varieties of the same species at the present

1 This is *Columba oenas,* the stock dove of Europe and western Asia, one of three species in the genus *Columba* native to Europe. The rock pigeon, *C. livia,* is the spectacularly diverse domesticated pigeon.

2 Darwin may overstate the hopelessness of recognizing the parental form of two highly divergent descendant species. Even in his day, skilled comparative anatomists would have a reasonably close idea of where to place highly derived groups (albeit working within the framework of "affinities," not evolutionary relationships). A "nearly perfect chain" of intermediate links is not necessary either to know where to place divergent forms or to map out evolutionary pathways.

3 Recall Darwin's principle of divergence. Descendant lineages become more and more dissimilar (divergent) over time, and in the process compete less directly. With that model in mind, we can see that the scenario given in this passage would have to be rare in Darwin's view, for he is talking about situations where a parental form remains essentially unchanged, while giving rise to a lineage that diverges substantially. Within-lineage evolution is now termed "anagenesis," while evolution by lineage splitting is termed "cladogenesis."

Darwin's view of divergence is not fully accepted today. In the modern understanding of the process it is not especially rare to have both lineages that largely retain ancestral traits and those that have diverged greatly from the ancestral form. Can it be said, as Darwin puts it here, that one living species *descends from* another living species? In principle yes, but we cannot know how similar or dissimilar the living species is genetically to its ancient relative. Could they reproduce?

4 Note the careful wording of this paragraph; the differences between species of a genus are of the same magnitude as the differences between varieties of a species—no gaps or leaps (*natura non facit saltum,* recall), but continuity. Darwin traces the chain of continuity into the distant past, pointing out that the number of transitional forms must have been immense. This grand vision of universal descent requires a grand vision of time, the subject of the next section.

day; and these parent-species, now generally extinct, have in their turn been similarly connected with more ancient species; and so on backwards, always converging to the common ancestor of each great class. So that the number of intermediate and transitional links, between all living and extinct species, must have been inconceivably great. But assuredly, if this theory be true, such have lived upon this earth.

1 *On the lapse of Time.*—Independently of our not finding fossil remains of such infinitely numerous connecting links, it may be objected, that time will not have sufficed for so great an amount of organic change, all changes having been effected very slowly through
2 natural selection. It is hardly possible for me even to recall to the reader, who may not be a practical geologist, the facts leading the mind feebly to comprehend the lapse of time. He who can read Sir Charles Lyell's grand work on the Principles of Geology, which the future historian will recognise as having produced a revolution in natural science, yet does not admit how incomprehensibly vast have been the past periods of time, may at once close this volume. Not that it suffices to study the Principles of Geology, or to read special treatises by different observers on separate formations, and to mark how each author attempts to give an inadequate idea of the duration of each formation or even each stratum. A man must for years examine for himself great piles of superimposed strata, and watch the sea at work grinding down old rocks and making fresh sediment, before he can hope to comprehend anything of the lapse of time, the monuments of which we see around us.

3 It is good to wander along lines of sea-coast, when formed of moderately hard rocks, and mark the

1 By the mid-nineteenth century scientists thought the age of the Earth stretched into the millions of years, though not as many millions as Darwin thought he needed. In the early 1860s William Thompson, later Lord Kelvin, calculated Earth's age at only about 100 million years (see p. 465). Note that in this section Darwin does not attempt to estimate the Earth's age; rather, he makes a case for how much time is needed for geological processes like erosion to yield the results we observe in particular cases. He can in that way establish that there is plenty of time available for natural selection to have generated the diversity of life—the Earth is inferred to be far older than the particular formations Darwin discusses.

2 In his book *Principles of Geology* (1830–1833), Charles Lyell did much to convince his peers of an ancient Earth—what the writer John McPhee memorably dubbed "deep time." The biggest stumbling block to accepting deep time is our inability to comprehend such time scales, which are akin to the equally mind-boggling spatial scales of the astronomers. One way to try to glimpse the immensity of geological time is to focus on something tangible. Darwin points to the products of erosion we see all around us: great beds of sedimentary strata that can only have been laid down at an infinitesimally slow rate. This passage in the *Origin* relates to a comment he made in the E notebook of 1838–1839: "No one but a practised geologist can really comprehend how old the world is, as the measurements refer not to revolutions of the sun & our lives, but to [a] period necessary to form heaps of pebbles &c &c" (Barrett et al. 1987, p. 432).

3 This is an effective example: the sea cliffs are eroded at their base. We can well imagine that this would take a long time indeed, but even more time is needed given that erosion is not continual, but occurs only twice daily when the tide is up. What is the origin of all that beach sand and near-shore sediment? The answer is rock formations worn down "atom by atom." Slowing erosion further still is the protective coating of algae and seaweed festooning the rocks!

process of degradation. The tides in most cases reach the cliffs only for a short time twice a day, and the waves eat intc them only when they are charged with sand or pebbles; for there is reason to believe that pure water can effect little or nothing in wearing away rock. At last the base of the cliff is undermined, huge fragments fall down, and these remaining fixed, have to be worn away, atom by atom, until reduced in size they can be rolled about by the waves, and then are more quickly ground into pebbles, sand, or mud. But how often do we see along the bases of retreating cliffs rounded boulders, all thickly clothed by marine productions, showing how little they are abraded and how seldom they are rolled about! Moreover, if we follow for a few miles any line of rocky cliff, which is undergoing degradation, we find that it is only here and there, along a short length or round a promontory, that the cliffs are at the present time suffering. The appearance of the surface and the vegetation show that elsewhere years have elapsed since the waters washed their base.

He who most closely studies the action of the sea on our shores, will, I believe, be most deeply impressed with the slowness with which rocky coasts are worn away. The observations on this head by Hugh Miller, and by that excellent observer Mr. Smith of Jordan Hill, are most impressive. With the mind thus impressed, let any one examine beds of conglomerate many thousand feet in thickness, which, though probably formed at a quicker rate than many other deposits, yet, from being formed of worn and rounded pebbles, each of which bears the stamp of time, are good to show how slowly the mass has been accumulated. Let him remember Lyell's profound remark, that the thickness and extent of sedimentary formations

1 A brilliant example with which to illustrate deep time: conglomerate rocks consist of pebbles embedded in a matrix of finer-grained material like sandstone or siltstone. This is sedimentary rock, and beds of the thickness Darwin describes here would require untold millennia to be deposited. But then consider: the pebbles in the conglomerate must themselves have been eroded out of parental rock still more remotely in time! Their size and well-worn character speak of countless *more* eons of erosion to generate the materials that are deposited as the sedimentary beds that, once consolidated, become the conglomerate rock we now see. Darwin eloquently notes that each pebble "bears the stamp of time"—deep time.

are the result and measure of the degradation which the earth's crust has elsewhere suffered. And what an amount of degradation is implied by the sedimentary
1 deposits of many countries! Professor Ramsay has given me the maximum thickness, in most cases from actual measurement, in a few cases from estimate, of each formation in different parts of Great Britain; and this is the result:—

	Feet.
Palæozoic strata (not including igneous beds) ..	57,154
Secondary strata	13,190
Tertiary strata	2,240

—making altogether 72,584 feet; that is, very nearly thirteen and three-quarters British miles. Some of these formations, which are represented in England by thin beds, are thousands of feet in thickness on the Continent. Moreover, between each successive formation, we have, in the opinion of most geologists, enormously long blank periods. So that the lofty pile of sedimentary rocks in Britain, gives but an inadequate idea of the time which has elapsed during their accumulation; yet what time this must have consumed! Good observers have estimated that sediment is deposited by the great Mississippi river at the rate of only 600 feet in a hundred thousand years. This estimate may be quite erroneous; yet, considering over what wide spaces very fine sediment is transported by the currents of the sea, the process of accumulation in any one area must be extremely slow.

2 But the amount of denudation which the strata have in many places suffered, independently of the rate of accumulation of the degraded matter, probably offers the best evidence of the lapse of time. I remember having been much struck with the evidence of denudation, when viewing volcanic islands, which have been

1 In 1858, while working on this chapter, Darwin wrote to the geologist Andrew Crombie Ramsay about the thickness of geological formations in the British Isles. Ramsay responded with a long catalog of data in December of that year (*Correspondence* 7: 225), which Darwin summarized in the condensed form seen on this page.

2 "Denudation" means erosion in modern terms; Darwin's point here is that the measurements of erosion give the best indication of the vast time scales inherent in geological processes. His observation of the erosional patterns in evidence on volcanic islands is astute: oceanic volcanic islands are typically shield volcanoes, which form by innumerable low-intensity eruptive lava flows over millions of years. Each fresh flow cools and hardens as a layer atop the previous one, which in turn is a layer atop the one before that, and so on. This layering is evident in cross section at the island's margin, where the sea has eroded vertical sea cliffs.

Following the trajectory indicated by the angle of the layers, Darwin could extrapolate how far seaward the land (layered lavas) formerly extended. His inspiration for this process is the south Atlantic island of St. Helena, which he describes and sketches in the red notebook (Barrett et al. 1987, pp. 31–32) and treats still further in the third of his *Beagle* voyage reports, *Geological Observations on South America* (1846, pp. 25–26).

worn by the waves and pared all round into perpendicular cliffs of one or two thousand feet in height; for the gentle slope of the lava-streams, due to their formerly liquid state, showed at a glance how far the hard, rocky beds had once extended into the open ocean. The same story is still more plainly told by faults,—those great cracks along which the strata have been upheaved on one side, or thrown down on the other, to the height or depth of thousands of feet; for since the crust cracked, the surface of the land has been so completely planed down by the action of the sea, that no trace of these vast dislocations is externally visible.

The Craven fault, for instance, extends for upwards of 30 miles, and along this line the vertical displacement of the strata has varied from 600 to 3000 feet. Prof. Ramsay has published an account of a downthrow in Anglesea of 2300 feet; and he informs me that he fully believes there is one in Merionethshire of 12,000 feet; yet in these cases there is nothing on the surface to show such prodigious movements; the pile of rocks on the one or other side having been smoothly swept away. The consideration of these facts impresses my mind almost in the same manner as does the vain endeavour to grapple with the idea of eternity.

I am tempted to give one other case, the well-known one of the denudation of the Weald. Though it must be admitted that the denudation of the Weald has been a mere trifle, in comparison with that which has removed masses of our palæozoic strata, in parts ten thousand feet in thickness, as shown in Prof. Ramsay's masterly memoir on this subject. Yet it is an admirable lesson to stand on the North Downs and to look at the distant South Downs; for, remembering that at no great distance to the west the northern and southern escarpments meet and close, one can safely picture to

1 The Craven Fault, located in Yorkshire, is a spectacular example of faulting in which a clear sequence of rock strata appears repeated. The repeat is due to vertical slippage along a fault line. Here is an especially clear description taken from the website of the Ingleborough Hall Outdoor Education Center, located within Yorkshire Dales National Park; it describes an exposure of the north Craven Fault at a site called Nappa Scar, where a walking path lies atop the down-faulted strata:

> Above the path in the base of the cliff you can see the older Silurian rocks overlain by a mixture of pebbles cemented together by limestone. This matrix is called a conglomerate. The pebbles, made up of Silurian greywacke and slate, get progressively smaller the farther up you look. The conglomerate is in turn overlain by Carboniferous limestone. It is possible to see a repeat of this sequence of greywacke/slate, conglomerate, and limestone below the path, having been down-faulted. The throw of this is only 10 or 20 metres, providing evidence that in this locality the North Craven Fault is a series of small step faults rather than one big fault.

Here the faulting is only 10 to 20 meters, but stretches as much as 1,000 meters in places, Darwin points out.

2 The Weald is a region in southern England stretching from Kent to Surrey and bounded by the North and South Downs, which are chalk ridges. The central wealden area consists of rolling sandstone hills older than the limestone of the Downs. The formation is understood today as an anticline, an upbuckled dome from which the chalk cap that would have formerly connected the North and South Downs has eroded, exposing the sandstones below. This is the denudation of the Weald: erosion of a prodigious amount of chalk. The formation actually extends under the English Channel and reemerges on the French coast.

oneself the great dome of rocks which must have covered up the Weald within so limited a period as since the latter part of the Chalk formation. The distance from the northern to the southern Downs is about 22 miles, and the thickness of the several formations is on an average about 1100 feet, as I am informed by Prof. Ramsay. But if, as some geologists suppose, a range of older rocks underlies the Weald, on the flanks of which the overlying sedimentary deposits might have accumulated in thinner masses than elsewhere, the above estimate would be erroneous; but this source of doubt probably would not greatly affect the estimate as applied to the western extremity of the district. If, then, we knew the rate at which the sea commonly wears away a line of cliff of any given height, we could measure the time requisite to have denuded the Weald. This, of course, cannot be done; but we may, in order to form some crude notion on the subject, assume that the sea would eat into cliffs 500 feet in height at the rate of one inch in a century. This will at first appear much too small an allowance; but it is the same as if we were to assume a cliff one yard in height to be eaten back along a whole line of coast at the rate of one yard in nearly every twenty-two years. I doubt whether any rock, even as soft as chalk, would yield at this rate excepting on the most exposed coasts; though no doubt the degradation of a lofty cliff would be more rapid from the breakage of the fallen fragments. On the other hand, I do not believe that any line of coast, ten or twenty miles in length, ever suffers degradation at the same time along its whole indented length; and we must remember that almost all strata contain harder layers or nodules, which from long resisting attrition form a breakwater at the base. Hence, under ordinary circumstances, I conclude that for a cliff 500 feet in height, a denudation

1 In March 1857 Ramsay provided Darwin with this estimate of the thickness of deposits in the Weald, which Darwin used to calculate how much time was needed to erode ("denude") these deposits. Darwin was being too bold in attempting to do so—perhaps knowing how fraught with assumption and error such estimates can be, even leading geologists like Lyell, who first proposed the model for the Weald formation accepted today, stopped short of proposing an actual time estimate. Lyell discussed the Weald at length in the third volume of *Principles of Geology,* and a nice geological map of the Weald prefaces the first chapter.

2 Darwin almost immediately regretted his estimate of 300 million years for the denudation of the Weald, which he based on the assumed erosion rate of one inch per century given in this sentence. Within a month a critique that he acknowledged to be "able and justly severe" appeared in the *Saturday Review* taking issue with his methodology. He qualified the rate accordingly in the second edition of the *Origin,* adding, "But perhaps it would be safer to allow two or three inches per century, and this would reduce the number of years to one hundred and fifty or one hundred million years" (Peckham 1959, p. 484). By the third edition he dropped the erosion time estimate altogether (Burchfield 1974, Herbert 2005). Even Darwin's revised estimate is likely too high. There are several competing current models for the formation and erosion of the Weald, but in general the deposits are understood as late Mesozoic and are estimated to have had an erosion rate of 1,350 meters in 66 million years, or about 20.4 meters per million years—a relatively low rate (Jones 1999).

of one inch per century for the whole length would be an ample allowance. At this rate, on the above data, the denudation of the Weald must have required 306,662,400 years; or say three hundred million years.

The action of fresh water on the gently inclined Wealden district, when upraised, could hardly have been great, but it would somewhat reduce the above estimate. On the other hand, during oscillations of level, which we know this area has undergone, the surface may have existed for millions of years as land, and thus have escaped the action of the sea: when deeply submerged for perhaps equally long periods, it would, likewise, have escaped the action of the coast-waves. So that in all probability a far longer period than 300 million years has elapsed since the latter part of the Secondary period.

I have made these few remarks because it is highly important for us to gain some notion, however imperfect, of the lapse of years. During each of these years, over the whole world, the land and the water has been peopled by hosts of living forms. What an infinite number of generations, which the mind cannot grasp, must have succeeded each other in the long roll of years! Now turn to our richest geological museums, and what a paltry display we behold! 1

On the poorness of our Palæontological collections.— 2
That our palæontological collections are very imperfect, is admitted by every one. The remark of that admirable palæontologist, the late Edward Forbes, should not be forgotten, namely, that numbers of our fossil species are known and named from single and often broken specimens, or from a few specimens collected on some one spot. Only a small portion of the surface of the earth has been geologically explored, and no part with

1 This is the real objective of Darwin's erosion-estimation exercise: an effort to give the reader a sense of the vastness of time, so necessary to his theory. In later *Origin* editions Darwin added several paragraphs here, including this attempt in the fifth edition to further help the reader visualize vast amounts of time: "Few of us, however, know what a million really means: Mr. Croll gives the following illustration: take a narrow strip of paper, 83 feet 4 inches in length, and stretch it along the wall of a large hall; then mark off at one end the tenth of an inch. This tenth of an inch will represent one hundred years, and the entire strip a million years" (Peckham 1959, p. 485). Taking this one step further, Darwin then points out how insignificant a hundred years is, yet how much change to plant and animal breeds can be effected in that time by artificial selection.

2 In this important section Darwin tries to explain why the fossil record is necessarily incomplete.

1 Given the diversity and abundance of species that are inferred to have once lived, we might expect the fossil record to be vast and respresentative. But several factors conspire to limit the range and number of fossils. Hard parts like bones and shells fossilize best, yet these are comparatively rare; poorly fossilizing soft parts are far more abundant. External conditions favorable to fossilization are also spotty: sedimentation occurs only in certain environments, and even in those experiencing deposition the rate is not constant. Some proportion of remains in the right place for fossilization are likely lost to degradation, too. The scant representatives of the few forms that succeed in becoming fossilized often run a gauntlet of destructive processes post-fossilization. No wonder the fossil record is "fragmentary in an extreme degree," as Darwin puts it on the next page.

2 Darwin named the family Chthamalidae, encrusting barnacles, and several species of the genus *Chthamalus.* These organisms, and the limpet *Chiton,* are common intertidal zone species.

sufficient care, as the important discoveries made every year in Europe prove. No organism wholly soft can be preserved. Shells and bones will decay and disappear when left on the bottom of the sea, where sediment is not accumulating. I believe we are continually taking a most erroneous view, when we tacitly admit to ourselves that sediment is being deposited over nearly the whole bed of the sea, at a rate sufficiently quick to embed and preserve fossil remains. Throughout an enormously large proportion of the ocean, the bright blue tint of the water bespeaks its purity. The many cases on record of a formation conformably covered, after an enormous interval of time, by another and later formation, without the underlying bed having suffered in the interval any wear and tear, seem explicable only on the view of the bottom of the sea not rarely lying for ages in an unaltered condition. The remains which do become embedded, if in sand or gravel, will when the beds are upraised generally be dissolved by the percolation of rain-water. I suspect that but few of the very many animals which live on the beach between high and low watermark are preserved. For instance, the several species of the Chthamalinæ (a subfamily of sessile cirripedes) coat the rocks all over the world in infinite numbers: they are all strictly littoral, with the exception of a single Mediterranean species, which inhabits deep water and has been found fossil in Sicily, whereas not one other species has hitherto been found in any tertiary formation: yet it is now known that the genus Chthamalus existed during the chalk period. The molluscan genus Chiton offers a partially analogous case.

With respect to the terrestrial productions which lived during the Secondary and Palæozoic periods, it is superfluous to state that our evidence from fossil remains is fragmentary in an extreme degree. For instance, not a land shell is known belonging to either of these vast periods, with one exception discovered by Sir C. Lyell in the carboniferous strata of North America. In regard to mammiferous remains, a single glance at the historical table published in the Supplement to Lyell's Manual, will bring home the truth, how accidental and rare is their preservation, far better than pages of detail. Nor is their rarity surprising, when we remember how large a proportion of the bones of tertiary mammals have been discovered either in caves or in lacustrine deposits; and that not a cave or true lacustrine bed is known belonging to the age of our secondary or palæozoic formations.

But the imperfection in the geological record mainly results from another and more important cause than any of the foregoing; namely, from the several formations being separated from each other by wide intervals of time. When we see the formations tabulated in written works, or when we follow them in nature, it is difficult to avoid believing that they are closely consecutive. But we know, for instance, from Sir R. Murchison's great work on Russia, what wide gaps there are in that country between the superimposed formations; so it is in North America, and in many other parts of the world. The most skilful geologist, if his attention had been exclusively confined to these large territories, would never have suspected that during the periods which were blank and barren in his own country, great piles of sediment, charged with new and peculiar forms of life, had elsewhere been accumulated. And if in each separate territory, hardly any idea can be formed of the length of time which has elapsed between the consecutive formations, we may infer that this could nowhere be ascertained. The frequent

1 For Lyell and Darwin, the Carboniferous mollusk example showed both that the fossil record is fragmentary, and that groups that seem to appear suddenly in this or that formation probably had a more gradual introduction farther down the formation. No mollusks at all were then known from the Carboniferous strata of Europe, leading some to assert that they had been created in the north-temperate latitudes after this period. Lyell and Dawson's 1852 discovery of fossil mollusks in North American Carboniferous strata (in Nova Scotia, to be exact; see p. 296) demolished this claim.

2 "Lyell's Manual" is the *Manual of Elementary Geology* (1852). This was essentially the fourth edition of Lyell's *Elements of Geology,* which was originally to have been presented as a fourth volume of *Principles of Geology.* In 1838 Lyell decided to publish this volume separately as *Elements,* thinking it might be a useful field guide.

3 Caves have yielded a remarkable diversity of mammal bones, probably because these areas are sought out as shelters or dens. In one spectacular find, heaps of fossil bones were found in Kirkdale Cave, Yorkshire, which was evidently occupied by many generations of hyenas. This discovery led the Oxford geologist William Buckland to write his famous book *Reliquiae Diluvianae* (Evidence of the Flood) in 1823.

4 The single biggest problem with the fossil record is the vast gaps in time between formations. Two spatially contiguous strata do not necessarily represent temporal contiguity. Oscillations in the elevation and subsidence of the Earth's surface will result in periods of deposition separated by intervals of time. Compounding the problem is the erosional loss of deposits during periods of uplift.

5 Here Darwin is referencing Sir Roderick Impey Murchison's books *On the Geological Structure of the Northern and Central Regions of Russia in Europe* (1841) and *Geology of Russia in Europe and the Ural Mountains* (1845).

1 Darwin was struck by the elevation of the western coast of South America, and he made detailed geological observations (with Lyellian eyes) to document it. On the *Beagle* voyage in 1834 he undertook a study of the elevation of Patagonia and established that the coastal landscape consists of a series of step-like terraces. He found abundant evidence for elevation all along the west coast, as reported here in chapter XVI of *Voyage of the Beagle:*

> I spent some days in examining the step-formed terraces of shingle, first noticed by Captain B. Hall, and believed by Mr. Lyell to have been formed by the sea, during the gradual rising of the land. This certainly is the true explanation, for I found numerous shells of existing species on these terraces. Five narrow, gently sloping, fringe-like terraces rise one behind the other, and where best developed are formed of shingle: they front the bay, and sweep up both sides of the valley. At Guasco, north of Coquimbo, the phenomenon is displayed on a much grander scale, so as to strike with surprise even some of the inhabitants. The terraces are there much broader, and may be called plains, in some parts there are six of them, but generally only five; they run up the valley for thirty-seven miles from the coast. These step-formed terraces or fringes closely resemble those in the valley of S. Cruz, and, except in being on a smaller scale, those great ones along the whole coast-line of Patagonia. They have undoubtedly been formed by the denuding power of the sea, during long periods of rest in the gradual elevation of the continent.

and great changes in the mineralogical composition of consecutive formations, generally implying great changes in the geography of the surrounding lands, whence the sediment has been derived, accords with the belief of vast intervals of time having elapsed between each formation.

But we can, I think, see why the geological formations of each region are almost invariably intermittent; that is, have not followed each other in close sequence. Scarcely any fact struck me more when examining many hundred miles of the South American coasts, which have been upraised several hundred feet within the recent period, than the absence of any recent deposits sufficiently extensive to last for even a short geological period. Along the whole west coast, which is inhabited by a peculiar marine fauna, tertiary beds are so scantily developed, that no record of several successive and peculiar marine faunas will probably be preserved to a distant age. A little reflection will explain why along the rising coast of the western side of South America, no extensive formations with recent or tertiary remains can anywhere be found, though the supply of sediment must for ages have been great, from the enormous degradation of the coast-rocks and from muddy streams entering the sea. The explanation, no doubt, is, that the littoral and sub-littoral deposits are continually worn away, as soon as they are brought up by the slow and gradual rising of the land within the grinding action of the coast-waves.

We may, I think, safely conclude that sediment must be accumulated in extremely thick, solid, or extensive masses, in order to withstand the incessant action of the waves, when first upraised and during subsequent oscillations of level. Such thick and extensive accumulations of sediment may be formed in two ways; either, in profound depths of the sea, in which case, judging from the researches of E. Forbes, we may conclude that the bottom will be inhabited by extremely few animals, and the mass when upraised will give a most imperfect record of the forms of life which then existed; or, sediment may be accumulated to any thickness and extent over a shallow bottom, if it continue slowly to subside. In this latter case, as long as the rate of subsidence and supply of sediment nearly balance each other, the sea will remain shallow and favourable for life, and thus a fossiliferous formation thick enough, when upraised, to resist any amount of degradation, may be formed.

I am convinced that all our ancient formations, which are rich in fossils, have thus been formed during subsidence. Since publishing my views on this subject in 1845, I have watched the progress of Geology, and have been surprised to note how author after author, in treating of this or that great formation, has come to the conclusion that it was accumulated during subsidence. I may add, that the only ancient tertiary formation on the west coast of South America, which has been bulky enough to resist such degradation as it has as yet suffered, but which will hardly last to a distant geological age, was certainly deposited during a downward oscillation of level, and thus gained considerable thickness.

All geological facts tell us plainly that each area has undergone numerous slow oscillations of level, and apparently these oscillations have affected wide spaces. Consequently formations rich in fossils and sufficiently thick and extensive to resist subsequent degradation, may have been formed over wide spaces during periods of subsidence, but only where the supply of sediment was sufficient to keep the sea shallow and to embed and

1 Darwin is referring to Edward Forbes's "azoic theory," which posited a depth limit to the distribution of marine life. Forbes developed his theory after conducting dredging studies in the Aegean Sea during a voyage of HMS *Beacon* in 1841 and 1842. This theory was disproved by the 1860s, when successively deeper dredging studies turned up weird and wonderful creatures from ever-greater depths (Anderson and Rice 2006).

2 Huttonian and Lyellian views of the earth had largely won the day by the 1850s. Darwin's geological work while on the *Beagle* voyage constituted a major contribution in support of Lyellian geology. It is sometimes forgotten that Darwin was hailed as a rising star in the field of geology in the years following the voyage; he read papers on the subject before the Geological Society, and three of the four books that grew out of the voyage were geological, treating coral reefs (1842), volcanic islands (1844), and South American geology (1846). Darwin wrote about evidence for subsidence in his contribution to *The Narrative of the Voyages of H. M. Ships* Adventure *and* Beagle in 1839, but greatly expanded his discussion of subsidence in 1845 when he reworked that contribution into the stand-alone *Journal of Researches,* later *Voyage of the Beagle.*

1 "Stations" mean habitats, a nineteenth-century usage consolidated by Augustin de Candolle in his essay on *géographie botanique* (1820). Darwin's point here is that new habitat is formed during uplift of the land, presenting conditions conducive to the production of new varieties and species. As the land subsides, however, this habitat is destroyed and populations migrate or are extirpated. Recalling his earlier argument that smaller populations generate fewer variations than larger populations, Darwin says here that populations in decline by the subsidence (if only local decline) will yield fewer new varieties and species. Fossilization will occur primarily during subsidence. Thus we have a paleontological catch-22: conditions favorable to the production of new species are not favorable to fossilization, while favorable fossil-forming conditions see a paucity of local species. This is, to Darwin, another important reason that the fossil record is fragmentary.

2 This is a serious issue for Darwin: that geological formations are fragmentary, with indeterminate time gaps between them, is well and good, but within any given thick formation should we not see the graduated intermediates the theory predicts? He confronts this difficult problem by arguing that animals come and go—immigrate and emigrate—and that even apparently continuous formations were not accumulated at a constant rate.

preserve the remains before they had time to decay. On the other hand, as long as the bed of the sea remained stationary, *thick* deposits could not have been accumulated in the shallow parts, which are the most favourable to life. Still less could this have happened during the alternate periods of elevation; or, to speak more accurately, the beds which were then accumulated will have been destroyed by being upraised and brought within the limits of the coast-action.

Thus the geological record will almost necessarily be rendered intermittent. I feel much confidence in the truth of these views, for they are in strict accordance with the general principles inculcated by Sir C. Lyell; and E. Forbes independently arrived at a similar conclusion.

One remark is here worth a passing notice. During periods of elevation the area of the land and of the adjoining shoal parts of the sea will be increased, and new stations will often be formed;—all circumstances most favourable, as previously explained, for the formation of new varieties and species; but during such periods there will generally be a blank in the geological record. On the other hand, during subsidence, the inhabited area and number of inhabitants will decrease (excepting the productions on the shores of a continent when first broken up into an archipelago), and consequently during subsidence, though there will be much extinction, fewer new varieties or species will be formed; and it is during these very periods of subsidence, that our great deposits rich in fossils have been accumulated. Nature may almost be said to have guarded against the frequent discovery of her transitional or linking forms.

From the foregoing considerations it cannot be doubted that the geological record, viewed as a whole, is extremely imperfect; but if we confine our attention to any one formation, it becomes more difficult to understand, why we do not therein find closely graduated varieties between the allied species which lived at its commencement and at its close. Some cases are on record of the same species presenting distinct varieties in the upper and lower parts of the same formation, but, as they are rare, they may be here passed over. Although each formation has indisputably required a vast number of years for its deposition, I can see several reasons why each should not include a graduated series of links between the species which then lived; but I can by no means pretend to assign due proportional weight to the following considerations.

Although each formation may mark a very long lapse of years, each perhaps is short compared with the period requisite to change one species into another. I am aware that two palæontologists, whose opinions are worthy of much deference, namely Bronn and Woodward, have concluded that the average duration of each formation is twice or thrice as long as the average duration of specific forms. But insuperable difficulties, as it seems to me, prevent us coming to any just conclusion on this head. When we see a species first appearing in the middle of any formation, it would be rash in the extreme to infer that it had not elsewhere previously existed. So again when we find a species disappearing before the uppermost layers have been deposited, it would be equally rash to suppose that it then became wholly extinct. We forget how small the area of Europe is compared with the rest of the world; nor have the several stages of the same formation throughout Europe been correlated with perfect accuracy. ◀1

With marine animals of all kinds, we may safely infer a large amount of migration during climatal and other changes; and when we see a species first appearing in any formation, the probability is that it

1 Darwin's immigration-emigration explanation for missing fossils seems unsatisfactory, though it is a reasonable argument for making sense of this puzzle.

only then first immigrated into that area. It is well known, for instance, that several species appeared somewhat earlier in the palæozoic beds of North America than in those of Europe; time having apparently been required for their migration from the American to the European seas. In examining the latest deposits of various quarters of the world, it has everywhere been noted, that some few still existing species are common in the deposit, but have become extinct in the immediately surrounding sea; or, conversely, that some are now abundant in the neighbouring sea, but are rare or absent in this particular deposit. It is an excellent lesson to reflect on the ascertained amount of migration of the inhabitants of Europe during the Glacial period, which forms only a part of one whole geological period; and likewise to reflect on the great changes of level, on the inordinately great change of climate, on the prodigious lapse of time, all included within this same glacial period. Yet it may be doubted whether in any quarter of the world, sedimentary deposits, *including fossil remains*, have gone on accumulating within the same area during the whole of this period. It is not, for instance, probable that sediment was deposited during the whole of the glacial period near the mouth of the Mississippi, within that limit of depth at which marine animals can flourish; for we know what vast geographical changes occurred in other parts of America during this space of time. When such beds as were deposited in shallow water near the mouth of the Mississippi during some part of the glacial period shall have been upraised, organic remains will probably first appear and disappear at different levels, owing to the migration of species and to geographical changes. And in the distant future, a geologist examining these beds, might be tempted to conclude that the average duration of life

1 Unbeknownst to Darwin, there were no American and European seas per se in the Paleozoic, owing to the very different distribution of land masses constituting proto–North America and proto-Eurasia. By the late Paleozoic, in fact, these land masses were contiguous.

2 The glacial theory was advanced in the early nineteenth century by several people, including Jean de Charpentier in Switzerland and Jens Esmark in Scandinavia. Early on, the hypothesized scope of glaciation was limited: alpine areas and far northern fjords, for example. Louis Agassiz became a convert to the idea of an earlier epoch of extensive alpine glaciation in 1837, and then took the idea much further to hypothesize panboreal glaciation. He set out in earnest to document the evidence, and published the important *Études sur les glaciers* in 1840. Agassiz thought of glaciers as "God's great plough," replacing, in his view, the Noachian flood as the most recent global catastrophe. At first Darwin was skeptical of the glacial theory, but by the 1840s he, too, accepted it (but not Agassiz's ice age idea of global catastrophe).

of the embedded fossils had been less than that of the glacial period, instead of having been really far greater, that is extending from before the glacial epoch to the present day.

In order to get a perfect gradation between two forms in the upper and lower parts of the same formation, the deposit must have gone on accumulating for a very long period, in order to have given sufficient time for the slow process of variation; hence the deposit will generally have to be a very thick one; and the species undergoing modification will have had to live on the same area throughout this whole time. But we have seen that a thick fossiliferous formation can only be accumulated during a period of subsidence; and to keep the depth approximately the same, which is necessary in order to enable the same species to live on the same space, the supply of sediment must nearly have counterbalanced the amount of subsidence. But this same movement of subsidence will often tend to sink the area whence the sediment is derived, and thus diminish the supply whilst the downward movement continues. In fact, this nearly exact balancing between the supply of sediment and the amount of subsidence is probably a rare contingency; for it has been observed by more than one palæontologist, that very thick deposits are usually barren of organic remains, except near their upper or lower limits.

It would seem that each separate formation, like the whole pile of formations in any country, has generally been intermittent in its accumulation. When we see, as is so often the case, a formation composed of beds of different mineralogical composition, we may reasonably suspect that the process of deposition has been much interrupted, as a change in the currents of the sea and a supply of sediment of a different nature will

1 The point of this paragraph is well taken: conditions for a truly continuous record of organisms living in a given area are rare. Between the movement of species and the changes in climate and other factors that affect sedimentation rates, it is unlikely that a formation will capture examples of the same continuous lineage long enough to display evolutionary changes.

2 Darwin thus concludes that individual formations have temporal gaps within them, despite the appearance of being a continuous record, just as there are temporal gaps among several successive formations.

generally have been due to geographical changes requiring much time. Nor will the closest inspection of a formation give any idea of the time which its deposition has consumed. Many instances could be given of beds only a few feet in thickness, representing formations, elsewhere thousands of feet in thickness, and which must have required an enormous period for their accumulation; yet no one ignorant of this fact would have suspected the vast lapse of time represented by the thinner formation. Many cases could be given of the lower beds of a formation having been upraised, denuded, submerged, and then re-covered by the upper beds of the same formation,—facts, showing what wide, yet easily overlooked, intervals have occurred in its accumulation. ▶1 In other cases we have the plainest evidence in great fossilised trees, still standing upright as they grew, of many long intervals of time and changes of level during the process of deposition, which would never even have been suspected, had not the trees ▶2 chanced to have been preserved: thus, Messrs. Lyell and Dawson found carboniferous beds 1400 feet thick in Nova Scotia, with ancient root-bearing strata, one above the other, at no less than sixty-eight different levels. Hence, when the same species occur at the bottom, middle, and top of a formation, the probability is that they have not lived on the same spot during the whole period of deposition, but have disappeared and reappeared, perhaps many times, during the same geological period. So that if such species were to undergo a considerable amount of modification during any one geological period, a section would not probably include all the fine intermediate gradations which must on my theory have existed between them, but abrupt, though perhaps very slight, changes of form.

It is all-important to remember that naturalists have

1 There are many petrified tree sites worldwide, but preservation of trees in an upright position is rare. An especially spectacular example of a well-preserved ancient forest of upright trees was unearthed in August 2007 in Hungary. Archaeologists found sixteen cypress (*Taxodium*) trees, part of a Miocene swamp forest that became buried in sand about seven million years ago (Holden 2007).

2 Darwin likely drew on Sir John William Dawson's book *Acadian Geology: An Account of the Geological Structure and Mineral Resources of Nova Scotia* (1855). Dawson was a leading Canadian paleontologist and geologist and a founder of the Royal Society of Canada. Lyell visited Dawson in 1852 to study a famous site along the Bay of Fundy in Nova Scotia, a Carboniferous formation called the Joggins Section (named a UNESCO World Heritage Site in 2008). This formation contains a veritable forest of fossil tree lycopods; on that expedition the two also discovered several fossil reptiles, including an "air-breathing reptile," as Dawson called it, dubbed *Dendrerpeton acadianum,* and several species of small reptiles named *Hylonomus* (including two named in honor of the codiscoverers: *Hylonomus lyelli* and *H. dawsonii*). These animals inhabited or became trapped in hollowed lycopod stumps, which might have acted like pitfall traps; Lyell and Dawson even found cases of multiple skeletons in a single trunk. Shortly after the *Origin* came out, Lyell told Darwin that Dawson had just found a fossil tree with four fossil reptile species; Darwin wrote back to Lyell, "What a marvellous geological Noah's ark that fossil tree in N. America was!" (*Correspondence* 7: 441). Dawson's book *Air Breathers of the Coal Period* (1863) later detailed the exciting findings from this site.

no golden rule by which to distinguish species and varieties; they grant some little variability to each species, but when they meet with a somewhat greater amount of difference between any two forms, they rank both as species, unless they are enabled to connect them together by close intermediate gradations. And this from the reasons just assigned we can seldom hope to effect in any one geological section. Supposing B and C to be two species, and a third, A, to be found in an underlying bed; even if A were strictly intermediate between B and C, it would simply be ranked as a third and distinct species, unless at the same time it could be most closely connected with either one or both forms by intermediate varieties. Nor should it be forgotten, as before explained, that A might be the actual progenitor of B and C, and yet might not at all necessarily be strictly intermediate between them in all points of structure. So that we might obtain the parent-species and its several modified descendants from the lower and upper beds of a formation, and unless we obtained numerous transitional gradations, we should not recognise their relationship, and should consequently be compelled to rank them all as distinct species.

It is notorious on what excessively slight differences many palæontologists have founded their species; and they do this the more readily if the specimens come from different sub-stages of the same formation. Some experienced conchologists are now sinking many of the very fine species of D'Orbigny and others into the rank of varieties; and on this view we do find the kind of evidence of change which on my theory we ought to find. Moreover, if we look to rather wider intervals, namely, to distinct but consecutive stages of the same great formation, we find that the embedded fossils, though almost universally ranked as specifically different,

1 Earlier in the *Origin* Darwin argued that we must look to the fossil record to find true intermediates between two extant species. His comment here is puzzling, then, when he suggests that even if an intermediate (A) is found between two species (B and C) of a more recent bed, it will not be recognized as such. This seems unlikely, but his point is well taken: intermediate series linking more recent forms are not likely to occur even within the same (apparently continuous) formation.

2 Darwin is referring to the French naturalist Alcide d'Orbigny's work on fossil mollusks, *Prodrome de paléontologie stratigraphique universelle des animaux mollesques et rayonnés* (1850–1852).

3 This observation resonates with another made by A. R. Wallace in his 1855 Sarawak Law paper: species similarity in time (as indicated by fossils) parallels species similarity in space, with the greatest similarity occurring between forms in proximity.

1 Darwin's model for the localized origin and subsequent spread of an incipient species would indeed seem to make it unlikely that forms in the early stages of divergence would be fossilized. It is a probabilistic argument: incipiently, such forms are neither populous nor widespread, so their chances of becoming preserved in the fossil record are very small. Still, this is a weak argument. Under Darwin's gradual-divergence model incipient species become ascendant and supplant parental forms; they then yield varieties that subsequently diverge and supplant them, and so on. Localized and limited forms become more widespread and dominant in Darwin's model and thus have a greater chance of being preserved in the fossil record. Critics might therefore expect to see a chain of such forms—it hardly seems necessary to have the earliest stages of diverging varieties preserved, too, in order to see the transition.

2 Here again Darwin seems to think that every single transitional step must be preserved to convince his critics, and he defensively explains why we will not find "numerous, fine, intermediate, fossil links" connecting species. Granted, paleontology was still a young science at the time Darwin wrote the *Origin*, and the fossil record was notoriously "gappy."

yet are far more closely allied to each other than are the species found in more widely separated formations; but to this subject I shall have to return in the following chapter.

1 One other consideration is worth notice: with animals and plants that can propagate rapidly and are not highly locomotive, there is reason to suspect, as we have formerly seen, that their varieties are generally at first local; and that such local varieties do not spread widely and supplant their parent-forms until they have been modified and perfected in some considerable degree. According to this view, the chance of discovering in a formation in any one country all the early stages of transition between any two forms, is small, for the successive changes are supposed to have been local or confined to some one spot. Most marine animals have a wide range; and we have seen that with plants it is those which have the widest range, that oftenest present varieties; so that with shells and other marine animals, it is probably those which have had the widest range, far exceeding the limits of the known geological formations of Europe, which have oftenest given rise, first to local varieties and ultimately to new species; and this again would greatly lessen the chance of our being able to trace the stages of transition in any one geological formation.

2 It should not be forgotten, that at the present day, with perfect specimens for examination, two forms can seldom be connected by intermediate varieties and thus proved to be the same species, until many specimens have been collected from many places; and in the case of fossil species this could rarely be effected by palæontologists. We shall, perhaps, best perceive the improbability of our being enabled to connect species by numerous, fine, intermediate, fossil links, by asking

ourselves whether, for instance, geologists at some future period will be able to prove, that our different breeds of cattle, sheep, horses, and dogs have descended from a single stock or from several aboriginal stocks; or, again, whether certain sea-shells inhabiting the shores of North America, which are ranked by some conchologists as distinct species from their European representatives, and by other conchologists as only varieties, are really varieties or are, as it is called, specifically distinct. This could be effected only by the future geologist discovering in a fossil state numerous intermediate gradations; and such success seems to me improbable in the highest degree.

Geological research, though it has added numerous species to existing and extinct genera, and has made the intervals between some few groups less wide than they otherwise would have been, yet has done scarcely anything in breaking down the distinction between species, by connecting them together by numerous, fine, intermediate varieties; and this not having been effected, is probably the gravest and most obvious of all the many objections which may be urged against my views. Hence it will be worth while to sum up the foregoing remarks, under an imaginary illustration. The Malay Archipelago is of about the size of Europe from the North Cape to the Mediterranean, and from Britain to Russia; and therefore equals all the geological formations which have been examined with any accuracy, excepting those of the United States of America. I fully agree with Mr. Godwin-Austen, that the present condition of the Malay Archipelago, with its numerous large islands separated by wide and shallow seas, probably represents the former state of Europe, when most of our formations were accumulating. The Malay Archipelago is one of the richest regions of the

1 Paleontologists are regularly finding forms that help fill the gaps between divergent taxa. The fossil record is still gappy, of course, but in the intervening century and a half since the *Origin,* a bounty of intermediate forms have been found in many groups. Nothing approaches the detailed chain of transition that Darwin lamented not having, but this is likely to be impossible in any case.

2 Darwin's informative "imaginary example" was first developed by Lyell in his address to the Geological Society of London in 1851. That parts of Europe once existed as a complex of islands created by incursions of the sea is clear from the enormous marine deposits in Britain as well as on the continent: sandstones, siltstones, limestones, and other formations teeming with marine fossils testify to this fact. Shallow seas in tropical and subtropical regions are ideal locales for preservation, since such seas have abundant life and experience much sedimentation. The Malay Archipelago, present-day Indonesia and the Malay peninsula, would seem to parallel that ancient European landscape.

3 Darwin's reference is to Robert Alfred Cloyne Godwin-Austen's *The Natural History of the European Seas* (1859), which was coauthored with Edward Forbes. This book was largely written by Forbes before his untimely death in 1854, then completed and edited by his friend Godwin-Austen. Joseph Hooker evidently supplied Darwin with an advance copy: he read the printer's proofs and wrote to Darwin in 1856, "I have been just reading Ed. Forbes first 112 pages of his little work on the European seas, all that is which Van Voorst had printed before Forbes death: would you like to see it?" (*Correspondence* 6: 175).

4 "The present is the key to the past" is the geologist's mantra. Darwin selected one of the most biologically rich tropical regions in the world, a region teeming with life and in which conditions favorable to sedimentation (and therefore fossilization) predominate. Yet even here, he argues, relatively few of the terrestrial species will be preserved for the reasons given in the next few paragraphs.

1 Missing species as a result of the rare confluence of all the right conditions for preservation is one thing, but quite another is, again, the apparent lack of transitional forms even in groups that are well represented in sizable formations. Darwin suggests here that if the average duration of species in geological time is longer than the average time span of continuous geological formations, complete transitions will be rare or absent.

whole world in organic beings; yet if all the species were to be collected which have ever lived there, how imperfectly would they represent the natural history of the world!

But we have every reason to believe that the terrestrial productions of the archipelago would be preserved in an excessively imperfect manner in the formations which we suppose to be there accumulating. I suspect that not many of the strictly littoral animals, or of those which lived on naked submarine rocks, would be embedded; and those embedded in gravel or sand, would not endure to a distant epoch. Wherever sediment did not accumulate on the bed of the sea, or where it did not accumulate at a sufficient rate to protect organic bodies from decay, no remains could be preserved.

In our archipelago, I believe that fossiliferous formations could be formed of sufficient thickness to last to an age, as distant in futurity as the secondary formations lie in the past, only during periods of subsidence. These periods of subsidence would be separated from each other by enormous intervals, during which the area would be either stationary or rising; whilst rising, each fossiliferous formation would be destroyed, almost as soon as accumulated, by the incessant coast-action, as we now see on the shores of South America. During the periods of subsidence there would probably be much extinction of life; during the periods of elevation, there would be much variation, but the geological record would then be least perfect.

1 It may be doubted whether the duration of any one great period of subsidence over the whole or part of the archipelago, together with a contemporaneous accumulation of sediment, would *exceed* the average duration of the same specific forms; and these contingencies are

indispensable for the preservation of all the transitional gradations between any two or more species. If such gradations were not fully preserved, transitional varieties would merely appear as so many distinct species. It is, also, probable that each great period of subsidence would be interrupted by oscillations of level, and that slight climatal changes would intervene during such lengthy periods; and in these cases the inhabitants of the archipelago would have to migrate, and no closely consecutive record of their modifications could be preserved in any one formation.

Very many of the marine inhabitants of the archipelago now range thousands of miles beyond its confines; and analogy leads me to believe that it would be chiefly these far-ranging species which would oftenest produce new varieties; and the varieties would at first generally be local or confined to one place, but if possessed of any decided advantage, or when further modified and improved, they would slowly spread and supplant their parent-forms. When such varieties returned to their ancient homes, as they would differ from their former state, in a nearly uniform, though perhaps extremely slight degree, they would, according to the principles followed by many palæontologists, be ranked as new and distinct species.

If then, there be some degree of truth in these remarks, we have no right to expect to find in our geological formations, an infinite number of those fine transitional forms, which on my theory assuredly have connected all the past and present species of the same group into one long and branching chain of life. We ought only to look for a few links, some more closely, some more distantly related to each other; and these links, let them be ever so close, if found in different stages of the same formation, would, by most palæonto- ◀1

1 Darwin is making the best of a bad situation. Had long chains of close transitional forms abounded in the fossil record, the transmutation theory would likely have gained purchase far earlier than it did. The fossil record is indeed spotty, and was far more so in Darwin's time. He must convince his readers that the transitional chains that his theory says must have been universal will almost never, if ever, be preserved *in toto.* He is thus in the awkward position of asking an already critical audience to accept that the record will yield only a few linking forms here and there. Understandably, this was one of the major points of criticism for Darwin's theory. We will revisit this at the end of the chapter.

logists, be ranked as distinct species. But I do not pretend that I should ever have suspected how poor a record of the mutations of life, the best preserved geological section presented, had not the difficulty of our not discovering innumerable transitional links between the species which appeared at the commencement and close of each formation, pressed so hardly on my theory.

[1] *On the sudden appearance of whole groups of Allied Species.*—The abrupt manner in which whole groups of species suddenly appear in certain formations, has been urged by several palæontologists, for instance, by
[2] Agassiz, Pictet, and by none more forcibly than by Professor Sedgwick, as a fatal objection to the belief in the transmutation of species. If numerous species, belonging to the same genera or families, have really started into life all at once, the fact would be fatal to the theory of descent with slow modification through natural selection. For the development of a group of forms, all of which have descended from some one progenitor, must have been an extremely slow process; and the progenitors must have lived long ages before their modified descendants. But we continually over-rate the perfection of the geological record, and falsely infer, because certain genera or families have not been found beneath a certain stage, that they did not exist before that stage. We continually forget how large the world is, compared with the area over which our geological formations have been carefully examined; we forget that groups of species may elsewhere have long existed and have slowly multiplied before they invaded the ancient archipelagoes of Europe and of the United States. We do not make due allowance for the enormous intervals of time, which have

1 Here is another empirical observation of the fossil record that is inconsistent with Darwin's model. Sudden or abrupt appearance suggests special creation, not transmutation! This subject is discussed more fully on pp. 306–309.

2 Darwin names three important geologists he knew would be hostile to his theory: Louis Agassiz, François Jules Pictet de la Rive, and his first geological mentor, Adam Sedgwick. All three were adherents of catastrophism, an idea championed by Cuvier in *Discours sur les révolutions de la surface du globe* (Paris, 1826). There were significant differences among them, however, reflecting the English and continental catastrophist schools. What they had in common was a shared Cuverian view that successive rounds of creation and extinction explained the patterns of faunal succession evident in the fossil record, while they differed on the need for miraculous intervention and on the rate of geological change in the past. The catastrophist school is often contrasted with uniformitarianism on these points (both terms were coined by Whewell in 1832): catastrophists are often mischaracterized as scriptural adherents blind to empirical evidence, but in fact they were stricter empiricists than Lyell, who for years rejected any kind of progressive change in the fossil record (Cannon 1960).

probably elapsed between our consecutive formations,—longer perhaps in some cases than the time required for the accumulation of each formation. These intervals will have given time for the multiplication of species from some one or some few parent-forms; and in the succeeding formation such species will appear as if suddenly created.

I may here recall a remark formerly made, namely that it might require a long succession of ages to adapt an organism to some new and peculiar line of life, for instance to fly through the air; but that when this had been effected, and a few species had thus acquired a great advantage over other organisms, a comparatively short time would be necessary to produce many divergent forms, which would be able to spread rapidly and widely throughout the world.

I will now give a few examples to illustrate these remarks; and to show how liable we are to error in supposing that whole groups of species have suddenly been produced. I may recall the well-known fact that in geological treatises, published not many years ago, the great class of mammals was always spoken of as having abruptly come in at the commencement of the tertiary series. And now one of the richest known accumulations of fossil mammals belongs to the middle of the secondary series; and one true mammal has been discovered in the new red sandstone at nearly the commencement of this great series. Cuvier used to urge that no monkey occurred in any tertiary stratum; but now extinct species have been discovered in India, South America, and in Europe even as far back as the eocene stage. The most striking case, however, is that of the Whale family; as these animals have huge bones, are marine, and range over the world, the fact of not a single bone of a whale having been discovered in

1 Here Darwin is suggesting that the abrupt appearance of groups in the fossil record could be a sampling artifact. It may take a long while for a group to become sufficiently abundant and widespread to be likely to be fossilized. The group thus appears rather suddenly in the fossil record but had been around a while before then.

2 The "one true mammal" of the Permo-Triassic New Red Sandstone that Darwin refers to here is probably *Chirotherium* ("hand animal"), named from fossilized tracks discovered in southern Germany. *Chirotherium* was much discussed in Darwin's day because it was thought to be the earliest mammalian fossil. This species is now believed to be an archosaur, an ancestral reptilian relative of dinosaurs. Darwin discusses this putative early mammal to make the point that groups, like mammals, that were thought to have appeared rather abruptly were in fact around longer than paleontologists supposed. Indeed, this is borne out: the Permo-Triassic is full of mammalian forerunners, the mammal-like reptiles (therapsids) (see Kemp 1982, Hopson 1987).

3 The earliest monkey fossil found to date was reported by Ni et al. (2004) from the early Eocene of Hunan Province, China. Recent years have seen exciting early cetacean discoveries. Like the primates, the earliest proto-whale fossils are early to mid-Eocene, and better transitional forms could not be found, for these early cetaceans can walk (e.g., *Ambulocetus* and *Pakicetus;* see Thewissen et al. 1996, Gingerich et al. 2001). These dates are a bit later than Darwin supposes here. In nineteenth-century terms, the "secondary" formations correspond to the modern Mesozoic, while the "tertiary" formations correspond to the modern Cenozoic. The Eocene, an early Cenozoic period, is more recent than the "upper greensand" Lyell discusses in his *Manual* (a late Mesozoic formation, Jurasso-Cretaceous). To date there are no late Mesozoic whale fossils.

any secondary formation, seemed fully to justify the belief that this great and distinct order had been suddenly produced in the interval between the latest secondary and earliest tertiary formation. But now we may read in the Supplement to Lyell's 'Manual,' published in 1858, clear evidence of the existence of whales in the upper greensand, some time before the close of the secondary period.

I may give another instance, which from having [1] passed under my own eyes has much struck me. In a memoir on Fossil Sessile Cirripedes, I have stated that, from the number of existing and extinct tertiary species; from the extraordinary abundance of the individuals of many species all over the world, from the Arctic regions to the equator, inhabiting various zones of depths from the upper tidal limits to 50 fathoms; from the perfect manner in which specimens are preserved in the oldest tertiary beds; from the ease with which even a fragment of a valve can be recognised; from all these circumstances, I inferred that had sessile cirripedes existed during the secondary periods, they would certainly have been preserved and discovered; and as not one species had been discovered in beds of this age, I concluded that this great group had been suddenly developed at the commencement of the tertiary series. This was a sore trouble to me, adding as I thought one more instance of the abrupt appearance of a great group of [2] species. But my work had hardly been published, when a skilful palæontologist, M. Bosquet, sent me a drawing of a perfect specimen of an unmistakeable sessile cirripede, which he had himself extracted from the chalk of Belgium. And, as if to make the case as striking as possible, this sessile cirripede was a Chthamalus, a very common, large, and ubiquitous genus, of which not one specimen has as yet been found even in any tertiary

1 Darwin produced four barnacle monographs between 1851 and 1854. This one is *A Monograph on the Fossil Balanidae and Verrucidae of Great Britain* (London, 1854).

2 The acorn barnacle that the paleontologist Joseph Augustin Hubert Bosquet sent Darwin—ironically, it was later named *Chthamalus darwini*—was putatively collected from the chalk (Cretaceous) of Belgium, but this turned out to be an error. Withers (1928, p. 44) noted that it was from the Recent period, "from the Mediterranean, presumably dumped in a field near Vaals, with the sweepings from a house." Fortunately for Darwin's argument, however, other Cretaceous barnacles have since been reported, so the acorn barnacles do not appear so suddenly after all. One of these made it into the sixth edition of the *Origin: Pyrgoma* (later *Brachylepas*) *cretacea*, described by Henry Woodward (Woodward 1901). More recently, *Pachydiadema cretaceum* from Sweden was described by Withers.

stratum. Hence we now positively know that sessile cirripedes existed during the secondary period; and these cirripedes might have been the progenitors of our many tertiary and existing species.

The case most frequently insisted on by palæontologists of the apparently sudden appearance of a whole group of species, is that of the teleostean fishes, low down in the Chalk period. This group includes the large majority of existing species. Lately, Professor Pictet has carried their existence one sub-stage further back; and some palæontologists believe that certain much older fishes, of which the affinities are as yet imperfectly known, are really teleostean. Assuming, however, that the whole of them did appear, as Agassiz believes, at the commencement of the chalk formation, the fact would certainly be highly remarkable; but I cannot see that it would be an insuperable difficulty on my theory, unless it could likewise be shown that the species of this group appeared suddenly and simultaneously throughout the world at this same period. It is almost superfluous to remark that hardly any fossil-fish are known from south of the equator; and by running through Pictet's Palæontology it will be seen that very few species are known from several formations in Europe. Some few families of fish now have a confined range; the teleostean fish might formerly have had a similarly confined range, and after having been largely developed in some one sea, might have spread widely. Nor have we any right to suppose that the seas of the world have always been so freely open from south to north as they are at present. Even at this day, if the Malay Archipelago were converted into land, the tropical parts of the Indian Ocean would form a large and perfectly enclosed basin, in which any great group of marine animals might be multiplied; and

1 The teleosts, subclass Teleostei, are the ray-finned fish. This group consists of most living fish species (some 20,000 species in about 40 orders) characterized by a symmetrical (homocercal) tail and a fully moveable upper jaw. "Teleost" means "complete bone," in reference to the fully bony (as opposed to cartilaginous) nature of the skeleton. It is ironic that this group should bear a name derived from the Greek *teleos,* also the root of teleology, an idea from which Darwin labored to free us.

2 Several extinct orders of early teleosts are known from the Mesozoic, including the Tselfatiformes, Leptolepidiformes, Pholidophoriformes, and Ichthyodectiformes. The group did not appear *in toto* at the commencement of the Cretaceous period, the view Darwin ascribes to Agassiz. In current thinking, the group likely arose in the early to mid-Triassic and is represented by several genera by the late Triassic. Most of the major teleost lineages and some of the modern ones are well represented by the mid- to late Jurassic.

Darwin tries to defuse the apparent problem of sudden appearance of teleosts in the Cretaceous by referring back to his earlier argument: progenitors of Cretaceous teleosts were around earlier, but perhaps elsewhere on earth, and only by the Cretaceous had they become widespread and abundant enough to begin to appear in fossiliferous formations. This argument is not very convincing, particularly to skeptics.

1 Darwin's main point is that we know precious little of the fossil record, after all, and so must withhold judgment about whether or how it supports transmutation. Darwin was confident that the fossils were out there awaiting discovery. "Absence of evidence is not evidence of absence" is a modern saying that he would have embraced in this context. In Darwin's lifetime discoveries were made that buttressed his claim (e.g., the dinosaur-bird link *Archaeopteryx*, discovered in 1861 and described by Owen in 1863). Still, arguing from negative evidence is not a position of strength, and it was to be expected that many of Darwin's harshest critics homed in on the weakness of the fossil evidence.

2 This matter is indeed much graver. The apparently sudden appearance of nearly the full panoply of animal phyla in the lowest fossiliferous strata certainly has the earmarks of special creation. This was a centerpiece of the argument for special creation by early paleontologists like Louis Agassiz. In a section entitled "Simultaneous existence in the earliest geological periods of all the great types of animals" in his 1851 *Essay on Classification,* Agassiz noted that "representatives of numerous families belonging to all the four great branches of the animal kingdom, are well known to have existed simultaneously in the oldest geological formations." Darwin does not shrink from consistent application of his theory of common descent: the argument that members of any given group share descent from a common ancestor must apply equally to the very earliest fossil species. In other words, there must have been progenitors from which these groups descended.

3 Trilobites are now-extinct arthropods that flourished in the shallow seas of the Paleozoic. *Nautilus* is a genus of squid, and *Lingula* is a genus of brachiopod. These last two groups are found in early- to mid-Paleozoic strata, and are also among the relatively few groups still extant today and exhibiting much the same morphology.

here they would remain confined, until some of the species became adapted to a cooler climate, and were enabled to double the southern capes of Africa or Australia, and thus reach other and distant seas.

1 From these and similar considerations, but chiefly from our ignorance of the geology of other countries beyond the confines of Europe and the United States; and from the revolution in our palæontological ideas on many points, which the discoveries of even the last dozen years have effected, it seems to me to be about as rash in us to dogmatize on the succession of organic beings throughout the world, as it would be for a naturalist to land for five minutes on some one barren point in Australia, and then to discuss the number and range of its productions.

On the sudden appearance of groups of Allied Species in
2 *the lowest known fossiliferous strata.*—There is another and allied difficulty, which is much graver. I allude to the manner in which numbers of species of the same group, suddenly appear in the lowest known fossiliferous rocks. Most of the arguments which have convinced me that all the existing species of the same group have descended from one progenitor, apply with nearly equal force to the earliest known species. For instance, I
3 cannot doubt that all the Silurian trilobites have descended from some one crustacean, which must have lived long before the Silurian age, and which probably differed greatly from any known animal. Some of the most ancient Silurian animals, as the Nautilus, Lingula, &c., do not differ much from living species; and it cannot on my theory be supposed, that these old species were the progenitors of all the species of the orders to which they belong, for they do not present characters in any degree intermediate between them.

If, moreover, they had been the progenitors of these orders, they would almost certainly have been long ago supplanted and exterminated by their numerous and improved descendants.

Consequently, if my theory be true, it is indisputable that before the lowest Silurian stratum was deposited, long periods elapsed, as long as, or probably far longer than, the whole interval from the Silurian age to the present day; and that during these vast, yet quite unknown, periods of time, the world swarmed with living creatures.

To the question why we do not find records of these vast primordial periods, I can give no satisfactory answer. Several of the most eminent geologists, with Sir R. Murchison at their head, are convinced that we see in the organic remains of the lowest Silurian stratum the dawn of life on this planet. Other highly competent judges, as Lyell and the late E. Forbes, dispute this conclusion. We should not forget that only a small portion of the world is known with accuracy. [1] M. Barrande has lately added another and lower stage to the Silurian system, abounding with new and peculiar species. Traces of life have been detected in the Longmynd beds beneath Barrande's so-called primordial zone. The presence of phosphatic nodules and bituminous matter in some of the lowest azoic rocks, [2] probably indicates the former existence of life at these periods. But the difficulty of understanding the absence of vast piles of fossiliferous strata, which on my theory no doubt were somewhere accumulated before the Silurian epoch, is very great. If these most ancient beds had been wholly worn away by denudation, or obliterated by metamorphic action, we ought to find only small remnants of the formations next succeeding them in age, and these ought to be very generally in

1 Barrande worked extensively on the Paleozoic fossils of Bohemia. He referred to fossiliferous rocks underlying Silurian strata near Prague as the "primordial zone," strata that were termed "Cambrian" by Sedgwick. Darwin was excited to learn that fossil-bearing rocks were found even lower than these. In June 1859 he wrote to Andrew Ramsay, "Is it certain that traces of organic remains have been found in the Longmynd Beds? And secondly is it . . . certain that these Beds are lower than Barrande's primordial zone?" (*Correspondence* 7: 311). The Longmynd Beds did not turn out to be earlier than the "primordial zone" and are now recognized as an early Cambrian formation found in Shropshire. But this did not solve the problem of sudden appearance; it just pushed it back. These earlier forms were simpler, however, supporting the idea that the suddenness of appearance of groups in the Cambrian is illusory.

2 Murchison dubbed the lowest formations "Azoic," the lifeless era, believing they formed under conditions too hot to support life. Today the early formations are called "Precambrian," itself divided into three eons: Hadean (~4.5–3.8 b.y.), Archaean (~3.8–2.5 b.y.), and Proterozoic (~2.5–0.5 b.y.).

A lifeless era capped by strata bearing complex taxa would be inconsistent with Darwin's model, so naturally he was keen on any evidence for life before the Cambrian. In the fourth *Origin* edition of 1866 Darwin inserted a sentence here relating an exciting new find: "*within the last year the great discovery of the Eozoon in the Laurentian formation of Canada has been made; . . . it is impossible to feel any doubt regarding its organic nature*" (Peckham 1959, p. 515). These concentric layered siliceous and calcareous nodules found in Ontario were thought to be fossilized foraminifera. They were dubbed *Eozoön,* "Dawn animal," in 1865 papers by Sir William Logan and William Benjamin Carpenter, but the organic nature of the structures was disputed almost immediately. The question was finally settled in the early twentieth century, in favor of an inorganic origin. Darwin hung on to the end while acknowledging dissention: in the sixth edition of *Origin* he altered the sentence to read that the reality of *Eozoön* was only "generally admitted" (Peckham 1959, p. 515; see O'Brien 1970).

a metamorphosed condition. But the descriptions which we now possess of the Silurian deposits over immense territories in Russia and in North America, do not support the view, that the older a formation is, the more it has suffered the extremity of denudation and metamorphism.

[1] The case at present must remain inexplicable; and may be truly urged as a valid argument against the views here entertained. To show that it may hereafter receive some explanation, I will give the following hypothesis. From the nature of the organic remains, which do not appear to have inhabited profound depths, in the several formations of Europe and of the United States; and from the amount of sediment, miles in thickness, of which the formations are composed, we may infer that from first to last large islands or tracts of land, whence the sediment was derived, occurred in the neighbourhood of the existing continents of Europe and North America. But we do not know what was the state of things in the intervals between the successive formations; whether Europe and the United States during these intervals existed as dry land, or as a submarine surface near land, on which sediment was not deposited, or again as the bed of an open and unfathomable sea.

Looking to the existing oceans, which are thrice as extensive as the land, we see them studded with many islands; but not one oceanic island is as yet known to afford even a remnant of any palæozoic or secondary
[2] formation. Hence we may perhaps infer, that during the palæozoic and secondary periods, neither continents nor continental islands existed where our oceans now extend; for had they existed there, palæozoic and secondary formations would in all probability have been accumulated from sediment derived from their wear and

1 Today, 150 years after the *Origin*'s publication, scientists have found considerable evidence of life in the Precambrian. Fossil prokaryotic mats called stromatolites (e.g., Allwood et al. 2006) are known from every continent. These were recognized as fossils only in the mid-twentieth century, once living mats were discovered (e.g., at Shark Bay, Australia). More complex organisms first appear in the Vendian Period of the Precambrian, 650 to 543 million years ago, marked by fossils known as the Vendian biota or Ediacaran fauna (named for the odd creatures of the Ediacara sandstone of Australia, discovered in 1946).

Nonetheless, Darwin is correct that there are very few intact sedimentary strata from this most ancient period. In the following paragraphs he attempts an explanation, but he is only partially correct. Geologists realize today that the rarity of strata may be attributed to the inexorable erosion, recycling, and metamorphism that has taken place over the eons. Moreover, the time scales involved are an order of magnitude greater than that for which Darwin dared hope, so there has been that much more time for degradation and metamorphism.

2 The Lyellian view of the earth, imbibed by Darwin, was one of oscillation of level. Elevation in one area was compensated by subsidence in another area. A compelling case for the idea of lateral movement of continents, or continental drift, was made in 1912 upon publication of Alfred Wegener's book *The Origin of Continents and Oceans.* Even then the theory was not accepted for another half century.

tear; and would have been at least partially upheaved by the oscillations of level, which we may fairly conclude must have intervened during these enormously long periods. If then we may infer anything from these facts, we may infer that where our oceans now extend, oceans have extended from the remotest period of which we have any record; and on the other hand, that where continents now exist, large tracts of land have existed, subjected no doubt to great oscillations of level, since the earliest silurian period. The coloured map appended to my volume on Coral Reefs, led me to conclude that the great oceans are still mainly areas of subsidence, the great archipelagoes still areas of oscillations of level, and the continents areas of elevation. But have we any right to assume that things have thus remained from eternity? Our continents seem to have been formed by a preponderance, during many oscillations of level, of the force of elevation; but may not the areas of preponderant movement have changed in the lapse of ages? At a period immeasurably antecedent to the silurian epoch, continents may have existed where oceans are now spread out; and clear and open oceans may have existed where our continents now stand. Nor should we be justified in assuming that if, for instance, the bed of the Pacific Ocean were now converted into a continent, we should there find formations older than the silurian strata, supposing such to have been formerly deposited; for it might well happen that strata which had subsided some miles nearer to the centre of the earth, and which had been pressed on by an enormous weight of superincumbent water, might have undergone far more metamorphic action than strata which have always remained nearer to the surface. The immense areas in some parts of the world, for instance in South America, of bare metamorphic rocks, which

1 Darwin's coral reef book (1842) included a fold-out map of the reefs color-coded, with red indicating areas of stability or elevation, and areas of subsidence in blue. Herbert (2005, p. 198) noted that the late marine geologist Henry William Menard sounded surprised when he wrote, "At times, Darwin seems to be concerned with the origin of coral reefs only so far as it gives evidence for regional subsidence of the sea floor." Herbert points out that that was precisely the case.

2 This, too, is very Lyellian. Lyell argued that it made more sense that the elevation of the land changes, even raising land out of the abyssal ocean, than that land was exposed by subsiding waters. In *Elements of Geology* he asserted that "were we to embrace the doctrine which ascribes the elevated position of marine formations, and the depression of certain freshwater strata, to oscillations in the level of the waters instead of the land, we should be compelled to admit that the ocean has been sometimes every where much shallower than at present, and at others more than three miles deeper" (Lyell 1865, p. 47). The seas do not rise and fall per se, in Lyell's view; rather, the land is raised and lowered.

3 This is a reasonable way to make sense of highly metamorphosed early rocks. Nineteenth-century geologists had a limited conception of the forces that could metamorphose rocks, but they correctly surmised that heat and pressure are involved. In the oscillating-earth model of Lyell and Darwin, heat and pressure are both provided by extremely deep submersion. In current understanding the very oldest rocks (~3.8 to 4 billion years old) are exposed cores of ancient mountain chains, metamorphosed in the extreme during mountain-building processes (e.g., Acasta Gneiss in the Canadian Shield). Darwin is correct that any sign of life in the oldest of rocks is likely to have been obliterated in bouts of metamorphism.

must have been heated under great pressure, have always seemed to me to require some special explanation; and we may perhaps believe that we see in these large areas, the many formations long anterior to the silurian epoch in a completely metamorphosed condition.

The several difficulties here discussed, namely our not finding in the successive formations infinitely numerous transitional links between the many species which now exist or have existed; the sudden manner in which whole groups of species appear in our European formations; the almost entire absence, as at present known, of fossiliferous formations beneath the Silurian strata, [1] are all undoubtedly of the gravest nature. We see this in the plainest manner by the fact that all the most eminent palæontologists, namely Cuvier, Owen, Agassiz, Barrande, Falconer, E. Forbes, &c., and all our greatest geologists, as Lyell, Murchison, Sedgwick, &c., have unanimously, often vehemently, maintained the immutability of species. But I have reason to believe that one great authority, Sir Charles Lyell, from further reflexion entertains grave doubts on this subject. I feel how rash it is to differ from these great authorities, to whom, with others, we owe all our knowledge. Those who think the natural geological record in any degree perfect, and who do not attach much weight to the facts and arguments of other kinds given in this volume, will undoubtedly at once re- [2] ject my theory. For my part, following out Lyell's metaphor, I look at the natural geological record, as a history of the world imperfectly kept, and written in a changing dialect; of this history we possess the last volume alone, relating only to two or three countries. Of this volume, only here and there a short chapter has

1 This is a formidable "who's who" of geology and paleontology, each likely to oppose the idea that species can change. Darwin alludes to Lyell being more receptive to the idea; indeed, Lyell was one of very few people with whom Darwin shared his theory. Lyell was impressed with Darwin's geological work on the *Beagle* voyage, and upon Darwin's return the two became friends. Lyell quickly became Darwin's mentor as well, introducing him to the leading lights in the English geological world. Darwin later wrote, "I always feel as if my books came half out of Lyell's brains and that I never acknowledge this sufficiently (*Correspondence* 3: 54).

Lyell was intensely interested in "the species question," and he filled seven notebooks on the subject between 1855 and 1861 (Wilson 1970). The date of the first notebook is significant: Lyell was long interested in the question of species origins and transmutation, but in November 1855 he read Wallace's Sarawak Law paper. The clarity of Wallace's insight had a profound effect on Lyell; soon afterward he urged Darwin to make his views on species origins known, given that Wallace was converging on the very same ideas. Darwin responded, initiating *Natural Selection* in 1856 (Stauffer 1975).

2 Darwin borrowed this metaphor from Lyell's language metaphor; by "following out" he means extending it. Lyell concluded volume III, chapter III, of *Principles* with a discussion on "cause of violations of continuity" to explain the lack of a continuous and complete record of life, concluding with an archaeological analogy. Lyell imagines that two buried cities are found at the foot of Mt. Vesuvius, the lower one Greek and the upper one Italian. An archaeologist would "reason very hastily" if he claimed the transition between the two had been abrupt. If the archaeologist afterward located three buried cities at distinct strata, the lowest Greek, the middle Roman, and the upper Italian, Lyell says, this would underscore the error of his former opinion of a sudden shift from Greek to Italian. And just as the Roman period interjected between the Greek and the Italian here, "so many other dialects may have been spoken in succession, and the passage from the Greek to the Italian may have been very gradual."

been preserved; and of each page, only here and there a few lines. Each word of the slowly-changing language, in which the history is supposed to be written, being more or less different in the interrupted succession of chapters, may represent the apparently abruptly changed forms of life, entombed in our consecutive, but widely separated, formations. On this view, the difficulties above discussed are greatly diminished, or even disappear.

1 Just as the occasional eruptions of Vesuvius preserve these human monuments at haphazard points in time, there is no continuous record of each and every step in human history, at least not in any one locale. "So in Geology," Lyell continued, "[we cannot] assume that it is the plan of nature to preserve, in every region of the globe, an unbroken series of monuments to commemorate the vicissitudes of the organic creation." For emphasis he ends this discussion by concluding that "we must shut our eyes to the whole economy of the existing causes . . . if we fail to perceive *that such is not the plan of Nature*" (Lyell 1833, pp. 33–34; emphasis in original).

Darwin's version of Lyell's metaphor imagines a flawed historical account of the world, written over time in a changing language; many volumes are missing, and of the few in hand only some chapters remain, and of those, only a few sentences. The metaphor is evocative of just how fragmentary the fossil record is, preserving precious little of the superabundance of life that must have flourished over the eons.

The tenor of Darwin's conclusion changed over time. By the fifth edition, the penultimate sentence was reworked to read: "Each word of the slowly-changing language, more or less different in the successive chapters, may represent the forms of life, which are entombed in our consecutive formations, and which falsely appear to us to have been abruptly introduced." This is a somewhat more forceful statement than the original version in the first edition, declaring as false the sudden appearance of species in the fossil record.

Having anticipated and addressed, if not defused, some of the graver problems that the fossil record presents for his model of descent with modification, Darwin turns here to empirical patterns to support his view. This chapter presents several key observations that are strongly suggestive of descent with modification—in particular the temporal sequence of fossil forms of a given group and their relationship to species that are still living in the same areas. Recall that this observation and its symmetry with spatial relationships of species (biogeography) gave Darwin and Wallace their first great insights into transmutation. Wallace's 1855 paper on the subject caught Lyell's attention and prompted the geologist to urge Darwin to publish his views. Darwin soon began to write his big species book, but Wallace was now hot on the trail of the idea and followed this paper with his seminal 1858 Ternate paper. The paleontological patterns discussed in this chapter thus constitute one of the two central lines of evidence that inspired both Darwin and Wallace. Yet the interpretation of the fossil record's empirical pattern that Darwin gives here did not meet with universal approval. Louis Agassiz in particular argued that the observed patterns compel a conclusion opposite to Darwin's.

---

1 By slow appearance of new species Darwin means that the species found in a stratigraphic column are distributed throughout that column; they do not all appear simultaneously at its inception and disappear together at its termination. This speaks to gradualism.

2 The works Darwin refers to here are Rudolph Amandus Philippi's treatment of the mollusks of Sicily, *Enumeratio Molluscorum Sicilia* (1844), and Heinrich Georg Bronn's *Letkaea Geognostica* (1851–1856). The paragraph ends with an explicit statement that appearance and disappearance are not abrupt. It implies that a distributed, gradual appearance and disappearance is more consistent with Darwin's model of descent with modification than with special creation. What follows is a series of empirical observations about species appearance and disappearance, with an eye toward emphasizing this point.

CHAPTER X.

ON THE GEOLOGICAL SUCCESSION OF ORGANIC BEINGS.

On the slow and successive appearance of new species — On their different rates of change — Species once lost do not reappear — Groups of species follow the same general rules in their appearance and disappearance as do single species — On Extinction — On simultaneous changes in the forms of life throughout the world — On the affinities of extinct species to each other and to living species — On the state of development of ancient forms — On the succession of the same types within the same areas — Summary of preceding and present chapters.

LET us now see whether the several facts and rules relating to the geological succession of organic beings, better accord with the common view of the immutability of species, or with that of their slow and gradual modification, through descent and natural selection.

1 New species have appeared very slowly, one after another, both on the land and in the waters. Lyell has shown that it is hardly possible to resist the evidence on this head in the case of the several tertiary stages; and every year tends to fill up the blanks between them, and to make the percentage system of lost and new forms more gradual. In some of the most recent beds, though undoubtedly of high antiquity if measured by years, only one or two species are lost forms, and only one or two are new forms, having here appeared for the first time, either locally, or, as far as we know, on
2 the face of the earth. If we may trust the observations of Philippi in Sicily, the successive changes in the marine inhabitants of that island have been many and most gradual. The secondary formations are more broken; but, as Bronn has remarked, neither the appearance

nor disappearance of their many now extinct species has been simultaneous in each separate formation.

Species of different genera and classes have not changed at the same rate, or in the same degree. In the oldest tertiary beds a few living shells may still be found in the midst of a multitude of extinct forms. Falconer has given a striking instance of a similar fact, in an existing crocodile associated with many strange and lost mammals and reptiles in the sub-Himalayan deposits. The Silurian Lingula differs but little from the living species of this genus; whereas most of the other Silurian Molluscs and all the Crustaceans have changed greatly. The productions of the land seem to change at a quicker rate than those of the sea, of which a striking instance has lately been observed in Switzerland. There is some reason to believe that organisms, considered high in the scale of nature, change more quickly than those that are low: though there are exceptions to this rule. The amount of organic change, as Pictet has remarked, does not strictly correspond with the succession of our geological formations; so that between each two consecutive formations, the forms of life have seldom changed in exactly the same degree. Yet if we compare any but the most closely related formations, all the species will be found to have undergone some change. When a species has once disappeared from the face of the earth, we have reason to believe that the same identical form never reappears. The strongest apparent exception to this latter rule, is that of the so-called "colonies" of M. Barrande, which intrude for a period in the midst of an older formation. and then allow the pre-existing fauna to reappear; but Lyell's explanation, namely, that it is a case of temporary migration from a distinct geographical province, seems to me satisfactory.

1 Observation no. 1: different species of a stratigraphic column do not change at the same rate or to the same degree. This is demonstrated by the occurrence of a few modern, even extant, mollusk species in Tertiary beds otherwise dominated by extinct forms. The Falconer reference is to Hugh Falconer's 1831 discovery of fossilized bones of an extant crocodile along with those of an assemblage of other animals in the Siwálik Hills of India, which he published in 1859.

2 Observation no. 2: more complex organisms appear to change more than less complex organisms. Different species change to different degrees as well as different rates. This observation sets up an important point that will be developed later, namely, that diverse and unrelated species changing in lockstep would be more consistent with supernatural agency than with natural law, but that is not what is observed.

3 Observation no. 3: once a species is gone from the fossil record, it is gone for good. This, too, is expected on Darwin's model, according to which the formation of new species goes hand-in-hand with the extinction of others.

Joachim Barrande's "colonies" concept posed a challenge to this theory. Barrande, in *Systême Silurien du Centre de la Bohême* (1852–1881), argued that certain rock series characteristic of a geological period reappear in older formations, bearing their fossils. Sets of the same species thus reappeared at different times in the fossil record, he asserted. He termed these assemblages of anachronistic rocks "colonies." Geologists challenged Barrande's interpretation of these rocks, but he spiritedly defended his interpretation in a series of books entitled *Défense des Colonies,* published over a twenty-two-year period, in which new colony formations were named after Barrande's critics. Barrande's colonies became one of the great debates in nineteenth-century geology and did not survive the test of time. (Neither, by the way, did Lyell's explanation, which Darwin favors here.) Today the so-called colonies are interpreted as younger Silurian rocks reverse-thrust into sequences of older Ordovician rocks in a complex folded structure, giving the appearance of reoccurrence of rock series and their fossils (Kríz and Pojeta 1974).

1 Darwin believes that species change according to natural law, and he expressly rejects any role for supernatural agency, inner driving forces tending toward complexity, divine guidance, etc. There is, therefore, no "fixed law of development" that controls when, how, and to what degree species change over time.

2 Here Darwin presents further observations in support of the assertion that there is no "fixed law of development." Species change at different rates and to different degrees; unique species of island systems like Madeira differ in varying degrees from their mainland relatives, and more complex organisms seem to change faster than do less complex organisms. (Darwin's information on Madeira came from the work of Thomas Vernon Wollaston.) This last observation is debatable; change might simply be more apparent in more complex forms, with a greater number of anatomical features capable of varying. Or perhaps these forms are better represented in the fossil record, making possible a more detailed picture of their change.

1 These several facts accord well with my theory. I believe in no fixed law of development, causing all the inhabitants of a country to change abruptly, or simultaneously, or to an equal degree. The process of modification must be extremely slow. The variability of each species is quite independent of that of all others. Whether such variability be taken advantage of by natural selection, and whether the variations be accumulated to a greater or lesser amount, thus causing a greater or lesser amount of modification in the varying species, depends on many complex contingencies,—on the variability being of a beneficial nature, on the power of intercrossing, on the rate of breeding, on the slowly changing physical conditions of the country, and more especially on the nature of the other inhabitants with which the varying species comes into
2 competition. Hence it is by no means surprising that one species should retain the same identical form much longer than others; or, if changing, that it should change less. We see the same fact in geographical distribution; for instance, in the land-shells and coleopterous insects of Madeira having come to differ considerably from their nearest allies on the continent of Europe, whereas the marine shells and birds have remained unaltered. We can perhaps understand the apparently quicker rate of change in terrestrial and in more highly organised productions compared with marine and lower productions, by the more complex relations of the higher beings to their organic and inorganic conditions of life, as explained in a former chapter. When many of the inhabitants of a country have become modified and improved, we can understand, on the principle of competition, and on that of the many all-important relations of organism to organism, that any form which does not become in some degree modified and improved,

will be liable to be exterminated. Hence we can see why all the species in the same region do at last, if we look to wide enough intervals of time, become modified; for those which do not change will become extinct.

In members of the same class the average amount of change, during long and equal periods of time, may, perhaps, be nearly the same; but as the accumulation of long-enduring fossiliferous formations depends on great masses of sediment having been deposited on areas whilst subsiding, our formations have been almost necessarily accumulated at wide and irregularly intermittent intervals; consequently the amount of organic change exhibited by the fossils embedded in consecutive formations is not equal. Each formation, on this view, does not mark a new and complete act of creation, but only an occasional scene, taken almost at hazard, in a slowly changing drama.

We can clearly understand why a species when once lost should never reappear, even if the very same conditions of life, organic and inorganic, should recur. For though the offspring of one species might be adapted (and no doubt this has occurred in innumerable instances) to fill the exact place of another species in the economy of nature, and thus supplant it; yet the two forms—the old and the new—would not be identically the same; for both would almost certainly inherit different characters from their distinct progenitors. For instance, it is just possible, if our fantail-pigeons were all destroyed, that fanciers, by striving during long ages for the same object, might make a new breed hardly distinguishable from our present fantail; but if the parent rock-pigeon were also destroyed, and in nature we have every reason to believe that the parent-form will generally be supplanted and

1 While generally true, this claim overlooks so-called living fossils, species that, judging from the fossil record, appear to be essentially unchanged for vast stretches of time. Examples include coelacanths *(Latimeria)* and horseshoe crabs *(Limulus)*. Elsewhere, such as in the divergence of character diagram (chapter IV), Darwin does acknowledge that some species will remain unchanged, presumably because they are exceptionally well adapted and live in a constant environment.

2 As we have repeatedly seen, Darwin turns to domesticated organisms for analogy whenever possible. Having just explained why a species cannot arise more than once—owing to the uniqueness of the particular lineage of each species—he underscores this with an example from living pigeons. Varieties, too, are unique products of a particular lineage that cannot be repeated. Should the parent stock of a given variety go extinct, no other species could give rise to precisely that same variety. (Of course, the uniqueness principle would also hold that *exactly* the same variety could not be developed even from the same parent stock, owing to the vagaries of chance variations, etc.)

exterminated by its improved offspring, it is quite incredible that a fantail, identical with the existing breed, could be raised from any other species of pigeon, or even from the other well-established races of the domestic pigeon, for the newly-formed fantail would be almost sure to inherit from its new progenitor some slight characteristic differences.

Groups of species, that is, genera and families, follow the same general rules in their appearance and disappearance as do single species, changing more or less quickly, and in a greater or lesser degree. A group does not reappear after it has once disappeared; or its existence, as long as it lasts, is continuous. I am aware that there are some apparent exceptions to this rule, but the exceptions are surprisingly few, so [1] few, that E. Forbes, Pictet, and Woodward (though all strongly opposed to such views as I maintain) admit its truth; and the rule strictly accords with my theory. For as all the species of the same group have descended from some one species, it is clear that as long as any species of the group have appeared in the long succession of ages, so long must its members have continuously existed, in order to have generated either new and modified or the same old and unmodified forms. Species [2] of the genus Lingula, for instance, must have continuously existed by an unbroken succession of generations, from the lowest Silurian stratum to the present day.

We have seen in the last chapter that the species of a group sometimes falsely appear to have come in abruptly; and I have attempted to give an explanation of this fact, which if true would have been fatal to my views. But such cases are certainly exceptional; the general rule being a gradual increase in number, till the group reaches its maximum, and then, sooner or later, it gradually decreases. If the

1 Forbes, Pictet, and Woodward are naturalists and geologists who staunchly rejected the idea of transmutation. Continuity of lineages is of great importance to Darwin's theory; a demonstrable break in the ancestral lineage chain of an extant species would compromise the whole model, as only supernatural agency could create a species anew.

2 *Lingula* is a genus in the brachiopod family Lingulidae. Brachiopods are marine invertebrates that superficially resemble bivalves, but they are quite distinct, placed in their own phylum, Brachiopoda. Extant brachiopods live in deep, cold waters; very few taxa exist today, but they were extremely common in the Paleozoic. *Lingula* has long been thought to be represented from mid-Paleozoic to modern times, as Darwin suggests here, though Emig (2003) argues that the group is not as little changed as paleontologists often assume.

number of the species of a genus, or the number of the genera of a family, be represented by a vertical line of varying thickness, crossing the successive geological formations in which the species are found, the line will sometimes falsely appear to begin at its lower end, not in a sharp point, but abruptly; it then gradually thickens upwards, sometimes keeping for a space of equal thickness, and ultimately thins out in the upper beds, marking the decrease and final extinction of the species. This gradual increase in number of the species of a group is strictly conformable with my theory; as the species of the same genus, and the genera of the same family, can increase only slowly and progressively; for the process of modification and the production of a number of allied forms must be slow and gradual,—one species giving rise first to two or three varieties, these being slowly converted into species, which in their turn produce by equally slow steps other species, and so on, like the branching of a great tree from a single stem, till the group becomes large.

On Extinction.—We have as yet spoken only incidentally of the disappearance of species and of groups of species. On the theory of natural selection the extinction of old forms and the production of new and improved forms are intimately connected together. The old notion of all the inhabitants of the earth having been swept away at successive periods by catastrophes, is very generally given up, even by those geologists, as Elie de Beaumont, Murchison, Barrande, &c., whose general views would naturally lead them to this conclusion. On the contrary, we have every reason to believe, from the study of the tertiary formations, that species and groups of species gradually disappear, one after another, first from one spot, then from another, and

1 The device of using width or thickness to represent the relative size of lineages, or clades, was common in Darwin's day and became even more so in the twentieth century. Note that this does not depend on evolutionary thinking, though Darwin recognized that the patterns illustrated with the device clearly had great relevance to his evolutionary model. Darwin's point here is that the clades should begin at a sharp point and gradually increase in width (species numbers) before declining again, ultimately disappearing at a sharp point as well.

2 A remarkable thing about Darwin's theory is the integral part that extinction plays in it. Note his wording here: the extinction of old forms and the production of new forms are intimately tied together. By 1859 the idea of successive catastrophes punctuating earth history was very much on the wane but still had its adherents (this obsolete idea was becoming gradually extinct . . .). A key point Darwin makes in this section is that species and species groups disappear gradually, not suddenly. The gradual supplanting of species by their descendant species is an important element of his theory.

1 The idea that species decline and disappear more slowly than they appear and spread is an empirical question. Fossil forms do tend to slowly decline in abundance, with the exception of mass extinction events like that behind the demise of the ammonites. The terminus of Darwin's "secondary period" (parts of our Paleozoic and Mesozoic) is defined by one of the great mass extinctions: the Cretaceous-Tertiary event. Ammonites are so common in earlier rocks that their sudden absence makes for an excellent marker for this boundary.

2 The brief mention of these fossils belies the momentous effect they would have on Darwin's thinking. Later described by Owen, these extinct giant quadrupeds, so different from living forms, yet so similar, did much to convince Darwin of the reality of transmutation. Here is an account of the discovery in Darwin's *Beagle* diary for September 22, 1832: "We staid sometime on Punta Alta [on the east coast of South America] about 10 miles from the ship; here I found some rocks.—These are the first I have seen, and are very interesting from containing numerous shells & the bones of large animals." This was just the beginning; the next day, he "walked on to Punta Alta to look after fossils; & to my great joy I found the head of some large animal, imbedded in soft rock.—It took me nearly 3 hours to get it out: As far as I am able to judge, it is allied to the Rhinoceros.—I did not get it on board till some hours after it was dark" (Keynes 2001, pp. 106–107). As later recounted in *Voyage,* Darwin found the fossil remains of nine extinct mammals at this site.

3 Imagine the surprise of naturalists to discover that horses once abounded in South America but went extinct there. The continent did not see horses again for thousands if not millions of years, until Europeans reintroduced them. "Certainly it is a marvelous fact in the history of the Mammalia," Darwin enthused in *Voyage,* "that in South America a native horse should have lived and disappeared, to be succeeded in after-ages by the countless herds descended from the few introduced with the Spanish colonists!" "The existence in South America of a fossil horse, of the mastodon, possibly of an elephant, and of a hollow-horned ruminant . . . are highly interesting facts with re-

(continued)

finally from the world. Both single species and whole groups of species last for very unequal periods; some groups, as we have seen, having endured from the earliest known dawn of life to the present day; some having disappeared before the close of the palæozoic period. No fixed law seems to determine the length of time during which any single species or any single genus endures. There is reason to believe that the complete extinction of the species of a group is generally a slower process than their production: if the appearance and disappearance of a group of species be represented, as before, by a vertical line of varying thickness, the line is found to taper more gradually at its upper end, which marks the progress of extermination, than at its lower end, which marks the first appearance and increase in numbers of the species. In some cases, however, the extermination of whole groups of beings, as of ammonites towards the close of the secondary period, has been wonderfully sudden.

The whole subject of the extinction of species has been involved in the most gratuitous mystery. Some authors have even supposed that as the individual has a definite length of life, so have species a definite duration. No one I think can have marvelled more at the extinction of species, than I have done. When I found in La Plata the tooth of a horse embedded with the remains of Mastodon, Megatherium, Toxodon, and other extinct monsters, which all co-existed with still living shells at a very late geological period, I was filled with astonishment; for seeing that the horse, since its introduction by the Spaniards into South America, has run wild over the whole country and has increased in numbers at an unparalleled rate, I asked myself what could so recently have exterminated the former horse under conditions of life apparently so favourable. But how utterly groundless was my astonishment! Professor Owen soon perceived that the tooth, though so like that of the existing horse, belonged to an extinct species. Had this horse been still living, but in some degree rare, no naturalist would have felt the least surprise at its rarity; for rarity is the attribute of a vast number of species of all classes, in all countries. If we ask ourselves why this or that species is rare, we answer that something is unfavourable in its conditions of life; but what that something is, we can hardly ever tell. On the supposition of the fossil horse still existing as a rare species, we might have felt certain from the analogy of all other mammals, even of the slow-breeding elephant, and from the history of the naturalisation of the domestic horse in South America, that under more favourable conditions it would in a very few years have stocked the whole continent. But we could not have told what the unfavourable conditions were which checked its increase, whether some one or several contingencies, and at what period of the horse's life, and in what degree, they severally acted. If the conditions had gone on, however slowly, becoming less and less favourable, we assuredly should not have perceived the fact, yet the fossil horse would certainly have become rarer and rarer, and finally extinct;—its place being seized on by some more successful competitor.

It is most difficult always to remember that the increase of every living being is constantly being checked by unperceived injurious agencies; and that these same unperceived agencies are amply sufficient to cause rarity, and finally extinction. We see in many cases in the more recent tertiary formations, that rarity precedes extinction; and we know that this has been the progress of events with those animals which have

(continued)

spect to the geographical distribution of animals" (*Voyage,* chapter VII). Highly interesting indeed. Owen, by the way, examined Darwin's fossil horse teeth and determined that they indicated a new horse species, which he named *Equus curvidens* in 1840.

1 Rarity precedes extinction . . . usually. The disappearance of groups in mass extinctions can seem rather abrupt, as with the ammonites mentioned earlier. In general, however, extinction usually comes at the end of a long decline.

been exterminated, either locally or wholly, through man's agency. [1] I may repeat what I published in 1845, namely, that to admit that species generally become rare before they become extinct—to feel no surprise at the rarity of a species, and yet to marvel greatly when it ceases to exist, is much the same as to admit that sickness in the individual is the forerunner of death—to feel no surprise at sickness, but when the sick man dies, to wonder and to suspect that he died by some unknown deed of violence.

The theory of natural selection is grounded on the belief that each new variety, and ultimately each new species, is produced and maintained by having some advantage over those with which it comes into competition; and the consequent extinction of less-favoured forms almost inevitably follows. It is the same with our domestic productions: when a new and slightly improved variety has been raised, it at first supplants the less improved varieties in the same neighbourhood; when much improved it is transported far and near, like our short-horn cattle, and takes the place of other breeds in other countries. Thus the appearance of new forms and the disappearance of old forms, both natural and artificial, are bound together. In certain flourishing groups, the number of new specific forms which have been produced within a given time is probably greater than that of the old forms which have been exterminated; but we know that the number of species has not gone on indefinitely increasing, at least during the later geological periods, so that looking to later times we may believe that the production of new forms has caused the extinction of about the same number of old forms.

The competition will generally be most severe, as formerly explained and illustrated by examples, between the forms which are most like each other in all respects.

1 Darwin is referring to observations he made in the *Journal of Researches*, published in 1845 and later republished as the *Voyage of the Beagle*. The specific passage he refers to appears in chapter VIII of the *Journal:*

> why should we feel such great astonishment at the rarity being carried a step further to extinction? An action going on, on every side of us, and yet barely appreciable, might surely be carried a little further, without exciting our observation . . . To admit that species generally become rare before they become extinct—to feel no surprise at the comparative rarity of one species with another, and yet to call in some extraordinary agent and to marvel greatly when a species ceases to exist, appears to me much the same as to admit that sickness in the individual is the prelude to death—to feel no surprise at sickness—but when the sick man dies to wonder, and to believe that he died through violence.

Hence the improved and modified descendants of a species will generally cause the extermination of the parent-species; and if many new forms have been developed from any one species, the nearest allies of that species, *i. e.* the species of the same genus, will be the most liable to extermination. Thus, as I believe, a number of new species descended from one species, that is a new genus, comes to supplant an old genus, belonging to the same family. But it must often have happened that a new species belonging to some one group will have seized on the place occupied by a species belonging to a distinct group, and thus caused its extermination; and if many allied forms be developed from the successful intruder, many will have to yield their places; and it will generally be allied forms, which will suffer from some inherited inferiority in common. But whether it be species belonging to the same or to a distinct class, which yield their places to other species which have been modified and improved, a few of the sufferers may often long be preserved, from being fitted to some peculiar line of life, or from inhabiting some distant and isolated station, where they have escaped severe competition. For instance, a single species of Trigonia, a great genus of shells in the secondary formations, survives in the Australian seas; and a few members of the great and almost extinct group of Ganoid fishes still inhabit our fresh waters. Therefore the utter extinction of a group is generally, as we have seen, a slower process than its production. ◀1

With respect to the apparently sudden extermination of whole families or orders, as of Trilobites at the close of the palæozoic period and of Ammonites at the close of the secondary period, we must remember what has been already said on the probable wide intervals of time ◀2

1 Darwin suggests two ways in which groups on the decline may linger and avoid being driven extinct altogether. One way might be specialization, adopting "some peculiar line of life," in Darwin's words, perhaps like those remnant few brachiopods, once superabundant and now found only in the cold abyssal arctic seas. Another way to stave off being outcompeted might be isolation: to remain safe in some distant land that time forgot, to invoke the steamy tropical islands of science fiction movies where dinosaurs still roam. Relict species of once common groups are found on modern oceanic islands, like the tuataras of New Zealand. Darwin had this idea in mind when he wrote in notebook C, "When species rare we infer extermination, when group few in number of kind, extermination . . . therefore an isolated form probably a remnant" (Barrett et al. 1987, p. 297). The important point is that such factors as specialization and isolation might extend the existence of otherwise declining taxa, underscoring the fact that extinction is typically a long and drawn-out process.

2 Darwin is offering a logical alternative to sudden demise of these groups, pointing out that the intervals between formations are often a blank, a gap of indeterminate length of time. In these cases, though, extinction was apparently rather sudden. The trilobites disappeared in the great extinction that marks the end of the Permian period of the Paleozoic, and we have already seen that ammonites and other groups were lost in the extinction that marks the end of the Cretaceous. Ironically, the late twentieth-century realization that impacts by extraterrestrial bodies may have punctuated the history of the earth, precipitating mass extinctions of stunning proportions, has reintroduced a form of catastrophism into paleontology. Catastrophism of natural, not supernatural, origin of course, but catastrophism nonetheless.

1 This closing paragraph of the section invokes natural law in explaining species extinction. It also asserts that the empirical record of the extinction process—the patterns of decline and disappearance—is consistent with the action of natural selection.

2 Here is an empirical observation that would seem anything but consistent with the action of natural selection. Similar (sometimes identical) taxa are often found in the same formations regardless of where in the world those formations lie, suggesting that widely separated groups changed in the same general ways at the same rates—this is termed "parallelism." Different lineages' apparently changing simultaneously in similar ways is more suggestive of supernatural guidance than of change by natural law.

As will become evident, Darwin attempted to explain parallelism in terms of natural selection and the poor fine-scale resolution of the fossil record. It is worth noting here that in some cases parallelism is an artifact of continental drift, an idea that gained acceptance a century after the *Origin* was published. Originally unitary formations have been rent and the pieces carried apart on different plates as a result of plate tectonics; in such cases, precisely the same strata bearing precisely the same fossil series will be found in widely separated parts of the world, giving the appearance of parallelism. In fact, such correspondences in rock types and fossils across continents formed one of the important lines of evidence establishing the reality of former unity and subsequent separation of the continents.

between our consecutive formations; and in these intervals there may have been much slow extermination. Moreover, when by sudden immigration or by unusually rapid development, many species of a new group have taken possession of a new area, they will have exterminated in a correspondingly rapid manner many of the old inhabitants; and the forms which thus yield their places will commonly be allied, for they will partake of some inferiority in common.

1 Thus, as it seems to me, the manner in which single species and whole groups of species become extinct, accords well with the theory of natural selection. We need not marvel at extinction; if we must marvel, let it be at our presumption in imagining for a moment that we understand the many complex contingencies, on which the existence of each species depends. If we forget for an instant, that each species tends to increase inordinately, and that some check is always in action, yet seldom perceived by us, the whole economy of nature will be utterly obscured. Whenever we can precisely say why this species is more abundant in individuals than that; why this species and not another can be naturalised in a given country; then, and not till then, we may justly feel surprise why we cannot account for the extinction of this particular species or group of species.

2 *On the Forms of Life changing almost simultaneously throughout the World.*—Scarcely any palæontological discovery is more striking than the fact, that the forms of life change almost simultaneously throughout the world. Thus our European Chalk formation can be recognised in many distant parts of the world, under the most different climates, where not a fragment of the mineral chalk itself can be found; namely, in North

America, in equatorial South America, in Tierra del Fuego, at the Cape of Good Hope, and in the peninsula of India. For at these distant points, the organic remains in certain beds present an unmistakeable degree of resemblance to those of the Chalk. It is not that the same species are met with; for in some cases not one species is identically the same, but they belong to the same families, genera, and sections of genera, and sometimes are similarly characterised in such trifling points as mere superficial sculpture. Moreover other forms, which are not found in the Chalk of Europe, but which occur in the formations either above or below, are similarly absent at these distant points of the world. In the several successive palæozoic formations of Russia, Western Europe and North America, a similar parallelism in the forms of life has been observed by several authors: so it is, according to Lyell, with the several European and North American tertiary deposits. Even if the few fossil species which are common to the Old and New Worlds be kept wholly out of view, the general parallelism in the successive forms of life, in the stages of the widely separated palæozoic and tertiary periods, would still be manifest, and the several formations could be easily correlated.

These observations, however, relate to the marine inhabitants of distant parts of the world: we have not sufficient data to judge whether the productions of the land and of fresh water change at distant points in the same parallel manner. We may doubt whether they have thus changed: if the Megatherium, Mylodon, Macrauchenia, and Toxodon had been brought to Europe from La Plata, without any information in regard to their geological position, no one would have suspected that they had coexisted with still living sea-shells; but as these anomalous monsters coexisted with the Masto-

1 Fossils of the Chalk formation of Europe, calcium carbonate marine deposits of what is now known as the Cretaceous period ("Cretaceous" being derived from the Latin *creta,* "chalk"), are similar to Cretaceous formations on other continents. The same is also true of other Mesozoic formations.

2 These are genera of mammalian megafauna that Darwin discovered in South America. *Megatherium* and *Mylodon* are giant ground sloths, and *Macrauchenia* is a camel-like animal belonging to a unique group of extinct South American mammals, the Litopterns (though Darwin mistakenly thought it a type of llama). *Toxodon* is a giant rodent, a distant relative of the still-extant capybara (the largest rodent species alive today). Other long-extinct giant South American rodents have since come to light, including the 1,540-lb. *Phoberomys pattersoni* (Sánchez-Villagro et al. 2003) and the behemoth *Josephoartigasia monesi,* estimated to weigh in at over 2,200 lbs. (Rinderknecht and Blanco 2008).

1 Simultaneity is a relative thing: in the vast sweep of geological time, what does it mean for species to change simultaneously? A million or ten million years' time difference cannot be resolved in the fossil record.

2 Edouard de Verneuil and Adolphe d'Archiac were French paleontologists who made extensive contributions to the study of the Paleozoic. They authored several paleontological books and papers individually and collaboratively, and were jointly awarded the Geological Society's prestigious Wollaston Medal in 1853: "we rejoice, not less than their own countrymen, in the well-merited reputation which [de Verneuil and d'Archiac] have gained by [their] scientific labours," the citation from the venerable British society read. The parallelism that Darwin mentions was first described by De Verneuil in an 1847 paper entitled *Note sur le parallelisme des roches des dépots paléozoiques de l'Amérique septentrionale avec ceux de l'Europe.* Darwin is quoting their collaborative work *On the Fossils of the older Deposits in the Rhenish Provinces,* published in the *Transactions of the Geological Society of London* in 1842.

don and Horse, it might at least have been inferred that they had lived during one of the later tertiary stages.

1 When the marine forms of life are spoken of as having changed simultaneously throughout the world, it must not be supposed that this expression relates to the same thousandth or hundred-thousandth year, or even that it has a very strict geological sense; for if all the marine animals which live at the present day in Europe, and all those that lived in Europe during the pleistocene period (an enormously remote period as measured by years, including the whole glacial epoch), were to be compared with those now living in South America or in Australia, the most skilful naturalist would hardly be able to say whether the existing or the pleistocene inhabitants of Europe resembled most closely those of the southern hemisphere. So, again, several highly competent observers believe that the existing productions of the United States are more closely related to those which lived in Europe during certain later tertiary stages, than to those which now live here; and if this be so, it is evident that fossiliferous beds deposited at the present day on the shores of North America would hereafter be liable to be classed with somewhat older European beds. Nevertheless, looking to a remotely future epoch, there can, I think, be little doubt that all the more modern *marine* formations, namely, the upper pliocene, the pleistocene and strictly modern beds, of Europe, North and South America, and Australia, from containing fossil remains in some degree allied, and from not including those forms which are only found in the older underlying deposits, would be correctly ranked as simultaneous in a geological sense.

2 The fact of the forms of life changing simultaneously, in the above large sense, at distant parts of the world, has greatly struck those admirable observers, MM.

de Verneuil and d'Archiac. After referring to the parallelism of the palæozoic forms of life in various parts of Europe, they add, "If struck by this strange sequence, we turn our attention to North America, and there discover a series of analogous phenomena, it will appear certain that all these modifications of species, their extinction, and the introduction of new ones, cannot be owing to mere changes in marine currents or other causes more or less local and temporary, but depend on general laws which govern the whole animal kingdom." M. Barrande has made forcible remarks to precisely the same effect. It is, indeed, quite futile to look to changes of currents, climate, or other physical conditions, as the cause of these great mutations in the forms of life throughout the world, under the most different climates. We must, as Barrande has remarked, look to some special law. We shall see this more clearly when we treat of the present distribution of organic beings, and find how slight is the relation between the physical conditions of various countries, and the nature of their inhabitants. ◀1

This great fact of the parallel succession of the forms of life throughout the world, is explicable on the theory of natural selection. New species are formed by new varieties arising, which have some advantage over older forms; and those forms, which are already dominant, or have some advantage over the other forms in their own country, would naturally oftenest give rise to new varieties or incipient species; for these latter must be victorious in a still higher degree in order to be preserved and to survive. We have distinct evidence on this head, in the plants which are dominant, that is, which are commonest in their own homes, and are most widely diffused, having produced the greatest number of new varieties. It is also natural that the domi-

1 Apparent "parallelism" (simultaneous changes in widely separated geological formations) cannot be attributed to similarities in environmental changes. In dismissing that notion, Darwin alludes to the two biogeography chapters coming up. What is the "special law" of which Barrande speaks? Darwin thinks the special law is natural selection. In the next few paragraphs he describes how this might result in parallelism: widely distributed common species, already dominant, are most likely to give rise to new varieties that will themselves have characteristics that will ensure their dominance. Darwin then imagines that these forms that have already been varying and spreading will spread further still, giving rise to yet newer forms of varieties and species in the new areas. We see where this is going: parallelism is thus a process of groups being supplanted by related but better-adapted groups (hence the similarity of successive fossil fauna), spreading far and wide, the low temporal resolution of the fossil record giving the appearance of simultaneity to a process that actually unfolds over a span of thousands to millions of generations. This is summarized on p. 327.

nant, varying, and far-spreading species, which already have invaded to a certain extent the territories of other species, should be those which would have the best chance of spreading still further, and of giving rise in new countries to new varieties and species. [1] The process of diffusion may often be very slow, being dependent on climatal and geographical changes, or on strange accidents, but in the long run the dominant forms will generally succeed in spreading. The diffusion would, it is probable, be slower with the terrestrial inhabitants of distinct continents than with the marine inhabitants of the continuous sea. We might therefore expect to find, as we apparently do find, a less strict degree of parallel succession in the productions of the land than of the sea.

Dominant species spreading from any region might encounter still more dominant species, and then their triumphant course, or even their existence, would cease. We know not at all precisely what are all the conditions most favourable for the multiplication of new and dominant species; but we can, I think, clearly see that a number of individuals, from giving a better chance of the appearance of favourable variations, and that severe competition with many already existing forms, would be highly favourable, as would be the power of spreading into new territories. A certain amount of isolation, recurring at long intervals of time, would probably be also favourable, as before explained. One quarter of the world may have been most favourable for the production of new and dominant species on the land, and another for those in the waters of the sea. If two great regions had been for a long period favourably circumstanced in an equal degree, whenever their inhabitants met, the battle would be prolonged and severe; and some from one birthplace and some from the other might be victorious. But in the course of time, the

1 Note that Darwin's mechanism makes some predictions: first, marine forms should "diffuse" or spread more quickly than terrestrial forms, and so faunal shifts should be faster and parallelism more apparent with marine fossils. This is generally the case compared with terrestrial fossils, but it may be an artifact of a far more complete fossil record. Terrestrial forms on different continents are more isolated, though when opportunities for migration arise, change can be relatively rapid; consider, for example, the supplanting of the South American marsupial fauna by North American placental mammals once the Central American land bridge opened just a few million years ago (see D. R. Wallace 1997). And that brings up a second prediction, the effects of isolation (discussed in the following paragraph). Isolation should protect species from being supplanted, enabling them to hang on even while superior competitors spread and dominate outside their bastion. On a larger, continental scale, isolation means that parallel succession should be less common in the fossil record. Note that to modern evolutionary biologists isolation plays a starring role in speciation, but Darwin very much downplays it. He notes that it permits interesting oddities to evolve but sees it mostly in terms of protecting the isolated forms from being routed.

forms dominant in the highest degree, wherever produced, would tend everywhere to prevail. As they prevailed, they would cause the extinction of other and inferior forms; and as these inferior forms would be allied in groups by inheritance, whole groups would tend slowly to disappear; though here and there a single member might long be enabled to survive.

Thus, as it seems to me, the parallel, and, taken in a large sense, simultaneous, succession of the same forms of life throughout the world, accords well with the principle of new species having been formed by dominant species spreading widely and varying; the new species thus produced being themselves dominant owing to inheritance, and to having already had some advantage over their parents or over other species; these again spreading, varying, and producing new species. The forms which are beaten and which yield their places to the new and victorious forms, will generally be allied in groups, from inheriting some inferiority in common; and therefore as new and improved groups spread throughout the world, old groups will disappear from the world; and the succession of forms in both ways will everywhere tend to correspond. ◀1

There is one other remark connected with this subject worth making. I have given my reasons for believing that all our greater fossiliferous formations were deposited during periods of subsidence; and that blank intervals of vast duration occurred during the periods when the bed of the sea was either stationary or rising, and likewise when sediment was not thrown down quickly enough to embed and preserve organic remains. During these long and blank intervals I suppose that the inhabitants of each region underwent a considerable amount of modification and extinction, and that there was much migration from

1 This paragraph offers a succinct statement of how parallelism follows from the action of natural selection.

1 Here is a third prediction: it is conceivable that, here and there, the preserved strata of a geological period were deposited at slightly different times in two nearby locales. As a result, the species in the two sets of strata would be similar but not exactly the same. This idea is expressed more clearly at the bottom of this page and onto the next page, in the final paragraph of the section.

2 Sir Joseph Prestwich published a series of papers on the Eocene strata of the London and Hampshire basins between 1846 and 1857, including an 1855 study in which he compared the early Tertiary strata of England with those of France and Belgium. Prestwich was awarded the Geological Society's Wollaston Medal in 1849 for his contributions to the science of stratigraphy.

other parts of the world. As we have reason to believe that large areas are affected by the same movement, it is probable that strictly contemporaneous formations have often been accumulated over very wide spaces in the same quarter of the world; but we are far from having any right to conclude that this has invariably been the case, and that large areas have invariably been affected by the same movements. When two formations have been deposited in two regions during ▶1 nearly, but not exactly the same period, we should find in both, from the causes explained in the foregoing paragraphs, the same general succession in the forms of life; but the species would not exactly correspond; for there will have been a little more time in the one region than in the other for modification, extinction, and immigration.

I suspect that cases of this nature have occurred in ▶2 Europe. Mr. Prestwich, in his admirable Memoirs on the eocene deposits of England and France, is able to draw a close general parallelism between the successive stages in the two countries; but when he compares certain stages in England with those in France, although he finds in both a curious accordance in the numbers of the species belonging to the same genera, yet the species themselves differ in a manner very difficult to account for, considering the proximity of the two areas,—unless, indeed, it be assumed that an isthmus separated two seas inhabited by distinct, but contemporaneous, faunas. Lyell has made similar observations on some of the later tertiary formations. Barrande, also, shows that there is a striking general parallelism in the successive Silurian deposits of Bohemia and Scandinavia; nevertheless he finds a surprising amount of difference in the species. If the several formations in these regions have not been deposited during the same exact periods,—a formation in one region often corresponding with a blank interval in the other,—and if in both regions the species have gone on slowly changing during the accumulation of the several formations and during the long intervals of time between them; in this case, the several formations in the two regions could be arranged in the same order, in accordance with the general succession of the form of life, and the order would falsely appear to be strictly parallel; nevertheless the species would not all be the same in the apparently corresponding stages in the two regions.

On the Affinities of extinct Species to each other, and to living forms.—Let us now look to the mutual affinities of extinct and living species. They all fall into one grand natural system; and this fact is at once explained on the principle of descent. The more ancient any form is, the more, as a general rule, it differs from living forms. But, as Buckland long ago remarked, all fossils can be classed either in still existing groups, or between them. That the extinct forms of life help to fill up the wide intervals between existing genera, families, and orders, cannot be disputed. For if we confine our attention either to the living or to the extinct alone, the series is far less perfect than if we combine both into one general system. With respect to the Vertebrata, whole pages could be filled with striking illustrations from our great palæontologist, Owen, showing how extinct animals fall in between existing groups. Cuvier ranked the Ruminants and Pachyderms, as the two most distinct orders of mammals; but Owen has discovered so many fossil links, that he has had to alter the whole classification of these two orders; and has placed certain pachyderms in the same sub-order with ruminants: for example, he dissolves by fine gradations the apparently

1 By "affinities" Darwin means relationship, similarity. The patterns of affinity in the fossil record constitute a key observation that strongly supports the transmutation hypothesis. We have already mentioned Alfred Russel Wallace's recognition of this significant relationship, expressed lucidly in his 1855 Sarawak Law paper. The affinities of extinct organisms and their relationship to those still living are highly instructive. Several broad trends are worthy of note: like living species, extinct forms fall into the Natural System of Linnaeus; that is, they are readily arranged into genera, families, orders, etc. Many extinct forms are in varying degrees intermediate between extant forms, and the older they are (reckoned by being found in progressively lower strata), the more they differ from extant forms.

2 In 1836 the celebrated geologist William Buckland of Oxford authored one of the Bridgewater Treatises: *Geology and Mineralogy Considered with Reference to Natural Theology.* Darwin made a note to himself in the B notebook to read this work when he was ruminating on the dynamics of faunal changes over time: "In a decreasing population at any one moment fewer closely related; [therefore] ultimately few genera . . . & lastly perhaps some one single one.—Will this not account for the odd genera with few species which stand between great groups, which we are bound to consider the increasing ones." He then inserted the note, "Read Buckland," and took copious notes on Buckland's account of fossil assemblages (Barrett et al. 1987, p. 206).

3 Ironically, the antitransmutationist Richard Owen's analyses provided Darwin with much of the evidence that convinced him of transmutation. Consider *Toxodon,* one of Darwin's famous finds in South America. Owen described *Toxodon* in a paper read before the Geological Society on April 19, 1837. The paper bore this telling title: "A description of the Cranium of the *Toxodon Platensis* . . . referrible by its dentition to the *Rodentia,* but with affinities to the *Pachydermata* and the *Herbivorous Cetacea.*" By "referrible" Owen meant that it shared characteristics with these three groups. *Toxodon,* he wrote,

> manifests an additional step in the gradation of mammiferous forms leading from the *Rodentia,* through the *Pachydermata* to the

(continued)

(continued)

> *Cetacea;* a gradation of which the water-hog of South America *(Hydrochaerus capybara)* already indicates the commencement amongst existing *Rodentia,* of which order it is interesting to observe this species is the largest, while at the same time it is peculiar to the continent in which the remains of the gigantic *Toxodon* were discovered. (Owen 1837, p. 542)

To Darwin, these affinities were profoundly revealing.

The ruminant-pachyderm relationship is revealing in another way. Cuvier recognized the ungulate orders Pachydermata (a group containing elephants, nonruminant ungulates, and hyraxes) and Ruminantia, or ruminants. Pachydermata was considered something of a hodgepodge, but linking forms in the fossil record unite the group, as pointed out in this 1883 encyclopedia entry: "[Pachydermata] has been often described as less natural than any other of Cuvier's mammalian orders, as it consists of animals among which there are wide diversities . . . but it is now universally received by naturalists as indicating a real, though not a close affinity; and when we extend our view from existing to fossil species, numerous links present themselves" (*Chambers's Encyclopaedia,* 1883, volume VII, p. 182). The taxonomic rearrangement Darwin mentions in this paragraph refers to Owen's 1848 description of fossil teeth from *Hyopotamus* (now *Ancodon*), an extinct mammal intermediate between pigs and hippos.

1 This point is worth repeating; by "intermediate" Darwin does not mean literally intermediate in every point of anatomy. Rather, some traits will link the form with one group, while other traits will link the form to another group. In that sense fossil forms often "stand between" extant forms.

2 *Lepidosiren paradoxa,* the South American lungfish, is the sole member of the family Lepidosirenidae. These obligate air-breathers are in the class Sarcopterygii, a primitive group that includes the lobe-finned fish.

wide difference between the pig and the camel. In regard to the Invertebrata, Barrande, and a higher authority could not be named, asserts that he is every day taught that palæozoic animals, though belonging to the same orders, families, or genera with those living at the present day, were not at this early epoch limited in such distinct groups as they now are.

[1] Some writers have objected to any extinct species or group of species being considered as intermediate between living species or groups. If by this term it is meant that an extinct form is directly intermediate in all its characters between two living forms, the objection is probably valid. But I apprehend that in a perfectly natural classification many fossil species would have to stand between living species, and some extinct genera between living genera, even between genera belonging to distinct families. The most common case, especially with respect to very distinct groups, such as fish and reptiles, seems to be, that supposing them to be distinguished at the present day from each other by a dozen characters, the ancient members of the same two groups would be distinguished by a somewhat lesser number of characters, so that the two groups, though formerly quite distinct, at that period made some small approach to each other.

It is a common belief that the more ancient a form is, by so much the more it tends to connect by some of its characters groups now widely separated from each other. This remark no doubt must be restricted to those groups which have undergone much change in the course of geological ages; and it would be difficult to prove the truth of the proposition, for every now and
[2] then even a living animal, as the Lepidosiren, is discovered having affinities directed towards very distinct groups. Yet if we compare the older Reptiles and

Batrachians, the older Fish, the older Cephalopods, and the eocene Mammals, with the more recent members of the same classes, we must admit that there is some truth in the remark.

Let us see how far these several facts and inferences accord with the theory of descent with modification. As the subject is somewhat complex, I must request the [1] reader to turn to the diagram in the fourth chapter. We may suppose that the numbered letters represent genera, and the dotted lines diverging from them the species in each genus. The diagram is much too simple, too few genera and too few species being given, but this is unimportant for us. The horizontal lines may represent successive geological formations, and all the forms beneath the uppermost line may be considered as extinct. The three existing genera, a^{14}, q^{14}, p^{14}, will form a small family; b^{14} and f^{14} a closely allied family or sub-family; and o^{14} e^{14}, m^{14}, a third family. These three families, together with the many extinct genera on the several lines of descent diverging from the parent-form A, will form an order; for all will have inherited something in common from their ancient and common progenitor. On the principle of the continued tendency to divergence of character, which was formerly illustrated by this diagram, the more recent any form is, the more it will generally differ from its ancient progenitor. Hence we [2] can understand the rule that the most ancient fossils differ most from existing forms. We must not, however, assume that divergence of character is a necessary contingency; it depends solely on the descendants from a species being thus enabled to seize on many and different places in the economy of nature. Therefore it is quite possible, as we have seen in the case of some Silurian forms, that a species might go on being slightly

1 The diagram from chapter IV is versatile: whereas before the horizontal lines represented a count of successive generations, here we can think of them as geological strata. The groups on the uppermost line are extant; all the others are extinct forms. Each extant group (a^{14}–z^{14}) represents a genus, grouped into six families. The three families on the left, descended from A, fall in one order, while those descended from I fall in another order. The sole and largely unchanged F descendant is one of those anomalous "living fossils" that would likely be placed in its own order.

2 The key point here is that the older a geological formation is, the more its fossil organisms tend to differ from living forms. Darwin is careful not to be absolutist about this: "living fossils" as represented by lineage F do occur (note that this is a modern term not used by Darwin).

1 This suggests that people tend to be struck more by differences than by similarities. In the absence of certain uniting forms, the apparent differences between genera, say, are more strongly impressed upon the mind than are their similarities, and the forms are split out into so many families. If, however, linking fossil forms were found, the similarities would become all important and the tendency might be to lump the extant genera together into the same family. There is a psychological dimension to recognizing natural groupings.

2 Reference to the diagram is a useful way to underscore this important point concerning the nature of intermediate forms. Linking, intermediate forms are neither literally intermediate in all characters nor necessarily directly ancestral.

modified in relation to its slightly altered conditions of life, and yet retain throughout a vast period the same general characteristics. This is represented in the diagram by the letter F^{14}.

All the many forms, extinct and recent, descended from A, make, as before remarked, one order; and this order, from the continued effects of extinction and divergence of character, has become divided into several sub-families and families, some of which are supposed to have perished at different periods, and some to have endured to the present day.

1 By looking at the diagram we can see that if many of the extinct forms, supposed to be embedded in the successive formations, were discovered at several points low down in the series, the three existing families on the uppermost line would be rendered less distinct from each other. If, for instance, the genera a^1, a^5, a^{10}, f^8, m^3, m^6, m^9, were disinterred, these three families would be so closely linked together that they probably would have to be united into one great family, in nearly the same manner as has occurred with ruminants and pachyderms. Yet
2 he who objected to call the extinct genera, which thus linked the living genera of three families together, intermediate in character, would be justified, as they are intermediate, not directly, but only by a long and circuitous course through many widely different forms. If many extinct forms were to be discovered above one of the middle horizontal lines or geological formations —for instance, above No. VI.—but none from beneath this line, then only the two families on the left hand (namely, a^{14}, &c., and b^{14}, &c.) would have to be united into one family; and the two other families (namely, a^{14} to f^{14} now including five genera, and o^{14} to m^{14}) would yet remain distinct. These two families, however, would be less distinct from each other than they were before the

discovery of the fossils. If, for instance, we suppose the existing genera of the two families to differ from each other by a dozen characters, in this case the genera, at the early period marked VI., would differ by a lesser number of characters; for at this early stage of descent they have not diverged in character from the common progenitor of the order, nearly so much as they subsequently diverged. Thus it comes that ancient and extinct genera are often in some slight degree intermediate in character between their modified descendants, or between their collateral relations.

In nature the case will be far more complicated than is represented in the diagram; for the groups will have been more numerous, they will have endured for extremely unequal lengths of time, and will have been modified in various degrees. As we possess only the last volume of the geological record, and that in a very broken condition, we have no right to expect, except in very rare cases, to fill up wide intervals in the natural system, and thus unite distinct families or orders. All that we have a right to expect, is that those groups, which have within known geological periods undergone much modification, should in the older formations make some slight approach to each other; so that the older members should differ less from each other in some of their characters than do the existing members of the same groups; and this by the concurrent evidence of our best palæontologists seems frequently to be the case. 1

Thus, on the theory of descent with modification, the main facts with respect to the mutual affinities of the extinct forms of life to each other and to living forms, seem to me explained in a satisfactory manner. And they are wholly inexplicable on any other view. 2

On this same theory, it is evident that the fauna of any great period in the earth's history will be inter-

1 By "make some slight approach to each other," Darwin means "are more similar to each other." Thinking of the x-axis of his diagram as representing the sweep of morphological space, the ancestral forms of a branching lineage found in the lowermost strata do literally make an approach to each other, as they are closer to the point of common ancestry.

2 Another line of evidence for the "consilience" or responsibility-case aspects of Darwin's argument in the book: patterns of relationship manifested in the fossil record now join with those that Darwin has discussed previously (patterns of variation, hybridization, expressions of behavior) as yet another, independent body of information consistent with his theory and "wholly inexplicable" otherwise.

mediate in general character between that which preceded and that which succeeded it. Thus, the species which lived at the sixth great stage of descent in the diagram are the modified offspring of those which lived at the fifth stage, and are the parents of those which became still more modified at the seventh stage; hence they could hardly fail to be nearly intermediate in character between the forms of life above and below. We must, however, allow for the entire extinction of some preceding forms, and for the coming in of quite new forms by immigration, and for a large amount of modification, during the long and blank intervals between the successive formations. Subject to these allowances, the fauna of each geological period undoubtedly is intermediate in character, between the preceding and
[1] succeeding faunas. I need give only one instance, namely, the manner in which the fossils of the Devonian system, when this system was first discovered, were at once recognised by palæontologists as intermediate in character between those of the overlying carboniferous, and underlying Silurian system. But each fauna is not necessarily exactly intermediate, as unequal intervals of time have elapsed between consecutive formations.

It is no real objection to the truth of the statement, that the fauna of each period as a whole is nearly intermediate in character between the preceding and succeeding faunas, that certain genera offer exceptions
[2] to the rule. For instance, mastodons and elephants, when arranged by Dr. Falconer in two series, first according to their mutual affinities and then according to their periods of existence, do not accord in arrangement. The species extreme in character are not the oldest, or the most recent; nor are those which are intermediate in character, intermediate in age. But

1 The Devonian "system," now Period, was first proposed in 1839 by Adam Sedgwick and Roderick Murchison. It is named for the county of Devon in southern England, where curious corals of an intermediate morphology had been discovered by their fellow geologist William Lonsdale. A leading fossil coral specialist, Lonsdale wrote that the evidence of the corals and other fossils of these beds led him "to suggest that the South Devon Limestones are of an intermediate age between the Carboniferous and Silurian Systems, and consequently of the age of the Old Red Sandstone." He further noted, "It is necessary to add, that Mr. Murchison had shown that there is a regular passage from the Old Red Sandstone upwards into the Carboniferous system, and downwards into the Silurian, and that the suites of fossils of the two systems are perfectly distinct" (Lonsdale 1840, p. 727). The ensuing brouhaha over this proposed geological period has been styled the Devonian Controversy by historians of science; see Martin Rudwick's 1985 book of that title.

2 This passage probably refers to Hugh Falconer's study of elephants published in 1846 as part I of *Fauna Antiqua Sivalensis.* Falconer published a later paper on elephants, in 1863, in which he agreed with Darwin's assessment: "The inferences which I draw from these facts, are not opposed to one of the leading propositions of Darwin's theory. With him I have no faith in the opinion, that the Mammoth and other extinct Elephants made their appearance suddenly, after the type in which their fossil remains are presented to us. The most rational views seem to be, that they are in some shape the modified descendants of earlier progenitors." He goes on to praise Darwin: "Darwin has, beyond all his cotemporaries [sic], given an impulse to the philosophical investigation of the most backward and obscure branch of the Biological Sciences of his day; he has laid the foundations of a great edifice; but he need not be surprised if, in the progress of erection, the superstructure is altered by his successors, like the Duomo of Milan, from the roman to a different style of architecture."

supposing for an instant, in this and other such cases, that the record of the first appearance and disappearance of the species was perfect, we have no reason to believe that forms successively produced necessarily endure for corresponding lengths of time: a very ancient form might occasionally last much longer than a form elsewhere subsequently produced, especially in the case of terrestrial productions inhabiting separated districts. To compare small things with great: if the principal living and extinct races of the domestic pigeon were arranged as well as they could be in serial affinity, this arrangement would not closely accord with the order in time of their production, and still less with the order of their disappearance; for the parent rock-pigeon now lives; and many varieties between the rock-pigeon and the carrier have become extinct; and carriers which are extreme in the important character of length of beak originated earlier than short-beaked tumblers, which are at the opposite end of the series in this same respect.

Closely connected with the statement, that the organic remains from an intermediate formation are in some degree intermediate in character, is the fact, insisted on by all palæontologists, that fossils from two consecutive formations are far more closely related to each other, than are the fossils from two remote formations. Pictet gives as a well-known instance, the general resemblance of the organic remains from the several stages of the chalk formation, though the species are distinct in each stage. This fact alone, from its generality, seems to have shaken Professor Pictet in his firm belief in the immutability of species. He who is acquainted with the distribution of existing species over the globe, will not attempt to account for the close resemblance of the distinct species in closely consecutive

1 Reference to living pigeon varieties is an effective way to make this point: the serial ordering of forms does not necessarily, or even usually, correspond to the actual order of their appearance or disappearance. Serial ordering is something humans read into the fossils, seeking to arrange them according to some conscious or unconscious dictum to make sense of them. Later evolutionists might have attended more carefully to Darwin's passage here; there was a long-standing tendency to order fossil series in a linear, graduated manner that gave the appearance of an arrow-like evolutionary trajectory. This went hand-in-hand with the discredited concept of "orthogenesis"—directional evolution, with teleological overtones.

2 Darwin sent a copy of the *Origin* to the Swiss paleontologist François Jules Pictet de la Rive, whom we met in the previous chapter on p. 302. In his accompanying letter, Darwin argued that his theory explained a wide range of facts, "otherwise inexplicable," and mentioned other naturalists convinced of his view. From the statement here that Pictet's "firm belief in the immutability of species" has been "shaken," one might conclude that Darwin also persuaded Pictet. That is not clear from Pictet's reply in a letter dated February 19, 1860, now lost except for a passage that Lyell had copied into his notebooks: "Your book makes science young, clear, elevated, but no facts to prove that principle that slight modification multiplied by any factor of time, no matter how long, could reach the character of families, etc., or could produce profound modifications in organization" (Wilson 1970, p. 354). In other respects the letter must have been favorable, considering that Darwin chose to share it with Lyell. Furthermore, this passage in the *Origin* remained unchanged in all subsequent editions.

formations, by the physical conditions of the ancient areas having remained nearly the same. Let it be remembered that the forms of life, at least those inhabiting the sea, have changed almost simultaneously throughout the world, and therefore under the most different climates and conditions. Consider the prodigious vicissitudes of climate during the pleistocene period, which includes the whole glacial period, and note how little the specific forms of the inhabitants of the sea have been affected.

On the theory of descent, the full meaning of the fact of fossil remains from closely consecutive formations, though ranked as distinct species, being closely related, is obvious. As the accumulation of each formation has often been interrupted, and as long blank intervals have intervened between successive formations, we ought not to expect to find, as I attempted to show in the last chapter, in any one or two formations all the intermediate varieties between the species which appeared at the commencement and close of these periods; but we ought to find after intervals, very long as measured by years, but only moderately long as measured geologically, closely allied forms, or, as they have been called by some authors, representative species; and these we assuredly do find. We find, in short, such evidence of the slow and scarcely sensible mutation of specific forms, as we have a just right to expect to find.

1 *On the state of Development of Ancient Forms.*—There has been much discussion whether recent forms are more highly developed than ancient. I will not here enter on this subject, for naturalists have not as yet defined to each other's satisfaction what is meant by high and low forms. But in one particular sense the

1 This is a concept with which Darwin struggled. The mental legacy of Aristotle's *scala naturae* and the inevitable human sense of self-importance result in a tendency to categorize the world from the least to the most complex, with humans at the apex of complexity. There are various objective measures by which degrees of complexity can be gauged, but this should not be confused with being "highly evolved" or not. Darwin recognized that seemingly simple organisms were as "evolved" as any vertebrate in terms of being well adapted to their environment. Indeed, so-called living fossils could represent species so well-adapted to their environment that overt morphological change has not been selected for millions of years. In a note pinned into his copy of the *Vestiges of the Natural History of Creation,* Darwin wrote, "Never use the word [sic] higher and lower." His B notebook of 1838 is punctuated with such cautionary notes: "It is absurd to talk of one animal being higher than another.—We consider those, where the intellectual faculties most developed, as highest.—A bee doubtless would [consider] instincts [as a criterion]"; and: "When we talk of higher orders, we should always say, intellectually higher.—But who with the face of the earth covered with the most beautiful savannahs & forests dare to say that intellectuality is the only aim in this world" (Barrett et al. 1987, pp. 189, 233).

more recent forms must, on my theory, be higher than the more ancient; for each new species is formed by having had some advantage in the struggle for life over other and preceding forms. If under a nearly similar climate, the eocene inhabitants of one quarter of the world were put into competition with the existing inhabitants of the same or some other quarter, the eocene fauna or flora would certainly be beaten and exterminated; as would a secondary fauna by an eocene, and a palæozoic fauna by a secondary fauna. I do not doubt that this process of improvement has affected in a marked and sensible manner the organisation of the more recent and victorious forms of life, in comparison with the ancient and beaten forms; but I can see no way of testing this sort of progress. Crustaceans, for instance, not the highest in their own class, may have beaten the highest molluscs. From the extraordinary manner in which European productions have recently spread over New Zealand, and have seized on places which must have been previously occupied, we may believe, if all the animals and plants of Great Britain were set free in New Zealand, that in the course of time a multitude of British forms would become thoroughly naturalized there, and would exterminate many of the natives. On the other hand, from what we see now occurring in New Zealand, and from hardly a single inhabitant of the southern hemisphere having become wild in any part of Europe, we may doubt, if all the productions of New Zealand were set free in Great Britain, whether any considerable number would be enabled to seize on places now occupied by our native plants and animals. Under this point of view, the productions of Great Britain may be said to be higher than those of New Zealand. Yet the most skilful naturalist from an examination of the

1 This passage supports the idea of progressivism in the sense that more recent forms are thought to be better-adapted competitors than earlier forms; the former would be "higher" and the latter "lower" as gauged by competitive ability. Darwin expressed this belief to Hooker in a letter of December 1858, while still acknowledging that the terms were troublesome:

> I intend carefully to avoid this expression, for I do not think that any one has a definite idea what is meant by higher, except in classes which can loosely be compared with man. On our theory of Natural Selection, if the organisms of any area belonging to the Eocene or Secondary periods were put into competition with those now existing in the same area (or probably in any part of the world) they (i.e. the old ones) would be beaten hollow and be exterminated; if the theory be true, this must be so. (*Correspondence* 7: 228)

species of the two countries could not have foreseen this result.

[1] Agassiz insists that ancient animals resemble to a certain extent the embryos of recent animals of the same classes; or that the geological succession of extinct forms is in some degree parallel to the embryological development of recent forms. I must follow Pictet and Huxley in thinking that the truth of this doctrine is very far from proved. Yet I fully expect to see it hereafter confirmed, at least in regard to subordinate groups, which have branched off from each other within comparatively recent times. For this doctrine of Agassiz accords well with the theory of natural selection. In a future chapter I shall attempt to show that the adult differs from its embryo, owing to variations supervening at a not early age, and being inherited at a corresponding age. This process, whilst it leaves the embryo almost unaltered, continually adds, in the course of successive generations, more and more difference to the adult.

Thus the embryo comes to be left as a sort of picture, preserved by nature, of the ancient and less modified condition of each animal. This view may be true, and yet it may never be capable of full proof. Seeing, for instance, that the oldest known mammals, reptiles, and fish strictly belong to their own proper classes, though some of these old forms are in a slight degree less distinct from each other than are the typical members of the same groups at the present day, it would be vain to look for animals having the common embryological character of the Vertebrata, until beds far beneath the lowest Silurian strata are discovered — a discovery of which the chance is very small.

On the Succession of the same Types within the same

1 Today the ideas discussed in this paragraph are not thought to hold water except in the most general of senses. We can readily appreciate why Darwin argues they "accord well" with natural selection—though they really accord more with evolution in general. In the nineteenth century, after Darwin, the idea that organisms go through embryological stages that correspond to earlier evolutionary stages was expressed as Von Baer's Law (generalized features common to all members of a group of related species appear earlier in development than do features found in subgroups), or the "biogenetic law" of the German zoologist Ernst Haeckel (summed up by the motto "ontogeny recapitulates phylogeny"). We will revisit this in chapter XIII, but suffice it to say here that while Von Baer's Law is accepted today, ontogeny does not recapitulate phylogeny in any but the most general of senses. Rather than reflecting an embryological record of each stage in transmutation, we understand the homologies of, say, vertebrate embryos as stemming from common ancestry. In this sense we can understand why humans, chickens, and salamanders have such similar early embryos, complete with structures marking gill arches, which then become more dissimilar as they develop along their respective trajectories.

How did Agassiz make sense of such embryological patterns that seemed to shout transmutation? He argued that the commonalities stemmed from creation on a common body plan, not common ancestry, and moreover that embryos do not exhibit "stages" reflecting any archetype but their own. In other words, vertebrate embryos may early on exhibit features indicative of the common vertebrate body plan, even including gill structures, but never exhibit features of another type like the radiates. "The embryo of the vertebrate is a vertebrate from the beginning," Agassiz wrote in his *Essay on Classification,* "and does not exhibit at any time a correspondence with the Invertebrates. The embryos of Vertebrates do not pass in their development through other permanent types of animals" (Agassiz 1857).

areas, during the later tertiary periods.—Mr. Clift many years ago showed that the fossil mammals from the Australian caves were closely allied to the living marsupials of that continent. In South America, a similar relationship is manifest, even to an uneducated eye, in the gigantic pieces of armour like those of the armadillo, found in several parts of La Plata; and Professor Owen has shown in the most striking manner that most of the fossil mammals, buried there in such numbers, are related to South American types. This relationship is even more clearly seen in the wonderful collection of fossil bones made by MM. Lund and Clausen in the caves of Brazil. I was so much impressed with these facts that I strongly insisted, in 1839 and 1845, on this "law of the succession of types,"—on "this wonderful relationship in the same continent between the dead and the living." Professor Owen has subsequently extended the same generalisation to the mammals of the Old World. We see the same law in this author's restorations of the extinct and gigantic birds of New Zealand. We see it also in the birds of the caves of Brazil. Mr. Woodward has shown that the same law holds good with sea-shells, but from the wide distribution of most genera of molluscs, it is not well displayed by them. Other cases could be added, as the relation between the extinct and living land-shells of Madeira; and between the extinct and living brackish-water shells of the Aralo-Caspian Sea.

Now what does this remarkable law of the succession of the same types within the same areas mean? He would be a bold man, who after comparing the present climate of Australia and of parts of South America under the same latitude, would attempt to account, on the one hand, by dissimilar physical conditions for the dissimilarity of the inhabitants of these two continents,

1 Darwin may be referring to William Clift's 1830 paper "Report in regard to the fossil bones found in New Holland," published in the *Edinburgh New Philosophical Journal* (vol. 10, pp. 394–396).

2 The Danish naturalists Peter Wilhelm Lund and Peter Clausen were acclaimed for the fossil mammals and birds they unearthed from the extensive limestone caverns of central Brazil. In the words of Robert Jameson, editor of the *Edinburgh New Philosophical Journal:* "In South America no less than 800 caves were explored by those indefatigable naturalists, Lund and Clausen, and they obtained the bones of 101 species of mammalia belonging to 50 genera, a fauna more rich and varied than that now inhabiting the same country" (Jameson 1851, p. 27).

3 Darwin is quoting himself from *Voyage of the Beagle.* The "succession of types" and "wonderful relationship in the same continent between the dead and the living" are striking observations underscoring the essential link between the extinct forms found as fossils and the living forms of the same continent wandering above them. By Darwin's account, these observations helped spark his insight into transmutation. "In July [1837] opened first note Book on 'transmutation of Species'.—Had been greatly struck from about month of previous March—on character of S. American fossils—& species on Galapagos Archipelago.—These facts origin (especially latter) of all my views," he excitedly wrote in a pocket diary. This later became the opening line of the *Origin*'s introduction.

4 Richard Owen described the moa *Diornis novaezealandiae* from a single gigantic femur in the 1840s. An often-reproduced engraving from this period shows Owen standing with this femur next to a towering reconstructed *D. novaezealandiae* skeleton.

5 This comment presages an argument that appears in the next chapter: there is no relationship between similarity of climate and taxonomic affinity, whereas geographical proximity does correlate with taxonomic affinity.

and, on the other hand, by similarity of conditions, for the uniformity of the same types in each during the later tertiary periods. Nor can it be pretended that it is an immutable law that marsupials should have been chiefly or solely produced in Australia; or that Edentata and other American types should have been solely produced in South America. For we know that Europe in ancient times was peopled by numerous marsupials; and I have shown in the publications above alluded to, that in America the law of distribution of terrestrial mammals was formerly different from what it now is. North America formerly partook strongly of the present character of the southern half of the continent; and the southern half was formerly more closely allied, than it is at present, to the northern half. In a similar manner we know from Falconer and Cautley's discoveries, that northern India was formerly more closely related in its mammals to Africa than it is at the present time. Analogous facts could be given in relation to the distribution of marine animals.

On the theory of descent with modification, the great law of the long enduring, but not immutable, succession of the same types within the same areas, is at once explained; for the inhabitants of each quarter of the world will obviously tend to leave in that quarter, during the next succeeding period of time, closely allied though in some degree modified descendants. If the inhabitants of one continent formerly differed greatly from those of another continent, so will their modified descendants still differ in nearly the same manner and degree. But after very long intervals of time and after great geographical changes, permitting much inter-migration, the feebler will yield to the more dominant forms, and there will be nothing immutable in the laws of past and present distribution.

1 Hugh Falconer and Proby Thomas Cautley coauthored the *Fauna Antiqua Sivalensis* (1846), a treatise on the fossils of the Siwálik Hills in northern India.

2 In this lucid paragraph Darwin is suggesting that patterns like those described in this section make sense on his theory—distinct groups of organisms are generally separated geographically but are not *strictly* partitioned. They have clearly occupied different locales at different times in history, as evidenced by their fossils left behind. Why these changes? He suggests that changes to the earth's surface that permit intermigration are responsible. Such changes not only permit groups to migrate but also permit better-adapted immigrant groups to supplant the ones that are already there.

It may be asked in ridicule, whether I suppose that the megatherium and other allied huge monsters have left behind them in South America the sloth, armadillo, and anteater, as their degenerate descendants. This cannot for an instant be admitted. These huge animals have become wholly extinct, and have left no progeny. But in the caves of Brazil, there are many extinct species which are closely allied in size and in other characters to the species still living in South America; and some of these fossils may be the actual progenitors of living species. It must not be forgotten that, on my theory, all the species of the same genus have descended from some one species; so that if six genera, each having eight species, be found in one geological formation, and in the next succeeding formation there be six other allied or representative genera with the same number of species, then we may conclude that only one species of each of the six older genera has left modified descendants, constituting the six new genera. The other seven species of the old genera have all died out and have left no progeny. Or, which would probably be a far commoner case, two or three species of two or three alone of the six older genera will have been the parents of the six new genera; the other old species and the other whole genera having become utterly extinct. In failing orders, with the genera and species decreasing in numbers, as apparently is the case of the Edentata of South America, still fewer genera and species will have left modified blood-descendants.

Summary of the preceding and present Chapters.—I have attempted to show that the geological record is extremely imperfect; that only a small portion of the globe has been geologically explored with care; that

1 This is an important paragraph, in that it succinctly presents Darwin's understanding of the evolutionary process and ancestor-descendant relationships. The comment about "degenerate descendants" in the first sentence is a reference to the views of Buffon in the late eighteenth century, who held that most animals originated in the northerly latitudes of the Old World, migrating to the New World and "degenerating" along the way in response to the climate. The example Darwin uses here is that of the edentates—sloths, armadillos, and anteaters, Order Xenarthra or Edentata, although it is an exclusively New World group. This notion was long since discredited by Darwin's time, but it is a useful foil for expressing his vision for how extinct and extant species of the same general group are related. In faunal succession, just a few representatives of an otherwise sizable assemblage of groups are likely to leave descendants. Over time the earlier forms are supplanted, and these few descendant groups flourish and diversify such that the replacement of the ancestral groups is accomplished by the descendants of just a subset of them. Iterate this over time, each subsequent group descended from a subset of ancestral groups, such that the living species are descended from a subset of a subset of a subset, etc., of groups that flourished at different periods in the past.

2 The summary is divided into two parts paralleling the content of the two geological chapters. Both involve empirical observations of the fossil record—species distributions through the geological column. The first treats observations inimical to Darwin's theory, summarizing his attempts to explain away the problems, while the second treats observations consistent with it.

1 The problems and challenges presented by the geological record, as presented in the rest of this paragraph, boil down to (1) absence of innumerable recorded transitional forms; (2) gaps between groups, giving the appearance of substantial jumps between types of species, and (3) seemingly sudden appearance of species groups. Darwin's answer to these problems may be summarized as follows:

- There are more transitional forms in the fossil record than we currently know of; potentially informative fossils have yet to be found.
- We will never have a complete record of ancient life, since fossilization itself is selective: species with hard parts are much more likely to be preserved, as are those living in or near zones of active subsidence where sediment is accumulating. Erosion, too, literally eats gaps in the record.
- What's more, diversification has probably been greater during periods of elevation, with transitional forms flourishing at times and places where fossilization is unlikely.
- Sudden appearance of groups in the fossil record is illusory; it is explained by the duration of species over time tending to be longer than the average geological formation, and species migrate in and out of different locales.
- Finally, varieties, which are transitional forms, arise at first on a very localized spatial scale and are unlikely to be preserved.

1 only certain classes of organic beings have been largely preserved in a fossil state; that the number both of specimens and of species, preserved in our museums, is absolutely as nothing compared with the incalculable number of generations which must have passed away even during a single formation; that, owing to subsidence being necessary for the accumulation of fossiliferous deposits thick enough to resist future degradation, enormous intervals of time have elapsed between the successive formations; that there has probably been more extinction during the periods of subsidence, and more variation during the periods of elevation, and during the latter the record will have been least perfectly kept; that each single formation has not been continuously deposited; that the duration of each formation is, perhaps, short compared with the average duration of specific forms; that migration has played an important part in the first appearance of new forms in any one area and formation; that widely ranging species are those which have varied most, and have oftenest given rise to new species; and that varieties have at first often been local. All these causes taken conjointly, must have tended to make the geological record extremely imperfect, and will to a large extent explain why we do not find interminable varieties, connecting together all the extinct and existing forms of life by the finest graduated steps.

He who rejects these views on the nature of the geological record, will rightly reject my whole theory. For he may ask in vain where are the numberless transitional links which must formerly have connected the closely allied or representative species, found in the several stages of the same great formation. He may disbelieve in the enormous intervals of time which have elapsed between our consecutive formations; he may overlook how important a part migration must have played, when the formations of any one great region alone, as that of Europe, are considered; he may urge the apparent, but often falsely apparent, sudden coming in of whole groups of species. He may ask where are the remains of those infinitely numerous organisms which must have existed long before the first bed of the Silurian system was deposited: I can answer this latter question only hypothetically, by saying that as far as we can see, where our oceans now extend they have for an enormous period extended, and where our oscillating continents now stand they have stood ever since the Silurian epoch; but that long before that period, the world may have presented a wholly different aspect; and that the older continents, formed of formations older than any known to us, may now all be in a metamorphosed condition, or may lie buried under the ocean.

Passing from these difficulties, all the other great leading facts in palæontology seem to me simply to follow on the theory of descent with modification through natural selection. We can thus understand how it is that new species come in slowly and successively; how species of different classes do not necessarily change together, or at the same rate, or in the same degree; yet in the long run that all undergo modification to some extent. The extinction of old forms is the almost inevitable consequence of the production of new forms. We can understand why when a species has once disappeared it never reappears. Groups of species increase in numbers slowly, and endure for unequal periods of time; for the process of modification is necessarily slow, and depends on many complex contingencies. The dominant species of the larger dominant groups tend to leave many modified

1 Here Darwin transitions from the half-empty portion of the paleontological glass to the half-full portion. Think consilience: he summarizes an array of observations that collectively, he argues, point in the common direction of slow, steady descent with modification. There are fourteen observations:

1. New species appear slowly and successively.
2. Species of different groups do not change at the same rate or to the same degree.
3. Extinction is common.
4. Once species go extinct they do not reappear.
5. Species groups start out with few representatives and steadily grow in number over time; the different member species persist for unequal lengths of time.
6. Successive groups, subgroups, and sub-subgroups tend to be derived from initial dominant and widely distributed forms.
7. Extinction is a slow process of decline and eventual disappearance.
8. Once species *groups* disappear they do not reappear.
9. The successive spread of new dominant forms gives the appearance of simultaneous change in different parts of the world.
10. All life can be related by nested hierarchy into one grand classification system.
11. The earlier a fossil occurs, the more it tends to differ from later fossil or living forms.
12. Ancient and extinct species present examples of linking forms.
13. These extinct forms are not direct intermediates but share key elements of the later groups.
14. The fossils of consecutive formations are more closely related to each other than are those of formations widely separated in time.

descendants, and thus new sub-groups and groups are formed. As these are formed, the species of the less vigorous groups, from their inferiority inherited from a common progenitor, tend to become extinct together, and to leave no modified offspring on the face of the earth. But the utter extinction of a whole group of species may often be a very slow process, from the survival of a few descendants, lingering in protected and isolated situations. When a group has once wholly disappeared, it does not reappear; for the link of generation has been broken.

We can understand how the spreading of the dominant forms of life, which are those that oftenest vary, will in the long run tend to people the world with allied, but modified, descendants; and these will generally succeed in taking the places of those groups of species which are their inferiors in the struggle for existence. Hence, after long intervals of time, the productions of the world will appear to have changed simultaneously.

We can understand how it is that all the forms of life, ancient and recent, make together one grand system; for all are connected by generation. We can understand, from the continued tendency to divergence of character, why the more ancient a form is, the more it generally differs from those now living. Why ancient and extinct forms often tend to fill up gaps between existing forms, sometimes blending two groups previously classed as distinct into one; but more commonly only bringing them a little closer together. The more ancient a form is, the more often, apparently, it displays characters in some degree intermediate between groups now distinct; for the more ancient a form is, the more nearly it will be related to, and consequently resemble, the common progenitor of groups, since become widely divergent. Extinct forms are seldom directly intermediate between existing forms; but are intermediate only by a long and circuitous course through many extinct and very different forms. We can clearly see why the organic remains of closely consecutive formations are more closely allied to each other, than are those of remote formations; for the forms are more closely linked together by generation: we can clearly see why the remains of an intermediate formation are intermediate in character.

The inhabitants of each successive period in the world's history have beaten their predecessors in the race for life, and are, in so far, higher in the scale of nature; and this may account for that vague yet ill-defined sentiment, felt by many palæontologists, that organisation on the whole has progressed. If it should hereafter be proved that ancient animals resemble to a certain extent the embryos of more recent animals of the same class, the fact will be intelligible. The succession of the same types of structure within the same areas during the later geological periods ceases to be mysterious, and is simply explained by inheritance.

If then the geological record be as imperfect as I believe it to be, and it may at least be asserted that the record cannot be proved to be much more perfect, the main objections to the theory of natural selection are greatly diminished or disappear. On the other hand, all the chief laws of palæontology plainly proclaim, as it seems to me, that species have been produced by ordinary generation: old forms having been supplanted by new and improved forms of life, produced by the laws of variation still acting round us, and preserved by Natural Selection.

1 The apparently progressive change in the history of life, a gnawing "vague yet ill-defined sentiment," was troubling to Darwin. Small and less complex organisms seem to transition to large and more complex ones, but remember Darwin's cautionary notes to himself about "higher" and "lower." Seeing progress in the fossil record and the history of life may be seductive, but "progress" is a loaded term that carries social meanings, teleological overtones, and other burdens. To Darwin, natural selection favors variants with superior abilities to survive and reproduce. Over time, there would then be a *progressive* improvement in, say, competitive ability, and it is in this sense that he embraces the idea of progress.

Embryonic morphology and development were thought to have a profound connection with evolutionary change over time. Haeckel first proposed his "biogenetic law" in 1866, and by the fifth edition of *Origin* in 1869, Darwin incorporated it by rephrasing the relevant sentence from a tentative to a more definitive embrace of embryological evidence. Instead of opening the sentence with the cautious phrase "if it should hereafter be proved," he boldly proclaims, "Extinct and ancient animals resemble to a certain extent the embryos of the more recent animals belonging to the same classes" (Peckham 1959, p. 561). The biogenetic law is long since discredited, yet embryological evidence is still strongly supportive of Darwin's theory.

2 The two-part structure of this concluding paragraph seems to parallel that of the two geological chapters. It delivers a one-two punch: first, if the fossil record is as fragmentary as Darwin has argued, then the objections enumerated in the first chapter evaporate. Second, and more positively, the patterns ("laws") presented by paleontology are altogether consistent with—even proclaim—his theory. Darwin modified the wording of the final sentence slightly in the fifth edition, ending with ". . . improved forms of life, the products of Variation and Survival of the Fittest," rather than with ". . . produced by the laws of variation still acting round us, and preserved by Natural Selection." Why is this? Perhaps because by then natural selection as the main mechanism of evolution was much beleaguered, while "survival of the fittest" as promoted by Spencer drew less fire.

Biogeography constitutes the second of two major lines of evidence that convinced Darwin (and Wallace) of the reality of transmutation. Like paleontology, this subject is treated in two chapters. The difficulties of puzzling distribution patterns and means of transport are addressed in this chapter, while positive biogeographical evidence pointing to transmutation is summarized in the next chapter. Darwin's overarching goals here are twofold: he seeks (1) to marshal the evidence of biogeographical distributions to show that these are logically interpreted in terms of common descent; and (2) to come up with a plausible explanation for global distribution patterns. From a modern perspective, he is on the mark in achieving the first goal, but often far off it with respect to the second. His explanations were formulated prior to an understanding of plate tectonics. As a result, Darwin was wedded to a dispersal model that explains distributions solely in terms of immigration and emigration controlled by oscillations of land and climate, ideas that dominated the field well into the twentieth century. Among the important elements of this chapter, note especially Darwin's thinking on the role of isolation in species formation. It is decidedly downplayed, particularly considering how early on he was inspired by the island-specific birds and tortoises of the Galápagos. This topic is explored further in the next chapter.

---

1 The chapter starts out boldly, declaring three "great facts" in the distribution of species over the first few pages. These facts can be summarized as follows: (1) affinities of species in different regions are independent of physical conditions; (2) barriers play a role in species differences in different regions; and (3) species of a given region are related, despite great differences in climate, habitat, etc.

2 Climatic or environmental conditions do not predict species relationships. Run a north-south transect down the center of North America and Eurasia and, despite an identical range of conditions and microhabitats, the species are different. This fact lies at the heart of one of the arguments against special creation: why duplicate efforts and create many distinct and unrelated species to occupy otherwise identical habitats around the world?

CHAPTER XI.

GEOGRAPHICAL DISTRIBUTION.

Present distribution cannot be accounted for by differences in physical conditions — Importance of barriers — Affinity of the productions of the same continent — Centres of creation — Means of dispersal, by changes of climate and of the level of the land, and by occasional means — Dispersal during the Glacial period co-extensive with the world.

▶1 IN considering the distribution of organic beings over the face of the globe, the first great fact which strikes us is, that neither the similarity nor the dissimilarity of the inhabitants of various regions can be accounted for by their climatal and other physical conditions. Of late, almost every author who has studied the subject ▶2 has come to this conclusion. The case of America alone would almost suffice to prove its truth: for if we exclude the northern parts where the circumpolar land is almost continuous, all authors agree that one of the most fundamental divisions in geographical distribution is that between the New and Old Worlds; yet if we travel over the vast American continent, from the central parts of the United States to its extreme southern point, we meet with the most diversified conditions; the most humid districts, arid deserts, lofty mountains, grassy plains, forests, marshes, lakes, and great rivers, under almost every temperature. There is hardly a climate or condition in the Old World which cannot be paralleled in the New—at least as closely as the same species generally require; for it is a most rare case to find a group of organisms confined to any small spot, having conditions peculiar in only a slight degree; for instance, small areas in the Old World could be pointed out hotter than any in the New World, yet these are not inhabited by a peculiar fauna or flora. Notwithstanding this parallelism in the conditions of the Old and New Worlds, how widely different are their living productions!

In the southern hemisphere, if we compare large tracts of land in Australia, South Africa, and western South America, between latitudes 25° and 35°, we shall find parts extremely similar in all their conditions, yet it would not be possible to point out three faunas and floras more utterly dissimilar. Or again we may compare the productions of South America south of lat. 35° with those north of 25°, which consequently inhabit a considerably different climate, and they will be found incomparably more closely related to each other, than they are to the productions of Australia or Africa under nearly the same climate. Analogous facts could be given with respect to the inhabitants of the sea. ◀1

A second great fact which strikes us in our general review is, that barriers of any kind, or obstacles to free migration, are related in a close and important manner to the differences between the productions of various regions. We see this in the great difference of nearly all the terrestrial productions of the New and Old Worlds, excepting in the northern parts, where the land almost joins, and where, under a slightly different climate, there might have been free migration for the northern temperate forms, as there now is for the strictly arctic productions. We see the same fact in the great difference between the inhabitants of Australia, Africa, and South America under the same latitude: for these countries are almost as much isolated from each other as is possible. On each continent, also, we see the same fact; for on the opposite sides of ◀2

1 Here is another, ingenious way of making the point raised on the previous page. This time run two transects, latitudinal and longitudinal, the first spanning parts of continents between latitudes 25 and 35 degrees south, and another down the middle of one continent (say South America). Highly similar conditions prevail in the latitudinal band, yet the assemblages of plants and animals differ tremendously. Conversely, highly diverse conditions are met along the longitudinal transect in South America, yet the plants and animals living in these diverse conditions show strong taxonomic affinities. The bottom line is that geography, not climate, is the best predictor of biogeographical relationships.

Interestingly, Louis Agassiz earlier made essentially the same observation to argue against species origins by natural law. In section VIII of the first chapter of his *Essay on Classification* (originally published in 1851), Agassiz wrote:

> In the study of the geographical distribution of animals and plants and their relations to the conditions under which they live, too little importance is attached to the circumstances that representations of the most diversified types are everywhere found associated, under limited areas, under identical conditions. These combinations of numerous and most heterogeneous types, under all possible variations of climatic influences . . . seems to me to present the most insuperable objection to the supposition that the organized beings, so combined, could in any way have originated spontaneously by the working of any natural law.

2 Between or within continental areas, isolating barriers are implicated in the species differences observed. Darwin is suggesting not that barriers *promote* species diversification in the manner of allopatric speciation, but rather that they serve to contain and thus delineate different species groups by preventing intermigration.

lofty and continuous mountain-ranges, and of great deserts, and sometimes even of large rivers, we find different productions; though as mountain-chains, deserts, &c., are not as impassable, or likely to have endured so long as the oceans separating continents, the differences are very inferior in degree to those characteristic of distinct continents.

[1] Turning to the sea, we find the same law. No two marine faunas are more distinct, with hardly a fish, shell, or crab in common, than those of the eastern and western shores of South and Central America; yet these great faunas are separated only by the narrow, but impassable, isthmus of Panama. Westward of the shores of America, a wide space of open ocean extends, with not an island as a halting-place for emigrants; here we have a barrier of another kind, and as soon as this is passed we meet in the eastern islands of the Pacific, with another and totally distinct fauna. So that here three marine faunas range far northward and southward, in parallel lines not far from each other, under corresponding climates; but from being separated from each other by impassable barriers, either of land or open sea, they are wholly distinct. On the other hand, proceeding still further westward from the eastern islands of the tropical parts of the Pacific, we encounter no impassable barriers, and we have innumerable islands as halting-places, until after travelling over a hemisphere we come to the shores of Africa; and over this vast space we meet with no well-defined and distinct marine faunas. Although hardly one shell, crab or fish is common to the above-named three approximate faunas of Eastern and Western America and the eastern Pacific islands, yet many fish range from the Pacific into the Indian Ocean, and many shells are common to the eastern islands of the Pacific

1 This paragraph is a good example of the significant editing down of Darwin's initial project, *Natural Selection,* to produce the *Origin.* Marine examples receive far more detailed treatment in *Natural Selection,* where Darwin cites data from the published works of seven authorities, bearing on fish, mollusks, bryozoans, coelenterates, crustaceans, and algae (Stauffer 1975, pp. 555–557).

and the eastern shores of Africa, on almost exactly opposite meridians of longitude.

A third great fact, partly included in the foregoing statements, is the affinity of the productions of the same continent or sea, though the species themselves are distinct at different points and stations. It is a law of the widest generality, and every continent offers innumerable instances. Nevertheless the naturalist in travelling, for instance, from north to south never fails to be struck by the manner in which successive groups of beings, specifically distinct, yet clearly related, replace each other. He hears from closely allied, yet distinct kinds of birds, notes nearly similar, and sees their nests similarly constructed, but not quite alike, with eggs coloured in nearly the same manner. The plains near the Straits of Magellan are inhabited by one species of Rhea (American ostrich), and northward the plains of La Plata by another species of the same genus; and not by a true ostrich or emeu, like those found in Africa and Australia under the same latitude. On these same plains of La Plata, we see the agouti and bizcacha, animals having nearly the same habits as our hares and rabbits and belonging to the same order of Rodents, but they plainly display an American type of structure. We ascend the lofty peaks of the Cordillera and we find an alpine species of bizcacha; we look to the waters, and we do not find the beaver or musk-rat, but the coypu and capybara, rodents of the American type. Innumerable other instances could be given. If we look to the islands off the American shore, however much they may differ in geological structure, the inhabitants, though they may be all peculiar species, are essentially American. We may look back to past ages, as shown in the last chapter, and we find American types then prevalent on

1 The third "great fact" is the flip side of the second: the differences *between* isolated areas are complemented by the affinities of species *within* those areas. By "stations" Darwin means habitats. This usage originates with Augustin de Candolle's influential essay *De géographie botanique* (1820), a study of the factors underlying plant geographical distribution, an important early marriage of plant ecology and biogeography (Browne 1983).

2 The way species seemed to transition from one to another across what often seems an invisible barrier began to puzzle Darwin after the *Beagle* voyage. The rheas, or ostriches of South America, are a good example. One species is found from Brazil to roughly the Rio Negro in Argentina, another one to the south.

The Indians and gauchos of South America recognized these two ostrich species. The common one was locally called the *Astrevuz* (Spanish for ostrich), and its smaller-bodied relative the *Astrevuz petiso,* "small ostrich," which Darwin recorded as Astrevuz petisse. At the time the smaller one had not been formally described as a new species. Why did South America have two rheas when it would seem that one would do? In the Red Notebook of 1836–1837, Darwin mused, "When we see Astrevuz two species. certainly different. not insensible change.—Yet one is urged to look to common parent? why should two of the most closely allied species occur in the same country?" [sic] (Barrett et al. 1987, p. 70).

As an aside, Darwin collected the petisse species in Patagonia, realizing its significance even as he and his comrades were eating it around a campfire! John Gould later (in 1837) named it *Rhea darwinii* in Darwin's honor, but Alcide d'Orbigny had already given it a name earlier that very year. Darwin was mistaken when he wrote in chapter V of *Voyage,* "M. A. d'Orbigny, when at the Rio Negro, made great exertions to procure this bird, but never had the good fortune to succeed." D'Orbigny had succeeded only too well; today this species is called *Pterocnemia pennata.*

the American continent and in the American seas. [1] We see in these facts some deep organic bond, prevailing throughout space and time, over the same areas of land and water, and independent of their physical conditions. The naturalist must feel little curiosity, who is not led to inquire what this bond is.

This bond, on my theory, is simply inheritance, that cause which alone, as far as we positively know, produces organisms quite like, or, as we see in the case of varieties nearly like each other. The dissimilarity of the inhabitants of different regions may be attributed to modification through natural selection, and in a quite subordinate degree to the direct influence of different [2] physical conditions. The degree of dissimilarity will depend on the migration of the more dominant forms of life from one region into another having been effected with more or less ease, at periods more or less remote ;—on the nature and number of the former immigrants ;—and on their action and reaction, in their mutual struggles for life ;—the relation of organism to organism being, as I have already often remarked, the most important of all relations. Thus the high importance of barriers comes into play by checking migration ; as does time for the slow process of modification through [3] natural selection. Widely-ranging species, abounding in individuals, which have already triumphed over many competitors in their own widely-extended homes will have the best chance of seizing on new places, when they spread into new countries. In their new homes they will be exposed to new conditions, and will frequently undergo further modification and improvement ; and thus they will become still further victorious, and will produce groups of modified descendants. On this principle of inheritance with modification, we can understand how it is that sections of genera, whole genera,

1 The three "great facts" are provocative indeed. Many naturalists of the day were curious about such observations, but faunal replacement and biogeographically distinct flora and fauna were conceptualized in a framework of special creation, not transmutation. The leading naturalists speculated on the number and locale of "centers of creation" around the globe, and most paleontologists saw successive faunal replacement evident in the fossil record in terms of an unfolding plan. Charles Lyell struggled with this. "Ascending from invertebrate through the vertebrate to Man by the gradual development of the Brain," he wrote in his journal on June 29, 1859, "[w]hat matters it whether this development is effected in millions of years by a series of 30 millions of special or miraculous interferences worked according to a plan—each million producing species on the whole resembling those who preceded and follow, & that which followed being always nearer the existing & more remote from the older set" (Wilson 1970, p. 276).

2 In part this is the view accepted today. Intermigration has the effect of mixing and homogenizing, whether at the level of the population or of whole species assemblages. The other part of the modern view is the necessity of barriers in *permitting* genetic differentiation and speciation. In both Darwin's view and the modern take, the degree of dissimilarity is related to how little migration there has been linking two areas. But note the subtle difference in the role of barriers in the two views: Darwin sees their role solely in terms of preventing mixing and the supplanting of one set of species by another, whereas modern biologists emphasize the role of barriers in holding populations apart long enough for speciation to occur.

3 Recall Darwin's "success breeds success" model from chapter IV. Wide-ranging and abundant species are successful and are likely to continue being successful because they will produce more variations than their competitors, making them better able to respond to selection pressures.

and even families are confined to the same areas, as is so commonly and notoriously the case.

I believe, as was remarked in the last chapter, in no law of necessary development. As the variability of each species is an independent property, and will be taken advantage of by natural selection, only so far as it profits the individual in its complex struggle for life, so the degree of modification in different species will be no uniform quantity. If, for instance, a number of species, which stand in direct competition with each other, migrate in a body into a new and afterwards isolated country, they will be little liable to modification; for neither migration nor isolation in themselves can do anything. These principles come into play only by bringing organisms into new relations with each other, and in a lesser degree with the surrounding physical conditions. As we have seen in the last chapter that some forms have retained nearly the same character from an enormously remote geological period, so certain species have migrated over vast spaces, and have not become greatly modified.

On these views, it is obvious, that the several species of the same genus, though inhabiting the most distant quarters of the world, must originally have proceeded from the same source, as they have descended from the same progenitor. In the case of those species, which have undergone during whole geological periods but little modification, there is not much difficulty in believing that they may have migrated from the same region; for during the vast geographical and climatal changes which will have supervened since ancient times, almost any amount of migration is possible. But in many other cases, in which we have reason to believe that the species of a genus have been produced within comparatively recent times, there is great difficulty on this head. It

1 This is a clear statement of the undirected nature of evolutionary change. Variability of each species is independent of that of others, and natural selection acts on each species individually.

is also obvious that the individuals of the same species, though now inhabiting distant and isolated regions, must have proceeded from one spot, where their parents were first produced: for, as explained in the last chapter, it is incredible that individuals identically the same should ever have been produced through natural selection from parents specifically distinct.

[1] We are thus brought to the question which has been largely discussed by naturalists, namely, whether species have been created at one or more points of the earth's surface. Undoubtedly there are very many cases of extreme difficulty, in understanding how the same species could possibly have migrated from some one point to the several distant and isolated points, where now found. Nevertheless the simplicity of the view that each species was first produced within a single region captivates the mind. He who rejects it, rejects the [2] *vera causa* of ordinary generation with subsequent migration, and calls in the agency of a miracle. It is universally admitted, that in most cases the area inhabited by a species is continuous; and when a plant or animal inhabits two points so distant from each other, or with an interval of such a nature, that the space could not be easily passed over by migration, the fact is given as something remarkable and exceptional. The capacity of migrating across the sea is more distinctly limited in terrestrial mammals, than perhaps in any other organic beings; and, accordingly, we find no inexplicable cases of the same mammal inhabiting distant points of the world. No geologist will feel any difficulty in such cases as Great Britain having been formerly united to Europe, and consequently possessing the same quadrupeds. But if the same species can be produced at two separate points, why do we not find a single mammal common to Europe and Australia or South America? The conditions of life are

1 The debate over single vs. multiple "centers of creation" began almost as soon as explorers came upon new worlds stocked with altogether unfamiliar plants and animals. The striking differences between Europe and Australia, South America, or North America puzzled fifteenth- and sixteenth-century naturalists, especially at a time when the bible was considered an accurate record of historical events. How to reconcile the dramatic species differences between regions—let alone the unique assemblages of species confined to lone island archipelagoes around the world—with the Noachian flood? Clearly that event suggested a single center or origin, requiring, among many other things, migration from the ark's resting place. Some naturalists abandoned the biblical account, opting for a creator that made sets of species for each continent and island system. The technique of botanical arithmetic, promoted by Alexander von Humboldt and others in the early nineteenth century, aimed at quantifying what they took to be the degree of "creative force" or creative activity in different regions. Darwin's model of common descent argues for a single center of creation, though Darwin regarded the Australian fauna as distinct.

Note that this subject is critically important to Darwin, because if it could be shown that the same species originated simultaneously in more than one locale, his whole theory would be undermined. (Multiple centers are workable as long as the same species is not created twice.) Much of the rest of this chapter and the next one are aimed at demonstrating how patterns of species distributions support the idea of a single origin and are explained by dispersal.

2 The philosophical term *vera causa*—true or actual cause—originated with Isaac Newton in 1687. In the words of the nineteenth-century astronomer Sir John Herschel: "To such causes Newton has applied the term *verae causae;* that is, causes recognized as having a real existence in nature, and not being mere hypotheses or figments of the mind" (*Preliminary Discourse on the Study of Natural Philosophy,* part II, 1831).

nearly the same, so that a multitude of European animals and plants have become naturalised in America and Australia; and some of the aboriginal plants are identically the same at these distant points of the northern and southern hemispheres? The answer, as I believe, is, that mammals have not been able to migrate, whereas some plants, from their varied means of dispersal, have migrated across the vast and broken interspace. The great and striking influence which barriers of every kind have had on distribution, is intelligible only on the view that the great majority of species have been produced on one side alone, and have not been able to migrate to the other side. Some few families, many sub-families, very many genera, and a still greater number of sections of genera are confined to a single region; and it has been observed by several naturalists, that the most natural genera, or those genera in which the species are most closely related to each other, are generally local, or confined to one area. What a strange anomaly it would be, if, when coming one step lower in the series, to the individuals of the same species, a directly opposite rule prevailed; and species were not local, but had been produced in two or more distinct areas!

Hence it seems to me, as it has to many other naturalists, that the view of each species having been produced in one area alone, and having subsequently migrated from that area as far as its powers of migration and subsistence under past and present conditions permitted, is the most probable. Undoubtedly many cases occur, in which we cannot explain how the same species could have passed from one point to the other. But the geographical and climatal changes, which have certainly occurred within recent geological times, must have interrupted or rendered discontinuous the formerly continuous range of many species. So that we are reduced to consider whether the exceptions to

1 As with some of the geological observations, Darwin is in the position of arguing that we do not fully understand the factors underlying geographical distribution and, in recognizing our ignorance, should not be too quick to declare that species can originate in more than one location. This is well expressed in the following passage from *Natural Selection:*

> The advocates of multiple creations, may, in my opinion, bring forward the species more especially those found in Chile and New Zealand as a very strong case in favour of their view: but it should not be overlooked that they would find it very difficult to give any rational explanation of the community of these few species, for the great mass of organic productions, & all the external conditions are widely different in Chile & New Zealand; & it might well be asked, why should these few plants be identical & so vast a number of other productions widely different: it seems to me safer to rely on our ignorance of the means of diffusion. (Stauffer 1975, p. 564)

continuity of range are so numerous and of so grave a nature, that we ought to give up the belief, rendered probable by general considerations, that each species has been produced within one area, and has migrated thence as far as it could. It would be hopelessly tedious to discuss all the exceptional cases of the same species, now living at distant and separated points; nor do I for a moment pretend that any explanation could be
1 offered of many such cases. But after some preliminary remarks, I will discuss a few of the most striking classes of facts; namely, the existence of the same species on the summits of distant mountain-ranges, and at distant points in the arctic and antarctic regions; and secondly (in the following chapter), the wide distribution of fresh-water productions; and thirdly, the occurrence of the same terrestrial species on islands and on the mainland, though separated by hundreds of miles of open sea. If
2 the existence of the same species at distant and isolated points of the earth's surface, can in many instances be explained on the view of each species having migrated from a single birthplace; then, considering our ignorance with respect to former climatal and geographical changes and various occasional means of transport, the belief that this has been the universal law, seems to me incomparably the safest.

In discussing this subject, we shall be enabled at the same time to consider a point equally important for us, namely, whether the several distinct species of a genus, which on my theory have all descended from a common progenitor, can have migrated (undergoing modification during some part of their migration) from the area
3 inhabited by their progenitor. If it can be shown to be almost invariably the case, that a region, of which most of its inhabitants are closely related to, or belong to the same genera with the species of a second region,

1 Darwin's approach here is much the same as his approach to other difficulties he tackles in the book: take the bull by the horns and consider the most difficult cases up front. Here, he takes the most extreme cases of dispersion; if he can explain them, the lesser cases are explained too.

2 The argument here is for parsimony of explanation, a time-honored scientific principle. Darwin is suggesting that if he can show that a scenario of single origin and subsequent dispersal plausibly explains these difficult cases of distribution, it is most parsimonious to accept this explanation as universal even in those cases where the mechanism of dispersal is not altogether clear. The "incomparably safest" explanation is the parsimonious explanation.

3 In demonstrating that geographical patterns of species relationship support the idea that species likely originate in one locality and disperse from there, Darwin is also showing that descent from a common ancestor is the most reasonable and likely mechanism.

has probably received at some former period immigrants from this other region, my theory will be strengthened; for we can clearly understand, on the principle of modification, why the inhabitants of a region should be related to those of another region, whence it has been stocked. A volcanic island, for instance, upheaved and formed at the distance of a few hundreds of miles from a continent, would probably receive from it in the course of time a few colonists, and their descendants, though modified, would still be plainly related by inheritance to the inhabitants of the continent. Cases of this nature are common, and are, as we shall hereafter more fully see, inexplicable on the theory of independent creation. This view of the relation of species in one region to those in another, does not differ much (by substituting the word variety for species) from that lately advanced in an ingenious paper by Mr. Wallace, in which he concludes, that "every species has come into existence coincident both in space and time with a pre-existing closely allied species." And I now know from correspondence, that this coincidence he attributes to generation with modification.

The previous remarks on "single and multiple centres of creation" do not directly bear on another allied question,—namely whether all the individuals of the same species have descended from a single pair, or single hermaphrodite, or whether, as some authors suppose, from many individuals simultaneously created. With those organic beings which never intercross (if such exist), the species, on my theory, must have descended from a succession of improved varieties, which will never have blended with other individuals or varieties, but will have supplanted each other; so that, at each successive stage of modification and improvement, all the individuals of each variety will have descended from

1 Darwin is explicit here in his overarching point: the patterns we see are inexplicable on the supposition of special creation, but they are either predicted by, make the most sense under, or are at least consistent with his theory of descent with modification.

2 The ingenious paper that Darwin quotes is Wallace's Sarawak Law paper, published in September 1855. This paper prompted Charles Lyell to initiate his journals on the species question. Lyell immediately recognized the significance of Wallace's keen observations, and he advised Darwin to publish his own views. Darwin seemed unfazed: he summed up Wallace's paper with a marginal note that read, "nothing very new." The following spring, Darwin wrote to Hooker that he "had good talk with Lyell about my species work, & he urges me strongly to publish something" (*Correspondence* 6: 106). On May 14, 1856, Darwin got moving at last, recording in his pocket diary that he "began by Lyell's advice writing Species Sketch."

1 This is populational thinking: favorable variations spread through the population by intercrossing.

2 Lyell discussed means of dispersal at great length, dedicating chapters 5–7 of the second volume of *Principles of Geology* to the subject. He also, like Darwin (see pp. 357–358), conducted experiments such as the persistence of seeds in salt water (Wilson 1970, p. 52) and recorded accounts of unlikely means of dispersal. For example, in a letter to Darwin in May 1856, Lyell excitedly wrote, "I have just heard from Woodward that his friend Mr C. Prentice of Cheltenham caught that large & most powerful of our water beetles Hydrobius piceus with an ancylus fluviatilis adhering to him! . . . Here is a new light as to the way by which these sedentary mollusks may get transported from one river basin to another—That species of Ancylus seems to have got into Madeira before Man—How far can an Hydrobius fly with a favourable gale?" In this same letter Lyell further reported an instance of a water beetle with the egg case of a water spider under its wings, musing, "What unexpected means of migration will in time be found out" (*Correspondence* 6: 89).

[1] a single parent. But in the majority of cases, namely, with all organisms which habitually unite for each birth, or which often intercross, I believe that during the slow process of modification the individuals of the species will have been kept nearly uniform by intercrossing; so that many individuals will have gone on simultaneously changing, and the whole amount of modification will not have been due, at each stage, to descent from a single parent. To illustrate what I mean: our English race-horses differ slightly from the horses of every other breed; but they do not owe their difference and superiority to descent from any single pair, but to continued care in selecting and training many individuals during many generations.

Before discussing the three classes of facts, which I have selected as presenting the greatest amount of difficulty on the theory of "single centres of creation," I must say a few words on the means of dispersal.

[2] *Means of Dispersal.*—Sir C. Lyell and other authors have ably treated this subject. I can give here only the briefest abstract of the more important facts. Change of climate must have had a powerful influence on migration: a region when its climate was different may have been a high road for migration, but now be impassable; I shall, however, presently have to discuss this branch of the subject in some detail. Changes of level in the land must also have been highly influential: a narrow isthmus now separates two marine faunas; submerge it, or let it formerly have been submerged, and the two faunas will now blend or may formerly have blended: where the sea now extends, land may at a former period have connected islands or possibly even continents together, and thus have allowed terrestrial productions to pass from one to the other.

No geologist will dispute that great mutations of level, have occurred within the period of existing organisms. Edward Forbes insisted that all the islands in the ◀1 Atlantic must recently have been connected with Europe or Africa, and Europe likewise with America. Other authors have thus hypothetically bridged over every ocean, and have united almost every island to some mainland. If indeed the arguments used by Forbes are to be trusted, it must be admitted that scarcely a single island exists which has not recently been united to some continent. This view cuts the Gordian knot of the dispersal of the same species to the most distant points, and removes many a difficulty: but to the best of my judgment we are not authorized in admitting such enormous geographical changes within the period of existing species. It seems to me that we have abundant evidence of great oscillations of level in our continents; but not of such vast changes in their position and extension, as to have united them within the recent period to each other and to the several intervening oceanic islands. I freely admit the former existence of many islands, now buried beneath the sea, which may have served as halting places for plants and for many animals during their migration. In the coral-producing oceans such sunken islands are now marked, as I believe, by rings of coral or atolls standing over them. Whenever it is fully admitted, as I believe it will some day be, that each species has proceeded from a single birthplace, and when in the course of time we know something definite about the means of distribution, we shall be enabled to speculate with security on the former extension of the land. But I do not ◀2 believe that it will ever be proved that within the recent period continents which are now quite separate, have been continuously, or almost continuously, united

1 Recall that Darwin, like Lyell, was an adherent of the idea of long-term subsidence and elevation of continents as a dominant geological process governing earth history. In 1846 Forbes published his "Atlantis theory," proposing the existence of a vast land bridge in the Miocene, spanning the east Atlantic from the Azores to Ireland. Darwin (and others) rejected this idea, favoring long-distance dispersal, but many geologists seemed taken with Forbes's land bridge theory, and "continental extensions" became a popular subject of speculation that continued well into the nineteenth century. Consider Wallace's conclusion to a letter he wrote to the editor of the magazine *Natural Science* in 1892:

> I cannot forget that it has been, and still is with many writers, the practice to assume former continental extensions across the great oceans in order to explain difficulties in the distribution of single genera or families; that geologists of repute have claimed the Dolphin bank in the Atlantic trough as the relic of a chain of mountains comparable with the Andes; that oceanic islands have been recently claimed to be merely the tops of submerged mountains, which can only be properly compared with the highest points of continents, and that a geological critic so late as 1879 considered the idea that the oceans had always been in their present positions "a funny one." If such extreme views are now less common than they were, I hope that I may, without presumption, claim to have had some share in bringing about the change in scientific opinion now in progress. (Wallace 1892, p. 718)

2 Darwin is a bit restrained here; he had little patience with the continental extension school and was somewhat irked to find that Lyell was more sympathetic to the idea. A long letter to Lyell dated June 25, 1856, begins: "As you say you would like to hear my reasons for being most unwilling to believe in the continental extensions of late authors, I gladly write them; as, without I am convinced of my error, I shall have to give them condensed in my Essay, when I discuss single and multiple creation. I shall therefore be particularly glad to have your general opinion on

(continued)

(continued)
them. I may quite likely have persuaded myself in my wrath that there is more in them than there is" (*Correspondence* 6: 153).

Darwin always maintained that the "extensionists" overlooked dispersal in their enthusiasm for the more grandiose notion of lost continents. "I quite agree in admiration of Forbes's Essay," Darwin wrote to Lyell in November 1860, "yet, on my life, I think, it has done in some respects as much mischief as good. Those who believe in vast continental extensions will never investigate means of distribution. Good Heavens look at Heers map of Atlantis!! I thought his division & lines of travel of the British Plants very wild & with hardly any foundation" (*Correspondence* 8: 479). "Heer" is Oswald Heer, who proposed a map of Atlantis in 1855.

1 Miles Joseph Berkeley reported the results of his seed immersion experiments in the *Gardeners' Chronicle and Agricultural Gazette* in September 1855. Darwin reported similar experiments in the same journal throughout that year (Darwin 1855a–d), and he wrote to Berkeley in February 1856 asking if he could combine Berkeley's findings with his own for a paper. Berkeley consented, and Darwin wrote "On the action of sea-water on the germination of seeds," read before the Linnean Society that May (and the following year published in the Society's *Journal*—see Darwin 1857b).

The significance of such studies for Darwin's theories is obvious: committed to long-distance dispersal as a key element of biogeography, he sought to establish that plants and animals could plausibly survive transoceanic passage. Hence this discussion of long-distance seed dispersal extends another six pages! Darwin documents myriad seed-dispersal observations and studies in his notebooks. In one passage in the B notebook, he even speculates about a double benefit of seed dispersal by sea: "It would be curious experiment to know whether soaking seeds in salt water &c has any tendency to form varieties?" (Barrett et al. 1987, p. 200).

with each other, and with the many existing oceanic islands. Several facts in distribution,—such as the great difference in the marine faunas on the opposite sides of almost every continent,—the close relation of the tertiary inhabitants of several lands and even seas to their present inhabitants,—a certain degree of relation (as we shall hereafter see) between the distribution of mammals and the depth of the sea,—these and other such facts seem to me opposed to the admission of such prodigious geographical revolutions within the recent period, as are necessitated on the view advanced by Forbes and admitted by his many followers. The nature and relative proportions of the inhabitants of oceanic islands likewise seem to me opposed to the belief of their former continuity with continents. Nor does their almost universally volcanic composition favour the admission that they are the wrecks of sunken continents;—if they had originally existed as mountain-ranges on the land, some at least of the islands would have been formed, like other mountain-summits, of granite, metamorphic schists, old fossiliferous or other such rocks, instead of consisting of mere piles of volcanic matter.

I must now say a few words on what are called accidental means, but which more properly might be called occasional means of distribution. I shall here confine myself to plants. In botanical works, this or that plant is stated to be ill adapted for wide dissemination; but for transport across the sea, the greater or less facilities may be said to be almost wholly unknown. Until I tried, with Mr. Berkeley's aid, a few experiments, it was not even known how far seeds could resist the injurious action of sea-water. To my surprise I found that out of 87 kinds, 64 germinated after an immersion of 28 days, and a few survived an immersion of 137 days.

For convenience sake I chiefly tried small seeds, without the capsule or fruit; and as all of these sank in a few days, they could not be floated across wide spaces of the sea, whether or not they were injured by the salt-water. Afterwards I tried some larger fruits, capsules, &c., and some of these floated for a long time. It is well known what a difference there is in the buoyancy of green and seasoned timber; and it occurred to me that floods might wash down plants or branches, and that these might be dried on the banks, and then by a fresh rise in the stream be washed into the sea. Hencc I was led to dry stems and branches of 94 plants with ripe fruit, and to place them on sea water. The majority sank quickly, but some which whilst green floated for a very short time, when dried floated much longer; for instance, ripe hazel-nuts sank immediately, but when dried, they floated for 90 days and afterwards when planted they germinated; an asparagus plant with ripe berries floated for 23 days, when dried it floated for 85 days, and the seeds afterwards germinated: the ripe seeds of Helosciadium sank in two days, when dried they floated for above 90 days, and afterwards germinated. Altogether out of the 94 dried plants, 18 floated for above 28 days, and some of the 18 floated for a very much longer period. So that as $\frac{64}{87}$ seeds germinated after an immersion of 28 days; and as $\frac{18}{94}$ plants with ripe fruit (but not all the same species as in the foregoing experiment) floated, after being dried, for above 28 days, as far as we may infer anything from these scanty facts, we may conclude that the seeds of $\frac{14}{100}$ plants of any country might be floated by sea-currents during 28 days, and would retain their power of germination. In Johnston's Physical Atlas, the average rate of the several Atlantic currents is 33 miles per diem (some currents running at the rate of 60 miles ◀1

1 Fourteen percent of the plants of Europe is a considerable proportion. In the context of sea currents and transoceanic distances involved, Darwin concludes that it is more than plausible that a significant number of plant species are dispersed in this way. The book mentioned here is Alexander Keith Johnston's *Physical Atlas of Natural Phenomena,* first published in 1850.

1 In this paragraph Darwin summarizes a study by the botanist Martin Charles Martens of the University of Louvain in Belgium. Martens's experiment on seed germination following long-term immersion in sea water was published in 1857 in the *Bulletin de la Société Botanique de France.*

per diem); on this average, the seeds of $\frac{14}{100}$ plants belonging to one country might be floated across 924 miles of sea to another country; and when stranded, if blown to a favourable spot by an inland gale, they would germinate.

1 Subsequently to my experiments, M. Martens tried similar ones, but in a much better manner, for he placed the seeds in a box in the actual sea, so that they were alternately wet and exposed to the air like really floating plants. He tried 98 seeds, mostly different from mine; but he chose many large fruits and likewise seeds from plants which live near the sea; and this would have favoured the average length of their flotation and of their resistance to the injurious action of the salt-water. On the other hand he did not previously dry the plants or branches with the fruit; and this, as we have seen, would have caused some of them to have floated much longer. The result was that $\frac{18}{98}$ of his seeds floated for 42 days, and were then capable of germination. But I do not doubt that plants exposed to the waves would float for a less time than those protected from violent movement as in our experiments. Therefore it would perhaps be safer to assume that the seeds of about $\frac{10}{100}$ plants of a flora, after having been dried, could be floated across a space of sea 900 miles in width, and would then germinate. The fact of the larger fruits often floating longer than the small, is interesting; as plants with large seeds or fruit could hardly be transported by any other means; and Alph. de Candolle has shown that such plants generally have restricted ranges.

But seeds may be occasionally transported in another manner. Drift timber is thrown up on most islands, even on those in the midst of the widest oceans; and the natives of the coral-islands in the Pacific, procure stones for their tools, solely from the roots of drifted trees, these stones being a valuable royal tax. I find on examination, that when irregularly shaped stones are embedded in the roots of trees, small parcels of earth are very frequently enclosed in their interstices and behind them,—so perfectly that not a particle could be washed away in the longest transport: out of one small portion of earth thus *completely* enclosed by wood in an oak about 50 years old, three dicotyledonous plants germinated: I am certain of the accuracy of this observation. Again, I can show that the carcasses of birds, when floating on the sea, sometimes escape being immediately devoured; and seeds of many kinds in the crops of floating birds long retain their vitality: peas and vetches, for instance, are killed by even a few days' immersion in sea-water; but some taken out of the crop of a pigeon, which had floated on artificial salt-water for 30 days, to my surprise nearly all germinated.

Living birds can hardly fail to be highly effective agents in the transportation of seeds. I could give many facts showing how frequently birds of many kinds are blown by gales to vast distances across the ocean. We may I think safely assume that under such circumstances their rate of flight would often be 35 miles an hour; and some authors have given a far higher estimate. I have never seen an instance of nutritious seeds passing through the intestines of a bird; but hard seeds of fruit will pass uninjured through even the digestive organs of a turkey. In the course of two months, I picked up in my garden 12 kinds of seeds, out of the excrement of small birds, and these seemed perfect, and some of them, which I tried, germinated. But the following fact is more important: the crops of birds do not secrete gastric juice, and do not in the

1 In 1855 and 1856 Darwin was much interested in the question of how long seeds remain viable when buried. He engaged gardeners on the question though letters to the *Gardener's Chronicle,* recorded examples from the literature, and developed experiments. This case of the seeds bound to the oak roots sounds much like another one that he described in a letter to his son William: "We have today cut down & grubbed the big Beech tree by the roundabout: I find by the rings it is 77 years old: I am going to try whether there are any seeds in the earth from right under it, for they must have been buried for 77 years" (February 26, 1856; *Correspondence* 6: 45).

2 This is one of Darwin's more celebrated experiments; harvesting the seeds at its conclusion was certainly not for the squeamish. He summed it up in the postscript to a December 1856 letter to Joseph Hooker:

> P.S. I must tell you another of my *profound* experiments! Franky [son Francis Darwin] said to me, "why [should] not a bird be killed (by hawk, lightning, apoplexy, hail &c) with seeds in crop, & it would swim." No sooner said, than done: a pigeon has floated for 30 days in salt water with seeds in crop & they have grown splendidly & to my great surprise even tares (Leguminosae, so generally killed by sea-water) which the Bird had naturally eaten have grown well.—You will say gulls & dog-fish &c [would] eat up the carcase, & so they [would] 999 out of a thousand, but one might escape: I have seen dead land bird in sea-drift. (*Correspondence* 6: 304)

3 In a letter to Joseph Hooker in November 1856, Darwin wrote: "Lately I have been looking during few walks at excrement of small birds; I have found 6 kinds of seeds, which is more than I expected" (*Correspondence* 6: 266).

least injure, as I know by trial, the germination of seeds; now after a bird has found and devoured a large supply of food, it is positively asserted that all the grains do not pass into the gizzard for 12 or even 18 hours. A bird in this interval might easily be blown to the distance of 500 miles, and hawks are known to look out for tired birds, and the contents of their torn crops might thus readily get scattered. Mr. Brent informs me that a friend of his had to give up flying carrier-pigeons from France to England, as the hawks on the English
[1] coast destroyed so many on their arrival. Some hawks and owls bolt their prey whole, and after an interval of from twelve to twenty hours, disgorge pellets, which, as I know from experiments made in the Zoological Gardens, include seeds capable of germination. Some seeds of the oat, wheat, millet, canary, hemp, clover, and beet germinated after having been from twelve to twenty-one hours in the stomachs of different birds of prey; and two seeds of beet grew after having been thus retained for two days and fourteen hours. Fresh-
[2] water fish, I find, eat seeds of many land and water plants: fish are frequently devoured by birds, and thus the seeds might be transported from place to place. I forced many kinds of seeds into the stomachs of dead fish, and then gave their bodies to fishing-eagles, storks, and pelicans; these birds after an interval of many hours, either rejected the seeds in pellets or passed them in their excrement; and several of these seeds retained their power of germination. Certain seeds, however, were always killed by this process.

[3] Although the beaks and feet of birds are generally quite clean, I can show that earth sometimes adheres to them: in one instance I removed twenty-two grains of dry argillaceous earth from one foot of a partridge, and in this earth there was a pebble quite as large as

1 This experiment was conducted in the Zoological Society's gardens in October 1856. Darwin recorded it in his "Experimental Book": "Killed some sparrow [*on 14th], one with wheat inside, put in Oats, Canary seed, Tares, Cabbage & Clover—gave 3 Birds to small S. African Eagle. (Bateleur): bolted them; threw up pellet in 18 hours, [*ie on morning of 16th] charged with seed: planted these seeds on 19th."

He later wrote to Hooker: "The seeds which the Eagle had in stomach for 18 hours looked so fresh that I would have bet 5 to 1 they would all have grown; but some kinds were all killed & 2 oats 1 Canary seed, 1 Clover & 1 Beet alone came up!" (*Correspondence* 6: 266).

2 Darwin had a frustrating time working with live fish. In early May 1855 he tried an experiment feeding fish a variety of seeds: "They took them in mouth & kept them for some seconds . . . & then rejected them with force," he recorded. Dejected, Darwin lamented to Hooker that "everything has been going wrong with me lately; the fish at the Zoolog. Soc. ate up lots of soaked seeds, & in imagination they had in my mind been swallowed, fish & all, by a heron, had been carried a hundred miles, been voided on the banks of some other lake & germinated splendidly,—when lo & behold, the fish ejected vehemently, & with disgust equal to my own, all the seeds from their mouths" (*Correspondence* 5: 329). Two years later he is still at it, and we find him more upbeat in another letter to Hooker: "I find Fish will greedily eat seeds of aquatic grasses, & that millet seed put into Fish & given to Stork & then voided will germinate" (January 20, 1857; *Correspondence* 6: 324).

3 In August 1856 Darwin appealed to Thomas Campbell Eyton, who kept a menagerie at his Shropshire estate, for help collecting mud from the feet of water fowl: "Would you render me a little assistance in this line? . . . I want to know whether on a wet muddy day, whether birds feet are dirty: I am going to send my servant out with some keeper & he shall wash all the partridges feet & save the dirty water!! But I want especially to know whether herons or any waders . . . or water-birds when suddenly sprung have ever dirty feet or beaks?" (*Correspondence* 6: 211).

the seed of a vetch. Thus seeds might occasionally be transported to great distances; for many facts could be given showing that soil almost everywhere is charged with seeds. Reflect for a moment on the millions of quails which annually cross the Mediterranean; and can we doubt that the earth adhering to their feet would sometimes include a few minute seeds? But I shall presently have to recur to this subject.

As icebergs are known to be sometimes loaded with earth and stones, and have even carried brushwood, bones, and the nest of a land-bird, I can hardly doubt that they must occasionally have transported seeds from one part to another of the arctic and antarctic regions, as suggested by Lyell; and during the Glacial period from one part of the now temperate regions to another. In the Azores, from the large number of the species of plants common to Europe, in comparison with the plants of other oceanic islands nearer to the mainland, and (as remarked by Mr. H. C. Watson) from the somewhat northern character of the flora in comparison with the latitude, I suspected that these islands had been partly stocked by ice-borne seeds, during the Glacial epoch. At my request Sir C. Lyell wrote to M. Hartung to inquire whether he had observed erratic boulders on these islands, and he answered that he had found large fragments of granite and other rocks, which do not occur in the archipelago. Hence we may safely infer that icebergs formerly landed their rocky burthens on the shores of these mid-ocean islands, and it is at least possible that they may have brought thither the seeds of northern plants.

Considering that the several above means of transport, and that several other means, which without doubt remain to be discovered, have been in action year after year, for centuries and tens of thousands of

1 In a letter to Lyell dated July 5, 1856, Darwin expressed his dissatisfaction with the continental-extension hypothesis as a means of explaining the distribution of species. Lyell mused in his journal that icebergs could do the job: "Letter Darwin July 5, 1856 Icebergs & floating ice between latitudes 35 & 80 in each hemisphere may have been great agents of transporting species & have done much which is attributed to continental extension. Take the Glacial period as a unit & multiply this by all post-miocene Time & the result may afford an amount of geograph[ical] change capable with the additional aid of means of transport & migration of species of carrying them every where even across [the equator] by aid of cold periods & mountainous islands & floating ice & floating timber" (Wilson 1970, p. 116). Neither Darwin nor Lyell originated the idea of iceberg transport (see Mills 1983), but once Agassiz's glacial theory was accepted in the early 1840s, they embraced it as a potentially important means of dispersal. Darwin published a paper on the subject as early as 1841.

The Azores, an archipelago lying in the north Atlantic over a thousand miles off the coast of the Iberian Peninsula, presented a possible case study in iceberg transport to Darwin. "I have just had the innermost cockles of my heart rejoiced by a letter from Lyell," Darwin wrote to Hooker in April 1858. "I said to him (or he to me) that I believed from character of Flora of Azores, that icebergs must have been stranded there; & that I expected erratic boulders [would] be detected embeded between the upheaved lava-beds: & I got Lyell to write to Hartung to ask, & now H. says my question explains what had astounded him viz large boulders (& some polished) of Mica-schist, quartz, sandstone &c, some embedded & some 40 & 50 ft above level of sea, so that he had inferred that they had not been brought as ballast. Is this not beautiful?" (*Correspondence* 7: 82). Georg Hartung geologized with Lyell on the Canary Islands and Madeira in the 1850s and later published several papers and books on the geology of these islands. An "erratic boulder," by the way, is any sizable rock inconsistent in type with the surrounding native rocks of the landscape. Erratics are associated with glaciation, but Darwin thought they were transported by floating icebergs.

years, it would I think be a marvellous fact if many plants had not thus become widely transported. These means of transport are sometimes called accidental, but this is not strictly correct: the currents of the sea are not accidental, nor is the direction of prevalent gales of wind. It should be observed that scarcely any means of transport would carry seeds for very great distances; for seeds do not retain their vitality when exposed for a great length of time to the action of sea-water; nor could they be long carried in the crops or intestines of birds. These means, however, would suffice for occasional transport across tracts of sea some hundred miles in breadth, or from island to island, or from a continent to a neighbouring island, but not from one distant continent to another. The floras of distant continents would not by such means become mingled in any great degree; but would remain as distinct as we now see them to be. The currents, from their course, would never bring seeds from North America to Britain, though they might and do bring seeds from the West Indies to our western shores, where, if not killed by so long an immersion in salt-water, they could not endure our climate. Almost every year, one or two land-birds are blown across the whole Atlantic Ocean, from North America to the western shores of Ireland and England; but seeds could be transported by these wanderers only by one means, namely, in dirt sticking to their feet, which is in itself a rare accident. Even in this case, how small would the chance be of a seed falling on favourable soil, and coming to maturity! But it would be a great error to argue that because a well-stocked island, like Great Britain, has not, as far as is known (and it would be very difficult to prove this), received within the last few centuries, through occasional means

1 Transoceanic dispersal went into a long eclipse after the discovery of plate tectonics, which led to the acceptance of continental drift. One commentator even condemned the idea of long-distance dispersal as a means of explaining species distributions as "a science of the improbable, the rare, the mysterious, and the miraculous" (Nelson 1979). Vicariance stemming from continental drift is important, but nonetheless transoceanic dispersal is now thought to be relatively common, particularly on ecological and evolutionary time scales. On long time scales, improbable events are sooner or later realized. See de Queiroz (2005) for a thorough review.

of transport, immigrants from Europe or any other continent, that a poorly-stocked island, though standing more remote from the mainland, would not receive colonists by similar means. I do not doubt that out of twenty seeds or animals transported to an island, even if far less well-stocked than Britain, scarcely more than one would be so well fitted to its new home, as to become naturalised. But this, as it seems to me, is no valid argument against what would be effected by occasional means of transport, during the long lapse of geological time, whilst an island was being upheaved and formed, and before it had become fully stocked with inhabitants. On almost bare land, with few or no destructive insects or birds living there, nearly every seed, which chanced to arrive, would be sure to germinate and survive.

Dispersal during the Glacial period.—The identity of many plants and animals, on mountain-summits, separated from each other by hundreds of miles of lowlands, where the Alpine species could not possibly exist, is one of the most striking cases known of the same species living at distant points, without the apparent possibility of their having migrated from one to the other. It is indeed a remarkable fact to see so many of the same plants living on the snowy regions of the Alps or Pyrenees, and in the extreme northern parts of Europe; but it is far more remarkable, that the plants on the White Mountains, in the United States of America, are all the same with those of Labrador, and nearly all the same, as we hear from Asa Gray, with those on the loftiest mountains of Europe. Even as long ago as 1747, such facts led Gmelin to conclude that the same species must have been independently created at several distinct points; and we might have remained

1 One of the great discoveries of the nineteenth century was the existence of glacial periods. The "glacial theory" appears to have been first proposed by the German naturalist Karl Schimper, but Louis Agassiz is credited with championing it and convincing the scientific world of its validity in a series of papers and a book, *Etudes sur les Glaciers* (1840). That same year Agassiz traveled to Scotland with William Buckland seeking evidence of glaciation. Buckland was soon convinced of the theory, and the two communicated their findings to the Geological Society of London in a series of papers delivered in the early 1840s (e.g., Buckland 1841, 1842; Agassiz 1841).

2 In April 1855 Darwin wrote to Gray asking for help listing alpine species of northeastern North America and their relationship to those of Europe. Gray responded the next month, confirming that "[t]he top of White Mts. N. Hampshire is, as it were, a bit of Labrador (alt. 5000–6000 feet), and I do not believe there is a plant there which is not in Labrador, except two" (*Correspondence* 5: 334). Gray was intensely interested in the geographical distribution of plants and felt that puzzling distributions like the eastern North America–eastern Asia disjunction (see p. 371) shed light on species origins. He was an adherent of the "single creation" school of thought, in contrast to the "multiple creations" school championed by Agassiz, among others. It was precisely over this point that Agassiz and Gray squared off in a series of debates in 1858–1859. This issue was much discussed in Boston intellectual circles. When in June 1858 Henry David Thoreau pondered the occurrence of toads and frogs atop Mt. Monadnock in New Hampshire, for example, he seemed dismissive when he wrote, "Agassiz might say that they originated on the top" (Torrey 1968).

3 Johann Georg Gmelin's 1747 work is *Flora Sibirica,* which stemmed from a decade-long expedition to Siberia at the request of Russian Empress Anna Ivanovna. Darwin recorded reading this work in 1854 (Vorzimmer 1977).

in this same belief, had not Agassiz and others called vivid attention to the Glacial period, which, as we shall immediately see, affords a simple explanation of these facts. We have evidence of almost every conceivable kind, organic and inorganic, that within a very recent geological period, central Europe and North America suffered under an Arctic climate. The ruins of a house burnt by fire do not tell their tale more plainly, than do the mountains of Scotland and Wales, with their scored flanks, polished surfaces, and perched boulders, of the icy streams with which their valleys were lately filled. So greatly has the climate of Europe changed, that in Northern Italy, gigantic moraines, left by old glaciers, are now clothed by the vine and maize. Throughout a large part of the United States, erratic boulders, and rocks scored by drifted icebergs and coast-ice, plainly reveal a former cold period.

The former influence of the glacial climate on the distribution of the inhabitants of Europe, as explained with remarkable clearness by Edward Forbes, is substantially as follows. But we shall follow the changes more readily, by supposing a new glacial period to come slowly on, and then pass away, as formerly occurred. As the cold came on, and as each more southern zone became fitted for arctic beings and ill-fitted for their former more temperate inhabitants, the latter would be supplanted and arctic productions would take their places. The inhabitants of the more temperate regions would at the same time travel southward, unless they were stopped by barriers, in which case they would perish. The mountains would become covered with snow and ice, and their former Alpine inhabitants would descend to the plains. By the time that the cold had reached its maximum, we should have a uniform arctic fauna and flora, covering the central parts of Europe, as far

1 Buckland's 1842 paper convinced a skeptical Darwin, who revisited Wales that same year with Buckland's paper in hand and saw the landscape with new eyes. Darwin acknowledged that he had erred in his interpretation of the geology of sites like Cwm Idwal in Snowdonia, and he lost no time communicating his own observations: "I cannot imagine a more instructive and interesting lesson for anyone who wishes . . . to learn the effects produced by the passage of glaciers, than to ascend a mountain like one of those south of the upper lake of Llanberis [in Snowdonia]" (Darwin 1842). With this in mind we appreciate his memorable comment here that the ruins of a burnt house "do not tell their tale more plainly."

2 Forbes (1846) proposed that plants of northern Europe had been pushed south by the glaciers of the Pleistocene (a name that Lyell proposed and that Forbes equated with the cold period), and then recolonized northward again when the climate warmed. These cold-adapted species also moved up in elevation at the same time, explaining their occurrence on scattered high mountaintops. Darwin recognized that alpine plants found on far-flung island-like mountaintops seemed contrary to the idea that species differences generally coincide with geographical barriers, and he wrote as much in his 1842 *Sketch* (Darwin 1909, p. 30). He did not publish his idea, however, and Forbes got there first. Darwin later wrote in his autobiography:

> I was forestalled in only one important point, which my vanity has always made me regret, namely, the explanation by means of the Glacial period of the presence of the same species of plants and of some few animals on distant mountain summits and in the arctic regions. This view pleased me so much that I wrote it out in extenso, and I believe that it was read by Hooker some years before E. Forbes published his celebrated memoir . . . on the subject. In the very few points in which we differed, I still think that I was in the right. (F. Darwin 1896, pp. 71–72)

south as the Alps and Pyrenees, and even stretching into Spain. The now temperate regions of the United States would likewise be covered by arctic plants and animals, and these would be nearly the same with those of Europe; for the present circumpolar inhabitants, which we suppose to have everywhere travelled southward, are remarkably uniform round the world. We may suppose that the Glacial period came on a little earlier or later in North America than in Europe, so will the southern migration there have been a little earlier or later; but this will make no difference in the final result.

As the warmth returned, the arctic forms would retreat northward, closely followed up in their retreat by the productions of the more temperate regions. And as the snow melted from the bases of the mountains, the arctic forms would seize on the cleared and thawed ground, always ascending higher and higher, as the warmth increased, whilst their brethren were pursuing their northern journey. Hence, when the warmth had fully returned, the same arctic species, which had lately lived in a body together on the lowlands of the Old and New Worlds, would be left isolated on distant mountain-summits (having been exterminated on all lesser heights) and in the arctic regions of both hemispheres.

Thus we can understand the identity of many plants at points so immensely remote as on the mountains of the United States and of Europe. We can thus also understand the fact that the Alpine plants of each mountain-range are more especially related to the arctic forms living due north or nearly due north of them: for the migration as the cold came on, and the re-migration on the returning warmth, will generally have been due south and north. The Alpine plants, for example, of Scotland, as remarked by Mr. H. C. Watson, ◀1

1 The alpine flora of Europe and North America are still thought to be glacial relicts, though not all taxa responded in precisely the same way to glacial cycles. The plants of different mountain ranges were affected in different ways and degrees by glaciation, depending on the geography of the range in terms of latitude and orientation (see Nagy et al. 2003).

[1] and those of the Pyrenees, as remarked by Ramond, are more especially allied to the plants of northern Scandinavia; those of the United States to Labrador; those of the mountains of Siberia to the arctic regions of that country. These views, grounded as they are on the perfectly well-ascertained occurrence of a former Glacial period, seem to me to explain in so satisfactory a manner the present distribution of the Alpine and Arctic productions of Europe and America, that when in other regions we find the same species on distant mountain-summits, we may almost conclude without other evidence, that a colder climate permitted their former migration across the low intervening tracts, since become too warm for their existence.

If the climate, since the Glacial period, has ever been in any degree warmer than at present (as some geologists in the United States believe to have been the [2] case, chiefly from the distribution of the fossil Gnathodon), then the arctic and temperate productions will at a very late period have marched a little further north, and subsequently have retreated to their present homes; but I have met with no satisfactory evidence with respect to this intercalated slightly warmer period, since the Glacial period.

The arctic forms, during their long southern migration and re-migration northward, will have been exposed to nearly the same climate, and, as is especially to be noticed, they will have kept in a body together; consequently their mutual relations will not have been much disturbed, and, in accordance with the principles inculcated in this volume, they will not have been liable to much modification. But with our Alpine productions, left isolated from the moment of the returning warmth, first at the bases and ultimately on the summits of the mountains, the case will have been somewhat dif-

1 The mountaineer and naturalist Louis François Élisabeth Ramond was renowned for his exploration of the high Pyrenees. He published several books and papers on the mountains between 1789 and 1825, the last being the botanical work *Sur l'état de la végétation au sommet du Pic du Midi* (On the Condition of the Vegetation on the Summit of the Pic du Midi).

2 *Gnathodon* is a brackish water bivalve, abundant along the American Gulf Coast. There is one extant North American species, *G. cuneatus* (now *Rangia cuneata*), fossils of which are known from the Miocene. In the memoir of his second visit to the United States, Lyell commented on roads in New Orleans constructed of *Gnathodon cuneatus* shells, and noted extensive beds of these fossils as much as twenty miles inland of the Port of Mobile (Lyell 1849). Fossils of *G. cuneatus* have been found as far north as the New Jersey coast, and its former northerly distribution is probably what led Darwin to mention its presence as an indication of past warmer conditions.

ferent; for it is not likely that all the same arctic species will have been left on mountain ranges distant from each other, and have survived there ever since; they will, also, in all probability have become mingled with ancient Alpine species, which must have existed on the mountains before the commencement of the Glacial epoch, and which during its coldest period will have been temporarily driven down to the plains; they will, also, have been exposed to somewhat different climatal influences. Their mutual relations will thus have been in some degree disturbed; consequently they will have been liable to modification; and this we find has been the case; for if we compare the present Alpine plants and animals of the several great European mountain-ranges, though very many of the species are identically the same, some present varieties, some are ranked as doubtful forms, and some few are distinct yet closely allied or representative species.

In illustrating what, as I believe, actually took place during the Glacial period, I assumed that at its commencement the arctic productions were as uniform round the polar regions as they are at the present day. But the foregoing remarks on distribution apply not only to strictly arctic forms, but also to many sub-arctic and to some few northern temperate forms, for some of these are the same on the lower mountains and on the plains of North America and Europe; and it may be reasonably asked how I account for the necessary degree of uniformity of the sub-arctic and northern temperate forms round the world, at the commencement of the Glacial period. At the present day, the sub-arctic and northern temperate productions of the Old and New Worlds are separated from each other by the Atlantic Ocean and by the extreme northern part of the Pacific. During the Glacial period, when the in-

1 This passage is a good reminder of Darwin's thinking on the generation of variability. He attributes the species differences between plants atop different mountains to variation generated during the glacial cycles, when the plants were exposed to different environmental conditions and comingled with lowland species. Note his language: these plants were "liable to modification" as a result of disturbance of their "mutual relations" (i.e., ecology). The modern take on these same species differences is that population isolation leads to divergence, with or without natural selection. Note, too, that spatial isolation plays a role: these alpine productions are stranded atop mountains. Is Darwin suggesting in the final sentence of the paragraph that the species of different mountaintops differ from one another, each showing some variants in isolation of the others?

habitants of the Old and New Worlds lived further southwards than at present, they must have been still more completely separated by wider spaces of ocean. I believe the above difficulty may be surmounted by looking to still earlier changes of climate of an opposite nature. We have good reason to believe that during the newer Pliocene period, before the Glacial epoch, and whilst the majority of the inhabitants of the world were specifically the same as now, the climate was warmer than at the present day. Hence we may suppose that the organisms now living under the climate of latitude 60°, during the Pliocene period lived further north under the Polar Circle, in latitude 66°-67°; and that the strictly arctic productions then lived on the ▶1 broken land still nearer to the pole. Now if we look at a globe, we shall see that under the Polar Circle there is almost continuous land from western Europe, through Siberia, to eastern America. And to this continuity of the circumpolar land, and to the consequent freedom for intermigration under a more favourable climate, I attribute the necessary amount of uniformity in the sub-arctic and northern temperate productions of the Old and New Worlds, at a period anterior to the Glacial epoch.

Believing, from reasons before alluded to, that our continents have long remained in nearly the same relative position, though subjected to large, but partial oscillations of level, I am strongly inclined to extend the above view, and to infer that during some earlier and still warmer period, such as the older Pliocene period, a large number of the same plants and animals inhabited the almost continuous circumpolar land; and that these plants and animals, both in the Old and New Worlds, began slowly to migrate southwards as the climate became less warm, long before the com-

1 In current thinking, intermigration occurred even during the glacial maxima; an ice-free corridor along Beringia, the Bering land bridge, linked Siberia and North America. During glaciation, sea level was 300 or more feet lower than present levels, exposing land approximately 1,000 miles across its widest point.

mencement of the Glacial period. We now see, as I believe, their descendants, mostly in a modified condition, in the central parts of Europe and the United States. On this view we can understand the relationship, with very little identity, between the productions of North America and Europe,—a relationship which is most remarkable, considering the distance of the two areas, and their separation by the Atlantic Ocean. We can further understand the singular fact remarked on by several observers, that the productions of Europe and America during the later tertiary stages were more closely related to each other than they are at the present time; for during these warmer periods the northern parts of the Old and New Worlds will have been almost continuously united by land, serving as a bridge, since rendered impassable by cold, for the inter-migration of their inhabitants.

During the slowly decreasing warmth of the Pliocene period, as soon as the species in common, which inhabited the New and Old Worlds, migrated south of the Polar Circle, they must have been completely cut off from each other. This separation, as far as the more temperate productions are concerned, took place long ages ago. And as the plants and animals migrated southward, they will have become mingled in the one great region with the native American productions, and have had to compete with them; and in the other great region, with those of the Old World. Consequently we have here everything favourable for much modification,—for far more modification than with the Alpine productions, left isolated, within a much more recent period, on the several mountain-ranges and on the arctic lands of the two Worlds. Hence it has come, that when we compare the now living productions of the temperate regions of the New and Old Worlds, we find very few identical

1 Darwin is right for the wrong reasons here: Eurasia and North America were indeed linked into a continuous land mass, but during the cold periods when sea level dropped, not during the warm interglacials when Beringia was flooded (as today). The time period to which Darwin is referring—"later Tertiary stages"—corresponds to the last epochs of the Tertiary, the Miocene and Pliocene (named by Lyell in 1833). We now know that these stages span a far longer stretch of time than the Victorians imagined, from about twenty-three million years ago to nearly two million years ago. "Tertiary" is no longer used by geologists and is now divided into the Paleogene and the Neogene periods.

2 Since the time of Linnaeus, botanists have noticed the affinities of Old World and New World flora. Asa Gray was especially intrigued by this distribution pattern, and in 1846 he produced a paper titled *Analogy between the flora of Japan and that of the United States.* A decade later, Gray's work on a list of North American alpine species for Darwin snowballed into his seminal paper *Statistics of the flora of the northern United States* (1856–1857), and then an even more detailed memoir on the botany of Japan (Gray 1859). Gray's Harvard colleague W. G. Farlow called this paper "a masterpiece" in his remembrance of Gray read before the National Academy of Sciences on April 17, 1889. The close relationship between the flora (as well as arthropods and fungi) of eastern Asia and eastern North America is highly striking: the flora of eastern North America is more similar to that of eastern China and Japan than to that of western North America, with more than sixty-five genera exhibiting the disjunction. It is believed that this pattern is a relict of the extensive north-temperate forests of the Neogene, dating to the Miocene Epoch ~23.8 to 5.3 million years ago (Boufford and Spongberg 1983; Wen 1999).

1 James Dwight Dana was a geologist for the U.S. Exploring Expedition of 1838–1842 and subsequently worked on the crustaceans collected on the expedition.

2 Relationship without identity is an important point: Darwin is underscoring the puzzle of having related yet distinct species living under essentially identical conditions.

species (though Asa Gray has lately shown that more plants are identical than was formerly supposed), but we find in every great class many forms, which some naturalists rank as geographical races, and others as distinct species; and a host of closely allied or representative forms which are ranked by all naturalists as specifically distinct.

As on the land, so in the waters of the sea, a slow southern migration of a marine fauna, which during the Pliocene or even a somewhat earlier period, was nearly uniform along the continuous shores of the Polar Circle, will account, on the theory of modification, for many closely allied forms now living in areas completely sundered. Thus, I think, we can understand the presence of many existing and tertiary representative forms on the eastern and western shores of temperate North America; and the still more striking case of many closely allied crustaceans (as described in Dana's admirable work), of some fish and other marine animals, in the Mediterranean and in the seas of Japan,—areas now separated by a continent and by nearly a hemisphere of equatorial ocean.

These cases of relationship, without identity, of the inhabitants of seas now disjoined, and likewise of the past and present inhabitants of the temperate lands of North America and Europe, are inexplicable on the theory of creation. We cannot say that they have been created alike, in correspondence with the nearly similar physical conditions of the areas; for if we compare, for instance, certain parts of South America with the southern continents of the Old World, we see countries closely corresponding in all their physical conditions, but with their inhabitants utterly dissimilar.

But we must return to our more immediate subject, the Glacial period. I am convinced that Forbes's view may be largely extended. In Europe we have the plainest evidence of the cold period, from the western shores of Britain to the Oural range, and southward to the Pyrenees. We may infer, from the frozen mammals and nature of the mountain vegetation, that Siberia was similarly affected. Along the Himalaya, at points 900 miles apart, glaciers have left the marks of their former low descent; and in Sikkim, Dr. Hooker saw maize growing on gigantic ancient moraines. South of the equator, we have some direct evidence of former glacial action in New Zealand; and the same plants, found on widely separated mountains in this island, tell the same story. If one account which has been published can be trusted, we have direct evidence of glacial action in the south-eastern corner of Australia.

Looking to America; in the northern half, ice-borne fragments of rock have been observed on the eastern side as far south as lat. 36°–37°, and on the shores of the Pacific, where the climate is now so different, as far south as lat. 46°; erratic boulders have, also, been noticed on the Rocky Mountains. In the Cordillera of Equatorial South America, glaciers once extended far below their present level. In central Chile I was astonished at the structure of a vast mound of detritus, about 800 feet in height, crossing a valley of the Andes; and this I now feel convinced was a gigantic moraine, left far below any existing glacier. Further south on both sides of the continent, from lat. 41° to the southernmost extremity, we have the clearest evidence of former glacial action, in huge boulders transported far from their parent source.

We do not know that the Glacial epoch was strictly simultaneous at these several far distant points on opposite sides of the world. But we have good evidence in almost every case, that the epoch was included within

1 Hooker led an expedition to the Himalayas in 1847–1850, during which he and his party were taken prisoner for several weeks by the Rajah of Sikkim. He recounted this and other adventures, along with vivid descriptions of the geology and botany of the region, in his book *Himalayan Journal* (1854), which he dedicated to Darwin. Darwin's information on New Zealand was also provided by Hooker, from an earlier voyage. Hooker was assistant surgeon and naturalist on the HMS *Erebus*, which with its sister ship HMS *Terror* explored the southern oceans from 1839 to 1841. The botanical work from this voyage made Hooker's reputation: his collections became one of the two *Flora Antarctica* volumes (1844–1847), and he subsequently published the acclaimed books *Flora Novae-Zelandiae* (1853–1855) and *Flora Tasmaniae* (1855–1860).

2 There is no explicit mention of this detritus mound in either the *Voyage of the Beagle* or *Geological Observations on South America*, but Darwin was much struck by conical mounds and terraces of unconsolidated material throughout the Chilean valleys. He interpreted these as evidence of incursions of the sea. In March 1835, for example, he inspected the valley of the Maypu River in Chile: "All the main valleys in the Cordillera are characterized by . . . a fringe or terrace of shingle and sand, rudely stratified, and generally of considerable thickness . . . and which were undoubtedly deposited when the sea penetrated Chile" (*Voyage*, chapter XV).

In chapter IX of *Voyage*, Darwin reported finding immense erratic boulders, which he interpreted as iceberg-borne. After his conversion to the glacial theory, Darwin realized that the terraces, detritus piles, and erratic boulders of South America were deposited by glaciers. His iceberg theory was not unreasonable: he had observed gigantic icebergs loaded with rocky debris in the Straits of Magellan: "In [Eyre's] Sound, about fifty icebergs were seen at one time floating outwards . . . Some of the icebergs were loaded with blocks of no inconsiderable size, of granite and other rocks, different from the clay-slate of the surrounding mountains" (*Voyage*, chapter XI).

the latest geological period. We have, also, excellent evidence, that it endured for an enormous time, as measured by years, at each point. The cold may have come on, or have ceased, earlier at one point of the globe than at another, but seeing that it endured for long at each, and that it was contemporaneous in a geological sense, it seems to me probable that it was, during a part at least of the period, actually simultaneous throughout the world. Without some distinct evidence to the contrary, we may at least admit as probable that the glacial action was simultaneous on the eastern and western sides of North America, in the Cordillera under the equator and under the warmer temperate zones, and on both sides of the southern extremity of the continent. If this be admitted, it is difficult to avoid believing that the temperature of the whole world was at this period simultaneously cooler. But it would suffice for my purpose, if the temperature was at the same time lower along certain broad belts of longitude.

On this view of the whole world, or at least of broad longitudinal belts, having been simultaneously colder from pole to pole, much light can be thrown on the present distribution of identical and allied species. In America, Dr. Hooker has shown that between forty and fifty of the flowering plants of Tierra del Fuego, forming no inconsiderable part of its scanty flora, are common to Europe, enormously remote as these two points are; and there are many closely allied species. On the lofty mountains of equatorial America a host of peculiar species belonging to European genera occur. On the highest mountains of Brazil, some few European genera were found by Gardner, which do not exist in the wide intervening hot countries. So on the Silla of Caraccas the illustrious Humboldt long ago found species belong-

1 Darwin conjectures that the glacial period was a time of global cooling, not just localized drops in temperature. He supposes that such cooling would homogenize conditions worldwide and allow for the southerly migration of now northerly adapted species. With warming of the climate, some of these plants would be extirpated locally, but others would retreat to high-elevation sites as the lowlands became too warm. The scenario of worldwide cooling is supported both by recent data indicating that the past two northern and southern hemisphere glacial advances (22,000 and 150,000 years ago) occurred simultaneously (Kaplan et al. 2004), and by pollen records from northern and southern hemisphere glacial sites (Moreno et al. 2001).

2 Darwin was keenly interested in Hooker's botanical findings in Tierra del Fuego. He wrote to him in November 1843, soon after Hooker's return to England:

> I have long thought that some general sketch of the Flora of [Tierra del Fuego], stretching so far into the southern seas, would be very curious.—Do make comparative remarks on the species allied to the European species, for the advantage of Botanical Ignoramus'es like myself. It has always struck me as a curious point to find out, whether there are many European genera in T. del Fuego, which are not found along the ridge of the Cordillera; the separation in such cases wd be so enormous. (*Correspondence* 2: 408)

3 The English botanical explorer George Gardner gave an account of the botany of the Organ Mountains, Brazil, in *Travels in the Interior of Brazil* (1846).

4 The nearly 3,000-meter-high Silla de Caracas is a mountain in the coastal range of Venezuela, near the city of Caracas. Alexander von Humboldt and Aimé Bonpland ascended the mountain in the course of their explorations in South America, and Darwin read about the mountain's botany in volume III of Humboldt's *Personal Narrative.*

ing to genera characteristic of the Cordillera. On the mountains of Abyssinia, several European forms and some few representatives of the peculiar flora of the Cape of Good Hope occur. At the Cape of Good Hope a very few European species, believed not to have been introduced by man, and on the mountains, some few representative European forms are found, which have not been discovered in the intertropical parts of Africa. On the Himalaya, and on the isolated mountain-ranges of the peninsula of India, on the heights of Ceylon, and on the volcanic cones of Java, many plants occur, either identically the same or representing each other, and at the same time representing plants of Europe, not found in the intervening hot lowlands. A list of the genera collected on the loftier peaks of Java raises a picture of a collection made on a hill in Europe! Still more striking is the fact that southern Australian forms are clearly represented by plants growing on the summits of the mountains of Borneo. Some of these Australian forms, as I hear from Dr. Hooker, extend along the heights of the peninsula of Malacca, and are thinly scattered, on the one hand over India and on the other as far north as Japan.

On the southern mountains of Australia, Dr. F. Müller has discovered several European species; other species, not introduced by man, occur on the lowlands; and a long list can be given, as I am informed by Dr. Hooker, of European genera, found in Australia, but not in the intermediate torrid regions. In the admirable 'Introduction to the Flora of New Zealand,' by Dr. Hooker, analogous and striking facts are given in regard to the plants of that large island. Hence we see that throughout the world, the plants growing on the more lofty mountains, and on the temperate lowlands of the northern and southern hemispheres, are sometimes

1 Ferdinand von Müller published several books on the flora of Australia, among them *Fragmenta Phytographica Australiae* (1862–1881) and *Plants of Victoria* (1860–1865). The "Introduction" Darwin refers to by Hooker is a thirty-nine-page introductory essay to his *Flora Novae-Zelandiae* (Flora of New Zealand) in 1853. Intended as a prefatory work, this essay became an acclaimed treatment of botanical geography in its own right.

2 This is a significant point: the plants at high elevations in these far-flung locales are usually "specifically distinct"—i.e., different species—and only in a few cases "identically the same," i.e., the same species. They are related in "a most remarkable manner" because they are very close relatives yet live so far apart. The supposition is that these close relatives descended from a common ancestor in the recent past, during or prior to the last glacial period. This scenario is largely accepted today, with at least three key differences from Darwin's era: the time scales are longer than the Victorians imagined; we now recognize multiple glacial cycles opening and closing migration corridors; and continents are now known to move, adding another factor to the dynamic of species movements in deep time.

identically the same; but they are much oftener specifically distinct, though related to each other in a most remarkable manner.

This brief abstract applies to plants alone: some strictly analogous facts could be given on the distribution of terrestrial animals. In marine productions, similar cases occur; as an example, I may quote a remark by the highest authority, Prof. Dana, that "it is certainly a wonderful fact that New Zealand should have a closer resemblance in its crustacea to Great Britain, its antipode, than to any other part of the world." Sir J. Richardson, also, speaks of the reappearance on the shores of New Zealand, Tasmania, &c., of northern forms of fish. Dr. Hooker informs me that twenty-five species of Algæ are common to New Zealand and to Europe, but have not been found in the intermediate tropical seas.

It should be observed that the northern species and forms found in the southern parts of the southern hemisphere, and on the mountain-ranges of the intertropical regions, are not arctic, but belong to the northern temperate zones. As Mr. H. C. Watson has recently remarked, "In receding from polar towards equatorial latitudes, the Alpine or mountain floras really become less and less arctic." Many of the forms living on the mountains of the warmer regions of the earth and in the southern hemisphere are of doubtful value, being ranked by some naturalists as specifically distinct, by others as varieties; but some are certainly identical, and many, though closely related to northern forms, must be ranked as distinct species.

Now let us see what light can be thrown on the foregoing facts, on the belief, supported as it is by a large body of geological evidence, that the whole world, or a large part of it, was during the Glacial period simulta-

1 The three authorities cited on this page have all made previous appearances. James Dwight Dana was cited on p. 372, as here, in connection with his work on the crustaceans of the U.S. Exploring Expedition. The Scottish naturalist and arctic explorer John Richardson wrote *Fauna Boreali-Americana.* Hooker we have met several times. He did not work extensively on algae, but he did collect many algal species on the voyage of the *Erebus* and would also have been intimately familiar with the work of the Irish phycologist William Henry Harvey, who described Darwin's algal collections from the *Beagle* voyage. The plant biogeographer Hewett Cottrell Watson, finally, treated British alpine plants in *Cybele Britannica* (1847).

2 Arctic, temperate, and tropical plants are imagined to respond to the steadily cooling climate of the glacial period in different ways. This long paragraph describes what happens to the temperate and tropical species. Here is a summary: Darwin supposes that the tropical species have stricter thermal requirements and experience much extinction, with surviving species retreating to the narrow remaining suitable areas within the tropics. The temperate-adapted species, in contrast, are imagined to be more physiologically flexible and thus ultimately become geographically displaced to a much greater degree. As a result of this migration, these plants experience altered environmental conditions and in some cases comingle with the more cool-tolerant species of the tropics. The net effect of this process, Darwin imagines, is to engender variations that selection may act upon to adapt the temperate species to their new homes, and new species are formed. Later, when the climate warms up again, these northern-affinity species are driven back north out of the tropics, except for some representatives that move upward to the mountain peaks instead of northward. Those that had made it beyond the equatorial zone into the southern hemisphere get driven still further southward, to the south temperate zone, as the tropical species expand out of their refugia and once again dominate the tropical landscape. Thus we are left with northern-affinity species in both hemispheres, especially in the cool climates of the high elevations. The species of these hemispheres differ but are closely related.

neously much colder than at present. The Glacial period, as measured by years, must have been very long; and when we remember over what vast spaces some naturalised plants and animals have spread within a few centuries, this period will have been ample for any amount of migration. As the cold came slowly on, all the tropical plants and other productions will have retreated from both sides towards the equator, followed in the rear by the temperate productions, and these by the arctic; but with the latter we are not now concerned. The tropical plants probably suffered much extinction; how much no one can say; perhaps formerly the tropics supported as many species as we see at the present day crowded together at the Cape of Good Hope, and in parts of temperate Australia. As we know that many tropical plants and animals can withstand a considerable amount of cold, many might have escaped extermination during a moderate fall of temperature, more especially by escaping into the warmest spots. But the great fact to bear in mind is, that all tropical productions will have suffered to a certain extent. On the other hand, the temperate productions, after migrating nearer to the equator, though they will have been placed under somewhat new conditions, will have suffered less. And it is certain that many temperate plants, if protected from the inroads of competitors, can withstand a much warmer climate than their own. Hence, it seems to me possible, bearing in mind that the tropical productions were in a suffering state and could not have presented a firm front against intruders, that a certain number of the more vigorous and dominant temperate forms might have penetrated the native ranks and have reached or even crossed the equator. The invasion would, of course, have been greatly favoured by high land, and perhaps by a dry climate; for Dr. Falconer informs me that it is the damp with the heat of the tropics which is so destructive to perennial plants from a temperate climate. On the other hand, the most humid and hottest districts will have afforded an asylum to the tropical natives. The mountain-ranges north-west of the Himalaya, and the long line of the Cordillera, seem to have afforded two great lines of invasion: and it is a striking fact, lately communicated to me by Dr. Hooker, that all the flowering plants, about forty-six in number, common [1] to Tierra del Fuego and to Europe still exist in North America, which must have lain on the line of march. But I do not doubt that some temperate productions entered and crossed even the *lowlands* of the tropics at the period when the cold was most intense,—when arctic forms had migrated some twenty-five degrees of latitude from their native country and covered the land at the foot of the Pyrenees. At this period of extreme cold, I believe that the climate under the equator at the level of the sea was about the same with that now felt there at the height of six or seven thousand feet. During this the coldest period, I suppose that large spaces of the tropical lowlands were clothed with a mingled tropical and temperate vegetation, like that now growing with strange luxuriance at the base of the [2] Himalaya, as graphically described by Hooker.

Thus, as I believe, a considerable number of plants, a few terrestrial animals, and some marine productions, migrated during the Glacial period from the northern and southern temperate zones into the intertropical regions, and some even crossed the equator. As the warmth returned, these temperate forms would naturally ascend the higher mountains, being exterminated on the lowlands; those which had not reached the equator, would re-migrate northward or southward towards their former

1 The commonalities between the flora of Tierra del Fuego and that of Europe were first described by Hooker in *Flora Antarctica,* but he did not specifically address Fuegian plants that were also present in Europe. This is important to Darwin because he opposes continental extensions and multiple creations; oversea and overland dispersal are the only mechanisms he accepts. We have already seen the energy Darwin expended on dispersal by sea. Here he explores the possibility of overland dispersal made possible by corridors created during cool periods.

Darwin suggested to Gray in May 1856 that it would be useful "to compare the list of European plants in Tierra del Fuego (in Hooker) with those in N. America; for without multiple creation, I think we must admit that all now in T. del Fuego, must have traveled through N. America" (*Correspondence* 6: 92). Two months later, still puzzling over the Patagonia-Europe connection, he asked Gray if the "Alleghenies" (Appalachians) could provide a migration route to the southern part of North America: "[Are the] Alleghenies . . . sufficiently continuous so that the plants could travel from the north in the course of ages thus far south? I remember Bartram makes the same remark with respect to several trees on the Occone Mts." (*Correspondence* 6: 182). Darwin read the *Travels* of William Bartram, the American botanist who collected in the southern Appalachians in the 1770s. "Occone Mts." refers to Station Mountain in upstate South Carolina.

In October 1856 Darwin asked Hooker to comment on a draft manuscript containing material that was to become much of this chapter. This was part of a months-long exchange between the two, in which Darwin tried to convince Hooker of his hypothesis of pathways of species migration during the cooler epochs. Hooker was wedded to the idea of continental extensions to explain the distribution of the southern hemisphere flora, a notion that Darwin found anathema.

2 Hooker's *Himalayan Journals* (1854–1855) offered accounts of the landscape and its people and natural history. Hooker comments on the plant families and their distribution with an especially keen eye. In *Natural Selection* (Stauffer 1975, p. 549), Darwin related Hooker's account of the vegetation found at the base of the Himalayas, "where true Tropical form are mingled with such northern forms as Birches, Maples, whortle-berries, strawberries, &c."

homes; but the forms, chiefly northern, which had crossed the equator, would travel still further from their homes into the more temperate latitudes of the opposite hemisphere. Although we have reason to believe from geological evidence that the whole body of arctic shells underwent scarcely any modification during their long southern migration and re-migration northward, the case may have been wholly different with those intruding forms which settled themselves on the intertropical mountains, and in the southern hemisphere. These being surrounded by strangers will have had to compete with many new forms of life; and it is probable that selected modifications in their structure, habits, and constitutions will have profited them. Thus many of these wanderers, though still plainly related by inheritance to their brethren of the northern or southern hemispheres, now exist in their new homes as well-marked varieties or as distinct species.

It is a remarkable fact, strongly insisted on by Hooker in regard to America, and by Alph. de Candolle in regard to Australia, that many more identical plants and allied forms have apparently migrated from the north to the south, than in a reversed direction. We see, however, a few southern vegetable forms on the mountains of Borneo and Abyssinia. I suspect that this preponderant migration from north to south is due to the greater extent of land in the north, and to the northern forms having existed in their own homes in greater numbers, and having consequently been advanced through natural selection and competition to a higher stage of perfection or dominating power, than the southern forms. And thus, when they became commingled during the Glacial period, the northern forms were enabled to beat the less powerful southern forms. Just in the same manner as we see at the present day,

1 This is a key concluding statement, to which the argument of the previous several pages has been headed.

2 Hooker observed in the introduction to *Flora of New Zealand* that a far greater number of northern species are represented in the south than are southern species in the north; Alphonse de Candolle made the same point in 1855 in his important work *Géographie Botanique Raisonnée.* Darwin called this a "curious difficulty" and a "singular fact" in *Natural Selection* (Stauffer 1975, pp. 558, 559). His explanation is instructive: he relates this finding back to land area, population size, and competitive ability (perhaps borrowing from de Candolle, who, he pointed out, reported that species in the Russian Empire have larger ranges than their relatives from the same botanical families at the southern tip of Africa). Recall Darwin's idea that large population sizes lead to greater variability, which in turn leads to greater responsiveness to selection pressures. Such populations thus become better and better adapted, while small populations languish. The greater land area of the northern hemisphere is supposed to lead to larger species ranges. Species from areas with larger ranges are thus imagined to be better competitors than those from limited land areas, so when the two come together and compete, the northern species tend to displace the southern rather than vice versa.

Hooker wrote a famous review of the *Géographie Botanique* in which he defended transmutation, an idea de Candolle vigorously opposed. By then Hooker had been let in on Darwin's thinking. Though he was skeptical about the theory, he acknowledged that it was an intriguing hypothesis worthy of consideration. Darwin commented to Hooker: "I have read half your Review & like it very much. D.C. ought to be very much pleased; but I suppose the sugar is at the top & the sour at the bottom" (*Correspondence* 6: 203).

1 This passage, too, gives insight into Darwin's thinking. Islands also have limited land areas; recall from chapter IV (see pp. 81, 105–107) his explanation for why species on islands are so easily outcompeted and displaced by invaders from continental areas: small land area means smaller population, less variability, and poorer competitiveness. His explanation for why we see northern-affinity species on the mountains of the tropics turns on precisely this point. "A mountain is an island on the land," and so the endemic alpine species of the tropical mountains, limited in land area and so poor competitors, have been displaced by northern-affinity rather than southern-affinity species.

that very many European productions cover the ground in La Plata, and in a lesser degree in Australia, and have to a certain extent beaten the natives; whereas extremely few southern forms have become naturalised in any part of Europe, though hides, wool, and other objects likely to carry seeds have been largely imported into Europe during the last two or three centuries from La Plata, and during the last thirty or forty years from Australia. Something of the same kind must have occurred on the intertropical mountains: no doubt before the Glacial period they were stocked with endemic Alpine forms; but these have almost everywhere largely yielded to the more dominant forms, generated in the larger areas and more efficient workshops of the north. In many islands the native productions are nearly equalled or even outnumbered by the naturalised; and if the natives have not been actually exterminated, their numbers have been greatly reduced, and this is the first stage towards extinction. A mountain is an island on the land; and the intertropical mountains before the Glacial period must have been completely isolated; and I believe that the productions of these islands on the land yielded to those produced within the larger areas of the north, just in the same way as the productions of real islands have everywhere lately yielded to continental forms, naturalised by man's agency.

I am far from supposing that all difficulties are removed on the view here given in regard to the range and affinities of the allied species which live in the northern and southern temperate zones and on the mountains of the intertropical regions. Very many difficulties remain to be solved. I do not pretend to indicate the exact lines and means of migration, or the reason why certain species and not others have migrated;

why certain species have been modified and have given rise to new groups of forms, and others have remained unaltered. We cannot hope to explain such facts, until we can say why one species and not another becomes naturalised by man's agency in a foreign land; why one ranges twice or thrice as far, and is twice or thrice as common, as another species within their own homes.

I have said that many difficulties remain to be solved: some of the most remarkable are stated with admirable clearness by Dr. Hooker in his botanical works on the antarctic regions. These cannot be here discussed. I will only say that as far as regards the occurrence of identical species at points so enormously remote as Kerguelen Land, New Zealand, and Fuegia, I believe that towards the close of the Glacial period, icebergs, as suggested by Lyell, have been largely concerned in their dispersal. But the existence of several quite distinct species, belonging to genera exclusively confined to the south, at these and other distant points of the southern hemisphere, is, on my theory of descent with modification, a far more remarkable case of difficulty. For some of these species are so distinct, that we cannot suppose that there has been time since the commencement of the Glacial period for their migration, and for their subsequent modification to the necessary degree. The facts seem to me to indicate that peculiar and very distinct species have migrated in radi ating lines from some common centre; and I am inclined to look in the southern, as in the northern hemisphere, to a former and warmer period, before the commencement of the Glacial period, when the antarctic lands, now covered with ice, supported a highly peculiar and isolated flora. I suspect that before this flora was exterminated by the Glacial epoch, a few forms were

1 Botanical geography of the southern hemisphere was given its first detailed treatment by Hooker in *Flora Antarctica* (1844–1847). Two facts about plant distribution in the far south had to be explained. On the one hand, identical species exist in widely separated areas, and on the other, sets of species—even genera—in these same areas are exclusively southern. The regions Darwin mentions are circum-Antarctic: Tierra del Fuego at the tip of South America, New Zealand, and Kerguelen Land (a large volcanic island in the south Indian Ocean, discovered in 1772 by the French explorer Yves Joseph de Kerguélen-Trémarec, who named it Desolation Island).

Darwin repeats his reliance on icebergs to explain the first observation, whereas Hooker was an adherent of the idea that an extensive land mass linked South America, Australia, New Zealand, and several islands in the southern oceans. The second observation is a real puzzle. Why should a sizable number of plant species, genera, and even families be found exclusively in the southern hemisphere, with no affinity to northern forms? This suggests transmutation over vast time periods, yet the north-to-south plant migration during the glacial epoch that Darwin proposed in this chapter is a comparatively recent phenomenon, so the flora would not have nearly enough time post–glacial period to become so strikingly differentiated. In *Natural Selection* Darwin refers to this mystery as the most "extraordinary [case] as yet known" concerning the distribution of plants in the southern hemisphere (Stauffer 1975, p. 560).

2 To explain the unique southern flora, Darwin posited an ancient warm period in which forests covered the Antarctic region. The onset of the glacial period would have exterminated much of this flora, but vestiges of it are seen in the circum-Antarctic distribution: "According to all analogy," he wrote, "this Antarctic vegetation from its isolation would have been very peculiar, but would have been in some degree related to that of the two nearest continents, America and Australia" (Stauffer 1975, p. 579). It would have delighted Darwin to learn that Antarctica once supported an extensive flora and fauna related to those of South America and Australia (Riffenburgh 2007). The explanation, *(continued)* widely dispersed to various points of the southern hemisphere by occasional means of transport, and by the aid, as halting-places, of existing and now sunken islands, and perhaps at the commencement of the Glacial period, by icebergs. By these means, as I believe, the southern shores of America, Australia, New Zealand have become slightly tinted by the same peculiar forms of vegetable life.

1 Sir C. Lyell in a striking passage has speculated, in language almost identical with mine, on the effects of great alternations of climate on geographical distribution. I believe that the world has recently felt one of his great cycles of change; and that on this view, combined with modification through natural selection, a multitude of facts in the present distribution both of the same and of allied forms of life can be ex-
2 plained. The living waters may be said to have flowed during one short period from the north and from the south, and to have crossed at the equator; but to have flowed with greater force from the north so as to have freely inundated the south. As the tide leaves its drift in horizontal lines, though rising higher on the shores where the tide rises highest, so have the living waters left their living drift on our mountain-summits, in a line gently rising from the arctic lowlands to a great height under the equator. The various beings thus left stranded may be compared with savage races of man, driven up and surviving in the mountain-fastnesses of almost every land, which serve as a record, full of interest to us, of the former inhabitants of the surrounding lowlands.

(continued)
however, would have amazed him: these continents were once united as part of the Pangean supercontinent. The "continental extension" school Darwin railed against was a bit closer to the truth than were the strict dispersalists. In 1851 Hooker told Darwin, "I am becoming slowly more convinced of the probability of the southern flora being a fragmentary one—all that remains of a great Southern Continent" (*Correspondence* 5: 67). Hooker envisioned a continent spanning the south polar region, one that upon its disintegration and subsidence left fragments of land scattered throughout the region. In modern understanding, there was no land extension linking the continents of the southern hemisphere; rather, the continents themselves were adjoined and through the mechanism of plate tectonics gradually moved apart.

1 The passage Darwin mentions here may come from chapter 43 of the eighth (1850) edition of *Principles of Geology;* note Lyell's reference to the "great cycle of climate":

> It will follow . . . that as often as the climates of the globe are passing from the extreme of heat to that of cold—from the summer to the winter of the great year . . . the migratory movement [of species] will be directed constantly from the poles towards the equator . . . But when, on the contrary, a series of changes in the physical geography of the globe, or any other supposed cause, occasions an elevation of the general temperature,—when there is a passage from the winter to one of the vernal or summer seasons of the great cycle of climate,—then the order of the migratory movement is inverted. The different species of animals and plants direct their course from the equator towards the poles.

2 This is Darwin at his lyrical best. The image of an ebb and flow of whole communities of species in response to climatic changes captures his grand scenario for worldwide species distributions, especially the anomalous northern-affinity species "stranded" in the southern hemisphere and atop the loftiest mountains. Darwin's analogy with the "savage races of man" able to hang on only in remote mountain fastnesses is very much of his colonialist time. His analogy makes the point that these remnant populations provide a clue to the past.

CHAPTER XII.

GEOGRAPHICAL DISTRIBUTION—*continued.*

Distribution of fresh-water productions — On the inhabitants of oceanic islands — Absence of Batrachians and of terrestrial Mammals — On the relation of the inhabitants of islands to those of the nearest mainland — On colonisation from the nearest source with subsequent modification — Summary of the last and present chapters.

As lakes and river-systems are separated from each other by barriers of land, it might have been thought that fresh-water productions would not have ranged widely within the same country, and as the sea is apparently a still more impassable barrier, that they never would have extended to distant countries. But the case is exactly the reverse. Not only have many fresh-water species, belonging to quite different classes, an enormous range, but allied species prevail in a remarkable manner throughout the world. I well remember, when first collecting in the fresh waters of Brazil, feeling much surprise at the similarity of the fresh-water insects, shells, &c., and at the dissimilarity of the surrounding terrestrial beings, compared with those of Britain.

But this power in fresh-water productions of ranging widely, though so unexpected, can, I think, in most cases be explained by their having become fitted, in a manner highly useful to them, for short and frequent migrations from pond to pond, or from stream to stream; and liability to wide dispersal would follow from this capacity as an almost necessary consequence. We can here consider only a few cases. In regard to

The previous chapter addressed the apparent anomalies of geographical dispersion, offering scenarios of migration during ice ages and mechanisms of long-distance dispersal in answer to the difficulties such peculiarities present. Much of this chapter, in contrast, presents biogeographical evidence in support of common descent. Patterns of relationship among island flora and fauna with respect to species of the nearest mainland, as well as the peculiarities of island species themselves, are all strongly suggestive of transmutation in splendid isolation. As early as 1836, musing on the distribution and relationships of the curious birds and tortoises of the Galápagos, Darwin noted in his diary that "the zoology of Archipelagoes will be well worth examining; for such facts [would] undermine the stability of Species" (see Barlow 1963).

Bear in mind that Darwin's thinking about the importance of spatial isolation changed in important ways over time. In the transmutation notebooks of the late 1830s through the *Essay* of 1844, he placed great emphasis on geographical separation in the formation of new species, but his discovery of the principle of divergence (see chapter IV) in 1852 led him to place increasing emphasis on ecological and what he termed "partial" isolation—forms of isolation more compatible with his idea of the importance of large continuous areas. It is sometimes asserted that Darwin abandoned geographical isolation as a factor in speciation, but he did not so much drop isolation as refine it by supplementing ecological with geographical barriers. He always maintained that complete isolation was necessary for the development of reproductive incompatibility, which is the point of no return with regard to species distinctness (Sulloway 1979). Still, it is true that spatial isolation, which is considered essential today, took on secondary importance in Darwin's thinking.

1 This similarity may be superficial, reflecting the general morphological conservatism of these invertebrate groups.

1 Another mechanism to consider is stream capture, as drainages evolve by erosion.

2 "Inosculation" is a term coined in connection with plants, referring to the merging or growing together of different individuals. Here Darwin uses the term in a similar vein to describe the merging of river systems. Elsewhere, however, he and others in his time used the word in a different sense. William Sharp Macleay referred to "inosculating" (merging) species in describing his ill-fated Quinarian System of classification. Darwin sometimes used "inosculate" in the sense of one species changing into another or passing from one form to another.

fish, I believe that the same species never occur in the fresh waters of distant continents. But on the same continent the species often range widely and almost capriciously; for two river-systems will have some fish in common and some different. A few facts seem to favour the possibility of their occasional transport by accidental means; like that of the live fish not rarely dropped by whirlwinds in India, and the vitality of their ova when removed from the water. But I am inclined to attribute the dispersal of fresh-water fish mainly to slight changes within the recent period in the level of the land, having caused rivers to flow into each other. Instances, also, could be given of this having occurred during floods, without any change of level. We have evidence in the loess of the Rhine of considerable changes of level in the land within a very recent geological period, and when the surface was peopled by existing land and fresh-water shells. The wide difference of the fish on opposite sides of continuous mountain-ranges, which from an early period must have parted river-systems and completely prevented their inosculation, seems to lead to this same conclusion. With respect to allied fresh-water fish occurring at very distant points of the world, no doubt there are many cases which cannot at present be explained: but some fresh-water fish belong to very ancient forms, and in such cases there will have been ample time for great geographical changes, and consequently time and means for much migration. In the second place, salt-water fish can with care be slowly accustomed to live in fresh water; and, according to Valenciennes, there is hardly a single group of fishes confined exclusively to fresh water, so that we may imagine that a marine member of a fresh-water group might travel far along the shores of the sea, and subsequently become modified and adapted to the fresh waters of a distant land.

Some species of fresh-water shells have a very wide range, and allied species, which, on my theory, are descended from a common parent and must have proceeded from a single source, prevail throughout the world. Their distribution at first perplexed me much, as their ova are not likely to be transported by birds, and they are immediately killed by sea water, as are the adults. I could not even understand how some naturalised species have rapidly spread throughout the same country. But two facts, which I have observed—and no doubt many others remain to be observed—throw some light on this subject. When a duck suddenly emerges from a pond covered with duck-weed, I have twice seen these little plants adhering to its back; and it has happened to me, in removing a little duck-weed from one aquarium to another, that I have quite unintentionally stocked the one with fresh-water shells from the other. But another agency is perhaps more effectual: I suspended a duck's feet, which might represent those of a bird sleeping in a natural pond, in an aquarium, where many ova of fresh-water shells were hatching; and I found that numbers of the extremely minute and just hatched shells crawled on the feet, and clung to them so firmly that when taken out of the water they could not be jarred off, though at a somewhat more advanced age they would voluntarily drop off. These just hatched molluscs, though aquatic in their nature, survived on the duck's feet, in damp air, from twelve to twenty hours; and in this length of time a duck or heron might fly at least six or seven hundred miles, and would be sure to alight on a pool or rivulet, if blown across sea to an oceanic island or to any other distant point. Sir Charles Lyell also

1 This is another of Darwin's celebrated "little experiments." Since 1855 he had been investigating means of dispersal and entreating friends and colleagues to try various experiments as well. In 1857 he wrote to the zoologist Philip Henry Gosse:

> I have thought that perhaps in course of summer you would have an opportunity & would be so very kind as to try a *little* experiment for me.—I think I can tell best what I want, by telling what I have done. The wide distribution of same species of F. Water Molluscs has long been a great perplexity to me: I have just lately hatched a lot & it occurred to me that when first born they might perhaps have not acquired phytophagous habits, & might perhaps like nibbling at a Ducks-foot . . . I found when there were many very young Molluscs in a small vessel with aquatic plants, amongst which I placed a dried Ducks foot, that the little barely visible shells often crawled over it, & that they adhered so firmly that they [could] not be shaken off, & that the foot being kept out of water in a damp atmosphere, the little Molluscs survived well 10, 12 & 15 hours & a few even 24 hours.—And thus, I believe, it must be that [fresh water] shells get from pond to pond & even to islands out at sea . . . A Heron fishing for instance, & then startled might well on a rainy day carry a young mollusc for a long distance.—Now what I want to beg of you, is, that you would try an analogous experiment with some sea-molluscs. (*Correspondence* 6: 382)

1 informs me that a Dyticus has been caught with an Ancylus (a fresh-water shell like a limpet) firmly adhering to it; and a water-beetle of the same family, a Colymbetes, once flew on board the 'Beagle,' when forty-five miles distant from the nearest land: how much farther it might have flown with a favouring gale no one can tell.

With respect to plants, it has long been known what enormous ranges many fresh-water and even marsh-species have, both over continents and to the most remote oceanic islands. This is strikingly shown, as remarked by Alph. de Candolle, in large groups of terrestrial plants, which have only a very few aquatic members; for these latter seem immediately to acquire, as if in consequence, a very wide range. I think favour-
2 able means of dispersal explain this fact. I have before mentioned that earth occasionally, though rarely, adheres in some quantity to the feet and beaks of birds. Wading birds, which frequent the muddy edges of ponds, if suddenly flushed, would be the most likely to have muddy feet. Birds of this order I can show are the greatest wanderers, and are occasionally found on the most remote and barren islands in the open ocean; they would not be likely to alight on the surface of the sea, so that the dirt would not be washed off their feet; when making land, they would be sure to fly to their
3 natural fresh-water haunts. I do not believe that botanists are aware how charged the mud of ponds is with seeds: I have tried several little experiments, but will here give only the most striking case: I took in February three table-spoonfuls of mud from three different points, beneath water, on the edge of a little pond; this mud when dry weighed only $6\frac{3}{4}$ ounces; I kept it covered up in my study for six months, pulling up and counting each plant as it grew; the plants were

1 Darwin kept records of odd long-distance dispersal events, and his friends readily obliged with reports of interesting cases. Lyell's 1856 letter to Darwin about the *Hydrobius* water beetle with its *Ancylus* snail passenger is quoted on page 356. Darwin replied: "Your cases of possible transportal beat all that I have ever heard of; & if any body had put such cases hypothetically I [should] have laughed at them. I have known *Colymbetes* fly on board Beagle 45 miles from land, which, by the way, surprised Wollaston much" (*Correspondence* 6: 99, italics added). *Colymbetes* and *Dyticus* (properly spelled *Dytiscus*) are genera of predacious diving beetles, family Dytiscidae.

2 Darwin reported to Lyell in May 1856 that he found twenty-nine species of plant in a tablespoon of mud from a pond. "Hooker was surprised at this, & struck with it, when I showed him how much mud I had scraped off one Duck's feet," he wrote (*Correspondence* 6: 99).

3 Darwin tried this experiment again in 1857 with mud from different ponds. In the case reported here, he initially collected mud on February 10 from a pond along the road to Westerham, about five miles south of Downe. He then kept a running tally of the seedlings that germinated from that sample. By the time of his final entry on August 1, the mud had yielded a remarkable 537 seedlings.

of many kinds, and were altogether 537 in number; and yet the viscid mud was all contained in a breakfast cup! Considering these facts, I think it would be an inexplicable circumstance if water-birds did not transport the seeds of fresh-water plants to vast distances, and if consequently the range of these plants was not very great. The same agency may have come into play with the eggs of some of the smaller fresh-water animals.

Other and unknown agencies probably have also played a part. I have stated that fresh-water fish eat some kinds of seeds, though they reject many other kinds after having swallowed them; even small fish swallow seeds of moderate size, as of the yellow water-lily and Potamogeton. Herons and other birds, century after century, have gone on daily devouring fish; they then take flight and go to other waters, or are blown across the sea; and we have seen that seeds retain their power of germination, when rejected in pellets or in excrement, many hours afterwards. When I saw the [1] great size of the seeds of that fine water-lily, the Nelumbium, and remembered Alph. de Candolle's remarks on this plant, I thought that its distribution must remain quite inexplicable; but Audubon states that he found the seeds of the great southern water-lily (probably, according to Dr. Hooker, the Nelumbium luteum) in a heron's stomach; although I do not know the fact, yet analogy makes me believe that a heron flying to another pond and getting a hearty meal of fish, would probably reject from its stomach a pellet containing the seeds of the Nelumbium undigested; or the seeds might be dropped by the bird whilst feeding its young, in the same way as fish are known sometimes to be dropped.

In considering these several means of distribution,

1 The Old World water lily referred to here was formerly called *Nelumbium speciosum*, now synonymized with *Nelumbo nucifera*, the sacred lotus of legend. The species name *nucifera* refers to the large seeds of this lily (from the Latin *nuci*, "nut"), which are borne in a distinctive flat-topped, podlike receptacle. This is the sacred "bean" of ancient India and Egypt; it was this bean and not the leguminous kind that Pythagoras forswore, but in any case the pod is used more often for floral arrangements than for consumption these days. Audubon's observation refers to the eating habits of the Great Blue Heron (*Ardea herodias*, Ardeidae), an account of which appears in his *Ornithological Biography* (4 volumes, 1831–1839), the text written to accompany the *Birds of America* paintings (see Irmscher 1999, p. 389).

it should be remembered that when a pond or stream is first formed, for instance, on a rising islet, it will be unoccupied; and a single seed or egg will have a good chance of succeeding. Although there will always be a struggle for life between the individuals of the species, however few, already occupying any pond, yet as the number of kinds is small, compared with those on the land, the competition will probably be less severe between aquatic than between terrestrial species; consequently an intruder from the waters of a foreign country, would have a better chance of seizing on a place, than in the case of terrestrial colonists. We should, also, remember that some, perhaps many, fresh-water productions are low in the scale of nature, and that we have reason to believe that such low beings change or become modified less quickly than the high; and this will give longer time than the average for the migration of the same aquatic species. We should not forget the probability of many species having formerly ranged as continuously as fresh-water productions ever can range, over immense areas, and having subsequently become extinct in intermediate regions. But the wide distribution of fresh-water plants and of the lower animals, whether retaining the same identical form or in some degree modified, I believe mainly depends on the wide dispersal of their seeds and eggs by animals, more especially by fresh-water birds, which have large powers of flight, and naturally travel from one to another and often distant piece of water. Nature, like a careful gardener, thus takes her seeds from a bed of a particular nature, and drops them in another equally well fitted for them.

On the Inhabitants of Oceanic Islands.—We now come to the last of the three classes of facts, which I

1 This idea is probably based on the fossil records of invertebrates vs. vertebrates, in which invertebrate forms do not appear to change at the same rates as vertebrate forms in a given time period. There is no reason, genetically speaking, that invertebrates should evolve at a slower rate, and the pattern may be artifactual—shelly or wormy morphologies are superficially similar, masking actual species differences over time.

2 The remainder of the chapter is dedicated to oceanic islands and their instructive flora and fauna. The third class of biogeograpical facts was introduced on p. 349: species show an affinity (relationship) in the same continent or sea yet differ in species identity at particular locales within these areas. Nowhere is this more striking than on island systems, small worlds unto themselves that reflect at a local level the large-scale patterns of species affinity found at the global level.

have selected as presenting the greatest amount of difficulty, on the view that all the individuals both of the same and of allied species have descended from a single parent; and therefore have all proceeded from a common birthplace, notwithstanding that in the course of time they have come to inhabit distant points of the globe. I have already stated that I cannot honestly admit Forbes's view on continental extensions, which, if legitimately followed out, would lead to the belief that within the recent period all existing islands have been nearly or quite joined to some continent. This view would remove many difficulties, but it would not, I think, explain all the facts in regard to insular productions. In the following remarks I shall not confine ◀1 myself to the mere question of dispersal; but shall consider some other facts, which bear on the truth of the two theories of independent creation and of descent with modification.

The species of all kinds which inhabit oceanic islands ◀2 are few in number compared with those on equal continental areas: Alph. de Candolle admits this for plants, and Wollaston for insects. If we look to the large size and varied stations of New Zealand, extending over 780 miles of latitude, and compare its flowering plants, only 750 in number, with those on an equal area at the Cape of Good Hope or in Australia, we must, I think, admit that something quite independently of any difference in physical conditions has caused so great a difference in number. Even the uniform county of Cambridge has 847 plants, and the little island of Anglesea 764, but a few ferns and a few introduced plants are included in these numbers, and the comparison in some other respects is not quite fair. We have evidence that the barren island of Ascension aboriginally possessed under half-a-dozen flowering

1 The casual tone belies the profound impact that the observations reported on subsequent pages had on the development of Darwin's thinking—particularly relating to endemism and affinity of island species.

2 The occurrence of lower species numbers on islands relative to mainland areas has since been recognized as a key fact in island biogeography, one with important implications for conservation biology. The species-area relationship, as it is now termed, was first formally treated in the twentieth century by Olof Arhennius (1921) and Henry Allan Gleason (1922), and then by Frank Preston in the development of his "canonical distribution" of species numbers as a function of area (Preston 1962). Philip Darlington (1957) used Caribbean lizard data to show that a tenfold increase in island area results in a twofold increase in species richness. The implications of the species-area relationship for conservation are obvious: habitat loss and fragmentation make islands of formerly continuous habitat, so fewer species supported in insularized areas means steady species loss.

1 Islands have proportionally fewer species than comparable mainland areas, yet they could support more, or at least different and better-adapted, species.

2 The disproportionate number of endemic species on islands provided a clue to Darwin years before, and it was island-specific endemism that converted him to transmutation. Isolation was centrally important to the production of new species in Darwin's early view, but by the time he wrote the *Origin,* Darwin argued that large continental areas were most important overall in species production. This was because large areas supported greater population densities, leading to the variations and intense competition that drive divergence of character. Yet Darwin acknowledged that islands could promote the formation of new species on a limited, local scale by presenting immigrants with new competitors and niche opportunities while keeping out a significant number of potentially better-adapted competitors. Thus isolated areas become tenanted with an especially high proportion of unique species. As he put it in *Natural Selection:* "I do not doubt that over the world far more species have been produced in continuous than in isolated areas. But I believe that *in relation to the area* far more species have been manufactured in, for instance, isolated islands than in continuous mainland" (Stauffer 1975, p. 254, emphasis added).

3 This is an important observation. Given the chance element of colonizing remote islands, it makes sense that each island system would be unique in its particular combination of taxa.

plants; yet many have become naturalised on it, as they have on New Zealand and on every other oceanic island which can be named. In St. Helena there is reason to believe that the naturalised plants and animals have nearly or quite exterminated many [1] native productions. He who admits the doctrine of the creation of each separate species, will have to admit, that a sufficient number of the best adapted plants and animals have not been created on oceanic islands; for man has unintentionally stocked them from various sources far more fully and perfectly than has nature.

[2] Although in oceanic islands the number of kinds of inhabitants is scanty, the proportion of endemic species (*i. e.* those found nowhere else in the world) is often extremely large. If we compare, for instance, the number of the endemic land-shells in Madeira, or of the endemic birds in the Galapagos Archipelago, with the number found on any continent, and then compare the area of the islands with that of the continent, we shall see that this is true. This fact might have been expected on my theory, for, as already explained, species occasionally arriving after long intervals in a new and isolated district, and having to compete with new associates, will be eminently liable to modification, and will often produce groups of modified descendants. But [3] it by no means follows, that, because in an island nearly all the species of one class are peculiar, those of another class, or of another section of the same class, are peculiar; and this difference seems to depend on the species which do not become modified having immigrated with facility and in a body, so that their mutual relations have not been much disturbed. Thus in the Galapagos Islands nearly every land-bird, but only two out of the eleven marine birds, are peculiar; and it is obvious that

marine birds could arrive at these islands more easily than land-birds. Bermuda, on the other hand, which lies at about the same distance from North America as the Galapagos Islands do from South America, and which has a very peculiar soil, does not possess one endemic land bird; and we know from Mr. J. M. Jones's admirable account of Bermuda, that very many North American birds, during their great annual migrations, visit either periodically or occasionally this island. Madeira does not possess one peculiar bird, and many European and African birds are almost every year blown there, as I am informed by Mr. E. V. Harcourt. So that these two islands of Bermuda and Madeira have been stocked by birds, which for long ages have struggled together in their former homes, and have become mutually adapted to each other; and when settled in their new homes, each kind will have been kept by the others to their proper places and habits, and will consequently have been little liable to modification. Madeira, again, is inhabited by a wonderful number of peculiar land-shells, whereas not one species of sea-shell is confined to its shores: now, though we do not know how sea-shells are dispersed, yet we can see that their eggs or larvæ, perhaps attached to seaweed or floating timber, or to the feet of wading-birds, might be transported far more easily than land-shells, across three or four hundred miles of open sea. The different orders of insects in Madeira apparently present analogous facts.

Oceanic islands are sometimes deficient in certain classes, and their places are apparently occupied by the other inhabitants; in the Galapagos Islands reptiles, and in New Zealand gigantic wingless birds, take the place of mammals. In the plants of the Galapagos Islands, Dr. Hooker has shown that the proportional numbers of the different orders are very different from

1 Bermuda is volcanic in origin, an igneous seamount capped with coralline limestone and aeolian sandstone (sandstone formed by the consolidation of limestone-derived sand blown into dunes). The weathering of these rocks has produced an alkaline iron-rich red soil that is nutrient-poor despite the limestone. Bermuda boasts at least one endemic bird, the Bermuda petrel or cahow *(Pterodroma cahow)*, and an endemic subspecies of vireo: the Bermuda white-eyed vireo *(Vireo griseus bermudianus)*. Darwin's point is well taken, however, considering that some 360 birds have been recorded from the island. The low avian endemism is due to constant immigration by prevailing winds. The "admirable account" Darwin references is probably John Matthew Jones's work *The Naturalist in Bermuda* (London, 1859).

2 Edward William Vernon Harcourt published a guide to the Madeira Archipelago in 1851 entitled *A Sketch of Madeira*, and in 1855 he published a paper on Madeiran birds. Darwin corresponded with Harcourt in 1856 about the avifauna of the archipelago.

3 This echoes a point made on the previous page. The absence of some groups and over-representation of others from island group to island group—particularly with respect to the makeup and relative proportions of taxa found on the nearest mainland—is called "disharmony" by biogeographers. The disharmonic nature of island biota puzzled the early naturalists: why create a wholly unique complement of species in odd proportions on each island archipelago? The Hawaiian Islands, for example, have hundreds of fruitfly and long-horned beetle species, yet only two native butterflies. On the other hand, innumerable insect taxa are missing altogether.

what they are elsewhere. Such cases are generally accounted for by the physical conditions of the islands; but this explanation seems to me not a little doubtful. Facility of immigration, I believe, has been at least as important as the nature of the conditions.

1 Many remarkable little facts could be given with respect to the inhabitants of remote islands. For instance, in certain islands not tenanted by mammals, some of the endemic plants have beautifully hooked seeds; yet few relations are more striking than the adaptation of hooked seeds for transportal by the wool and fur of quadrupeds. 2 This case presents no difficulty on my view, for a hooked seed might be transported to an island by some other means; and the plant then becoming slightly modified, but still retaining its hooked seeds, would form an endemic species, having as useless an appendage as any rudimentary organ,—for instance, as the shrivelled wings under the soldered elytra of many insular beetles. 3 Again, islands often possess trees or bushes belonging to orders which elsewhere include only herbaceous species; now trees, as Alph. de Candolle has shown, generally have, whatever the cause may be, confined ranges. Hence trees would be little likely to reach distant oceanic islands; and an herbaceous plant, though it would have no chance of successfully competing in stature with a fully developed tree, when established on an island and having to compete with herbaceous plants alone, might readily gain an advantage by growing taller and taller and overtopping the other plants. If so, natural selection would often tend to add to the stature of herbaceous plants when growing on an island, to whatever order they belonged, and thus convert them first into bushes and ultimately into trees.

4 With respect to the absence of whole orders on

1 Darwin benefited not only from his own travels around the world but also from the legacy of centuries of European exploration. Naturalists had long marveled at "remarkable little facts" such as these. Individually they might be dismissed as curiosities or simply evidence of the mysteries of the creator, but collectively they make a powerful case for transmutation, for island species having descended from mainland relatives. Note Darwin's rhetorical approach as he goes through these examples. He repeatedly points out that the observations are consistent with his theory but are inexplicable under the supposition of special creation (except, of course, in unsatisfying terms such as "because the creator willed it so").

2 Vestigial structures reflect ancestry. Seed hooks are dispersal devices that adhere to the fur of passing mammals. This feature is an oddity in an environment with no mammals but bats.

3 One of the marvels of island systems is that plant groups that are usually herbaceous become arborescent, or tree-like. The Hawaiian Islands, for example, are home to a diversity of spectacular bird-pollinated arborescent lobelias (*Trematolobelia, Cyanea,* and others; Mabberley 1975), as well as shrubby violets in the Moloka'i highlands related to an herbaceous amphi-Beringian species, *Viola langsdorfii* (Ballard and Sytsma 2000). Darwin cites de Candolle to make the point that trees are less likely than herbs to be dispersed over the ocean (presumably owing to seed number and size—think dandelion versus oak tree seeds), yet on islands we find many tree-sized representatives of groups consisting mostly of herbaceous species. This could be the botanical equivalent of the gigantism seen in island forms of many ordinarily small-framed mammals.

4 Amphibians are especially sensitive to desiccation and salt water, and thus should be among the least likely organisms to survive oversea dispersal. Yet genetic evidence shows that some batrachians survived such unlikely journeys. Analysis of two mantellid frogs from Mayotte, in the Comoros Archipelago west of Madagascar, suggests two colonizations (Vences et al. 2003).

(continued)

oceanic islands, Bory St. Vincent long ago remarked that Batrachians (frogs, toads, newts) have never been found on any of the many islands with which the great oceans are studded. I have taken pains to verify this assertion, and I have found it strictly true. I have, however, been assured that a frog exists on the mountains of the great island of New Zealand; but I suspect that this exception (if the information be correct) may be explained through glacial agency. This general absence of frogs, toads, and newts on so many oceanic islands cannot be accounted for by their physical conditions; indeed it seems that islands are peculiarly well fitted for these animals; for frogs have been introduced into Madeira, the Azores, and Mauritius, and have multiplied so as to become a nuisance. But as these animals and their spawn are known to be immediately killed by sea-water, on my view we can see that there would be great difficulty in their transportal across the sea, and therefore why they do not exist on any oceanic island. But why, on the theory of creation, they should not have been created there, it would be very difficult to explain.

Mammals offer another and similar case. I have carefully searched the oldest voyages, but have not finished my search; as yet I have not found a single instance, free from doubt, of a terrestrial mammal (excluding domesticated animals kept by the natives) inhabiting an island situated above 300 miles from a continent or great continental island; and many islands situated at a much less distance are equally barren. The Falkland Islands, which are inhabited by a wolf-like fox, come nearest to an exception; but this group cannot be considered as oceanic, as it lies on a bank connected with the mainland; moreover, icebergs formerly brought boulders to its western shores, and they may

(continued)
A more recent study shows that eleutherodactyline frogs of Caribbean islands likely arose by ancient oversea dispersal events (Heinicke et al. 2007). Rafting is probably common in hurricane-prone regions (Censky et al. 1998, Calsbeek and Smith 2003). The New Zealand account, by the way, was correct: a rare group of frogs called Pepeketua (*Leiopelma* spp., family Leiopelmatidae) is endemic to New Zealand. They are unusual nocturnal frogs that have lost croak, ears (tympanum), and toe webbing.

1 Some introduced amphibians have become more than nuisances. The infamous cane toad *(Bufo marinus)*, a native of Central and South America, was introduced to many tropical islands such as Hawaii in the 1920s and 1930s as a means of controlling sugarcane pests. This enormous toad reproduces explosively, becoming a noxious pest owing to its voracious appetite and toxic skin. See Lever (2003) for accounts of this and other amphibian introductions.

2 The Falkland Islands fox, or Warrah (*Dusicyon australis*, formerly *Canis antarcticus*), has been extinct since 1876. Darwin anticipated its demise when he visited the Falklands in 1834:

> The only quadruped native to the islands is a large wolf-like fox *(Canis antarcticus)*, which is common to both East and West Falkland. I have no doubt it is a peculiar species, and confined to this archipelago . . . As far as I am aware, there is no other instance . . . of so small a mass of broken land, distant from a continent, possessing so large an aboriginal quadruped peculiar to itself. Their numbers have rapidly decreased. Within a very few years after these islands shall have become regularly settled, in all probability this fox will be classed with the dodo, as an animal which has perished from the face of the earth. (*Voyage*, chapter IX)

have formerly transported foxes, as so frequently now happens in the arctic regions. Yet it cannot be said that small islands will not support small mammals, for they occur in many parts of the world on very small islands, if close to a continent; and hardly an island can be named on which our smaller quadrupeds have not become naturalised and greatly multiplied. It cannot be said, on the ordinary view of creation, that there has not been time for the creation of mammals; many volcanic islands are sufficiently ancient, as shown by the stupendous degradation which they have suffered and by their tertiary strata: there has also been time for the production of endemic species belonging to other classes; and on continents it is thought that mammals appear and disappear at a quicker rate than other and lower animals. Though terrestrial mammals do not occur on oceanic islands, aërial mammals do occur on almost every island. New Zealand possesses two bats found nowhere else in the world: Norfolk Island, the Viti Archipelago, the Bonin Islands, the Caroline and Marianne Archipelagoes, and Mauritius, all possess their peculiar bats. Why, it may be asked, has the supposed creative force produced bats and no other mammals on remote islands? On my view this question can easily be answered; for no terrestrial mammal can be transported across a wide space of sea, but bats can fly across. Bats have been seen wandering by day far over the Atlantic Ocean; and two North American species either regularly or occasionally visit Bermuda, at the distance of 600 miles from the mainland. I hear from Mr. Tomes, who has specially studied this family, that many of the same species have enormous ranges, and are found on continents and on far distant islands. Hence we have only to suppose that such wandering species have been modi-

1 Saying, as Darwin does here, that small mammals are readily naturalized and greatly multiply on remote islands is an understatement. The introduction of such mammals as rats, rabbits, mice, cats, and mongooses to oceanic islands is one of the great ecological tragedies of recent centuries. Their success does underscore Darwin's point that, despite being able to thrive in island environments, these species do not *naturally* occur there.

2 Here is another telling observation that spoke volumes to Darwin: it is curious that the creator chose to stock remote islands with the only mammals capable of flight.

3 Robert Fisher Tomes was a bat enthusiast who published many papers on the vesper or evening bats (Vespertilionidae), the largest and best-known bat family with more than 300 species worldwide.

fied through natural selection in their new homes in relation to their new position, and we can understand the presence of endemic bats on islands, with the absence of all terrestrial mammals.

Besides the absence of terrestrial mammals in relation to the remoteness of islands from continents, there is also a relation, to a certain extent independent of distance, between the depth of the sea separating an island from the neighbouring mainland, and the presence in both of the same mammiferous species or of allied species in a more or less modified condition. Mr. Windsor Earl has made some striking observations on this head in regard to the great Malay Archipelago, which is traversed near Celebes by a space of deep ocean; and this space separates two widely distinct mammalian faunas. On either side the islands are situated on moderately deep submarine banks, and they are inhabited by closely allied or identical quadrupeds. No doubt some few anomalies occur in this great archipelago, and there is much difficulty in forming a judgment in some cases owing to the probable naturalisation of certain mammals through man's agency; but we shall soon have much light thrown on the natural history of this archipelago by the admirable zeal and researches of Mr. Wallace. I have not as yet had time to follow up this subject in all other quarters of the world; but as far as I have gone, the relation generally holds good. We see Britain separated by a shallow channel from Europe, and the mammals are the same on both sides; we meet with analogous facts on many islands separated by similar channels from Australia. The West Indian Islands stand on a deeply submerged bank, nearly 1000 fathoms in depth, and here we find American forms, but the species and even the genera are distinct. As the amount of modification in all cases depends to

1 Darwin is saying that there is a relationship between the depth of the ocean between mainlands and islands and the presence of distinct mammalian species on the islands. (The observation was not new, unsurprisingly. Lyell noted in 1837 that "quadrupeds found on islands situated near the continents generally form a part of the stock of animals belonging to the adjacent mainland"; *Principles*, vol. 3, p. 31.) Islands separated by a shallow sea connect with the mainland during periods of low sea level, undermining isolation and speciation. These are now termed "land-bridge islands." Isolation and divergence characterize populations on islands that rarely or never connect with the mainland, as predicted by transmutation theory. The reference to George Windsor Earl's observations concerns an 1845 paper that seemed to anticipate Wallace (see below). After Darwin read a draft of Wallace's paper on Malay zoogeography he wrote to ask: "Are you aware that Mr W. Earl published several years ago the view of distribution of animals in Malay Archipelago in relation to the depth of the sea between the islands?" (*Correspondence* 7: 323).

2 The biogeography of the Malay Archipelago (peninsular Malaysia, Indonesia, New Guinea, and the Philippines) is now virtually synonymous with Alfred Russel Wallace, who covered 14,000 miles in his eight years exploring and collecting in that vast galaxy of islands. Wallace's important study on the zoogeography of the archipelago (appearing in 1859) gave the first detailed description of the line of demarcation between the Australian and the Indo-Malayan biogeographical provinces, now called the Wallace Line, that Earl had noted for the mammals (Camerini 1993). In 1869 Wallace published his rousing travelogue *The Malay Archipelago*, in which he referred to Earl's paper as providing "a clew to the most radical contrast in the Archipelago." "By following it out in detail," Wallace continued, "I have arrived at the conclusion that we can draw a line among the islands, which shall so divide them that one-half shall truly belong to Asia, while the other shall no less certainly be allied to Australia" (*Malay Archipelago*, chapter I).

a certain degree on the lapse of time, and as during changes of level it is obvious that islands separated by shallow channels are more likely to have been continuously united within a recent period to the mainland than islands separated by deeper channels, we can understand the frequent relation between the depth of the sea and the degree of affinity of the mammalian inhabitants of islands with those of a neighbouring continent,—an inexplicable relation on the view of independent acts of creation.

[1] All the foregoing remarks on the inhabitants of oceanic islands,—namely, the scarcity of kinds—the richness in endemic forms in particular classes or sections of classes,—the absence of whole groups, as of batrachians, and of terrestrial mammals notwithstanding the presence of aërial bats,—the singular proportions of certain orders of plants,—herbaceous forms having been developed into trees, &c.,—seem to me to accord better with the view of occasional means of transport having been largely efficient in the long course of time, than with the view of all our oceanic islands having been formerly connected by continuous land with the nearest continent; for on this latter view the migration would probably have been more complete; and if modification be admitted, all the forms of life would have been more equally modified, in accordance with the paramount importance of the relation of organism to organism.

I do not deny that there are many and grave difficulties in understanding how several of the inhabitants of the more remote islands, whether still retaining the same specific form or modified since their arrival, could have reached their present homes. But the probability of many islands having existed as halting-places, of which not a wreck now remains, must not be over-

1 Darwin, as we have seen, was highly critical of Forbes's continental-extension theory postulating land bridges connecting continents and islands. This does not mean that Darwin rejects land bridges altogether, as we see in the foregoing comments regarding the Malay Archipelago—exposed continental shelves become land bridges. Islands separated from continents by the abyssal sea are another matter, Darwin argued. He is asserting that the wonderful island oddities summarized here are themselves evidence that such islands could never have been connected with mainland areas, for if they had, the wholesale migration of plants and animals would have resulted in a less disharmonic biota (to use a modern term) with far lower levels of endemism.

looked. I will here give a single instance of one of the cases of difficulty. Almost all oceanic islands, even the most isolated and smallest, are inhabited by land-shells, generally by endemic species, but sometimes by species found elsewhere. Dr. Aug. A. Gould has given several interesting cases in regard to the land-shells of the islands of the Pacific. Now it is notorious that land-shells are very easily killed by salt; their eggs, at least such as I have tried, sink in sea-water and are killed by it. Yet there must be, on my view, some unknown, but highly efficient means for their transportal. Would the just-hatched young occasionally crawl on and adhere to the feet of birds roosting on the ground, and thus get transported? It occurred to me that land-shells, when hybernating and having a membranous diaphragm over the mouth of the shell, might be floated in chinks of drifted timber across moderately wide arms of the sea. And I found that several species did in this state withstand uninjured an immersion in sea-water during seven days: one of these shells was the Helix pomatia, and after it had again hybernated I put it in sea-water for twenty days, and it perfectly recovered. As this species has a thick calcareous operculum, I removed it, and when it had formed a new membranous one, I immersed it for fourteen days in sea-water, and it recovered and crawled away: but more experiments are wanted on this head.

The most striking and important fact for us in regard to the inhabitants of islands, is their affinity to those of the nearest mainland, without being actually the same species. Numerous instances could be given of this fact. I will give only one, that of the Galapagos Archipelago, situated under the equator, between 500 and 600 miles from the shores of South America. Here

1 Augustus Addison Gould described the mollusks of the United States Exploring Expedition of 1838–1841, contributing volume XII of the expedition, *Mollusca and Shells* (Philadelphia, 1852).

2 Darwin kept a terrarium of snails—his "snailery," as he called it—for his dispersal experiments. In 1857 he conducted the immersion experiments mentioned here. He noted their progress to family and friends in letters; this comment to William Darwin Fox is typical: "I have just had a *Helix pomatia* withstand 14 days well in Salt-water; to my very great surprise" (*Correspondence* 6: 334).

Recall from p. 385 that Darwin also did experiments on the colonization of the feet of waterfowl by freshwater mollusks: "I have just lately hatched a lot [of snails] & it occurred to me that when first born they might perhaps have not acquired phytophagous habits, & might perhaps like nibbling at a Ducks-foot" (*Correspondence* 6: 382).

Readers of the *Origin* sometimes wrote to Darwin with bits of information they thought might be of interest. Several wrote with accounts of snails and bivalves adhering to the feet of waterfowl, and Darwin summarized these in letters to the journal *Nature* (Darwin 1878, 1882). As a result of these efforts, duck's feet became a well-established means of mollusk dispersal!

3 "Striking and important fact" indeed. The biogeographical patterns of remote islands bearing high proportions of unique species—species that also happen to exhibit a clear relationship to those of the nearest mainland—were revelatory to Darwin.

almost every product of the land and water bears the unmistakeable stamp of the American continent. There are twenty-six land birds, and twenty-five of these are [1] ranked by Mr. Gould as distinct species, supposed to have been created here; yet the close affinity of most of these birds to American species in every character, in their habits, gestures, and tones of voice, was manifest. So it is with the other animals, and with nearly all the plants, as shown by Dr. Hooker in his admirable memoir on the Flora of this archipelago. The naturalist, looking at the inhabitants of these volcanic islands in the Pacific, distant several hundred miles from the continent, yet feels that he is standing on American land. Why should this be so? why should the species which are supposed to have been created in the Galapagos Archipelago, and nowhere else, bear so plain a stamp of affinity to those created in America? There is nothing in the conditions of life, in the geological nature of the islands, in their height or climate, or in the proportions in which the several classes are associated together, which resembles closely the conditions of the South American coast: in fact there is a considerable dissimilarity in all these respects. On [2] the other hand, there is a considerable degree of resemblance in the volcanic nature of the soil, in climate, height, and size of the islands, between the Galapagos and Cape de Verde Archipelagos: but what an entire and absolute difference in their inhabitants! The inhabitants of the Cape de Verde Islands are related to those of Africa, like those of the Galapagos to America. I believe this grand fact can receive no sort of explanation on the ordinary view of independent creation; whereas on the view here maintained, it is obvious that the Galapagos Islands would be likely to receive colonists, whether by occasional means of transport or

1 The ornithologist John Gould described Darwin's bird collections from the Galápagos. Gould alerted Darwin to the curious relationship between the finches of different islands, and the group's relationship to South American birds (Gould 1837a). Today fourteen species in the genera *Geospiza, Camarhynchus, Certhidea,* and *Pinaroloxias* are recognized. The mockingbirds of the genus *Nesomimus* show island-specific endemism to an even greater degree than do the finches. Darwin collected three of the four recognized mockingbird species, which Gould (1837b) named genus *Orpheus* (now *Nesomimus*). Other unique "land-birds" include the Galápagos hawk *(Buteo galapagoensis),* Galápagos screech owl *(Asio flammeus galapagoensis),* large-billed flycatcher *(Myiarchus magnirostris),* and Galápagos dove *(Zenaida galapagoensis),* among others; of the twenty-nine species of land birds living in the Galápagos Islands today, twenty-two are endemic. Joseph Hooker did for the Galápagos plants what Gould had done for the birds. In 1846 he read several papers on the flora of the islands to the Linnean Society, which published them in its *Transactions* and *Proceedings* journals.

2 Darwin had direct experience with both archipelagoes. The Cape Verdes were the first and next-to-last experience Darwin had with tropical islands, as the *Beagle* visited the archipelago on the outbound and inbound legs of the voyage. Outbound the ship first stopped at Tenerife in the Canaries—the setting of stirring accounts by Darwin's hero Alexander von Humboldt—but to everyone's disappointment they were denied permission to land because the island was under quarantine. Eventually they landed at St. Jago, the main island of the Cape Verdes. In his *Beagle* diary Darwin recorded "treading on Volcanic rocks, hearing the notes of unknown birds, & seeing new insects fluttering about still newer flowers. It has been for me a glorious day, like giving to a blind man eyes" (Keynes 2001, p. 23). As Gordon Chancellor writes in his Introduction to Darwin's *Beagle* voyage field notebooks (http://darwin-online.org.uk/), "Darwin had by 1836 realised that the fact that the plants and animals on the Cape Verdes were of an African cast was because that is where

(continued)

by formerly continuous land, from America; and the Cape de Verde Islands from Africa; and that such colonists would be liable to modification;—the principle of inheritance still betraying their original birthplace.

Many analogous facts could be given: indeed it is an almost universal rule that the endemic productions of islands are related to those of the nearest continent, or of other near islands. The exceptions are few, and most of them can be explained. Thus the plants of Kerguelen Land, though standing nearer to Africa than to America, are related, and that very closely, as we know from Dr. Hooker's account, to those of America: but on the view that this island has been mainly stocked by seeds brought with earth and stones on icebergs, drifted by the prevailing currents, this anomaly disappears. New Zealand in its endemic plants is much more closely related to Australia, the nearest mainland, than to any other region: and this is what might have been expected; but it is also plainly related to South America, which, although the next nearest continent, is so enormously remote, that the fact becomes an anomaly. But this difficulty almost disappears on the view that both New Zealand, South America, and other southern lands were long ago partially stocked from a nearly intermediate though distant point, namely from the antarctic islands, when they were clothed with vegetation, before the commencement of the Glacial period. The affinity, which, though feeble, I am assured by Dr. Hooker is real, between the flora of the south-western corner of Australia and of the Cape of Good Hope, is a far more remarkable case, and is at present inexplicable: but this affinity is confined to the plants, and will, I do not doubt, be some day explained.

The law which causes the inhabitants of an archi-

(continued)
they had originated. Any Cape Verde 'centre of creation' was a fiction."

1 The relationship between island-specific species and those found on the nearest mainland had a profound impact on Darwin's thinking. Of the Galápagos he marveled, "The archipelago is a little world within itself, or rather a satellite attached to America" (*Voyage*, chapter 17). In one of his Galápagos field notebooks, he mused, "The Thenca [mockingbirds] very tame & curious in these [islands]. I certainly recognise S. America in ornithology, would a botanist?" (Darwin 1835). Later, from Australia, Darwin notes in a letter to Henslow: "I last wrote to you from Lima, since which time I have done disgracefully little in Nat. History; or rather I should say since the Galapagos Islands, where I worked hard.—Amongst other things, I collected every plant, which I could see in flower, & as it was the flowering season I hope my collection may be of some interest to you.—*I shall be very curious to know whether the Flora belongs to America, or is peculiar. I paid also much attention to the Birds, which I suspect are very curious*" (*Correspondence* 1: 485, emphasis added). Remember that Darwin fully embraced transmutation only in the months following his arrival home (Sulloway 1982a), but here we see an idea beginning to take form.

2 This biogeographical mystery has been solved. The botanical affinities of Australia, southern Africa, and southern South America are understood in part by continental drift. It is an interesting twist of history that southern hemisphere biogeography led Hooker to adhere to the continental-extension theory of Forbes (much to Darwin's chagrin), yet though that theory was discredited, Pangaea solves the problem in much the same way.

pelago, though specifically distinct, to be closely allied to those of the nearest continent, we sometimes see displayed on a small scale, yet in a most interesting manner, within the limits of the same archipelago.
1 Thus the several islands of the Galapagos Archipelago are tenanted, as I have elsewhere shown, in a quite marvellous manner, by very closely related species; so that the inhabitants of each separate island, though mostly distinct, are related in an incomparably closer degree to each other than to the inhabitants of any other part of the world. And this is just what might have been expected on my view, for the islands are situated so near each other that they would almost certainly receive immigrants from the same original source, or from each other. But this dissimilarity between the endemic inhabitants of the islands may be used as an argument against my views; for it may be asked, how has it happened in the several islands situated within sight of each other, having the same geological nature, the same height, climate, &c., that many of the immigrants should have been differently modified, though only in a small degree. This long appeared to me a great difficulty: but it arises in chief part from the deeply-seated error of considering the physical conditions of a country as the most important for its inhabitants; whereas it cannot, I think, be disputed that the nature of the other inhabitants, with which each has to compete, is at least as important, and generally a far more important element of
2 success. Now if we look to those inhabitants of the Galapagos Archipelago which are found in other parts of the world (laying on one side for the moment the endemic species, which cannot be here fairly included, as we are considering how they have come to be modified since their arrival), we find a considerable amount

1 Darwin was first clued into this fact by the English-born vice-governor of the islands, Nicholas O. Lawson, who commented that he could readily tell the island of origin of a given giant tortoise just by looking at it. Soon after leaving the islands, Darwin wrote in his ornithological notebook:

> When I recollect the fact, that from the form of the body, shape of scales & general size, the Spaniards can at once pronounce from which [island] any tortoise may have been brought:—when I see these Islands in sight of each other and possessed of but a scanty stock of animals, tenanted by these [mockingbirds] but slightly differing in structure & filling the same place in Nature, I must suspect they are only varieties . . . If there is the slightest foundation for these remarks, the Zoology of Archipelagoes will be well worth examining; for such facts would undermine the stability of species. (quoted in Barlow 1963)

He later wrote in *Voyage* that "by far the most remarkable feature in the natural history of this archipelago [is] that the different islands to a considerable extent are inhabited by a different set of beings" (*Voyage,* chapter 17).

2 The *non-endemic* species differences among islands underscore the occasional and random nature of colonization. Additionally, there is competition: arriving on a vacant island is a very different situation from arriving on a "fully tenanted" one. Degree of modification might depend, too, on how much time has elapsed since the ancestral individuals arrived, an idea captured in this 1837 entry in the B notebook: "if species (1) may be derived from form (2). &c. Then (remembering Lyells arguments of transportal) island near continents might have some species same as nearest land, which were late arrivals others old ones [sic] . . . Hence the type would be of the continent though species all different . . . On this idea of propagation of species we can see why a form [i.e., genus] peculiar to continents; all bred in from one parent" (Barrett et al. 1987, p. 173).

of difference in the several islands. This difference might indeed have been expected on the view of the islands having been stocked by occasional means of transport—a seed, for instance, of one plant having been brought to one island, and that of another plant to another island. Hence when in former times an immigrant settled on any one or more of the islands, or when it subsequently spread from one island to another, it would undoubtedly be exposed to different conditions of life in the different islands, for it would have to compete with different sets of organisms: a plant, for instance, would find the best-fitted ground more perfectly occupied by distinct plants in one island than in another, and it would be exposed to the attacks of somewhat different enemies. If then it varied, natural selection would probably favour different varieties in the different islands. Some species, however, might spread and yet retain the same character throughout the group, just as we see on continents some species spreading widely and remaining the same.

The really surprising fact in this case of the Galapagos Archipelago, and in a lesser degree in some analogous instances, is that the new species formed in the separate islands have not quickly spread to the other islands. But the islands, though in sight of each other, are separated by deep arms of the sea, in most cases wider than the British Channel, and there is no reason to suppose that they have at any former period been continuously united. The currents of the sea are rapid and sweep across the archipelago, and gales of wind are extraordinarily rare; so that the islands are far more effectually separated from each other than they appear to be on a map. Nevertheless a good many species, both those found in other parts of the world and those confined to the archipelago, are common to the several islands, and we may infer from certain facts that these have probably spread from some one island to the others. But we often take, I think, an erroneous view of the probability of closely allied species invading each other's territory, when put into free intercommunication. Undoubtedly if one species has any advantage whatever over another, it will in a very brief time wholly or in part supplant it; but if both are equally well fitted for their own places in nature, both probably will hold their own places and keep separate for almost any length of time. Being familiar with the fact that many species, naturalised through man's agency, have spread with astonishing rapidity over new countries, we are apt to infer that most species would thus spread; but we should remember that the forms which become naturalised in new countries are not generally closely allied to the aboriginal inhabitants, but are very distinct species, belonging in a large proportion of cases, as shown by Alph. de Candolle, to distinct genera. In the Galapagos Archipelago, many even of the birds, though so well adapted for flying from island to island, are distinct on each; thus there are three closely-allied species of mocking-thrush, each confined to its own island. Now let us suppose the mocking-thrush of Chatham Island to be blown to Charles Island, which has its own mocking-thrush: why should it succeed in establishing itself there? We may safely infer that Charles Island is well stocked with its own species, for annually more eggs are laid there than can possibly be reared; and we may infer that the mocking-thrush peculiar to Charles Island is at least as well fitted for its home as is the species peculiar to Chatham Island. Sir C. Lyell and Mr. Wollaston have communicated to me a remarkable fact bearing on this subject; namely, that Madeira and the adjoining islet of

1 It is important to appreciate the role that competition plays in species distributions. The likelihood of successfully spreading to another island depends on the competitors already present on that island.

2 In fact, many of these "weedy" species are adapted to disturbance, and thus easily follow in the footsteps of humans. Their close association with human activity makes them very familiar to us, and in that way perhaps they give the false impression that just about any species can spread with such facility. Note the comment about the relationship between naturalized and native species—Darwin states that they are "not closely allied." The implication is that invading forms closely related to native forms would encounter stronger competitive resistance, on the supposition that their ecological requirements would be more similar. It is difficult to generalize thus, particularly since, as just pointed out, so many invading species are generalists that thrive under conditions of disturbance.

3 The "mocking-thrushes" are now called mockingbirds, family Mimidae. The Ecuadoran names for the islands are now used. Chatham Island (now San Cristóbal) is home to *Nesomimus melanotis*—among the first Galápagos birds Darwin encountered when he disembarked from the *Beagle* on September 17, 1835. *Nesomimus trifasciatus* was found on Charles Island (Floreana). The Floreana population of *N. trifasciatus* has been extirpated since about 1880, and only a few hundred individuals of this critically endangered species persist on small nearby islets.

4 Porto Santo is a small island about 23 miles northeast of Madeira, in the eastern Atlantic. This otherwise unremarkable island was for a time the home of Christopher Columbus, who married the governor's daughter. The rock Darwin refers to is probably the columnar-jointed lava that was quarried from Pico de Ana Ferreira, the island's highest point (Czajkowski 2002). Wollaston, Lyell, and other naturalists were struck with the fact that Porto Santo and Madeira share many plant species in com-

(continued)

Porto Santo possess many distinct but representative land-shells, some of which live in crevices of stone; and although large quantities of stone are annually transported from Porto Santo to Madeira, yet this latter island has not become colonised by the Porto Santo species: nevertheless both islands have been colonised by some European land-shells, which no doubt had some advantage over the indigenous species. From these considerations I think we need not greatly marvel at the endemic and representative species, which inhabit the several islands of the Galapagos Archipelago, not having universally spread from island to island. In many other instances, as in the several districts of the same continent, pre-occupation has probably played an important part in checking the commingling of species under the same conditions of life. Thus, the south-east and south-west corners of Australia have nearly the same physical conditions, and are united by continuous land, yet they are inhabited by a vast number of distinct mammals, birds, and plants.

The principle which determines the general character of the fauna and flora of oceanic islands, namely, that the inhabitants, when not identically the same, yet are plainly related to the inhabitants of that region whence colonists could most readily have been derived,—the colonists having been subsequently modified and better fitted to their new homes,—is of the widest application throughout nature. We see this on every mountain, in every lake and marsh. For Alpine species, excepting in so far as the same forms, chiefly of plants, have spread widely throughout the world during the recent Glacial epoch, are related to those of the surrounding lowlands;—thus we have in South America, Alpine humming-birds, Alpine rodents, Alpine plants, &c., all of strictly American forms, and it is obvious

(continued)
mon, yet their snail fauna is remarkably divergent. Lyell exclaimed to Wollaston in an April 1856 letter: "How can there be only 11 or 12 per cent of Pulmoniferous mollusks common to Madeira & Porto Santo if, as Dr. Hooker says, nearly all the plants are common to the two islands?" He went on to muse in his journal that same day that "the land shells of all the 100 British Isles are the same & agree with Germany & France except a few Portuguese . . . It is only very ancient islands which have their land shells peculiar . . . Hence a geologist may infer the antiquity of islands by the isolation of their shells" (L. G. Wilson 1970, pp. 51–52).

1 Many habitats are island-like, from lakes and ponds to caves to alpine zones. Wherever Darwin looked at island environments he saw the same general pattern: closest relationship to species of the nearest "mainland." The boreal zone of the southern Appalachians, for example, is home to the endemic Fraser fir (*Abies fraseri*) and Carolina hemlock (*Tsuga caroliniana*); both are close relatives of firs and hemlock of the northeastern boreal zone, and not of species of high elevations of the Rockies, say, or Europe.

that a mountain, as it became slowly upheaved, would naturally be colonised from the surrounding lowlands. So it is with the inhabitants of lakes and marshes, excepting in so far as great facility of transport has given the same general forms to the whole world. We see this same principle in the blind animals inhabiting the caves of America and of Europe. Other analogous facts could be given. And it will, I believe, be universally found to be true, that wherever in two regions, let them be ever so distant, many closely allied or representative species occur, there will likewise be found some identical species, showing, in accordance with the foregoing view, that at some former period there has been intercommunication or migration between the two
[1] regions. And wherever many closely-allied species occur, there will be found many forms which some naturalists rank as distinct species, and some as varieties; these doubtful forms showing us the steps in the process of modification.

[2] This relation between the power and extent of migration of a species, either at the present time or at some former period under different physical conditions, and the existence at remote points of the world of other species allied to it, is shown in another and more general way. Mr. Gould remarked to me long ago, that in those genera of birds which range over the world, many of the species have very wide ranges. I can hardly doubt that this rule is generally true, though it would be difficult to prove it. Amongst mammals, we see it strikingly displayed in Bats, and in a lesser degree in the Felidæ and Canidæ. We see it, if we compare the distribution of butterflies and beetles. So it is with most fresh-water productions, in which so many genera range over the world, and many individual species have
[3] enormous ranges. It is not meant that in world-

1 Here Darwin is linking his observations on range and distribution back to the model of divergence. Recall his argument from chapter IV that *widely ranging* groups will tend to be the most speciose and have the most varieties on average. In keeping with this model, wherever lots of "closely-allied" (related) species are found, so too will linking species—those telltale "doubtful forms" that are so suggestive of transmutation.

2 The argument here is that species belonging to wide-ranging genera *themselves* have large ranges, which bears directly on Darwin's principle of divergence, as we saw in chapter IV. Darwin was collecting information on this idea in the late 1830s and 1840s. In 1839 he had several conversations with Gould in the British Museum (now Natural History Museum), notes of which are found in the C transmutation notebook (Barrett et al. 1987).

3 This is an important qualifier: the existence of at least some far-ranging species in far-ranging genera is consistent with Darwin's argument for common descent—the large species range shows there is potential for long-distance colonization, which might lead to exposure to new conditions and eventually to speciation. This point is emphasized in the last sentence of this paragraph.

ranging genera all the species have a wide range, or even that they have on an *average* a wide range; but only that some of the species range very widely; for the facility with which widely-ranging species vary and give rise to new forms will largely determine their average range. For instance, two varieties of the same species inhabit America and Europe, and the species thus has an immense range; but, if the variation had been a little greater, the two varieties would have been ranked as distinct species, and the common range would have been greatly reduced. Still less is it meant, that a species which apparently has the capacity of crossing barriers and ranging widely, as in the case of certain powerfully-winged birds, will necessarily range widely; for we should never forget that to range widely implies not only the power of crossing barriers, but the more important power of being victorious in distant lands in the struggle for life with foreign associates. But on the view of all the species of a genus having dedescended from a single parent, though now distributed to the most remote points of the world, we ought to find, and I believe as a general rule we do find, that some at least of the species range very widely; for it is necessary that the unmodified parent should range widely, undergoing modification during its diffusion, and should place itself under diverse conditions favourable for the conversion of its offspring, firstly into new varieties and ultimately into new species.

In considering the wide distribution of certain genera, we should bear in mind that some are extremely ancient, and must have branched off from a common parent at a remote epoch; so that in such cases there will have been ample time for great climatal and geographical changes and for accidents of transport; and consequently for the migration of some of the species into all

1 Put another way, it would be anomalous to have widely dispersed species in a genus if none of those species showed any capacity for broad dispersal. Remember that Darwin lived in a time when the stability of continents and ocean basins was a given, so high "dispersability" was essential for getting around the globe (continental extensionists notwithstanding).

1 Plate tectonics gives another good reason that we might expect to see very early life forms (as reckoned by the fossil record) broadly distributed over the globe—they were carried there by long-moving plates. Turning this around, such observations provided one of the lines of evidence (in consilience fashion!) for establishing continental drift as early as the late nineteenth century. The early fossil gymnosperm *Glossopteris,* for example, is found in South America, Africa, Antarctica, India, and Australia. The rock formations in which the fossils are found match one another nicely, helping to reconstruct Gondwanaland in much the same way that the different parts of a picture on puzzle pieces provide the clues to reconstruct the puzzle. In fact, the name Gondwanaland was coined in 1861 by the Austrian geologist Eduard Suess for the region of India where he found many *Glossopteris* fossils, which he took to be a clue to the former link of the southern continents.

2 We do not see long sentences like this one very often. Think of it as "one long sentence" that embodies Darwin's approach in the "one long argument" of the *Origin:* the patterns in evidence all speak to the same process, namely, descent with modification. His constrained passion is evident in phrasing like "... utterly inexplicable ..."

quarters of the world, where they may have become slightly modified in relation to their new conditions. There is, also, some reason to believe from geological evidence that organisms low in the scale within each great class, generally change at a slower rate than the higher forms; and consequently the lower forms will have had a better chance of ranging widely and of still retaining the same specific character. This fact, together with the seeds and eggs of many low forms being very minute and better fitted for distant transportation, probably accounts for a law which has long been observed, and which has lately been admirably discussed by Alph. de Candolle in regard to plants, namely, that the lower any group of organisms is, the more widely it is apt to range.

The relations just discussed,—namely, low and slowly-changing organisms ranging more widely than the high,—some of the species of widely-ranging genera themselves ranging widely,—such facts, as alpine, lacustrine, and marsh productions being related (with the exceptions before specified) to those on the surrounding low lands and dry lands, though these stations are so different—the very close relation of the distinct species which inhabit the islets of the same archipelago,—and especially the striking relation of the inhabitants of each whole archipelago or island to those of the nearest mainland,—are, I think, utterly inexplicable on the ordinary view of the independent creation of each species, but are explicable on the view of colonisation from the nearest and readiest source, together with the subsequent modification and better adaptation of the colonists to their new homes.

Summary of last and present Chapters.—In these chapters I have endeavoured to show, that if we make due allowance for our ignorance of the full effects of all the changes of climate and of the level of the land, which have certainly occurred within the recent period, and of other similar changes which may have occurred within the same period; if we remember how profoundly ignorant we are with respect to the many and curious means of occasional transport,—a subject which has hardly ever been properly experimentised on; if we bear in mind how often a species may have ranged continuously over a wide area, and then have become extinct in the intermediate tracts, I think the difficulties in believing that all the individuals of the same species, wherever located, have descended from the same parents, are not insuperable. And we are led to this conclusion, which has been arrived at by many naturalists under the designation of single centres of creation, by some general considerations, more especially from the importance of barriers and from the analogical distribution of sub-genera, genera, and families.

With respect to the distinct species of the same genus, which on my theory must have spread from one parent-source; if we make the same allowances as before for our ignorance, and remember that some forms of life change most slowly, enormous periods of time being thus granted for their migration, I do not think that the difficulties are insuperable; though they often are in this case, and in that of the individuals of the same species, extremely grave.

As exemplifying the effects of climatal changes on distribution, I have attempted to show how important has been the influence of the modern Glacial period, which I am fully convinced simultaneously affected the whole world, or at least great meridional belts. As showing how diversified are the means of occasional transport, I have discussed at some little length the means of dispersal of fresh-water productions.

1 Arguing that all members of the same species descended from the same ancestral parents might seem unnecessary today, but at the time it was supposed that a creator who made species to begin with could just as easily create individuals *in situ*, anywhere on the globe; recall Thoreau's poke at Agassiz, who, he half-jokingly mused, would have suggested that the toads atop Mt. Monadnock (in New Hampshire) were created there rather than arriving by hopping (see commentary on p. 365). This is a lower-level argument for material vs. supernatural agency: did species become widely distributed by various probable and improbable means of dispersal over the eons, or were they placed where we see them by divine fiat?

2 Note Darwin's rhetorical strategy here: he is applying the argument given above, about individuals of the same species, to species of the same genus.

1 Migration with subsequent modification and multiplication of species—that, in essence, is the process underlying the "grand facts" of biogeography. At least eleven "grand facts" are presented in this paragraph, five on this page and another six on the next.

"The high importance of barriers" refers to the correspondence between distinctness of flora and fauna between regions and the barriers separating those regions. Recall that though Darwin changed his thinking on the importance of barriers in generating new species, demoting them relative to the large populations and direct competition experienced on continuous land areas, he always recognized that barriers are key to preventing (or at least slowing) intermigration, promoting the growing distinctiveness of species assemblages in different areas over time. This relates to the next grand fact, the "localization of species, genera," etc.

By the same token, it is a grand fact that the species assemblage of an area will resemble the extinct species of that same area (but less so the older the fossil assemblage is): species "linked together by affinity . . . are likewise linked to the extinct beings . . . of the same continent." Species affinities in both space and time are the key observation of the landmark Sarawak Law paper of 1855, though note that Wallace is not credited. Finally, it is a grand fact that species affinities correspond to geography and not to environment. Areas "having nearly the same physical conditions" are typically "inhabited by very different forms of life." This relates to the vagaries of which species colonized a given area ancestrally, the time that had elapsed since then, what other species may or may not have also colonized the area, etc.

If the difficulties be not insuperable in admitting that in the long course of time the individuals of the same species, and likewise of allied species, have proceeded from some one source; then I think all the grand leading facts of geographical distribution are explicable on the theory of migration (generally of the more dominant forms of life), together with subsequent modification and the multiplication of new forms. We can thus understand the high importance of barriers, whether of land or water, which separate our several zoological and botanical provinces. We can thus understand the localisation of sub-genera, genera, and families; and how it is that under different latitudes, for instance in South America, the inhabitants of the plains and mountains, of the forests, marshes, and deserts, are in so mysterious a manner linked together by affinity, and are likewise linked to the extinct beings which formerly inhabited the same continent. Bearing in mind that the mutual relations of organism to organism are of the highest importance, we can see why two areas having nearly the same physical conditions should often be inhabited by very different forms of life; for according to the length of time which has elapsed since new inhabitants entered one region; according to the nature of the communication which allowed certain forms and not others to enter, either in greater or lesser numbers; according or not, as those which entered happened to come in more or less direct competition with each other and with the aborigines; and according as the immigrants were capable of varying more or less rapidly, there would ensue in different regions, independently of their physical conditions, infinitely diversified conditions of life,—there would be an almost endless amount of organic action and reaction,—and we should find, as we do find, some groups of beings greatly, and some only slightly modified,—some developed in great force, some existing in scanty numbers—in the different great geographical provinces of the world.

On these same principles, we can understand, as I have endeavoured to show, why oceanic islands should have few inhabitants, but of these a great number should be endemic or peculiar; and why, in relation to the means of migration, one group of beings, even within the same class, should have all its species endemic, and another group should have all its species common to other quarters of the world. We can see why whole groups of organisms, as batrachians and terrestrial mammals, should be absent from oceanic islands, whilst the most isolated islands possess their own peculiar species of aërial mammals or bats. We can see why there should be some relation between the presence of mammals, in a more or less modified condition, and the depth of the sea between an island and the mainland. We can clearly see why all the inhabitants of an archipelago, though specifically distinct on the several islets, should be closely related to each other, and likewise be related, but less closely, to those of the nearest continent or other source whence immigrants were probably derived. We can see why in two areas, however distant from each other, there should be a correlation, in the presence of identical species, of varieties, of doubtful species, and of distinct but representative species. 1

As the late Edward Forbes often insisted, there is a striking parallelism in the laws of life throughout time and space: the laws governing the succession of forms in past times being nearly the same with those governing at the present time the differences in different areas. We see this in many facts. The endurance of each species and group of species is continuous in time; for the exceptions to the rule are so few, that they may 2

1 The remaining six "grand facts" are given in this paragraph, all pertaining to oceanic islands. These striking observations speak for themselves and need no summary by me. Darwin's overarching point is that these curiosities of oceanic islands virtually scream out transmutation; they make sense only on the supposition that island endemics descended from chance ancestral colonists from the nearest mainland.

2 The naturalist Edward Forbes died prematurely in 1854, just shy of his fortieth year. The "parallelism" that Darwin mentions here refers to Forbes's ideas about the relationships between past and present species as discussed in his "polarity" theory (1854) and his 1846 monograph on the biogeography of the British Isles in relation to glacial events and other aspects of geological history (recall Forbes's adherence to extensive land bridges). Darwin may have rejected Forbes's quasi-supernatural polarity theory, but he greatly admired Forbes: he was one of the men Darwin suggested to Emma as an editor to publish the sketch of his species theory in the event of his untimely death (F. Darwin 1909).

1 This concise discussion of species relationships in time and space seems to ring of uniformity: the land is continuously occupied, the species upon it slowly changing. "All the world's a stage," in this biological version of Shakespeare. A more accurate theatrical metaphor was given by the noted ecologist G. Evelyn Hutchinson in the title of his 1965 book *The Ecological Theater and the Evolutionary Play.*

The importance of biogeography in providing crucial insights into the reality of evolutionary change cannot be over-emphasized. From the time of the earliest voyages to new lands, the striking similarities and differences in flora and fauna provoked wonder, and the more philosophically minded—those interested in the "species question"—instinctively realized that geographical distribution somehow held the key to solving the mystery. This is reflected in a remark made by the ornithologist John Gould, who wrote in the introduction to his *Birds of Australia,* "I cannot conclude . . . without mentioning the very remarkable manner in which many of the Australian Birds represent other nearly allied species belonging to the Old World, as if some particular law existed in reference to the subject, the species so represented being evidently destined to fulfill the same offices in either hemisphere" (Gould 1848, vol. I, p. xvi).

Darwin clearly sees the grand story told by biogeography; with understatement he writes here: "On my theory these several relations throughout time and space are intelligible."

fairly be attributed to our not having as yet discovered in an intermediate deposit the forms which are therein absent, but which occur above and below: so in space, it certainly is the general rule that the area inhabited by a single species, or by a group of species, is continuous; and the exceptions, which are not rare, may, as I have attempted to show, be accounted for by migration at some former period under different conditions or by occasional means of transport, and by the species having become extinct in the intermediate tracts. Both in time and space, species and groups of species have their points of maximum development. Groups of species, belonging either to a certain period of time, or to a certain area, are often characterised by trifling characters in common, as of sculpture or colour. In looking to the long succession of ages, as in now looking to distant provinces throughout the world, we find that some organisms differ little, whilst others belonging to a different class, or to a different order, or even only to a different family of the same order, differ greatly. In both time and space the lower members of each class generally change less than the higher; but there are in both cases marked exceptions to the rule. On my theory these several relations throughout time and space are intelligible; for whether we look to the forms of life which have changed during successive ages within the same quarter of the world, or to those which have changed after having migrated into distant quarters, in both cases the forms within each class have been connected by the same bond of ordinary generation; and the more nearly any two forms are related in blood, the nearer they will generally stand to each other in time and space; in both cases the laws of variation have been the same, and modifications have been accumulated by the same power of natural selection.

CHAPTER XIII.

MUTUAL AFFINITIES OF ORGANIC BEINGS: MORPHOLOGY: EMBRYOLOGY: RUDIMENTARY ORGANS.

CLASSIFICATION, groups subordinate to groups — Natural system — Rules and difficulties in classification, explained on the theory of descent with modification — Classification of varieties — Descent always used in classification — Analogical or adaptive characters — Affinities, general, complex and radiating — Extinction separates and defines groups — MORPHOLOGY, between members of the same class, between parts of the same individual — EMBRYOLOGY, laws of, explained by variations not supervening at an early age, and being inherited at a corresponding age — RUDIMENTARY ORGANS; their origin explained — Summary.

FROM the first dawn of life, all organic beings are found to resemble each other in descending degrees, so that they can be classed in groups under groups. This classification is evidently not arbitrary like the grouping of the stars in constellations. The existence of groups would have been of simple signification, if one group had been exclusively fitted to inhabit the land, and another the water; one to feed on flesh, another on vegetable matter, and so on; but the case is widely different in nature; for it is notorious how commonly members of even the same sub-group have different habits. In our second and fourth chapters, on Variation and on Natural Selection, I have attempted to show that it is the widely ranging, the much diffused and common, that is the dominant species belonging to the larger genera, which vary most. The varieties, or incipient species, thus produced ultimately become converted, as I believe, into new and distinct species; and these, on the principle of inheritance, tend to produce other new and dominant

This last "applications" chapter is about the unity of life: the taxonomic relationship, structure, and development of organisms. The overarching argument is that the patterns in evidence in these areas are not only consistent with descent with modification but even flow from it. In fact, Darwin became so struck with morphological affinity and embryological development that they usurped biogeography in his thinking as the single area with the most convincing evidence for transmutation. The late eighteenth and nineteenth centuries saw the blossoming of comparative anatomy, and indeed the comparative approach was seen not only as revolutionizing the understanding of species relationships but also as providing greater insights into the plan of creation. To Darwin, however, comparative anatomy told a different story. The idea of unchanging archetypes simply created thus was inelegant, an explanation that failed to explain much. To him the unity of life, including oddities and puzzles like rudimentary or atrophied structures, spoke resoundingly of transmutation.

1 Just as stars are joined in a constellation if they happen to appear close to one another in our line of sight, at one time superficial resemblance sufficed to group some organisms. The old grouping called *Vermes,* for example, contained a zoo of small, squiggly, worm-like animals, from nematodes to earthworms to certain insects. Comparative anatomy provided a more accurate picture of relationships. Darwin's point about the different habits of members of the same subgroup may seem trivial, but it harmonizes well with his principle of divergence: recall that in Darwin's model, the most similar varieties (and by extension, species) are driven to diverge by natural selection. This is evident in the sentences that follow.

1 The divergence of character diagram was introduced in chapter IV, following page 117. The branching and rebranching of the tree of life yield a pattern of "groups subordinate to groups," as Darwin puts it here; in other words, a nested hierarchy. Note the versatility of this diagram—we have seen it twice before, first when it was introduced to explain the divergence principle, and again in chapter X, on geological succession. Here we see that the same process that drives divergence and yields a predictable distribution of species through the fossil record also explains the intuitive classification system based on nested degrees of relationship. Note that the leftmost branch in the diagram, yielding taxa a^{14} through f^{14}, can be conceptualized as nested taxonomic groupings.

2 Here a^{14}, q^{14}, etc., are depicted as genera, say, G_1 through G_6. The three leftmost genera (G_1–G_3) are united by common descent into a common family or subfamily, say, F_1, while the two genera on the right (G_5 and G_6) have a more recent common ancestor and could, accordingly, be united in a related family or subfamily (F_2). These two families are related by common ancestry (node at a^5) and could be grouped together into the same class (C_1). The genera o^{14}, e^{14}, and m^{14} could constitute a single family within a single (monotypic) class that we would designate C_2. Darwin points out that all descendents of A, consisting of these two classes, would be united into a common order, while all the descendents of I on the right side of the diagram would be united into a different order. We could continue by identifying even more inclusive groups, until we united the entire array of species into, say, a common phylum or kingdom. Notice that the end result is nested sets, groups subordinate to groups.

species. Consequently the groups which are now large, and which generally include many dominant species, tend to go on increasing indefinitely in size. I further attempted to show that from the varying descendants of each species trying to occupy as many and as different places as possible in the economy of nature, there is a constant tendency in their characters to diverge. This conclusion was supported by looking at the great diversity of the forms of life which, in any small area, come into the closest competition, and by looking to certain facts in naturalisation.

I attempted also to show that there is a constant tendency in the forms which are increasing in number and diverging in character, to supplant and exterminate the less divergent, the less improved, and preceding forms. [1] I request the reader to turn to the diagram illustrating the action, as formerly explained, of these several principles; and he will see that the inevitable result is that the modified descendants proceeding from one progenitor become broken up into groups subordinate to groups. [2] In the diagram each letter on the uppermost line may represent a genus including several species; and all the genera on this line form together one class, for all have descended from one ancient but unseen parent, and, consequently, have inherited something in common. But the three genera on the left hand have, on this same principle, much in common, and form a sub-family, distinct from that including the next two genera on the right hand, which diverged from a common parent at the fifth stage of descent. These five genera have also much, though less, in common; and they form a family distinct from that including the three genera still further to the right hand, which diverged at a still earlier period. And all these genera, descended from (A), form an order distinct from the genera descended from (I). So that we here have many species descended from a single progenitor grouped into genera; and the genera are included in, or subordinate to, sub-families, families, and orders, all united into one class. Thus, the grand fact in natural history of the subordination of group under group, which, from its familiarity, does not always sufficiently strike us, is in my judgment fully explained.

Naturalists try to arrange the species, genera, and families in each class, on what is called the Natural System. But what is meant by this system? Some authors look at it merely as a scheme for arranging together those living objects which are most alike, and for separating those which are most unlike; or as an artificial means for enunciating, as briefly as possible, general propositions,—that is, by one sentence to give the characters common, for instance, to all mammals, by another those common to all carnivora, by another those common to the dog-genus, and then by adding a single sentence, a full description is given of each kind of dog. The ingenuity and utility of this system are indisputable. But many naturalists think that something more is meant by the Natural System; they believe that it reveals the plan of the Creator; but unless it be specified whether order in time or space, or what else is meant by the plan of the Creator, it seems to me that nothing is thus added to our knowledge. Such expressions as that famous one of Linnæus, and which we often meet with in a more or less concealed form, that the characters do not make the genus, but that the genus gives the characters, seem to imply that something more is included in our classification, than mere resemblance. I believe that something more is included; and that propinquity of descent,—the only known cause of the similarity of organic beings,—is the bond, hidden as it is by various degrees of modifi-

1 The "Natural System" of classification takes a comprehensive approach to identifying taxonomic affinity, utilizing data from the entire organism (in the case of plants, say, fruits, flowers, leaves, roots, etc., even habitat). The Natural System has its origins in Linnaeus: "The true beginning and end of botany is the natural system," he wrote, in which "all plants exhibit mutual affinities, as territories on a geographical map" (*Philosophia Botanica,* 1751, quoted in Müller-Wille 2007). In practice this was difficult to achieve, and simpler taxonomic systems (like Linnaeus's sexual system of classifying plants on the basis of numbers and arrangement of stamens and pistils) were adopted as proxies. These in turn were superceded by the work of the French botanist Antoine Laurent de Jussieu, whose much more integrative system eventually became the basis for modern botanical systematics.

Darwin suggests here that the Natural System works so well because it captures affinities that reflect common descent. This is ironic, since most of its adherents believed that the system simply reflected the plan of the creator. Darwin points out that something more than mere resemblance underlies the classification, and that something is true genealogical relationship. He suggests that this idea is inherent in a Linnaean aphorism, also from *Philosophia Botanica:* "The characters do not make the genus; but the genus gives the characters."

The latest revolution in systematics is the use of descent—phylogeny—to inform classification. *PhyloCode,* the International Code of Phylogenetic Nomenclature, is a new system that uses phylogenetic data to define placement at different levels of the taxonomic hierarchy (see http://www.ohiou.edu/phylocode/).

cation, which is partially revealed to us by our classifications.

Let us now consider the rules followed in classification, and the difficulties which are encountered on the view that classification either gives some unknown plan of creation, or is simply a scheme for enunciating general propositions and of placing together the forms most like each other. [1] It might have been thought (and was in ancient times thought) that those parts of the structure which determined the habits of life, and the general place of each being in the economy of nature, would be of very high importance in classification. Nothing can be more false. No one regards the external similarity of a mouse to a shrew, of a dugong to a whale, of a whale to a fish, as of any importance. These resemblances, though so intimately connected with the whole life of the being, are ranked as merely "adaptive or analogical characters;" but to the consideration of these resemblances we shall have to recur. It may even be given as a general rule, that the less any part of the organisation is concerned with special habits, the more important it becomes for classification. As an instance: [2] Owen, in speaking of the dugong, says, "The generative organs being those which are most remotely related to the habits and food of an animal, I have always regarded as affording very clear indications of its true affinities. We are least likely in the modifications of these organs to mistake a merely adaptive for an essential character." So with plants, how remarkable it is that the organs of vegetation, on which their whole life depends, are of little signification, excepting in the first main divisions; whereas the organs of reproduction, with their product the seed, are of paramount importance!

We must not, therefore, in classifying, trust to resemblances in parts of the organisation, however important

1 Darwin makes a curious point in this paragraph. Obvious organs or structures, those that play a role in procuring food, etc., might have been thought important for classification. The opposite is true: they are among the least useful traits. Why? Consider this observation in the context of Darwin's remark about domesticated organisms. Recall his point in chapter I about the "valued" and "unvalued" parts of a domesticated species like an apple tree. Parts under the most selection, like the fruits, vary significantly, while the "unvalued" parts like the leaves are left pretty much alone, and thus reveal relationship better than do the fruits. So, too, might seemingly trivial details of structure reveal relationship better than do obviously modified organs like limbs. In the next few pages Darwin drives home the point that obscure or trivial characters are often of great value in classification (i.e., revealing affinity). This seemingly obvious point underscores Darwin's core belief in variability and the idea that every character, however minor, has an evolutionary history.

2 The dugong *(Dugong dugon)* is the sole remaining species of the family Dugongidae. These sirenians are widely distributed in the eastern hemisphere, while their close relatives the manatees are found in the western hemisphere. Darwin is quoting an 1839 paper by Owen from the *Annals of Natural History.* The reproductive organs reveal affinity, Owen says, while structures used for swimming and feeding are less useful. This is echoed in the botanical example—in fact, the Linnaean sexual system of plant classification is based on this very idea, namely, that affinities are best seen in flower and fruit structures as opposed to vegetative structures. Darwin copied out the Owen quote in a notebook (the E transmutation notebook; see Barrett et al. 1987, p. 421), adding the suggestive comment, "How little clear meaning has this to what it might have." Darwin had the habit, by the way, of tearing out pages of his early notebooks when he later needed the information they contained. This is one of those pages that has been recovered; quite a few are lost.

they may be for the welfare of the being in relation to the outer world. Perhaps from this cause it has partly arisen, that almost all naturalists lay the greatest stress on resemblances in organs of high vital or physiological importance. No doubt this view of the classificatory importance of organs which are important is generally, but by no means always, true. But their importance for classification, I believe, depends on their greater constancy throughout large groups of species; and this constancy depends on such organs having generally been subjected to less change in the adaptation of the species to their conditions of life. That the mere physiological importance of an organ does not determine its classificatory value, is almost shown by the one fact, that in allied groups, in which the same organ, as we have every reason to suppose, has nearly the same physiological value, its classificatory value is widely different. No naturalist can have worked at any group without being struck with this fact; and it has been most fully acknowledged in the writings of almost every author. It will suffice to quote the highest authority, Robert Brown, who in speaking of certain organs in the Proteaceæ, says their generic importance, "like that of all their parts, not only in this but, as I apprehend, in every natural family, is very unequal, and in some cases seems to be entirely lost." Again in another work he says, the genera of the Connaraceæ "differ in having one or more ovaria, in the existence or absence of albumen, in the imbricate or valvular æstivation. Any one of these characters singly is frequently of more than generic importance, though here even when all taken together they appear insufficient to separate Cnestis from Connarus." To give an example amongst insects, in one great division of the Hymenoptera, the antennæ, as Westwood has remarked, are most constant in structure;

1 The Scottish botanist Robert Brown sailed as a naturalist on Matthew Flinders's Australian expedition on HMS *Investigator* and became an expert on southern hemisphere flora. He read a paper on the southern hemisphere plant family Proteaceae before the Linnean Society in 1809 (published in the *Transactions* in 1810), and named the family Connaraceae in 1818. The type genus *Connarus* is a pantropical group of about 100 species; *Cnestis,* also Connaraceae, is an Old World genus of about 85 species.

2 The entomologist John Obadiah Westwood, professor and keeper of the Hope collection of insects at Oxford, made a special study of Hymenoptera (ants, bees, wasps, and their relatives). Darwin may have referred to Westwood's 1835 paper "Characters of new genera and species of hymenopterous insects" (*Proceedings of the Zoological Society of London* 3: 51–72).

1 Rudimentary organs are weakly developed or reduced structures. Either they are no longer under selection or selection is favoring their reduction. In cases such as the ones Darwin mentions here, the rudimentary structure is expressed perhaps fleetingly in development or persists in juveniles but is later lost. Rudimentary organs themselves are powerful evidence for descent. Here Darwin is making the point that seemingly trivial structures can provide important clues to true taxonomic affinity—owing, of course, to common descent.

2 This is *Ornithorhynchus anatinus,* the duck-billed platypus of eastern Australia and Tasmania. The Latin name is wonderfully descriptive: *Ornithorhynchus* means "bird-beaked," and the species name is derived from *anat,* the Latin word for "duck."

in another division they differ much, and the differences are of quite subordinate value in classification; yet no one probably will say that the antennæ in these two divisions of the same order are of unequal physiological importance. Any number of instances could be given of the varying importance for classification of the same important organ within the same group of beings.

1 Again, no one will say that rudimentary or atrophied organs are of high physiological or vital importance; yet, undoubtedly, organs in this condition are often of high value in classification. No one will dispute that the rudimentary teeth in the upper jaws of young ruminants, and certain rudimentary bones of the leg, are highly serviceable in exhibiting the close affinity between Ruminants and Pachyderms. Robert Brown has strongly insisted on the fact that the rudimentary florets are of the highest importance in the classification of the Grasses.

Numerous instances could be given of characters derived from parts which must be considered of very trifling physiological importance, but which are universally admitted as highly serviceable in the definition of whole groups. For instance, whether or not there is an open passage from the nostrils to the mouth, the only character, according to Owen, which absolutely distinguishes fishes and reptiles—the inflection of the angle of the jaws in Marsupials—the manner in which the wings of insects are folded—mere colour in certain Algæ—mere pubescence on parts of the flower in grasses—the nature of the dermal covering, as hair or
2 feathers, in the Vertebrata. If the Ornithorhynchus had been covered with feathers instead of hair, this external and trifling character would, I think, have been considered by naturalists as important an aid in determining the degree of affinity of this strange creature to

birds and reptiles, as an approach in structure in any one internal and important organ.

The importance, for classification, of trifling characters, mainly depends on their being correlated with several other characters of more or less importance. The value indeed of an aggregate of characters is very evident in natural history. Hence, as has often been remarked, a species may depart from its allies in several characters, both of high physiological importance and of almost universal prevalence, and yet leave us in no doubt where it should be ranked. Hence, also, it has been found, that a classification founded on any single character, however important that may be, has always failed; for no part of the organisation is universally constant. The importance of an aggregate of characters, even when none are important, alone explains, I think, that saying of Linnæus, that the characters do not give the genus, but the genus gives the characters; for this saying seems founded on an appreciation of many trifling points of resemblance, too slight to be defined. Certain plants, belonging to the Malpighiaceæ, bear perfect and degraded flowers; in the latter, as A. de Jussieu has remarked, "the greater number of the characters proper to the species, to the genus, to the family, to the class, disappear, and thus laugh at our classification." But when Aspicarpa produced in France, during several years, only degraded flowers, departing so wonderfully in a number of the most important points of structure from the proper type of the order, yet M. Richard sagaciously saw, as Jussieu observes, that this genus should still be retained amongst the Malpighiaceæ. This case seems to me well to illustrate the spirit with which our classifications are sometimes necessarily founded.

Practically when naturalists are at work, they do

1 This is still true today. Systematists use as much data as possible in identifying relationships and phylogenetic branching order—they often prefer a "total evidence" approach that combines morphological and genetic data. This idea lies at the heart of Linnaeus's "Natural System." Artificial systems, in contrast, tend to rely on a few characters for classification. Darwin points out that the utility of a trivial character depends on whether it is correlated with several other characters; correlation means the character can serve as a proxy for looking at multiple characters.

2 The de Jussieus were a distinguished French family of botanists and naturalists, including brothers Antoine de Jussieu (1686–1758) and Bernard de Jussieu (1699–1777), their nephew Antoine Laurent de Jussieu (1748–1836), and his son Adrien-Henri Laurent de Jussieu (1797–1853). Darwin is citing Adrien-Henri Laurent, who succeeded his father as Professor of Botany at the Muséum National d'Histoire Naturelle in Paris. He treated the Malpighiaceae (the Barbados cherry family) in his 1843 treatise *Monographie des Malpighiacées,* as well as in earlier works.

3 *Aspicarpa* (common name asphead) is a New World genus of trees in the family Malpighiaceae. Aspheads cultivated in France for some years produced only nonfunctional flowers with reduced stamens and pistils—the very characters considered important in taxonomic placement within the Malpighiaceae. Darwin's point here is that the French botanists A.-H. L. Jussieu and L. C. Richard (who named the genus *Aspicarpa* in 1816) nonetheless recognized the plants for what they are.

not trouble themselves about the physiological value of the characters which they use in defining a group, or in allocating any particular species. If they find a character nearly uniform, and common to a great number of forms, and not common to others, they use it as one of high value; if common to some lesser number, they use it as of subordinate value. This principle has been broadly confessed by some naturalists to be the true one; and by none more clearly than by [1] that excellent botanist, Aug. St. Hilaire. If certain characters are always found correlated with others, though no apparent bond of connexion can be discovered between them, especial value is set on them. As in most groups of animals, important organs, such as those for propelling the blood, or for aërating it, or those for propagating the race, are found nearly uniform, they are considered as highly serviceable in classification; but in some groups of animals all these, the most important vital organs, are found to offer characters of quite subordinate value.

We can see why characters derived from the embryo should be of equal importance with those derived from the adult, for our classifications of course include all [2] ages of each species. But it is by no means obvious, on the ordinary view, why the structure of the embryo should be more important for this purpose than that of the adult, which alone plays its full part in the economy of nature. Yet it has been strongly urged by those great naturalists, Milne Edwards and Agassiz, that embryonic characters are the most important of any in the classification of animals; and this doctrine has very [3] generally been admitted as true. The same fact holds good with flowering plants, of which the two main divisions have been founded on characters derived from the embryo,—on the number and position of the em-

1 This "excellent botanist" is the Frenchman Augustin de Saint-Hilaire (1799–1853), who made extensive collections of Brazilian flora. Darwin is referring to Saint-Hilaire's treatment of plant morphology: *Leçons de Botanique, Comprénant Principalement la Morphologie Végetale* (Paris, 1841), which he read in 1846 (Vorzimmer 1977).

2 In 1844 Henri Milne-Edwards published an important essay on classification discussing the value of comparative embryology in revealing systematic affinities. Darwin immediately saw its significance: "This is the most profound paper I have ever seen on Affinities," he wrote on his copy (Cambridge University Library ms. CUL-DAR 72. 117–151). Agassiz concurred in his *Essay on Classification:* "Embryology furnishes also the best measure of the true affinities existing between animals," he wrote. "Knowledge of the embryonic changes of certain animals gave the first clue to their true affinities, while in other cases it has furnished a very welcome confirmation of relationships" (Agassiz 1857, p. 86). Darwin wrote in his *Essay* of 1844: "The cause of the greater value of characters, drawn from the early stages of life, can . . . be in a considerable degree explained, on the theory of descent, although inexplicable on the views of the creationist" (F. Darwin 1909, p. 201).

3 In botany, the use of cotyledons (seed-leaves) as a fundamental character of plant classification was first proposed by Bernard de Jussieu and subsequently published by his nephew Antoine Laurent in 1789 in *Genera Plantatum* (they coined the terms "monocot" and "dicot"). The basic division of plants by number of cotyledons is still recognized today.

bryonic leaves or cotyledons, and on the mode of development of the plumule and radicle. In our discussion ◀1 on embryology, we shall see why such characters are so valuable, on the view of classification tacitly including the idea of descent.

Our classifications are often plainly influenced by ◀2 chains of affinities. Nothing can be easier than to define a number of characters common to all birds; but in the case of crustaceans, such definition has hitherto been found impossible. There are crustaceans at the opposite ends of the series, which have hardly a character in common; yet the species at both ends, from being plainly allied to others, and these to others, and so onwards, can be recognised as unequivocally belonging to this, and to no other class of the Articulata.

Geographical distribution has often been used, though ◀3 perhaps not quite logically, in classification, more especially in very large groups of closely allied forms. Temminck insists on the utility or even necessity of this practice in certain groups of birds; and it has been followed by several entomologists and botanists.

Finally, with respect to the comparative value of the ◀4 various groups of species, such as orders, sub-orders, families, sub-families, and genera, they seem to be, at least at present, almost arbitrary. Several of the best botanists, such as Mr. Bentham and others, have strongly insisted on their arbitrary value. Instances could be given amongst plants and insects, of a group of forms, first ranked by practised naturalists as only a genus, and then raised to the rank of a sub-family or family; and this has been done, not because further research has detected important structural differences, at first overlooked, but because numerous allied species, with slightly different grades of difference, have been subsequently discovered.

1 In the fifth edition of the *Origin* Darwin added a sentence here to clarify his point: "We shall immediately see why these characters possess so high a value in classification, namely, from the natural system being genealogical in its arrangement" (Peckham 1959, p. 655).

2 By its very structure, the nested hierarchy used in classification inherently recognizes that organisms are related in "chains of affinities." The example of crustaceans to illustrate long series with species at either end that barely resemble one another was carefully chosen. Recall that Darwin authored several monographs on barnacles, a group with highly modified morphology compared with, say, their relatives the lobsters, shrimp, and water fleas. The ability to link species with such extreme morphologies in a chain of affinity is *itself* strongly suggestive of transmutation.

3 Coenraad Jacob Temminck was a distinguished Dutch ornithologist and author of many ornithological works. Darwin may have consulted Temminck's treatment of bird classification, *Observations sur la classification méthodique des oiseaux et remarques sur l'analyse d'une nouvelle ornithologie élémentaire* (Amsterdam and Paris, 1817).

4 Darwin is saying that *at present*—i.e., in his day—there is little information inherent in comparative taxonomic ranks owing to the arbitrariness of these ranks. In delineating higher taxa, systematists often differ on which characters to use as the basis for classification. Characters sufficiently important or informative to one systematist might be deemed uninformative or even misleading by another. Darwin is arguing here that with a genealogical approach to classification, in contrast, comparative information is inherent. This is evident in the final sentence of the paragraph, where he suggests that knowledge of intermediates informs classification. Once intermediates are discovered, a fuller picture of relationships emerges, allowing a closer approach to a true genealogical classification.

1 What makes the Natural System natural is genealogy, in Darwin's view. Take careful note of his wording in this paragraph and the next. Naturalists have not realized it, but what they have been seeking as the fundamental principle of classification is a "community of descent": common descent is a hidden bond that underlies a truly natural system of classification. All true classifications are genealogical, not mere groupings of similar forms.

Scholars have attempted to interpret Darwin's views on classification in modern terms, citing him as the standard-bearer for different schools of thought in systematics (i.e., evolutionary taxonomy vs. Hennigian cladistics). This effort is misguided. Darwin was concerned with convincing naturalists of the reality of common descent and (in this chapter) how it provides a framework for classification based on genealogy. There is a wide gulf between ideal classification and classification in practice, however; in reality, genealogical relationships are obscured by phenomena like adaptive convergence and extinction of intermediates. Darwin's own efforts at higher classification underscore this fact: his barnacle classification is not strictly genealogical but uses general similarity of groups to inform classification wherever necessary. Padian (1999, p. 357) nicely encapsulates Darwin's approach in theory and practice, to borrow from the title of his paper: "Darwin regarded *genealogy* as the overriding principle by which organisms should be ordered; he saw the *arrangement* of organisms as basically genealogical but varying in placement according to the evolutionary differentiation of features and the work of extinction in eliminating intermediate forms . . .; and *classification* as an expression of genealogy, ironically made possible in part because extinction had created visible divisions in the natural genealogical flow" (emphasis in original). Why is it ironic that extinction is useful in this way? On the one hand knowledge of the complete genealogical tree is lost by extinction, but on the other hand extinction helps to delineate groups more sharply, thereby making it easier to classify them into discrete taxa.

All the foregoing rules and aids and difficulties in classification are explained, if I do not greatly deceive myself, on the view that the natural system is founded on descent with modification; that the characters which naturalists consider as showing true affinity between any two or more species, are those which have been inherited from a common parent, and, in so far, all true classification is genealogical; that community of descent is the hidden bond which naturalists have been unconsciously seeking, and not some unknown plan of creation, or the enunciation of general propositions, and the mere putting together and separating objects more or less alike.

But I must explain my meaning more fully. I believe that the *arrangement* of the groups within each class, in due subordination and relation to the other groups, must be strictly genealogical in order to be natural; but that the *amount* of difference in the several branches or groups, though allied in the same degree in blood to their common progenitor, may differ greatly, being due to the different degrees of modification which they have undergone; and this is expressed by the forms being ranked under different genera, families, sections, or orders. The reader will best understand what is meant, if he will take the trouble of referring to the diagram in the fourth chapter. We will suppose the letters A to L to represent allied genera, which lived during the Silurian epoch, and these have descended from a species which existed at an unknown anterior period. Species of three of these genera (A, F, and I) have transmitted modified descendants to the present day, represented by the fifteen genera (a^{14} to z^{14}) on the uppermost horizontal line. Now all these modified descendants from a single species, are represented as related in blood or descent to the same degree; they may metaphorically be called cousins to the same millionth degree; yet they differ widely and in different degrees from each other. The forms descended from A, now broken up into two or three families, constitute a distinct order from those descended from I, also broken up into two families. Nor can the existing species, descended from A, be ranked in the same genus with the parent A; or those from I, with the parent I. But the existing genus F^{14} may be supposed to have been but slightly modified; and it will then rank with the parent-genus F; just as some few still living organic beings belong to Silurian genera. So that the amount or value of the differences between organic beings all related to each other in the same degree in blood, has come to be widely different. Nevertheless their genealogical *arrangement* remains strictly true, not only at the present time, but at each successive period of descent. All the modified descendants from A will have inherited something in common from their common parent, as will all the descendants from I; so will it be with each subordinate branch of descendants, at each successive period. If, however, we choose to suppose that any of the descendants of A or of I have been so much modified as to have more or less completely lost traces of their parentage, in this case, their places in a natural classification will have been more or less completely lost,—as sometimes seems to have occurred with existing organisms. All the descendants of the genus F, along its whole line of descent, are supposed to have been but little modified, and they yet form a single genus. But this genus, though much isolated, will still occupy its proper intermediate position; for F originally was intermediate in character between A and I, and the several genera descended from these two genera will

1 All current living species descending from a distant common ancestor are related to one another in the same degree, in the sense that their lineages have the same starting point. However, these descendant species might differ greatly in degree of modification from the ancestral species, and from each other, depending on the vagaries of natural selection and other historical factors. These differences necessitate their placement in different taxonomic groups. Their general taxonomic arrangement will still be genealogical, Darwin emphasizes, but the similarities and differences among them can be expressed by placing them in separate higher taxa.

1 Darwin is the first person to represent species relationships (as opposed to relationships between higher taxonomic groups) in the form of a branching tree, a pattern that immediately lends itself to arranging species in groups, subgroups, etc. Darwin sees this arrangement as natural, whereas forcing species into a linear series is decidedly *un*natural.

2 Darwin realized that language evolution provides an excellent parallel for species evolution. The process is not the same—languages do not experience natural selection based on variations, and they are certainly not heritable—yet they change slowly and diverge as a result of isolation, with dialects developing into bona fide languages by "insensible gradations." Degree of difference varies widely among language lineages, and some ancestral languages have given rise to many descendant forms, whereas others have generated few or none. The net result is that languages have a natural arrangement of their own, an arrangement that is genealogical. Darwin borrowed the language metaphor from geology; in an 1850 letter to Lyell he wrote, "Your metaphor of the pebbles of preexisting languages, reminds me that I heard Sir J. Herschel at the Cape say, how he wished someone [would] treat languages, as you had Geology, & study the existing causes of change & apply the deductions to old languages" (*Correspondence* 4: 319). (Recall that Darwin met Herschel in South Africa in June 1836 while on the *Beagle* voyage.)

have inherited to a certain extent their characters. ▶1 This natural arrangement is shown, as far as is possible on paper, in the diagram, but in much too simple a manner. If a branching diagram had not been used, and only the names of the groups had been written in a linear series, it would have been still less possible to have given a natural arrangement; and it is notoriously not possible to represent in a series, on a flat surface, the affinities which we discover in nature amongst the beings of the same group. Thus, on the view which I hold, the natural system is genealogical in its arrangement, like a pedigree; but the degrees of modification which the different groups have undergone, have to be expressed by ranking them under different so-called genera, sub-families, families, sections, orders, and classes.

▶2 It may be worth while to illustrate this view of classification, by taking the case of languages. If we possessed a perfect pedigree of mankind, a genealogical arrangement of the races of man would afford the best classification of the various languages now spoken throughout the world; and if all extinct languages, and all intermediate and slowly changing dialects, had to be included, such an arrangement would, I think, be the only possible one. Yet it might be that some very ancient language had altered little, and had given rise to few new languages, whilst others (owing to the spreading and subsequent isolation and states of civilisation of the several races, descended from a common race) had altered much, and had given rise to many new languages and dialects. The various degrees of difference in the languages from the same stock, would have to be expressed by groups subordinate to groups; but the proper or even only possible arrangement would still be genealogical; and this would be strictly natural, as it would connect together all languages, extinct and modern, by the closest affinities, and would give the filiation and origin of each tongue.

In confirmation of this view, let us glance at the classification of varieties, which are believed or known to have descended from one species. These are grouped under species, with sub-varieties under varieties; and with our domestic productions, several other grades of difference are requisite, as we have seen with pigeons. The origin of the existence of groups subordinate to groups, is the same with varieties as with species, namely, closeness of descent with various degrees of modification. Nearly the same rules are followed in classifying varieties, as with species. Authors have insisted on the necessity of classing varieties on a natural instead of an artificial system; we are cautioned, for instance, not to class two varieties of the pine-apple together, merely because their fruit, though the most important part, happens to be nearly identical; no one puts the swedish and common turnips together, though the esculent and thickened stems are so similar. Whatever part is found to be most constant, is used in classing varieties: thus the great agriculturist Marshall says the horns are very useful for this purpose with cattle, because they are less variable than the shape or colour of the body, &c.; whereas with sheep the horns are much less serviceable, because less constant. In classing varieties, I apprehend if we had a real pedigree, a genealogical classification would be universally preferred; and it has been attempted by some authors. For we might feel sure, whether there had been more or less modification, the principle of inheritance would keep the forms together which were allied in the greatest number of points. In tumbler pigeons, though some sub-varieties differ from the others

1 The production of domesticated varieties by artificial selection provided a potent analogy for natural selection, as we saw in chapter I. In this paragraph that analogy is continued: people have intuitively classified domesticated varieties genealogically, Darwin suggests, as opposed to grouping them by similarity of traits, which can arise by convergence. Turnips are a case in point. The Swedish turnip, or rutabaga *(Brassica napobrassica)*, originated as a cross between cabbage *(Brassica oleracea)* and the common turnip *(Brassica rapa)*. In a discussion of the origins and variability of *Brassica* vegetables in the first volume of *Variation*, Darwin mentions turnip-like convergence among species: "In the production of large, fleshy, turnip-like stems, we have a case of analogous variation in three forms which are generally considered as distinct species. But scarcely any modification seems so easily acquired as a succulent enlargement of the stem or root—that is, a store of nutriment laid up for the plant's own future use" (Darwin 1883, ch. IX).

2 Darwin read several books by William Marshall. A discussion of horns is given in a book Darwin did *not* record in his reading list, but it nicely expresses Marshall's views on the use of horns in identifying cattle breeds:

> The doctrine of horns has long appeared to me . . . as a craft convenient to *leading-breeders,* in establishing their respective *systems.* . . . The horn has been mentioned as a permanent specific character of cattle. Hence in varieties it may have its use as a criterion. Thus supposing a male and female of superior form and flesh, and with horns resembling each other (as nearly as the horns of males and females of the same variety naturally do), no matter whether short or long, sharp or clubbed, rising or falling; and supposing a variety to be established from this parentage, it is highly probable that the horns of the parents would continue for a while to be a characteristic of the *true breed,* and might by inferior judges be depended upon, *in some degree,* as a criterion. (Marshall 1788, pp. 189–190; emphases in original)

in the important character of having a longer beak, yet all are kept together from having the common habit of tumbling; but the short-faced breed has nearly or quite lost this habit; nevertheless, without any reasoning or thinking on the subject, these tumblers are kept in the same group, because allied in blood and alike in some other respects. If it could be proved that the Hottentot had descended from the Negro, I think he would be classed under the Negro group, however much he might differ in colour and other important characters from negroes.

With species in a state of nature, every naturalist has in fact brought descent into his classification; for he includes in his lowest grade, or that of a species, the two sexes; and how enormously these sometimes differ in the most important characters, is known to every naturalist: scarcely a single fact can be predicated in common of the males and hermaphrodites of certain cirripedes, when adult, and yet no one dreams of separating them. The naturalist includes as one species the several larval stages of the same individual, however much they may differ from each other and from the adult; as he likewise includes the so-called alternate generations of Steenstrup, which can only in a technical sense be considered as the same individual. He includes monsters; he includes varieties, not solely because they closely resemble the parent-form, but because they are descended from it. He who believes that the cowslip is descended from the primrose, or conversely, ranks them together as a single species, and gives a single definition. As soon as three Orchidean forms (Monochanthus, Myanthus, and Catasetum), which had previously been ranked as three distinct genera, were known to be sometimes produced on the same spike, they were immediately included as a single species.

1 "Hottentot" was the name given to the pastoral Khoikhoi people of southwestern Africa by the European colonists; it is considered offensive today. The Khoikhoi are one of several ethnically distinct southern African groups, along with the Nguni, Basotho, Zulu, Khoisan, and Xhosa. The Khoikhoi possess several distinctive physical characteristics, including steatopygia (significant fat deposits in and around the buttocks) and elongated labia minora. In Darwin's day these features led naturalists to identify the Khoikhoi as a distinct "race" and to speculate as to their relationship to the more widespread Bantu ethnolinguistic group, which white Europeans collectively referred to as "negro." This is the context for the idea of the Hottentot "descend[ing] from the Negro." (This sentence was dropped from the fifth and sixth editions.)

2 This is a curious line of argument: one demonstration that true affinity and not mere outward similarity is the guiding principle of classification is given by cases in which individuals differ morphologically (sexually dimorphic species, larval vs. adult forms, etc.) yet are recognized as members of the same species. This is also true of animals that alternate between morphological forms in their development—dubbed "alternation of generations" in 1845 by Johannes Japetus Steenstrup. Darwin is suggesting that these cases are, in essence, an intuitive acknowledgment of the primacy of descent, insofar as the distinctive morphs are by definition related by common ancestry.

Alternation of generations also occurs in plants, as do cases of such strikingly different flower morphs that early naturalists assumed they belonged to different taxa. The case of the orchids given near the bottom of this page is interesting because Darwin played a role in establishing that these three "genera" are really floral morphs of a single species. In 1854 the English explorer Sir Robert Schomburgk reported finding flowers of all three occurring on a single plant! The name *Catasetum* had taxonomic priority, so the others became synonyms of that genus. Darwin became keenly interested in this floral polymorphism, and in 1861 (the same year his orchid book came out) he read a paper to the Linnean Society announcing that the three morphs were in fact male, female, and hermaphroditic flowers.

But it may be asked, what ought we to do, if it could be proved that one species of kangaroo had been produced, by a long course of modification, from a bear? Ought we to rank this one species with bears, and what should we do with the other species? The supposition is of course preposterous; and I might answer by the *argumentum ad hominem*, and ask what should be done if a perfect kangaroo were seen to come out of the womb of a bear? According to all analogy, it would be ranked with bears; but then assuredly all the other species of the kangaroo family would have to be classed under the bear genus. The whole case is preposterous; for where there has been close descent in common, there will certainly be close resemblance or affinity.

As descent has universally been used in classing together the individuals of the same species, though the males and females and larvæ are sometimes extremely different; and as it has been used in classing varieties which have undergone a certain, and sometimes a considerable amount of modification, may not this same element of descent have been unconsciously used in grouping species under genera, and genera under higher groups, though in these cases the modification has been greater in degree, and has taken a longer time to complete? I believe it has thus been unconsciously used; and only thus can I understand the several rules and guides which have been followed by our best systematists. We have no written pedigrees; we have to make out community of descent by resemblances of any kind. Therefore we choose those characters which, as far as we can judge, are the least likely to have been modified in relation to the conditions of life to which each species has been recently exposed. Rudimentary structures on this view are as good as, or even sometimes better than, other parts of the organisation. We

1 The issue at the heart of this passage—not Darwin's most elegant example—is, again, classification by similarity of structure vs. true affinity. Darwin is saying that just as one would unhesitatingly classify kangaroos with bears if a kangaroo were demonstrably born of a bear, so too would these animals have to be recognized as close relatives (and classified together) if it could be shown that kangaroos descended from bears through transmutation. Their relationship is genealogical in both cases. Earlier Darwin argued with Huxley over classification: Huxley held that similarity of structure was sufficient to inform classification, while Darwin argued that it must be genealogical. In a letter to Huxley dated just days after the *Origin* was published, Darwin wrote, "On classification I fear we shall split. Did you perceive argumentum ad hominem Huxley. about Kangaroo & Bear" (*Correspondence* 7: 398). Huxley apparently did not reply, and Darwin decided to drop the example in the next edition of the *Origin.*

2 Resemblance is a good starting point in determining "community of descent" but is notoriously problematic when it comes to the nuts-and-bolts of deciding branching order of evolutionary trees. This field was revolutionized with the advent of cladistics, the basis for phylogenetic systematics. The cladistic approach, pioneered by the German entomologist Willi Hennig (1913–1976), was first presented in 1950 but became widely known with the 1966 English translation of Hennig's book *Phylogenetic Systematics.* Cladistics permits reconstruction of branching order by identifying successive subsets of shared-derived (synapomorphic) characters among species. Real sets of character trait data are rarely so straightforward, and different techniques are used to infer the most likely tree out of what can amount to a "forest" of possibilities. Just which technique to use has been an acrimonious issue among systematists, as has the extent to which branching order, once identified, should be used to define our classifications. All agree with Darwin that genealogy should form the basis of classification, but just how rigidly one should adhere to branching order to identify higher taxa is a point of contention (see Hull 1988, Kitching et al. 1998, Schmitt 2003).

1 This is an important principle used by systematists today. It is a parsimony argument: a pattern seen by one piece of evidence may not be itself compelling, but when multiple pieces of evidence exhibit the same pattern we have greater confidence that we are drawing the correct inference. The following paragraph is a beautifully written expression of how even trifling characters can reveal common descent.

care not how trifling a character may be—let it be the mere inflection of the angle of the jaw, the manner in which an insect's wing is folded, whether the skin be covered by hair or feathers—if it prevail throughout many and different species, especially those having very different habits of life, it assumes high value; for we can account for its presence in so many forms with such different habits, only by its inheritance from a common parent. We may err in this respect in regard to single points of structure, but when several characters, let them be ever so trifling, occur together throughout a large group of beings having different habits, we may feel almost sure, on the theory of descent, that these characters have been inherited from a common ancestor. And we know that such correlated or aggregated characters have especial value in classification.

We can understand why a species or a group of species may depart, in several of its most important characteristics, from its allies, and yet be safely classed with them. This may be safely done, and is often done, as long as a sufficient number of characters, let them be ever so unimportant, betrays the hidden bond of community of descent. Let two forms have not a single character in common, yet if these extreme forms are connected together by a chain of intermediate groups, we may at once infer their community of descent, and we put them all into the same class. As we find organs of high physiological importance—those which serve to preserve life under the most diverse conditions of existence—are generally the most constant, we attach especial value to them; but if these same organs, in another group or section of a group, are found to differ much, we at once value them less in our classification. We shall hereafter, I think, clearly see why embryological characters are of such high classificatory importance.

Geographical distribution may sometimes be brought usefully into play in classing large and widely-distributed genera, because all the species of the same genus, inhabiting any distinct and isolated region, have in all probability descended from the same parents.

We can understand, on these views, the very important distinction between real affinities and analogical or adaptive resemblances. Lamarck first called attention to this distinction, and he has been ably followed by Macleay and others. The resemblance, in the shape of the body and in the fin-like anterior limbs, between the dugong, which is a pachydermatous animal, and the whale, and between both these mammals and fishes, is analogical. Amongst insects there are innumerable instances: thus Linnæus, misled by external appearances, actually classed an homopterous insect as a moth. We see something of the same kind even in our domestic varieties, as in the thickened stems of the common and swedish turnip. The resemblance of the greyhound and racehorse is hardly more fanciful than the analogies which have been drawn by some authors between very distinct animals. On my view of characters being of real importance for classification, only in so far as they reveal descent, we can clearly understand why analogical or adaptive character, although of the utmost importance to the welfare of the being, are almost valueless to the systematist. For animals, belonging to two most distinct lines of descent, may readily become adapted to similar conditions, and thus assume a close external resemblance; but such resemblances will not reveal—will rather tend to conceal their blood-relationship to their proper lines of descent. We can also understand the apparent paradox, that the very same characters are analogical when one class or order is compared with another, but give true affinities when the members of

1 Some taxonomists, particularly those wedded to the "single center of creation" idea, used biogeography to help delineate species. In the creationist view, species were created in one place, though their descendants could subsequently migrate to other regions. Note that Darwin agrees with this basic scenario but believes that the species originate by a different means.

2 "Real affinities" are homologies in today's terms, and "analogical or adaptive resemblances" are termed analogies. Darwin is mistaken in attributing the distinction between the two to Lamarck. The homology concept was first clearly expressed by the French zoologist Pierre Belon (1517–1564), in a comparison of bird and human skeletons published in 1555. Lamarck made implicit use of the concepts of homology and analogy, though he used neither term. His "first law" of 1815 saw homology as a byproduct of transmutation of species up the scale of nature *(escalator naturae),* while analogies represented convergence in acquired characteristics. William Sharpe Macleay's quinary system of classification, published a dozen years later, was not evolutionary but also recognized the difference between homology and analogy (what he termed "affinity" and "parallelism," respectively). Richard Owen, however, gets credit for crafting the first clear definitions of these terms in the glossary of a treatise on invertebrate zoology. In Owen's formulation, homology is the "same organ in different animals under every variety of form and function" (Owen 1843, p. 379), while analogy is "a part or organ in one animal which has the same function as another part or organ in a different animal" (Owen 1843, p. 374). See Boyden (1943) for a concise history of the terms.

3 Note from the foregoing discussion that the concepts of homology and analogy long predated transmutationist thinking. One of Darwin's contributions was providing an evolutionary explanation for homologies and analogies—these take on a whole new meaning in the context of common descent, and in this way we can understand Darwin's assertion that homologies are important in classification insofar as they reveal affinities, while analogies are nearly worthless.

1 By "numerical parallelism" Darwin means multiple cases of convergence that give the false impression of occurring in a certain mathematical pattern. Hence, the proponents of the quinary system (Macleay, William Swainson, and followers) thought that all classification could be expressed in multiples of five. There were similar, though less well known, systems, such as the seven-based (septenary) system proposed by the entomologist Edward Newman in the 1830s.

the same class or order are compared one with another: thus the shape of the body and fin-like limbs are only analogical when whales are compared with fishes, being adaptations in both classes for swimming through the water; but the shape of the body and fin-like limbs serve as characters exhibiting true affinity between the several members of the whale family; for these cetaceans agree in so many characters, great and small, that we cannot doubt that they have inherited their general shape of body and structure of limbs from a common ancestor. So it is with fishes.

[1] As members of distinct classes have often been adapted by successive slight modifications to live under nearly similar circumstances,—to inhabit for instance the three elements of land, air, and water,—we can perhaps understand how it is that a numerical parallelism has sometimes been observed between the sub-groups in distinct classes. A naturalist, struck by a parallelism of this nature in any one class, by arbitrarily raising or sinking the value of the groups in other classes (and all our experience shows that this valuation has hitherto been arbitrary), could easily extend the parallelism over a wide range; and thus the septenary, quinary, quaternary, and ternary classifications have probably arisen.

As the modified descendants of dominant species, belonging to the larger genera, tend to inherit the advantages, which made the groups to which they belong large and their parents dominant, they are almost sure to spread widely, and to seize on more and more places in the economy of nature. The larger and more dominant groups thus tend to go on increasing in size; and they consequently supplant many smaller and feebler groups. Thus we can account for the fact that all organisms, recent and extinct, are included under a few great orders, under still fewer classes, and all in one great natural system. As showing how few the higher groups are in number, and how widely spread they are throughout the world, the fact is striking, that the discovery of Australia has not added a single insect belonging to a new order; and that in the vegetable kingdom, as I learn from Dr. Hooker, it has added only two or three orders of small size.

In the chapter on geological succession I attempted to show, on the principle of each group having generally diverged much in character during the long-continued process of modification, how it is that the more ancient forms of life often present characters in some slight degree intermediate between existing groups. A few old and intermediate parent-forms having occasionally transmitted to the present day descendants but little modified, will give to us our so-called osculant or aberrant groups. The more aberrant any form is, the greater must be the number of connecting forms which on my theory have been exterminated and utterly lost. And we have some evidence of aberrant forms having suffered severely from extinction, for they are generally represented by extremely few species; and such species as do occur are generally very distinct from each other, which again implies extinction. The genera Ornithorhynchus and Lepidosiren, for example, would not have been less aberrant had each been represented by a dozen species instead of by a single one; but such richness in species, as I find after some investigation, does not commonly fall to the lot of aberrant genera. We can, I think, account for this fact only by looking at aberrant forms as failing groups conquered by more successful competitors, with a few members preserved by some unusual coincidence of favourable circumstances.

Mr. Waterhouse has remarked that, when a member

1 An "aberrant" species could be highly derived or retain primarily primitive characteristics. The two examples Darwin mentions here are primitive cases. *Lepidosiren paradoxa,* the South American lungfish (see p. 330), belongs to the lobe-finned fish lineage, which is basal to terrestrial vertebrates. *Ornithorhynchus anatinus,* the duck-billed platypus (see p. 416), is a monotreme, a group that diverged from marsupials and eutherian mammals over 160 million years ago (see the latest phylogenies of these groups in the Tree of Life project website, tolweb.org).

1 This assessment of the relationship between the "bizcacha"—now better known as the plains viscacha, *Lagostomus maximus*—and marsupials is incorrect. The viscacha, of the South American family Chinchillidae, is not more closely related to marsupials than to other rodents; in fact, most studies suggest that marsupials and rodents are only distantly related (Kullberg et al. 2006). Any points of similarity are thus convergent, which means that Darwin was closer to the truth in his *original* discussion of these rodents in the 1844 *Essay*, which he treated under the heading "analogical affinity":

> . . . of all Rodents the Bizcacha, by certain peculiarities in its reproductive system, approaches nearest to the Marsupials; of all Marsupials the Phascolomys [wombat], on the other hand, appears to approach in the form of its teeth and intestines nearest to the Rodents; but there is no special relation between these two genera; the Bizcacha is no nearer related to the Phascolomys than to any other Marsupial . . . nor again is the Phascolomys, in the points of structure in which it approaches the Rodents, any nearer related to the Bizcacha than to any other Rodent. Other examples might have been chosen, but I have given (from Waterhouse) this example as it illustrates . . . the difficulty of determining what are analogical or adaptive and what real affinities; it seems that the teeth of the Phascolomys though *appearing closely* to resemble those of a Rodent are found to be built on the Marsupial type; and it is thought that these teeth and consequently the intestines may have been adapted to the peculiar life of this animal and therefore may not show any real relation. (F. Darwin 1909, p. 203, emphasis in original)

George Robert Waterhouse discussed the anatomy and general biology of the viscacha in volume II of his book *Natural History of the Mammalia* (London, 1848), and the rodent-like dentition of wombats in volume I (1846). He had earlier argued (1843) that the similarities between wombats and rodents were purely convergent, as Darwin indicates here.

belonging to one group of animals exhibits an affinity to a quite distinct group, this affinity in most cases is general and not special: thus, according to Mr. Waterhouse, of all Rodents, the bizcacha is most nearly related to Marsupials; but in the points in which it approaches this order, its relations are general, and not to any one marsupial species more than to another. As the points of affinity of the bizcacha to Marsupials are believed to be real and not merely adaptive, they are due on my theory to inheritance in common. Therefore we must suppose either that all Rodents, including the bizcacha, branched off from some very ancient Marsupial, which will have had a character in some degree intermediate with respect to all existing Marsupials; or that both Rodents and Marsupials branched off from a common progenitor, and that both groups have since undergone much modification in divergent directions. On either view we may suppose that the bizcacha has retained, by inheritance, more of the character of its ancient progenitor than have other Rodents; and therefore it will not be specially related to any one existing Marsupial, but indirectly to all or nearly all Marsupials, from having partially retained the character of their common progenitor, or of an early member of the group. On the other hand, of all Marsupials, as Mr. Waterhouse has remarked, the phascolomys resembles most nearly, not any one species, but the general order of Rodents. In this case, however, it may be strongly suspected that the resemblance is only analogical, owing to the phascolomys having become adapted to habits like those of a Rodent. The elder De Candolle has made nearly similar observations on the general nature of the affinities of distinct orders of plants.

On the principle of the multiplication and gradual divergence in character of the species descended from a common parent, together with their retention by inheritance of some characters in common, we can understand the excessively complex and radiating affinities by which all the members of the same family or higher group are connected together. For the common parent of a whole family of species, now broken up by extinction into distinct groups and sub-groups, will have transmitted some of its characters, modified in various ways and degrees, to all ; and the several species will consequently be related to each other by circuitous lines of affinity of various lengths (as may be seen in the diagram so often referred to), mounting up through many predecessors. As it is difficult to show the blood-relationship between the numerous kindred of any ancient and noble family, even by the aid of a genealogical tree, and almost impossible to do this without this aid, we can understand the extraordinary difficulty which naturalists have experienced in describing, without the aid of a diagram, the various affinities which they perceive between the many living and extinct members of the same great natural class.

Extinction, as we have seen in the fourth chapter, has played an important part in defining and widening the intervals between the several groups in each class. We may thus account even for the distinctness of whole classes from each other—for instance, of birds from all other vertebrate animals—by the belief that many ancient forms of life have been utterly lost, through which the early progenitors of birds were formerly connected with the early progenitors of the other vertebrate classes. There has been less entire extinction of the forms of life which once connected fishes with batrachians. There has been still less in some other classes, as in that of the Crustacea, for here the most wonderfully diverse forms are still tied

1 By dropping out intermediate forms, extinction has the effect of sharpening the apparent distinctiveness of remaining forms that once would have been linked by a graduated series. This is what Darwin means by "defining and widening" the intervals between taxonomic groups. The affinities of highly derived groups like birds or whales would have been difficult to identify in the absence of fossil intermediate or transitional forms. Darwin mentions birds here; the reptilian affinities of that group were suspected but confirmed in a most spectacular way with the discovery in 1861 of *Archaeopteryx lithographica*, the famous "missing link" found in the Jurassic-age Solnhofen limestone of Germany. Note that this species is not thought to be a *direct* link between reptiles and birds; it is, rather, in an offshoot lineage of early birds. Think of it as a "collateral relative" (to use one of Darwin's phrases) of the direct ancestors, one that has retained many saurian features.

1 This is an interesting thought experiment. Imagine if all extinct species were alive, notwithstanding how crowded the earth would be: if every intermediate, linking form between every group were present they could be arranged in an almost continuous series. The ever-useful divergence of character diagram following page 117 helps us visualize this: simply raise each species at each node in the branching trees (all letters with superscript numerals) up to the top line, where a^{14} through z^{14} are found. If we recall that the x-axis in this diagram represents morphological divergence (see ch. IV), the result of this exercise is a crowded series, albeit with some remaining gaps between the larger sets of taxa (i.e., between all descendants of A and all descendants of I). Note how the intermediates link a^{14} and m^{14} in a continuous series. And yet, as Darwin points out, a natural classification would still be possible because all these species fall into nested subsets reflecting different degrees of similarity and difference from one another.

together by a long, but broken, chain of affinities. Extinction has only separated groups: it has by no means made them; for if every form which has ever lived on this earth were suddenly to reappear, though it would be quite impossible to give definitions by which each group could be distinguished from other groups, as all would blend together by steps as fine as those between the finest existing varieties, nevertheless a natural classification, or at least a natural arrangement, would be possible. We shall see this by turning to the diagram: the letters, A to L, may represent eleven Silurian genera, some of which have produced large groups of modified descendants. Every intermediate link between these eleven genera and their primordial parent, and every intermediate link in each branch and sub-branch of their descendants, may be supposed to be still alive; and the links to be as fine as those between the finest varieties. In this case it would be quite impossible to give any definition by which the several members of the several groups could be distinguished from their more immediate parents; or these parents from their ancient and unknown progenitor. Yet the natural arrangement in the diagram would still hold good; and, on the principle of inheritance, all the forms descended from A, or from I, would have something in common. In a tree we can specify this or that branch, though at the actual fork the two unite and blend together. We could not, as I have said, define the several groups; but we could pick out types, or forms, representing most of the characters of each group, whether large or small, and thus give a general idea of the value of the differences between them. This is what we should be driven to, if we were ever to succeed in collecting all the forms in any class which have lived throughout all time and space. We shall certainly never succeed in making so perfect a collection: nevertheless, in certain classes, we are tending in this direction; and Milne Edwards has lately insisted, in an able paper, on the high importance of looking to types, whether or not we can separate and define the groups to which such types belong.

Finally, we have seen that natural selection, which results from the struggle for existence, and which almost inevitably induces extinction and divergence of character in the many descendants from one dominant parent-species, explains that great and universal feature in the affinities of all organic beings, namely, their subordination in group under group. We use the element of descent in classing the individuals of both sexes and of all ages, although having few characters in common, under one species; we use descent in classing acknowledged varieties, however different they may be from their parent; and I believe this element of descent is the hidden bond of connexion which naturalists have sought under the term of the Natural System. On this idea of the natural system being, in so far as it has been perfected, genealogical in its arrangement, with the grades of difference between the descendants from a common parent, expressed by the terms genera, families, orders, &c., we can understand the rules which we are compelled to follow in our classification. We can understand why we value certain resemblances far more than others; why we are permitted to use rudimentary and useless organs, or others of trifling physiological importance; why, in comparing one group with a distinct group, we summarily reject analogical or adaptive characters, and yet use these same characters within the limits of the same group. We can clearly see how it is that all living and extinct forms can be grouped together in one great system; and how the several members of each class are connected together by the most complex and radiating

1 It may seem odd for Darwin to endorse "looking to types," given that he was no advocate of the unchanging (Platonic) archetype concept of Owen and Agassiz. But the idea has its value: "type" can simply refer to basic body plan. In fact, Huxley adopted the term "archetype" from Owen but redefined it. He came up with the idea of using embryology to help define anatomical uniformity of classes of organisms, an approach that Darwin embraced: "I am very much obliged for your Paper on the mollusca," he wrote to Huxley in April 1853. "I have read it all with much interest; but it [would] be ridiculous in me to make any remarks on a subject on which I am so utterly ignorant; but I can see its high importance. The discovery of the type or 'idea' (in your sense, for I detest the word as used by Owen, Agassiz & Co) of each great class, I cannot doubt is one of the very highest ends of Natural History: & certainly most interesting to the worker" (*Correspondence* 5: 133).

2 This is the crux of this concluding paragraph on classification. Common descent is implicitly used by naturalists, Darwin argues; he urges his readers to recognize this "hidden bond" underlying classification. Common descent is what makes the Natural System natural and at once explains why the various rules and practices used in classification work so well. The concluding sentence on this topic in the 1844 *Essay* expresses this nicely: "These terms of affinity, relations, families, adaptive characters, &c., which naturalists cannot avoid using, though metaphorically, cease being so, and are full of plain signification" (F. Darwin 1909, p. 213).

[1] lines of affinities. We shall never, probably, disentangle the inextricable web of affinities between the members of any one class; but when we have a distinct object in view, and do not look to some unknown plan of creation, we may hope to make sure but slow progress.

[2] *Morphology.*—We have seen that the members of the same class, independently of their habits of life, resemble each other in the general plan of their organisation. This resemblance is often expressed by the
[3] term "unity of type;" or by saying that the several parts and organs in the different species of the class are homologous. The whole subject is included under the general name of Morphology. This is the most interesting department of natural history, and may be said to be its very soul. What can be more curious than that the hand of a man, formed for grasping, that of a mole for digging, the leg of the horse, the paddle of the porpoise, and the wing of the bat, should all be constructed on the same pattern, and should include the same bones, in the same relative positions? Geoffroy St. Hilaire has insisted strongly on the high importance of relative connexion in homologous organs:
[4] the parts may change to almost any extent in form and size, and yet they always remain connected together in the same order. We never find, for instance, the bones of the arm and forearm, or of the thigh and leg, transposed. Hence the same names can be given to the homologous bones in widely different animals. We see the same great law in the construction of the mouths of insects: what can be more different than the immensely long spiral proboscis of a sphinx-moth, the curious folded one of a bee or bug, and the great jaws of a beetle?—yet all these organs, serving for such dif-

1 This final sentence of the classification section is worth noting for its dismissal of the then-conventional view that classification signifies nothing more than a plan of creation—which was long the philosophical underpinning of classification. Linnaeus's *Systema Naturae* (1758) opens with an epigraph paraphrasing Psalm 103, verse 24: "O Lord, How manifold are your works! How wisely have you made them! How full is the earth with your riches!" In Linnaeus's own words: "God created, Linnaeus arranged."

2 The anatomical structure of organisms forms another important line of evidence for Darwin in support of his theory. The term "morphology" was first used by Goethe in 1796. He defined it in an 1817 treatise as "the science of the form *(Gestalt)*, formation *(Bildung)* and transformation *(Umbildung)* of organic beings" (quoted in Opitz 2004, p. 6). The comparative study of morphology had strong philosophical/theological underpinnings, and it is unsurprising that many of Darwin's most severe critics were the great morphologists of the day. Darwin's evolutionary interpretation of patterns evident in comparative morphology—what he calls "the most interesting department of natural history—meant these naturalists had been barking up the wrong explanatory tree their whole careers!

3 "Unity of type"—Owen's "homology"—was a philosophical position stemming from the German idealistic *Naturphilosophie* tradition. Recall from chapter VI the Cuvier-Geoffroy debates. Geoffroy and followers focused on adaptive function: human arms, bird and bat wings, and whale flukes, for example, were seen as contrivances to allow adaptation to "conditions of existence." Cuvier, Owen, Agassiz, and their followers held that "unity of type" was of primary importance and adaptations secondary. Bird, bat, and whale limbs are modifications of a single body plan, and the task of comparative morphology is discovering that plan to find the form upon which the creator designed the organisms; adaptations were incidental.

4 Darwin would have been amazed at the depth of understanding achieved in so many areas of biology, but perhaps no ad-
(continued)

ferent purposes, are formed by infinitely numerous modifications of an upper lip, mandibles, and two pairs of maxillæ. Analogous laws govern the construction of the mouths and limbs of crustaceans. So it is with the flowers of plants.

Nothing can be more hopeless than to attempt to explain this similarity of pattern in members of the same class, by utility or by the doctrine of final causes. The hopelessness of the attempt has been expressly admitted by Owen in his most interesting work on the 'Nature of Limbs.' [1] On the ordinary view of the independent creation of each being, we can only say that so it is;—that it has so pleased the Creator to construct each animal and plant.

The explanation is manifest on the theory of the natural selection of successive slight modifications,—each modification being profitable in some way to the modified form, but often affecting by correlation of growth other parts of the organisation. In changes of this nature, there will be little or no tendency to modify the original pattern, or to transpose parts. The bones of a limb might be shortened and widened to any extent, and become gradually enveloped in thick membrane, so as to serve as a fin; or a webbed foot might have all its bones, or certain bones, lengthened to any extent, and the membrane connecting them increased to any extent, so as to serve as a wing: yet in all this great amount of modification there will be no tendency to alter the framework of bones or the relative connexion of the several parts. If we suppose that the ancient progenitor, the archetype as it may be called, of all mammals, had its limbs constructed on the existing general pattern, for whatever purpose they served, we can at once perceive the plain signification of the homologous construction of the limbs throughout the whole class. So with the mouths of insects, we have only to

(continued)
vance would have amazed him more than the genetics and development of organisms. The body plan of animals is determined by a set of "homeobox" developmental genes. Mutations in these genes can yield oddities like transposed limbs and repeated structures, a fact that reveals their homology at the most fundamental genetic level.

In modern terms, the principle here is modification of structures in serial homology (called the Law of Repetition by Goethe). One or more rounds of duplication of an ancestral structure yield multiple copies that might then be modified in different, complementary ways. Thus the mouthpart modifications of insects, which Darwin mentions here, are serially homologous appendages that amount to variations on a theme: the labrum, mandibles, labium, and maxillae are each modified in different ways to yield chewing, sucking, piercing, etc., mouthpart sets (Jockusch et al. 2004).

1 In 1849 Owen published *On the Nature of Limbs: A Discourse*, in which he developed his concept of the archetype. Darwin wrote to him in 1850: "I am in the middle of the 'Limbs' with uncommon interest—The manner in which you work out the toes strikes me as quite beautiful. Whoever would have thought that a great Cart-horse walked on four fingers!" (*Correspondence* 4: 295). Owen would not have taken kindly to Darwin's dismissal of the "ordinary view of independent creation"—a view that Owen shared in some measure. It should be pointed out, however, that Owen was somewhat receptive to transmutationist thinking as early as the 1840s but for complex reasons rejected Darwin's model. Later in life, maybe sensing that he was on the losing side of history, he actually proposed an evolutionary scheme of his own, the "derivation" or "filiation" hypothesis (see Owen 1868, chapter XI). This ill-conceived model was one of transmutation driven by predestination, but it firmly rejected natural selection and gradualism (MacLeod 1965, Padian 1997).

suppose that their common progenitor had an upper lip, mandibles, and two pair of maxillæ, these parts being perhaps very simple in form; and then natural selection will account for the infinite diversity in structure and function of the mouths of insects. Nevertheless, it is conceivable that the general pattern of an organ might become so much obscured as to be finally lost, by the atrophy and ultimately by the complete abortion of certain parts, by the soldering together of other parts, and by the doubling or multiplication of others,—variations which we know to be within the limits of possibility. In the paddles of the extinct gigantic sea-lizards, and [1] in the mouths of certain suctorial crustaceans, the general pattern seems to have been thus to a certain extent obscured.

[2] There is another and equally curious branch of the present subject; namely, the comparison not of the same part in different members of a class, but of the different parts or organs in the same individual. Most physiologists believe that the bones of the skull are homologous with—that is correspond in number and in relative connexion with—the elemental parts of a certain number of vertebræ. The anterior and posterior limbs in each member of the vertebrate and articulate classes are plainly homologous. We see the same law in comparing the wonderfully complex jaws and legs in [3] crustaceans. It is familiar to almost every one, that in a flower the relative position of the sepals, petals, stamens, and pistils, as well as their intimate structure, are intelligible on the view that they consist of metamorphosed leaves, arranged in a spire. In monstrous plants, we often get direct evidence of the possibility of one organ being transformed into another; and we can actually see in embryonic crustaceans and in many other animals, and in flowers, that organs, which when mature

1 What in the world is a suctorial crustacean, you may be wondering. The evocative term refers to a hodgepodge of parasitic crustaceans that adhere to the bodies of fish and other marine animals, including other crustaceans. They imbibe blood and other fluids by means of a proboscis sometimes called the suctorial cone, a mouthpart modification unlike that found in any other crustacean group. Significantly, parasitic taxa differ in the anatomical details of these mouthparts: variations on a theme.

2 This is the other side of the serial homology coin. Structures of the same individual can be homologous. Thus mouthparts, legs, and antennae of insects are modifications of a common ancestral structure, as are the mouthparts and legs of crustaceans. With vertebrates, the homology of cranial bones with the vertebrae is accepted today in very general terms (Gans 1993, Forey and Janvier 1994) but was hotly debated in the early nineteenth century. The idea originated with the German naturalist Lorenz Oken, who presented his theory in an address entitled "Signification of the bones of the skull" in 1807. Owen later championed the idea in his book *On the Archetype and Homologies of the Vertebrate Skeleton* (London, 1848).

3 Horticulturists were familiar with "monstrous plants," or sports (e.g., cases in which a leaf develops where a petal or thorn should be). Sporting is so common that some later critics proposed a now-defunct theory of saltational evolution (from Latin *saltus*, "jumping") based on mutations. The most famous exponent of this idea was the Dutch botanist Hugo de Vries (1848–1935), author of *Species and Varieties, Their Origin by Mutation* (Chicago, 1905). De Vries thought that species arose not gradually but by radical shifts in morphology brought about by mutations. His treatise concludes with the memorable statement: "Natural selection may explain the survival of the fittest, but it cannot explain the arrival of the fittest."

become extremely different, are at an early stage of growth exactly alike.

How inexplicable are these facts on the ordinary view of creation! Why should the brain be enclosed in a box composed of such numerous and such extraordinarily shaped pieces of bone? As Owen has remarked, the benefit derived from the yielding of the separate pieces in the act of parturition of mammals, will by no means explain the same construction in the skulls of birds. Why should similar bones have been created in the formation of the wing and leg of a bat, used as they are for such totally different purposes? Why should one crustacean, which has an extremely complex mouth formed of many parts, consequently always have fewer legs; or conversely, those with many legs have simpler mouths? Why should the sepals, petals, stamens, and pistils in any individual flower, though fitted for such widely different purposes, be all constructed on the same pattern? ◀1

On the theory of natural selection, we can satisfactorily answer these questions. In the vertebrata, we see a series of internal vertebræ bearing certain processes and appendages; in the articulata, we see the body divided into a series of segments, bearing external appendages; and in flowering plants, we see a series of successive spiral whorls of leaves. An indefinite repetition of the same part or organ is the common characteristic (as Owen has observed) of all low or little-modified forms; therefore we may readily believe that the unknown progenitor of the vertebrata possessed many vertebræ; the unknown progenitor of the articulata, many segments; and the unknown progenitor of flowering plants, many spiral whorls of leaves. We have formerly seen that parts many times repeated are eminently liable to vary in number and structure; consequently it is quite probable that

1 In this lyrical passage Darwin touches on an apparent difficulty for those inclined to see evidence of design in organisms. The skull sutures of mammals—perhaps especially humans—might be seen as a sign of benevolent design, permitting the compression of the skull during birth. As even Owen was aware, however, this feature could not have been designed for human parturition since such sutures are also found in birds and reptiles, which are born from eggs. Darwin refers to Owen's remark in his treatise *On the Archetype and Homologies of the Vertebrate Skeleton:*

> We may admit that the multiplied points of ossification in the skull of the human foetus facilitate, and were designed to facilitate, childbirth; yet something more than such a final purpose lies beneath the fact, that most of those osseous centres represent permanently distinct bones in the cold-blooded vertebrates. The cranium of the bird, which is composed in the adult of a single bone, is ossified from the same number of points as in the human embryo, without the possibility of a similar purpose being subserved thereby, in the extrication of the chick from the fractured egg-shell. (Owen 1848b, p. 73)

Owen also considered the homology of repeated flower parts, writing in 1855 of the "law of vegetative or irrelative repetition," which explained the "multiplication of organs performing the same function." Darwin heavily annotated this and other works by Owen.

1 Many mollusks are so modified that identifying homologies is very difficult. Bivalves and gastropods posed a special challenge: bivalves (clams, mussels, etc.) have no obvious head and tail end, and gastropods (snail, slugs, etc.) exhibit *torsion* in development, which involves a 180° twisting of the mantle and viscera such that the posterior end becomes situated above the head facing forward. Darwin found them perplexing. Huxley sent him a copy of an article he wrote attempting to establish molluskan homologies; Darwin wrote back in gratitude: "Many thanks for your paper . . . I have read it with attention & interest, for I had often wondered how a gasteropod [sic], a bivalve, & cephalopod [could] be brought to same type. . . . An acephalous Mollusc has always looked to me a complete mystery, & I really know no more about it, than a man does, who has only eat oyster patties; the relation of the animal to the shell & crust being about the same in my eyes" (*Correspondence* 5: 281).

2 We saw on p. 436 that Owen, after Goethe and Oken, argued for the homology of cranial plates and vertebrae. Huxley, who squabbled endlessly with Owen, took issue with the idea and attacked it at length in his Croonian Lecture *On The Theory of the Vertebrate Skull.* Huxley argued on the basis of embryology that the two were not homologous. Owen's philosophy, Huxley said, was "introduc[ing] the phraseology and mode of thought of an obsolete and scholastic realism into biology." He continued: "The fallacy involved in the vertebral theory of the skull is like that which . . . infested our notions of the relations between fishes and mammals. The mammal was imagined to be a modified fish, whereas, in truth, fish and mammal start from a common point, and each follows its own road thence" (Huxley 1858, p. 432). This is what Darwin is getting at when he speaks of skull and vertebrae having been metamorphosed from a "common element."

natural selection, during a long-continued course of modification, should have seized on a certain number of the primordially similar elements, many times repeated, and have adapted them to the most diverse purposes. And as the whole amount of modification will have been effected by slight successive steps, we need not wonder at discovering in such parts or organs, a certain degree of fundamental resemblance, retained by the strong principle of inheritance.

1 In the great class of molluscs, though we can homologise the parts of one species with those of another and distinct species, we can indicate but few serial homologies; that is, we are seldom enabled to say that one part or organ is homologous with another in the same individual. And we can understand this fact; for in molluscs, even in the lowest members of the class, we do not find nearly so much indefinite repetition of any one part, as we find in the other great classes of the animal and vegetable kingdoms.

2 Naturalists frequently speak of the skull as formed of metamorphosed vertebræ: the jaws of crabs as metamorphosed legs; the stamens and pistils of flowers as metamorphosed leaves; but it would in these cases probably be more correct, as Professor Huxley has remarked, to speak of both skull and vertebræ, both jaws and legs, &c.,—as having been metamorphosed, not one from the other, but from some common element. Naturalists, however, use such language only in a metaphorical sense: they are far from meaning that during a long course of descent, primordial organs of any kind—vertebræ in the one case and legs in the other—have actually been modified into skulls or jaws. Yet so strong is the appearance of a modification of this nature having occurred, that naturalists can hardly avoid employing language having this plain signification. On my view

these terms may be used literally; and the wonderful fact of the jaws, for instance, of a crab retaining numerous characters, which they would probably have retained through inheritance, if they had really been metamorphosed during a long course of descent from true legs, or from some simple appendage, is explained.

Embryology.—It has already been casually remarked that certain organs in the individual, which when mature become widely different and serve for different purposes, are in the embryo exactly alike. The embryos, also, of distinct animals within the same class are often strikingly similar: a better proof of this cannot be given, than a circumstance mentioned by Agassiz, namely, that having forgotten to ticket the embryo of some vertebrate animal, he cannot now tell whether it be that of a mammal, bird, or reptile. The vermiform larvæ of moths, flies, beetles, &c., resemble each other much more closely than do the mature insects; but in the case of larvæ, the embryos are active, and have been adapted for special lines of life. A trace of the law of embryonic resemblance, sometimes lasts till a rather late age: thus birds of the same genus, and of closely allied genera, often resemble each other in their first and second plumage; as we see in the spotted feathers in the thrush group. In the cat tribe, most of the species are striped or spotted in lines; and stripes can be plainly distinguished in the whelp of the lion. We occasionally though rarely see something of this kind in plants: thus the embryonic leaves of the ulex or furze, and the first leaves of the phyllodineous acaceas, are pinnate or divided like the ordinary leaves of the leguminosæ.

The points of structure, in which the embryos of widely different animals of the same class resemble each other, often have no direct relation to their condi-

1 Darwin may have singled out the crab example because crabs exhibit a series of appendages from walking legs to mouthparts, making the homology of mouthparts with walking legs especially obvious. Decapods, the order of crabs and lobsters, have an eight-segmented thorax. The first three thoracic segments (thoracomeres) are fused with the head segments to form a cephalothorax. Anterior to the thoracomeres, on the head proper, are the mandibles and maxillae. Posterior to the thoracomeres are the five independent thoracic segments. All segments of the thorax have a pair of appendages: the thoracomeres each bear maxillipeds, and the independent segments each bear walking legs (the name Decapoda, "ten feet," alludes to these five pairs of legs). The maxillipeds, then, are mouthparts that are literally and structurally intermediate between the biting-and-chewing mouthparts of the head and the walking legs.

2 Darwin opened the section on embryology in his 1844 *Essay* with a stronger statement of its significance: "The unity of type in the great classes is shown in another and very striking manner, namely, in the stages through which the embryo passes in coming to maturity" (F. Darwin 1909, p. 218).

3 Darwin erroneously attributes this anecdote to Agassiz; in fact, the German embryologist Karl Ernst von Baer related this story in his important embryological treatise *Über Entwickelungsgeschichte der Thiere* (Konigsberg, 1828). Darwin corrected his error in the third edition.

4 A botanical example: *Ulex europaeus,* also known as gorse, furze, and Spanish broom, is a dry-adapted legume native to southwestern Europe and northwest Africa. It has prickly, almost thorny, narrow leaves. Acacias (genus *Acacia*) are legumes widely distributed throughout the tropical and warm-temperate regions of the world. One large group of acacias has well-developed phyllodes, broadened and flattened petioles that function as leaves, in addition to the "true" pinnately compound leaves so common in the family. Darwin's point here is that the first or "embryonic" leaves to develop in *Ulex* and phyllode-

(continued)

[1] tions of existence. We cannot, for instance, suppose that in the embryos of the vertebrata the peculiar loop-like course of the arteries near the branchial slits are related to similar conditions,—in the young mammal which is nourished in the womb of its mother, in the egg of the bird which is hatched in a nest, and in the spawn of a frog under water. We have no more reason to believe in such a relation, than we have to believe that the same bones in the hand of a man, wing of a bat, and fin of a porpoise, are related to similar conditions of life. No one will suppose that the stripes on the whelp of a lion, or the spots on the young blackbird, are of any use to these animals, or are related to the conditions to which they are exposed.

[2] The case, however, is different when an animal during any part of its embryonic career is active, and has to provide for itself. The period of activity may come on earlier or later in life; but whenever it comes on, the adaptation of the larva to its conditions of life is just as perfect and as beautiful as in the adult animal. From such special adaptations, the similarity of the larvæ or active embryos of allied animals is sometimes much obscured; and cases could be given of the larvæ of two species, or of two groups of species, differing quite as much, or even more, from each other than do their adult parents. In most cases, however, the larvæ, though active, still obey more or less closely the law of com-
[3] mon embryonic resemblance. Cirripedes afford a good instance of this: even the illustrious Cuvier did not perceive that a barnacle was, as it certainly is, a crustacean; but a glance at the larva shows this to be the case in an unmistakeable manner. So again the two main divisions of cirripedes, the pedunculated and sessile, which differ widely in external appearance, have larvæ in all their several stages barely distinguishable.

(continued)
bearing *Acacia* are pinnately compound like most leguminous species, thus exhibiting taxonomic affinity early in development and becoming more specialized later in development.

1 The branchial arteries carry blood to and from the gills in fish; the Greek *branchium* translates, in fact, as "gill." The oxygenated blood from the gills is returned by the trematic loops, which surround each gill opening. Birds, amphibians, reptiles, and mammals exhibit gill arches in embryonic development and retain the associated looping of the arterial system. Why, Darwin asks, should organisms that are embryonically nourished in such different ways all retain essentially the same anatomical arrangement as fish? The implication is that this represents the stamp of history and reflects the common descent of these vertebrates.

2 Darwin uses the term "embryonic" in the broad sense of all pre-adult, juvenile stages.

3 Barnacles are highly modified crustaceans, subclass Cirripedia (a name derived from the feathery appearance of the legs, which function as modified filter-feeding structures). Cuvier classified barnacles as mollusks, placing them within their own molluskan class, the Cirrhopoda (note the similarity of the names). Crustaceans have a nonfeeding larval stage called the *nauplius* (plural *nauplii*). Some groups pass through this stage in embryonic development (e.g., lobsters), while others have an active nauplius capable of swimming upon hatching (e.g., most shrimp). Many nauplii use their antennae for swimming—itself an oddity. Interestingly, the term "nauplius" is derived from the Greek term for a type of mollusk; it was introduced in 1785 by the zoologist O. F. Müller, who concurred with Cuvier's classification.

Evidence from barnacles appears repeatedly in this section. Darwin's four-volume treatise on barnacles (1851–1854) is a study in the application of his conviction that homology and embryology best reveal species relationships and thus inform classification. Struck by the work of Milne-Edwards (1844), Darwin embraced the idea that the most general features of a
(continued)

The embryo in the course of development generally rises in organisation: I use this expression, though I am aware that it is hardly possible to define clearly what is meant by the organisation being higher or lower. But no one probably will dispute that the butterfly is higher than the caterpillar. In some cases, however, the mature animal is generally considered as lower in the scale than the larva, as with certain parasitic crustaceans. To refer once again to cirripedes: the larvæ in the first stage have three pairs of legs, a very simple single eye, and a probosciformed mouth, with which they feed largely, for they increase much in size. In the second stage, answering to the chrysalis stage of butterflies, they have six pairs of beautifully constructed natatory legs, a pair of magnificent compound eyes, and extremely complex antennæ; but they have a closed and imperfect mouth, and cannot feed: their function at this stage is, to search by their well-developed organs of sense, and to reach by their active powers of swimming, a proper place on which to become attached and to undergo their final metamorphosis. When this is completed they are fixed for life: their legs are now converted into prehensile organs; they again obtain a well-constructed mouth; but they have no antennæ, and their two eyes are now reconverted into a minute, single, and very simple eye-spot. In this last and complete state, cirripedes may be considered as either more highly or more lowly organised than they were in the larval condition. But in some genera the larvæ become developed either into hermaphrodites having the ordinary structure, or into what I have called complemental males: and in the latter, the development has assuredly been retrograde; for the male is a mere sack, which lives for a short time, and is destitute of mouth, stomach, or other organ of importance, excepting for reproduction.

(continued)

group appear earliest in development and are highly informative for revealing affinities. Owen, in contrast, emphasized homologies of the adults. On that basis he put barnacles in their own class between annelids and crustaceans, while Darwin recognized them as crustaceans (see Appendix II in *Correspondence* vol. 4).

1 Darwin struggled with labels like "higher" and "lower" or "more" and "less" evolved. On the one hand, he knew that on his theory all organisms, even the seemingly simplest in structure, must be highly adapted to their conditions of life, and that we, big-brained bipedal mammals that we are, are biased in assessing complexity. "It is absurd to talk of one animal being higher than another," he wrote in the B notebook. "*We* consider those, where the cerebral structure/intellectual faculties most developed, as highest.—A bee doubtless would when the instincts were.—" (Barrett et al. 1987, p. 189, emphasis Darwin's). So is a bacterium "lower" or "less evolved" than a barnacle or a bird? Hardly; but on the other hand, Darwin was immersed in a culture of progressivist thinking and was steeped in a science imbued with "scale of nature" philosophy, both of which exerted an irresistible gravitational pull. With the advent of embryology, some zoologists used degree of developmental departure from a type or basic body plan to determine how highly developed an organism was.

2 But species like barnacles with complicated developmental morphs—each well adapted to particular circumstances or uses—presented a conundrum for those inclined to think in terms of "higher" and "lower" animals. First-stage larvae are nauplii, capable of swimming and feeding. In the second stage, they develop even more legs (natatory legs are swimming legs, from Latin *nato*, "swimming") plus antennae, but lose their mouth and do not feed. These seek a settling area, where they metamorphose to the sessile encrusting barnacle form with which most people are familiar; they get their mouth back, and their legs morph into feeding structures. So which stage is "highest" in development?

1 This is an insightful argument. Darwin is pointing out that things do not have to be thus; in fact, on the alternative model of special creation one might expect a more straightforward pattern in evidence. A developmental system in which the embryonic stages are just successively larger versions of the same form seems to suffice for some groups, like Owen's cuttlefish—why the complicated and radical morphological changes, not to mention the way seemingly unrelated groups (fish, birds, mammals, say) all seem to have the same starting point in embryonic morphology? It is possible to respond that development is thus because God made it that way, but Darwin would say that answer sidesteps the question and is tantamount to no answer at all.

2 Aphids are one of the few insect groups to have live births, and they reproduce asexually (parthenogenetically) to boot. Mother aphids thus produce perfectly formed miniatures of themselves; Huxley read a paper on this subject before the Linnean Society in 1857. By contrast, most insects (in particular those with complete development) exhibit a wormlike (vermiform) starting point, just as vertebrate embryos proceed from a virtually indistinguishable starting point.

Insects nicely illustrate Darwin's argument in a different manner, too. Three post-embryonic developmental patterns are evident in insects. *Ametabolous* insects (e.g., silverfish and bristletails) exhibit no metamorphosis; each stage appears similar to the previous one in nearly all respects. *Hemimetabolous* insects (e.g., grasshoppers, mayflies, true bugs) exhibit incomplete metamorphosis, with a gradual change in external appearance, notably in the development of wings. Feeding ecology of the juveniles is the same as that of the adults in both groups, while it differs radically between juveniles and adults in *Holometabolous* insects (e.g., ants, beetles, butterflies, true flies, etc.). These have complete development, exhibiting morphologically distinct stages from larva to pupa to adult. The feeding specialization of larvae vs. adults might explain the holometabolous pattern of development, but why should ametabola and hemimetabola differ when adult and juvenile ecologies are the same?

1 We are so much accustomed to see differences in structure between the embryo and the adult, and likewise a close similarity in the embryos of widely different animals within the same class, that we might be led to look at these facts as necessarily contingent in some manner on growth. But there is no obvious reason why, for instance, the wing of a bat, or the fin of a porpoise, should not have been sketched out with all the parts in proper proportion, as soon as any structure became visible in the embryo. And in some whole groups of animals and in certain members of other groups, the embryo does not at any period differ widely from the adult: thus Owen has remarked in regard to cuttle-fish, "there is no metamorphosis; the cephalopodic character is manifested long before the parts of the embryo are completed;" and again in spiders, "there is nothing worthy to be called a metamorphosis." The larvæ of insects, whether adapted to the most diverse and active habits, or quite inactive, being fed by their parents or placed in the midst of proper nutriment, yet nearly all pass through a similar worm-like stage of development;
2 but in some few cases, as in that of Aphis, if we look to the admirable drawings by Professor Huxley of the development of this insect, we see no trace of the vermiform stage.

How, then, can we explain these several facts in embryology,—namely the very general, but not universal difference in structure between the embryo and the adult;—of parts in the same indivividual embryo, which ultimately become very unlike and serve for diverse purposes, being at this early period of growth alike;—of embryos of different species within the same class, generally, but not universally, resembling each other;—of the structure of the embryo not being closely related to its conditions of existence, except when the embryo becomes at any period of life active and has to provide for itself;—of the embryo apparently having sometimes a higher organisation than the mature animal, into which it is developed. I believe that all these facts can be explained, as follows, on the view of descent with modification.

It is commonly assumed, perhaps from monstrosities often affecting the embyro at a very early period, that slight variations necessarily appear at an equally early period. But we have little evidence on this head—indeed the evidence rather points the other way; for it is notorious that breeders of cattle, horses, and various fancy animals, cannot positively tell, until some time after the animal has been born, what its merits or form will ultimately turn out. We see this plainly in our own children; we cannot always tell whether the child will be tall or short, or what its precise features will be. The question is not, at what period of life any variation has been caused, but at what period it is fully displayed. The cause may have acted, and I believe generally has acted, even before the embryo is formed; and the variation may be due to the male and female sexual elements having been affected by the conditions to which either parent, or their ancestors, have been exposed. Nevertheless an effect thus caused at a very early period, even before the formation of the embryo, may appear late in life; as when an hereditary disease, which appears in old age alone, has been communicated to the offspring from the reproductive element of one parent. Or again, as when the horns of cross-bred cattle have been affected by the shape of the horns of either parent. For the welfare of a very young animal, as long as it remains in its mother's womb, or in the egg, or as long as it is nourished and protected by its parent, it must be quite unimportant whether most of its characters are fully

1 The gist of Darwin's argument is that his theory can explain the empirical patterns observed of embryology. The seemingly disparate observations all point in the same direction—they are consilient, as William Whewell might have put it.

2 Embryonic development tends to proceed from more general features to more specific, or specialized, features. This fact lies at the heart of the simplistic idea that evolution was a process of simply adding on to ancestral life stages—one manifestation of which was recapitulation theory, which posited that organisms literally trace their evolutionary lineage in the process of development. This long-discredited idea was expressed in the famous aphorism "ontogeny recapitulates phylogeny." Later in this section (p. 449) Darwin will discuss recapitulation in a more nuanced way.

1 The point here is that there is much predictability in development; moreover, the timing of development of certain features is itself a trait that exhibits heritability and variation (which in turn means that it can be shaped by natural selection).

2 To be clear on the two principles: the first is that successive modifications appear later rather than earlier in development, as if they were "add-ons" (Darwin uses the word "supervening" to describe them); the second is that the age of appearance of traits tends to be the same in parents and offspring.

3 These studies were undertaken as part of Darwin's research applying his theory of descent to domesticated varieties. His correspondence provides a wonderful source of insight into the day-to-day mechanics—and frustrations—of this work. The following passage from a May 1855 letter to his cousin William Darwin Fox (*Correspondence* 5: 325) mentions both the puppies and the horses discussed in this paragraph: "[Should] an old wild Turkey ever die please remember me: I do not care for Baby turkey. Nor for a mastiff. Very many thanks for your offer.—I have puppies of Bull-dogs & Greyhound in salt.—& I have had Carthorse & Race Horse young colts carefully measured.—Whether I shall do any good I doubt: I am getting out of my depth."

acquired a little earlier or later in life. It would not signify, for instance, to a bird which obtained its food best by having a long beak, whether or not it assumed a beak of this particular length, as long as it was fed by its parents. Hence, I conclude, that it is quite possible, that each of the many successive modifications, by which each species has acquired its present structure, may have supervened at a not very early period of life; and some direct evidence from our domestic animals supports this view. But in other cases it is quite possible that each successive modification, or most of them, may have appeared at an extremely early period.

1 I have stated in the first chapter, that there is some evidence to render it probable, that at whatever age any variation first appears in the parent, it tends to reappear at a corresponding age in the offspring. Certain variations can only appear at corresponding ages, for instance, peculiarities in the caterpillar, cocoon, or imago states of the silk-moth; or, again, in the horns of almost full-grown cattle. But further than this, variations which, for all that we can see, might have appeared earlier or later in life, tend to appear at a corresponding age in the offspring and parent. I am far from meaning that this is invariably the case; and I could give a good many cases of variations (taking the word in the largest sense) which have supervened at an earlier age in the child than in the parent.

2 These two principles, if their truth be admitted, will, I believe, explain all the above specified leading facts in embryology. But first let us look at a few analogous
3 cases in domestic varieties. Some authors who have written on Dogs, maintain that the greyhound and bulldog, though appearing so different, are really varieties most closely allied, and have probably descended from

the same wild stock; hence I was curious to see how far their puppies differed from each other: I was told by breeders that they differed just as much as their parents, and this, judging by the eye, seemed almost to be the case; but on actually measuring the old dogs and their six-days old puppies, I found that the puppies had not nearly acquired their full amount of proportional difference. So, again, I was told that the foals of cart and race-horses differed as much as the full-grown animals; and this surprised me greatly, as I think it probable that the difference between these two breeds has been wholly caused by selection under domestication; but having had careful measurements made of the dam and of a three-days old colt of a race and heavy cart-horse, I find that the colts have by no means acquired their full amount of proportional difference.

As the evidence appears to me conclusive, that the several domestic breeds of Pigeon have descended from one wild species, I compared young pigeons of various breeds, within twelve hours after being hatched; I carefully measured the proportions (but will not here give details) of the beak, width of mouth, length of nostril and of eyelid, size of feet and length of leg, in the wild stock, in pouters, fantails, runts, barbs, dragons, carriers, and tumblers. Now some of these birds, when mature, differ so extraordinarily in length and form of beak, that they would, I cannot doubt, be ranked in distinct genera, had they been natural productions. But when the nestling birds of these several breeds were placed in a row, though most of them could be distinguished from each other, yet their proportional differences in the above specified several points were incomparably less than in the full-grown birds. Some characteristic points of difference—for instance, that of the width of mouth—could hardly be detected in the ◀1

1 In the mid-1850s Darwin amassed a sizable collection of pigeon varieties. He wrote to the naturalist Thomas Campbell Eyton, who had a museum of bird specimens, for advice on preparing the skeletons: "I am getting on with my collection of Pigeons, & now have pairs of ten varieties alive & shall on Saturday receive two or three more kinds." Initially he prepared his own skeletons, using this recipe found among his notes: "Skeletons 1/4 oz Caustic Potash to Pint of Water Silver Oxide 1/2 gram: twice a day, for month." He could not tolerate the stench, however, so in 1856 he began sending the skeletons out for preparation (see *Correspondence* 5: 508, and notes). Significantly, Darwin found that the young birds differed far less from one another than did the adults.

young. But there was one remarkable exception to this rule, for the young of the short-faced tumbler differed from the young of the wild rock-pigeon and of the other breeds, in all its proportions, almost exactly as much as in the adult state.

[1] The two principles above given seem to me to explain these facts in regard to the later embryonic stages of our domestic varieties. Fanciers select their horses, dogs, and pigeons, for breeding, when they are nearly grown up: they are indifferent whether the desired qualities and structures have been acquired earlier or later in life, if the full-grown animal possesses them. And the cases just given, more especially that of pigeons, seem to show that the characteristic differences which give value to each breed, and which have been accumulated by man's selection, have not generally first appeared at an early period of life, and have been inherited by the offspring at a corresponding not early period. But the case of the short-faced tumbler, which when twelve hours old had acquired its proper proportions, proves that this is not the universal rule; for here the characteristic differences must either have appeared at an earlier period than usual, or, if not so, the differences must have been inherited, not at the corresponding, but at an earlier age.

Now let us apply these facts and the above two principles—which latter, though not proved true, can be shown to be in some degree probable—to species [2] in a state of nature. Let us take a genus of birds, descended on my theory from some one parent-species, and of which the several new species have become modified through natural selection in accordance with their diverse habits. Then, from the many slight successive steps of variation having supervened at a rather late age, and having been inherited at a corresponding

1 Following the approach he has taken throughout the book, Darwin turns to domesticated varieties to see if the rules or patterns he thinks he discerns hold true. Domestication is a microcosm or exemplar of the natural process of transmutation by natural selection, in Darwin's view. If the embryological patterns exhibited by domestic pigeon varieties parallel those seen in groups of related species in nature, this furthers the argument that domestication can be taken as an analog of the transmutation process in nature. And sure enough, pigeon varieties begin to show the traits that make their breed distinctive later rather than earlier in development. This observation formed an important part of the evidence Darwin cited to Lyell when he revealed his transmutation theory to him in April 1856. Lyell later recounted Darwin's observations in his journal: "The young pigeons are more of the normal type than the old of each variety. Embryology, therefore, leads to the opinion that you get nearer the type in going nearer to the foetal archetype & in like manner in Time we may get back nearer to the archetype of each genus & family & class" (Wilson 1970, p. 54).

2 Developmentally, the net result of descent from a common ancestral type is that the young will resemble each other more than they resemble the adults, since the successive variations are add-ons. To cite a familiar bird example mentioned earlier by Darwin, a common trait of the thrush family Turdidae is a speckled or spotted breast, a trait exhibited by virtually all juveniles but retained into adulthood only in some groups. Adult wood thrushes *(Hylocichla)*, for example, have this trait, while adult American robins *(Turdus migratorius)* have no speckled breast, though their young are speckled in the same manner as the young of *Hylocichla* and other turdid species.

age, the young of the new species of our supposed genus will manifestly tend to resemble each other much more closely than do the adults, just as we have seen in the case of pigeons. We may extend this view to whole families or even classes. The fore-limbs, for instance, which served as legs in the parent-species, may become, by a long course of modification, adapted in one descendant to act as hands, in another as paddles, in another as wings; and on the above two principles—namely of each successive modification supervening at a rather late age, and being inherited at a corresponding late age—the fore-limbs in the embryos of the several descendants of the parent-species will still resemble each other closely, for they will not have been modified. But in each individual new species, the embryonic fore-limbs will differ greatly from the fore-limbs in the mature animal; the limbs in the latter having undergone much modification at a rather late period of life, and having thus been converted into hands, or paddles, or wings. Whatever [1] influence long-continued exercise or use on the one hand, and disuse on the other, may have in modifying an organ, such influence will mainly affect the mature animal, which has come to its full powers of activity and has to gain its own living; and the effects thus produced will be inherited at a corresponding mature age. Whereas the young will remain unmodified, or be modified in a lesser degree, by the effects of use and disuse.

In certain cases the successive steps of variation might supervene, from causes of which we are wholly ignorant, at a very early period of life, or each step might be inherited at an earlier period than that at which it first appeared. In either case (as with the short-faced tumbler) the young or embryo would closely

1 Shades of Lamarck: in this sentence we see Darwin explicitly admit the possibility of inheritance of acquired characteristics. He suggests that use and disuse affect the adults more than the juveniles. Older animals are imagined to have had more time (and need) to accumulate modifications of some structure compared with young animals, so this is perhaps another reason the modifications appear later rather than earlier in development.

1 Darwin put his finger on the explanation that is accepted today. Note the use of "causality" language so characteristic of Darwin's century. Scientists no longer speak of final causation per se, but the nearest thing to it in the modern scientific lexicon is "ultimate" causation. The evolutionary biologist Ernst Mayr first used the terms "ultimate" and "proximate" causation in the context of, respectively, evolutionary vs. more immediate mechanistic explanations for biological phenomena (Mayr 1961).

2 This rather vague statement about further explanation was finally dropped in the fourth edition.

3 There are many examples of "retrograde" development in insects, particularly in cases where the juveniles need to be mobile while the adults do not. The barnacle-like scale insects, family Coccidae, are a good example: these sap-sucking insects can form sizable encrustations on plants, and many are serious agricultural and ornamental plant pests. The immatures form the primary dispersal stage, while the sessile females develop a shell-like carapace that becomes cemented to the plant. Under the carapace the insects tap their proboscis into the plant. The eyes, wings, antennae, etc., of such females have been lost or are greatly reduced, while the immatures have the fully developed eyes and legs necessary for dispersal. Bagworm moths, family Psychidae, are another familiar example. Caterpillars of both sexes are active leaf feeders with fully developed sensory and locomotory appendages. The caterpillars construct structures of silk and leaf debris that they carry around with them—convenient shelters into which to withdraw in times of danger. The female caterpillar "degenerates" in metamorphosis to a grub-like form devoid of all external organs and never leaves her bag. Winged males locate females via sex pheromones and mate with them from outside the structure. Which is "higher" or more complex? In traditional terms the question is moot.

resemble the mature parent-form. We have seen that this is the rule of development in certain whole groups of animals, as with cuttle-fish and spiders, and with a few members of the great class of insects, as with Aphis. [1] With respect to the final cause of the young in these cases not undergoing any metamorphosis, or closely resembling their parents from their earliest age, we can see that this would result from the two following contingencies; firstly, from the young, during a course of modification carried on for many generations, having to provide for their own wants at a very early stage of development, and secondly, from their following exactly the same habits of life with their parents; for in this case, it would be indispensable for the existence of the species, that the child should be modified at a very early age in the same manner with its parents, in [2] accordance with their similar habits. Some further explanation, however, of the embryo not undergoing any metamorphosis is perhaps requisite. If, on the other hand, it profited the young to follow habits of life in any degree different from those of their parent, and consequently to be constructed in a slightly different manner, then, on the principle of inheritance at corresponding ages, the active young or larvæ might easily be rendered by natural selection different to any conceivable extent from their parents. Such differences might, also, become correlated with successive stages of development; so that the larvæ, in the first stage, might differ greatly from the larvæ in the second stage, as we [3] have seen to be the case with cirripedes. The adult might become fitted for sites or habits, in which organs of locomotion or of the senses, &c., would be useless; and in this case the final metamorphosis would be said to be retrograde.

As all the organic beings, extinct and recent, which

have ever lived on this earth have to be classed together, and as all have been connected by the finest gradations, the best, or indeed, if our collections were nearly perfect, the only possible arrangement, would be genealogical. [1] Descent being on my view the hidden bond of connexion which naturalists have been seeking under the term of the natural system. On this view we can understand how it is that, in the eyes of most naturalists, the structure of the embryo is even more important for classification than that of the adult. For the embryo is the animal in its less modified state ; and in so far it reveals the structure of its progenitor. In two groups of animal, however much they may at present differ from each other in structure and habits, if they pass through the same or similar embryonic stages, we may feel assured that they have both descended from the same or nearly similar parents, and are therefore in that degree closely related. Thus, community in [2] embryonic structure reveals community of descent. It will reveal this community of descent, however much the structure of the adult may have been modified and obscured ; we have seen, for instance, that cirripedes can at once be recognised by their larvæ as belonging to the great class of crustaceans. As the embryonic [3] state of each species and group of species partially shows us the structure of their less modified ancient progenitors, we can clearly see why ancient and extinct forms of life should resemble the embryos of their descendants,—our existing species. Agassiz believes this to be a law of nature ; but I am bound to confess that I only hope to see the law hereafter proved true. It can be proved true in those cases alone in which the ancient state, now supposed to be represented in many embryos, has not been obliterated, either by the successive variations in a long course of modification having super-

1 Although Darwin suggests here and elsewhere in the *Origin* that the best classification should be based on genealogy, he did not follow this approach in his work on barnacle taxonomy, despite having fully worked out his theory by then. Perhaps doing so would have tipped his hand prematurely (Padian 1999). The important point here is that embryology helps reveal this genealogical arrangement—revealing more of affinities than do the adults, embryos provide a window into that "hidden bond of connexion" that is common descent.

2 Darwin's embryology was essentially that of Karl Ernst von Baer, who is credited with establishing the techniques and conceptual framework of modern embryology. Although von Baer was an adherent of the archetype concept, his recognition that development consists of embryonic differentiation from a common form or body plan set the stage for naturalists like Darwin to see branching differentiation in an evolutionary context.

3 What does Darwin mean by ancient and extinct forms resembling the embryos of their descendants? He is referring to Agassiz's idea of a threefold parallelism among individual development (embryology), relationships within a type (phylogeny), and geological history (paleontology). In his work on fossil fish (1844), for example, Agassiz stated that "the successive creations [of life on earth] have passed through phases of development analogous to those through which the embryo passes during its growth." He went on to argue that "the embryo of the fish during its development, the class of existing fishes in its numerous families, and the fish type in its planetary history traverse in all respects analogous phases" (quoted in Ospovat 1976). One could interpret parallelism in terms of common descent, as Darwin did. This exciting idea was one of those Darwin shared with Lyell in 1856. "Darwin thinks that Agassiz's embryology has something in it, or that the order of development in individuals & of similar types in time may be connected," Lyell reported in his journal (Wilson 1970, p. 54).

1 Darwin is revolutionizing the science of embryology here! The field had already shifted from viewing development in terms of recapitulation of a linear chain of being (in its most extreme form) to the idea of branching development from common types. To strict recapitulationists of the early nineteenth century, the embryo literally manifests a sequence of organisms lower down in its series as it develops. They supposed that each stage was created, but to adherents of the archetype concept (Owen, Agassiz, et al.) this was dangerously close to transmutationist thinking. They rejected strict recapitulation, arguing instead that the animal kingdom was divided into four *completely* separate types. Organisms showed fundamental affinity to their type in early development but bore no relationship to other types. These scientists advocated the embryology of von Baer, who argued in 1828 that, *within types,* embryonic development represents divergence from a common form—the archetype. Darwin rejects "chain of being" thinking *and* the archetype idea, though development from common forms is closer to his descent theory. He reinterpreted von Baerian embryology, prophetically closing this section with the comment that embryology "rises greatly in interest" when seen in the light of descent (see Ospovat 1976).

2 All the examples of "rudimentary" organs given in this paragraph are fascinating. In the case of the "bastard-wing" (alula) of birds, Darwin predicted to Lyell in 1859 that the structure would be confirmed as a reduced digit if early bird fossils were found. He received exciting news in January 1863, when Falconer wrote to him about the discovery of a nearly complete fossil *Archaeopteryx* skeleton. Falconer suggested that it was just the sort of linking form Darwin had predicted. Darwin wrote back: "I much wish to hear . . . which digits are developed; when examining birds two or three years ago, I distinctly remember writing to Lyell that some day a fossil bird would be found with . . . the bastard wing and other part both well developed" (*Correspondence* 11: 11). Indeed, the alula is well developed in this and other dinosaur-bird transitional fossils.

vened at a very early age, or by the variations having been inherited at an earlier period than that at which they first appeared. It should also be borne in mind, that the supposed law of resemblance of ancient forms of life to the embryonic stages of recent forms, may be true, but yet, owing to the geological record not extending far enough back in time, may remain for a long period, or for ever, incapable of demonstration.

1 Thus, as it seems to me, the leading facts in embryology, which are second in importance to none in natural history, are explained on the principle of slight modifications not appearing, in the many descendants from some one ancient progenitor, at a very early period in the life of each, though perhaps caused at the earliest, and being inherited at a corresponding not early period. Embryology rises greatly in interest, when we thus look at the embryo as a picture, more or less obscured, of the common parent-form of each great class of animals.

2 *Rudimentary, atrophied, or aborted organs.*—Organs or parts in this strange condition, bearing the stamp of inutility, are extremely common throughout nature. For instance, rudimentary mammæ are very general in the males of mammals: I presume that the "bastard-wing" in birds may be safely considered as a digit in a rudimentary state: in very many snakes one lobe of the lungs is rudimentary; in other snakes there are rudiments of the pelvis and hind limbs. Some of the cases of rudimentary organs are extremely curious; for instance, the presence of teeth in fœtal whales, which when grown up have not a tooth in their heads; and the presence of teeth, which never cut through the gums, in the upper jaws of our unborn calves. It has even been stated on good authority that rudiments of teeth can be detected

in the beaks of certain embryonic birds. Nothing can be plainer than that wings are formed for flight, yet in how many insects do we see wings so reduced in size as to be utterly incapable of flight, and not rarely lying under wing-cases, firmly soldered together!

The meaning of rudimentary organs is often quite unmistakeable: for instance there are beetles of the same genus (and even of the same species) resembling each other most closely in all respects, one of which will have full-sized wings, and another mere rudiments of membrane; and here it is impossible to doubt, that the rudiments represent wings. Rudimentary organs sometimes retain their potentiality, and are merely not developed: this seems to be the case with the mammæ of male mammals, for many instances are on record of these organs having become well developed in full-grown males, and having secreted milk. So again there are normally four developed and two rudimentary teats in the udders of the genus Bos, but in our domestic cows the two sometimes become developed and give milk. In individual plants of the same species the petals sometimes occur as mere rudiments, and sometimes in a well-developed state. In plants with separated sexes, the male flowers often have a rudiment of a pistil; and Kölreuter found that by crossing such male plants with an hermaphrodite species, the rudiment of the pistil in the hybrid offspring was much increased in size; and this shows that the rudiment and the perfect pistil are essentially alike in nature.

An organ serving for two purposes, may become rudimentary or utterly aborted for one, even the more important purpose; and remain perfectly efficient for the other. Thus in plants, the office of the pistil is to allow the pollen-tubes to reach the ovules protected in the ovarium at its base. The pistil consists of a stigma

1 The "unmistakeable" meaning of rudimentary organs is that they are in fact derived from functional versions of the same organ. The cases given in this paragraph demonstrate variability in these organs, a necessary ingredient if they are to become gradually modified. Note that in some cases the organs vary between species of the same genus, and in others they vary among individuals of the same species.

The significance of rudimentary organs did not escape the notice of Alfred Russel Wallace. In his Sarawak Law paper, Wallace wrote:

> Another important series of facts, quite in accordance with, and even necessary deductions from, the law now developed, are those of rudimentary organs . . . To every thoughtful naturalist the question must arise, What are these for? What have they to do with the great laws of creation? Do they not teach us something of the system of Nature? If each species has been created independently, and without any necessary relations with pre-existing species, what do these rudiments, these apparent imperfections mean? There must be a cause for them; they must be the necessary results of some great natural law. (Wallace 1855, p. 195)

2 Today it is thought that gene families evolve in precisely the same way. Genes can become duplicated through various mechanisms (e.g., unequal crossing over, gene conversion, etc.), and once redundancy occurs, the extra copies may be modified by natural selection. Globin genes are a good example of this. There are five alpha-globin gene loci (plus two non-functional—vestigial!—pseudogenes) on chromosome 16. Chromosome 11 bears six beta-globin loci plus one pseudogene. The related myoglobin gene is found on chromosome 22 (Ohta 2003).

supported on the style; but in some Compositæ, the male florets, which of course cannot be fecundated, have a pistil, which is in a rudimentary state, for it is not crowned with a stigma; but the style remains well developed, and is clothed with hairs as in other compositæ, for the purpose of brushing the pollen out of the surrounding anthers. Again, an organ may become rudimentary for its proper purpose, and be used for a distinct object: in certain fish the swim-bladder seems to be rudimentary for its proper function of giving buoyancy, but has become converted into a nascent breathing organ or lung. Other similar instances could be given.

Rudimentary organs in the individuals of the same species are very liable to vary in degree of development and in other respects. Moreover, in closely allied species, the degree to which the same organ has been rendered rudimentary occasionally differs much. This latter fact is well exemplified in the state of the wings of the female moths in certain groups. Rudimentary organs may be utterly aborted; and this implies, that we find in an animal or plant no trace of an organ, which analogy would lead us to expect to find, and which is occasionally found in monstrous individuals of the species. Thus in the snapdragon (antirrhinum) we generally do not find a rudiment of a fifth stamen; but this may sometimes be seen. In tracing the homologies of the same part in different members of a class, nothing is more common, or more necessary, than the use and discovery of rudiments. This is well shown in the drawings given by Owen of the bones of the leg of the horse, ox, and rhinoceros.

It is an important fact that rudimentary organs, such as teeth in the upper jaws of whales and ruminants, can often be detected in the embryo, but afterwards wholly disappear. It is also, I believe, a universal

1 "Compositae" is an obsolete botanical name now replaced by family Asteraceae. These plants, which include daisies, asters, and sunflowers, are still called composites, however, because the "flower" is actually a composite of many tiny florets. These are often differentiated into disk florets with undeveloped petals in the center, and ray florets with one long petal along the margin. Collectively, the ray florets give the appearance of a single flower encircled by petals. Darwin undertook a lengthy study of flower form and male-female flower differentiation that was published as a book in 1877.

2 Darwin's point is well taken here, but the modern understanding of swimbladders and their relationship to lungs differs from his. This topic was discussed on p. 190; see also essay seven in Gould (1993).

3 The discovery of rudimentary teeth in the dental grooves of fetal baleen whales (Balaenidae) is attributed to Geoffroy St. Hilaire but was described most fully after the *Origin* was published, notably by the Danish cetologists D. F. Eschricht and J. Reinhart, who published a monograph on the right whale (then called the Greenland whale) in 1861. Baleen whales were known as "whalebone whales," an ironic name since "whalebone" (baleen) is not bone tissue but keratin, the same material as hair and nails.

rule, that a rudimentary part or organ is of greater size relatively to the adjoining parts in the embryo, than in the adult; so that the organ at this early age is less rudimentary, or even cannot be said to be in any degree rudimentary. Hence, also, a rudimentary organ in the adult, is often said to have retained its embryonic condition.

I have now given the leading facts with respect to rudimentary organs. In reflecting on them, every one must be struck with astonishment: for the same reasoning power which tells us plainly that most parts and organs are exquisitely adapted for certain purposes, tells us with equal plainness that these rudimentary or atrophied organs, are imperfect and useless. In works on natural history rudimentary organs are generally said to have been created "for the sake of symmetry," or in order "to complete the scheme of nature;" but this seems to me no explanation, merely a re-statement of the fact. Would it be thought sufficient to say that because planets revolve in elliptic courses round the sun, satellites follow the same course round the planets, for the sake of symmetry, and to complete the scheme of nature? An eminent physiologist accounts for the presence of rudimentary organs, by supposing that they serve to excrete matter in excess, or injurious to the system; but can we suppose that the minute papilla, which often represents the pistil in male flowers, and which is formed merely of cellular tissue, can thus act? Can we suppose that the formation of rudimentary teeth which are subsequently absorbed, can be of any service to the rapidly growing embryonic calf by the excretion of precious phosphate of lime? When a man's fingers have been amputated, imperfect nails sometimes appear on the stumps: I could as soon believe that these vestiges of nails have appeared, not from unknown laws

1 Darwin is contemptuous of such "special pleading." As early as 1842 he wrote that "these wondrous facts, of parts created for no use in past and present time, all can by my theory receive simple explanation; or they receive none and we must be content with some such empty metaphor, as that of De Candolle, who compares creation to a well covered table, and says abortive organs may be compared to the dishes (some should be empty) placed symmetrically!" (F. Darwin 1909, p. 47). The "eminent physiologist," in contrast, is positing a function for putatively nonfunctional rudimentary organs. This is a legitimate hypothesis that can be tested; the hypothesis of excreting excess matter may not hold water, as Darwin shows, but there are other cases in which vestigial structures are shown to have some uses.

It is worth cautioning against the common error of mistaking a rudimentary or vestigial state with nonfunctionality; a rudimentary structure is reduced or degenerated relative to its ancestral condition, but it may or may not retain some—or even different—functionality. The human appendix is a case in point. This structure is homologous with the end of the gastric caecum, where it appears to play a digestive role housing cellulytic enzyme-producing bacteria, a useful thing for herbivores. In *The Descent of Man* (1871) Darwin attributed the rudimentary state of the human appendix to diet change. Evolutionary biologists accept this interpretation today. It has been suggested that the appendix currently plays a role in the immune system or in the maintenance of our bacterial gut flora (Bollinger et al. 2007); it may have these or other functions but is viewed nonetheless as vestigial relative to its considerable development in other primates (see, for example, Romer and Parsons 1986). Perhaps worse than superfluous, the appendix has its costs: appendicitis can lead to certain death without surgical intervention.

of growth, but in order to excrete horny matter, as that [1] the rudimentary nails on the fin of the manatee were formed for this purpose.

On my view of descent with modification, the origin of rudimentary organs is simple. We have plenty of [2] cases of rudimentary organs in our domestic productions,—as the stump of a tail in tailless breeds,—the vestige of an ear in earless breeds,—the reappearance of minute dangling horns in hornless breeds of cattle, more especially, according to Youatt, in young animals, —and the state of the whole flower in the cauliflower. We often see rudiments of various parts in monsters. But I doubt whether any of these cases throw light on the origin of rudimentary organs in a state of nature, further than by showing that rudiments can be produced; for I doubt whether species under nature ever [3] undergo abrupt changes. I believe that disuse has been the main agency; that it has led in successive generations to the gradual reduction of various organs, until they have become rudimentary,—as in the case of the eyes of animals inhabiting dark caverns, and of the wings of birds inhabiting oceanic islands, which have seldom been forced to take flight, and have ultimately lost the power of flying. Again, an organ useful under certain conditions, might become injurious under others, as with the wings of beetles living on small and exposed islands; and in this case natural selection would continue slowly to reduce the organ, until it was rendered harmless and rudimentary.

Any change in function, which can be effected by insensibly small steps, is within the power of natural selection; so that an organ rendered, during changed habits of life, useless or injurious for one purpose, might easily be modified and used for another purpose. Or an organ might be retained for one alone of its

1 Florida manatees (*Trichechus manatus,* family Trichachidae) sport rudimentary nails on the second, third, and fourth digits, while their southerly relatives the Amazonian manatees *(T. inunguis)* have lost these nail rudiments (Berta et al. 2005).

2 The domestication analogy is never far from mind; for those of Darwin's readers who accepted his argument that domestic varieties are produced by a process of gradual selection, the ability to render traits rudimentary by selectively breeding them away is a particularly potent demonstration of what natural selection might effect. It is perhaps no accident that this observation segues into a statement about rudimentary structures in so-called monsters. These are individuals with striking morphological mutations (the equivalent in plants would be "sports"). Darwin is quick to downplay their importance because many readers were probably familiar with such cases, where change seems more saltational than gradual. A rudimentary tail, for example, is often caused by a single genetic mutation, as in Manx and Japanese bobtail cat breeds. Such changes can occur literally in a generation, while Darwin is trying to make a case for gradual change.

3 Curiously, Darwin cites disuse as the main agency in the reduction in organs. This is rather Lamarckian, though his mechanism of the effects of disuse differs somewhat from Larmarck's. Although natural selection is not mentioned here, it acts hand-in-hand with use and disuse; Darwin's basic belief was that use and disuse can affect the variability of a trait, which natural selection might then act upon.

former functions. An organ, when rendered useless, may well be variable, for its variations cannot be checked by natural selection. At whatever period of life disuse or selection reduces an organ, and this will generally be when the being has come to maturity and to its full powers of action, the principle of inheritance at corresponding ages will reproduce the organ in its reduced state at the same age, and consequently will seldom affect or reduce it in the embryo. Thus we can understand the greater relative size of rudimentary organs in the embryo, and their lesser relative size in the adult. But if each step of the process of reduction were to be inherited, not at the corresponding age, but at an extremely early period of life (as we have good reason to believe to be possible) the rudimentary part would tend to be wholly lost, and we should have a case of complete abortion. The principle, also, of economy, explained in a former chapter, by which the materials forming any part or structure, if not useful to the possessor, will be saved as far as is possible, will probably often come into play; and this will tend to cause the entire obliteration of a rudimentary organ.

As the presence of rudimentary organs is thus due to the tendency in every part of the organisation, which has long existed, to be inherited—we can understand, on the genealogical view of classification, how it is that systematists have found rudimentary parts as useful as, or even sometimes more useful than, parts of high physiological importance. Rudimentary organs may be compared with the letters in a word, still retained in the spelling, but become useless in the pronunciation, but which serve as a clue in seeking for its derivation. On the view of descent with modification, we may conclude that the existence of organs in a rudimentary, imperfect, and useless condition, or quite aborted, far

1 This is a clever observation insofar as functionality of a trait, and therefore the relevance of use and disuse, applies to individuals having "full powers of action"—i.e., post-natal individuals, alive and kicking. Selection (or just disuse) will reduce the trait at the relevant life stage, but if this stage is always post-natal, this nicely explains why the traits in question should be well developed in embryos and then decline.

The principle of economy, discussed in chapter V (see p. 147), refers to the idea of compensation, whereby under-development of some organs is compensated by over-development of others.

2 Once again language and its evolution provide an analogy for organic evolution (see also pp. 40, 160, 310–311, 422). Here Darwin makes the point that vestiges of earlier states of words are like rudimentary organs. Consider, for example, the "o" in "leopard" (Thomson 1932, p. 320). In antiquity, the leopard (*Felis pardus*) was said to be the product of a cross between a male "pard" (perhaps a panther) and a lioness (Albertus Magnus gives a thorough discussion in book 22 of *De Animalibus* [On Animals]; see Kitchell and Resnick 1999). Hence the word is derived from *leo* (Latin for "lion") + *pard* (Greek for "panther" or, confusingly, "leopard"). The "o" in "leopard" is not pronounced but gives us a clue that the word's origin pertains to lions. A phonetic spelling like "lepard" would lose this clue, and in fact could mislead a sleuthing linguist into supposing that the name pertains to rabbits: *lepus* is Latin for "hare."

1 In this final section of the chapter Darwin succeeds in tying together four great groups of observations: classification, morphology, embryology, and rudimentary or vestigial structures. He argues that the patterns in evidence are more than simply explained or accountable by his theory: they are actually expected. It is worth quoting his closing statement on the subject from the 1844 *Essay:*

> Finally then I must repeat that these wonderful facts of organs formed with traces of exquisite care, but now either absolutely useless or adapted to ends wholly different from their ordinary end, being present and forming part of the structure of almost every inhabitant of this world, both in long-past and present times—being best developed and often only discoverable at a very early embryonic period, and being full of signification in arranging the long series of organic beings in a natural system—these wonderful facts not only receive a simple explanation on the theory of long-continued selection of many species from a few parent-stocks, but necessarily follow from this theory. If this theory be rejected, these facts remain quite inexplicable. (F. Darwin 1909, p. 238)

from presenting a strange difficulty, as they assuredly do on the ordinary doctrine of creation, might even have been anticipated, and can be accounted for by the laws of inheritance.

Summary.—In this chapter I have attempted to show, that the subordination of group to group in all organisms throughout all time; that the nature of the relationship, by which all living and extinct beings are united by complex, radiating, and circuitous lines of affinities into one grand system; the rules followed and the difficulties encountered by naturalists in their classifications; the value set upon characters, if constant and prevalent, whether of high vital importance, or of the most trifling importance, or, as in rudimentary organs, of no importance; the wide opposition in value between analogical or adaptive characters, and characters of true affinity; and other such rules;—all naturally follow on the view of the common parentage of those forms which are considered by naturalists as allied, together with their modification through natural selection, with its contingencies of extinction and divergence of character. In considering this view of classification, it should be borne in mind that the element of descent has been universally used in ranking together the sexes, ages, and acknowledged varieties of the same species, however different they may be in structure. If we extend the use of this element of descent,—the only certainly known cause of similarity in organic beings,—we shall understand what is meant by the natural system: it is genealogical in its attempted arrangement, with the grades of acquired difference marked by the terms varieties, species, genera, families, orders, and classes.

On this same view of descent with modification, all the great facts in Morphology become intelligible,—

whether we look to the same pattern displayed in the homologous organs, to whatever purpose applied, of the different species of a class; or to the homologous parts constructed on the same pattern in each individual animal and plant.

On the principle of successive slight variations, not necessarily or generally supervening at a very early period of life, and being inherited at a corresponding period, we can understand the great leading facts in Embryology; namely, the resemblance in an individual embryo of the homologous parts, which when matured will become widely different from each other in structure and function; and the resemblance in different species of a class of the homologous parts or organs, though fitted in the adult members for purposes as different as possible. Larvæ are active embryos, which have become specially modified in relation to their habits of life, through the principle of modifications being inherited at corresponding ages. On this same principle—and bearing in mind, that when organs are reduced in size, either from disuse or selection, it will generally be at that period of life when the being has to provide for its own wants, and bearing in mind how strong is the principle of inheritance—the occurrence of rudimentary organs and their final abortion, present to us no inexplicable difficulties; on the contrary, their presence might have been even anticipated. The importance of embryological characters and of rudimentary organs in classification is intelligible, on the view that an arrangement is only so far natural as it is genealogical.

Finally, the several classes of facts which have been considered in this chapter, seem to me to proclaim so plainly, that the inumerable species, genera, and families of organic beings, with which this world is ◀1

1 This final sentence is remarkable: Darwin asserts that the evidence of this chapter is so compelling that *it alone* would suffice to convince him of common descent even in the absence of any other data. He later wrote in his autobiography: "Hardly any point gave me so much satisfaction when I was at work on the Origin as the explanation of the wide difference in many classes between the embryo and the adult animal, and of the close resemblance of the embryos within the same class." Darwin thought that embryology itself provided the elusive "smoking gun" of incontrovertible evidence for transmutation, and he was disappointed that his friends and colleagues did not seem to realize this. He continues in his autobiography: "No notice of this point was taken, as far as I remember, in the early reviews of the Origin, and I recollect expressing my surprise on this head in a letter to Asa Gray." He wrote to Hooker just a month after the *Origin* came out: "Embryology is my pet bit in my book, & confound my friends not one has noticed this to me" (*Correspondence* 7: 431). To Darwin's chagrin, this continued to be the case. Here he is the following September, in the communication to Asa Gray mentioned in the autobiography: "It is curious how each one, I suppose weighs arguments in a different balance: embryology is to me by far strongest single class of facts in favour of change of form, & not one, I think, of my reviewers has alluded to this. Variations not coming on at a very early age, & being inherited at not very early corresponding period, explains, as it seems to me, the grandest of all facts in Nat. History, or rather in Zoology. viz the resemblance of Embryos" (*Correspondence* 8: 349).

peopled, have all descended, each within its own class or group, from common parents, and have all been modified in the course of descent, that I should without hesitation adopt this view, even if it were unsupported by other facts or arguments.

CHAPTER XIV.

RECAPITULATION AND CONCLUSION.

Recapitulation of the difficulties on the theory of Natural Selection — Recapitulation of the general and special circumstances in its favour — Causes of the general belief in the immutability of species — How far the theory of natural selection may be extended — Effects of its adoption on the study of Natural history — Concluding remarks.

As this whole volume is one long argument, it may be convenient to the reader to have the leading facts and inferences briefly recapitulated. ◀1

That many and grave objections may be advanced against the theory of descent with modification through natural selection, I do not deny. I have endeavoured to give to them their full force. Nothing at first can appear more difficult to believe than that the more complex organs and instincts should have been perfected, not by means superior to, though analogous with, human reason, but by the accumulation of innumerable slight variations, each good for the individual possessor. Nevertheless, this difficulty, though appearing to our imagination insuperably great, cannot be considered real if we admit the following propositions, namely,—that gradations in the perfection of any organ or instinct, which we may consider, either do now exist or could have existed, each good of its kind,—that all organs and instincts are, in ever so slight a degree, variable,—and, lastly, that there is a struggle for existence leading to the preservation of each profitable deviation of structure or instinct. The truth of these propositions cannot, I think, be disputed. ◀2

If this book is, as Darwin called it, an abstract of his theory, then this chapter represents an abstract of an abstract: the first two-thirds are a lucid synopsis of the argument of the entire book. It is fitting, too, that the manner of presentation is not *too* literal a recapitulation: Darwin dispenses with difficulties first, then summarizes the main arguments and observations in largely the same order as they appear in the book. The remaining third of the chapter is an exhilarating look to the future—our time, in effect. Like a prophet, Darwin foresees the transformation of all branches of the biological sciences. Indeed, his prose in the conclusion reads with a Biblical cadence—perhaps by design.

---

1 This famous line expresses the grand, holistic view of the *Origin* that Darwin hoped to express. It has been suggested that nearly *all* of Darwin's major works taken together can also be viewed as "one long argument." Stepping back, we can glimpse the thread of a continuous narrative: Darwin's geological works argue for gradualism and uniformity, the barnacle monographs explore the nature of species and varieties, and *Origin* posits a gradualist view of evolution by a process of selection among variations. Post-*Origin* Darwin wrote a series of books exploring the model of evolution by natural selection in different circumstances, from sexual selection and human evolution to orchid pollination and expression of emotions. His final work on the long-term effects of earthworm activity seems to bring him full circle: uniformitarian gradualism on a scale to which people can relate.

2 Paralleling his approach in the "consilience" or responsibility-case chapters (5–13), Darwin addresses difficulties first. He knew that his theory undermined beliefs that many of his contemporaries held dear, and, as he says on pp. 465–466, he was keenly aware of the magnitude of the problems his theory faced. Notice on the next several pages the phrases he uses: the difficulties and objections he knows will be raised are called "insuperably great," "extreme," "grave," and "obvious and forcible."

1 This has been a sticking point for many critics of Darwin. Under his model, all traits, from the most basic hair on the back of a fly to the most exquisitely complex organ like the eye or the ear, must have evolved by natural selection, or in correlation with other traits that so evolved. The process is stepwise, gradual, and slow. From the moment the *Origin* appeared, adherents of the argument from design emphasized the "designedness" of these same traits, contending that they are too complex to have evolved by gradual steps. This argument was resurrected for molecular-level traits in recent years by the so-called intelligent design adherents. Their assertion is that some traits, such as the bacterial flagellum or certain biochemical pathways, are "irreducibly complex"—meaning that they are only functional as operational wholes; fractional versions would be utterly useless and so could not have served as intermediate steps in some evolutionary scenario (e.g., Behe 1996). The argument, then, is not new, but the same wine in a different bottle, and the assertion of irreducible complexity for such traits has been refuted (see, e.g., Pallen and Matzke 2006 and Liu and Ochman 2007 on the evolution of the flagellum).

It is, no doubt, extremely difficult even to conjecture by what gradations many structures have been perfected, more especially amongst broken and failing groups of organic beings; but we see so many strange gradations in nature, as is proclaimed by the canon, "Natura non facit saltum," that we ought to be extremely cautious in saying that any organ or instinct, or any whole being, could not have arrived at its present state by many graduated steps. There are, it must be admitted, cases of special difficulty on the theory of natural selection; and one of the most curious of these is the existence of two or three defined castes of workers or sterile females in the same community of ants; but I have attempted to show how this difficulty can be mastered.

With respect to the almost universal sterility of species when first crossed, which forms so remarkable a contrast with the almost universal fertility of varieties when crossed, I must refer the reader to the recapitulation of the facts given at the end of the eighth chapter, which seem to me conclusively to show that this sterility is no more a special endowment than is the incapacity of two trees to be grafted together; but that it is incidental on constitutional differences in the reproductive systems of the intercrossed species. We see the truth of this conclusion in the vast difference in the result, when the same two species are crossed reciprocally; that is, when one species is first used as the father and then as the mother.

The fertility of varieties when intercrossed and of their mongrel offspring cannot be considered as universal; nor is their very general fertility surprising when we remember that it is not likely that either their constitutions or their reproductive systems should have been profoundly modified. Moreover, most of the varieties which have been experimentised on have been produced under domestication; and as domestication apparently tends to eliminate sterility, we ought not to expect it also to produce sterility.

The sterility of hybrids is a very different case from that of first crosses, for their reproductive organs are more or less functionally impotent; whereas in first crosses the organs on both sides are in a perfect condition. As we continually see that organisms of all kinds are rendered in some degree sterile from their constitutions having been disturbed by slightly different and new conditions of life, we need not feel surprise at hybrids being in some degree sterile, for their constitutions can hardly fail to have been disturbed from being compounded of two distinct organisations. This parallelism is supported by another parallel, but directly opposite, class of facts; namely, that the vigour and fertility of all organic beings are increased by slight changes in their conditions of life, and that the offspring of slightly modified forms or varieties acquire from being crossed increased vigour and fertility. So that, on the one hand, considerable changes in the conditions of life and crosses between greatly modified forms, lessen fertility; and on the other hand, lesser changes in the conditions of life and crosses between less modified forms, increase fertility.

Turning to geographical distribution, the difficulties encountered on the theory of descent with modification are grave enough. All the individuals of the same species, and all the species of the same genus, or even higher group, must have descended from common parents; and therefore, in however distant and isolated parts of the world they are now found, they must in the course of successive generations have passed from some one part to the others. We are often wholly unable [1]

1 The factors responsible for the biogeographical distribution of some groups are still debated today. Recall that several processes are relevant at different temporal scales, some or all of which may have had a hand in shaping distribution, depending on group. Continental drift effected through plate tectonics is a process found at one end of the temporal spectrum, as is mountain building, formation of epicontinental seas, etc. Further along the spectrum are global changes in sea level and shifts in climatic zones. These processes link areas (e.g., land bridges) or lead to isolation (e.g., forming a barrier to migration). Darwin and most of his contemporaries were wedded to a static-earth model, and as a result they had to rely on migration and long-distance dispersal as influenced by global climatic changes.

even to conjecture how this could have been effected. Yet, as we have reason to believe that some species have retained the same specific form for very long periods, enormously long as measured by years, too much stress ought not to be laid on the occasional wide diffusion of the same species; for during very long periods of time there will always be a good chance for wide migration by many means. A broken or interrupted range may often be accounted for by the extinction of the species in the intermediate regions. It cannot be denied that we are as yet very ignorant of the full extent of the various climatal and geographical changes which have affected the earth during modern periods; and such changes will obviously have greatly facilitated migration. As an example, I have attempted to show how potent has been the influence of the Glacial period on the distribution both of the same and of representative species throughout the world. We are as yet profoundly ignorant of the many occasional means of transport. With respect to distinct species of the same genus inhabiting very distant and isolated regions, as the process of modification has necessarily been slow, all the means of migration will have been possible during a very long period; and consequently the difficulty of the wide diffusion of species of the same genus is in some degree lessened.

As on the theory of natural selection an interminable number of intermediate forms must have existed, linking together all the species in each group by gradations as fine as our present varieties, it may be asked, Why do we not see these linking forms all around us? Why are not all organic beings blended together in an inextricable chaos? With respect to existing forms, we should remember that we have no right to expect (excepting in rare cases) to discover *directly* connecting links between them, but only between each and some extinct and supplanted form. Even on a wide area, which has during a long period remained continuous, and of which the climate and other conditions of life change insensibly in going from a district occupied by one species into another district occupied by a closely allied species, we have no just right to expect often to find intermediate varieties in the intermediate zone. For we have reason to believe that only a few species are undergoing change at any one period; and all changes are slowly effected. I have also shown that the intermediate varieties which will at first probably exist in the intermediate zones, will be liable to be supplanted by the allied forms on either hand; and the latter, from existing in greater numbers, will generally be modified and improved at a quicker rate than the intermediate varieties, which exist in lesser numbers; so that the intermediate varieties will, in the long run, be supplanted and exterminated.

On this doctrine of the extermination of an infinitude [1] of connecting links, between the living and extinct inhabitants of the world, and at each successive period between the extinct and still older species, why is not every geological formation charged with such links? Why does not every collection of fossil remains afford plain evidence of the gradation and mutation of the forms of life? We meet with no such evidence, and this is the most obvious and forcible of the many objections which may be urged against my theory. Why, again, do whole groups of allied species appear, though certainly they often falsely appear, to have come in suddenly on the several geological stages? Why do we not find great piles of strata beneath the Silurian system, stored with the remains of the progenitors of the Silurian groups of fossils? For certainly on my theory such

1 Although intermediate or linking forms of many groups have been found in the fossil record since Darwin's time (see examples in Martin 2004), this remains one of the more problematic areas for Darwin's model. In principle *all* extant species are linked through successive points of common ancestry, and so transitional forms between all these species must have existed. If transitional fossil series linking species were the rule rather than the exception, an evolutionary model of some kind would surely have been adopted at the dawn of paleontology in the early Enlightenment. As it is, the fossil record is patchy, with well-preserved fossil intermediates found for some groups and sizable gaps separating others. The vagaries of the formation and fate of fossils (what can be fossilized, suitable conditions for fossilization, and subsequent destruction of fossil strata) are the main reasons for gaps. They are also the main reason for the secondary problem Darwin mentions here: the apparently sudden appearance of suites of species in the fossil record, patterns evident in such phrases as the "Cambrian explosion."

strata must somewhere have been deposited at these ancient and utterly unknown epochs in the world's history.

I can answer these questions and grave objections only on the supposition that the geological record is far more imperfect than most geologists believe. It cannot be objected that there has not been time sufficient for any amount of organic change; for the lapse of time has been so great as to be utterly inappreciable by the human intellect. The number of specimens in all our museums is absolutely as nothing compared with the countless generations of countless species which certainly have existed. We should not be able to recognise a species as the parent of any one or more species if we were to examine them ever so closely, unless we likewise possessed many of the intermediate links between their past or parent and present states; and these many links we could hardly ever expect to discover, owing to the imperfection of the geological record. Numerous existing doubtful forms could be named which are probably varieties; but who will pretend that in future ages so many fossil links will be discovered, that naturalists will be able to decide, on the common view, whether or not these doubtful forms are varieties? As long as most of the links between any two species are unknown, if any one link or intermediate variety be discovered, it will simply be classed as another and distinct species. Only a small portion of the world has been geologically explored. Only organic beings of certain classes can be preserved in a fossil condition, at least in any great number. Widely ranging species vary most, and varieties are often at first local,—both causes rendering the discovery of intermediate links less likely. Local varieties will not spread into other and distant regions until they are considerably modified and improved; and when they do spread, if discovered in a geological formation, they will appear as if suddenly created there, and will be simply classed as new species. Most formations have been intermittent in their accumulation; and their duration, I am inclined to believe, has been shorter than the average duration of specific forms. Successive formations are separated from each other by enormous blank intervals of time; for fossiliferous formations, thick enough to resist future degradation, can be accumulated only where much sediment is deposited on the subsiding bed of the sea. During the alternate periods of elevation and of stationary level the record will be blank. During these latter periods there will probably be more variability in the forms of life; during periods of subsidence, more extinction.

With respect to the absence of fossiliferous formations 1
beneath the lowest Silurian strata, I can only recur to the hypothesis given in the ninth chapter. That the geological record is imperfect all will admit; but that it is imperfect to the degree which I require, few will be inclined to admit. If we look to long enough intervals of time, geology plainly declares that all species have changed; and they have changed in the manner which my theory requires, for they have changed slowly and in a graduated manner. We clearly see this in the fossil remains from consecutive formations invariably being much more closely related to each other, than are the fossils from formations distant from each other in time.

Such is the sum of the several chief objections and difficulties which may justly be urged against my theory; and I have now briefly recapitulated the answers and explanations which can be given to them. I have felt these difficulties far too heavily during many years to

x 3

1 One area of significant criticism concerned the age of the earth, in particular whether there was time enough for the evolutionary changes that Darwin advocated. In the sixth edition, Darwin inserted a sentence here addressing an attack mounted by the renowned physicist William Thompson, Lord Kelvin. Kelvin published papers in 1862 and 1863 arguing that the earth was only about 93 million years old, a date based on the thermodynamic properties of the cooling globe. His argument was amplified by the engineer Fleeming Jenkin in an important review of the *Origin* published in 1867. Kelvin and Jenkin took on quite a lot: besides arguing that there was insufficient time for Darwin's hypothesized transmutations, they railed against Lyellian uniformity and the efficacy of natural selection. The argument intensified over a period of years, and Darwin was finally inclined to address the criticism in this section's litany of difficulties: "With respect to the lapse of time not having been sufficient since our planet was consolidated for the assumed amount of organic change, and this objection, as urged by Sir William Thompson, is probably one of the gravest as yet advanced, I can only say, firstly, that we do not know at what rate species change as measured by years, and secondly, that many philosophers are not as yet willing to admit that we know enough of the constitution of the universe and of the interior of our globe to speculate with safety on its past duration" (Peckham 1959, p. 728). This problem was not solved in Darwin's lifetime, but the discovery of radioactivity at the turn of the twentieth century revealed the grave error of Kelvin's calculations, vindicating Darwin's and Lyell's estimates of an earth hundreds of millions of years old. Today the age has been determined with considerable precision, and the planet, it turns out, is far older than Darwin would have imagined, at approximately 4.5 billion years (see discussion by Burchfield 1974).

1 Doubts and difficulties weighed heavily on Darwin's mind since the first inkling of the reality of transmutation occurred to him in 1837. Still, as weighty as the difficulties are, he saw beyond them to the more positive evidence in support of the idea. Here is an expression of that confidence from a July 1857 letter to Asa Gray: "as an honest man I must tell you that I have come to the heteredox [sic] conclusion that there are no such things as independently created species—that species are only strongly defined varieties. I know that this will make you despise me.—I do not much underrate the many huge difficulties on this view, but yet it seems to me to explain too much, otherwise inexplicable, to be false" (*Correspondence* 6: 431).

In this passage Darwin makes the interesting point that the areas of the gravest difficulty are also those in which our ignorance is greatest; in other words, he is suggesting that the evidence against his theory is not so much *positive* as *negative* (as in missing data).

2 Beginning here Darwin offers a précis of the positive arguments in favor of his theory, starting with the domestication analogy and the variability of organisms. On the next page he considers struggle for existence; on page 468 he then puts these together, outlining natural and sexual selection. This recap continues through most of page 469.

1 doubt their weight. But it deserves especial notice that the more important objections relate to questions on which we are confessedly ignorant; nor do we know how ignorant we are. We do not know all the possible transitional gradations between the simplest and the most perfect organs; it cannot be pretended that we know all the varied means of Distribution during the long lapse of years, or that we know how imperfect the Geological Record is. Grave as these several difficulties are, in my judgment they do not overthrow the theory of descent with modification.

2 Now let us turn to the other side of the argument. Under domestication we see much variability. This seems to be mainly due to the reproductive system being eminently susceptible to changes in the conditions of life; so that this system, when not rendered impotent, fails to reproduce offspring exactly like the parent-form. Variability is governed by many complex laws,—by correlation of growth, by use and disuse, and by the direct action of the physical conditions of life. There is much difficulty in ascertaining how much modification our domestic productions have undergone; but we may safely infer that the amount has been large, and that modifications can be inherited for long periods. As long as the conditions of life remain the same, we have reason to believe that a modification, which has already been inherited for many generations, may continue to be inherited for an almost infinite number of generations. On the other hand we have evidence that variability, when it has once come into play, does not wholly cease; for new varieties are still occasionally produced by our most anciently domesticated productions.

Man does not actually produce variability; he only

unintentionally exposes organic beings to new conditions of life, and then nature acts on the organisation, and causes variability. But man can and does select the variations given to him by nature, and thus accumulate them in any desired manner. He thus adapts animals and plants for his own benefit or pleasure. He may do this methodically, or he may do it unconsciously by preserving the individuals most useful to him at the time, without any thought of altering the breed. It is certain that he can largely influence the character of a breed by selecting, in each successive generation, individual differences so slight as to be quite inappreciable by an uneducated eye. This process of selection has been the great agency in the production of the most distinct and useful domestic breeds. That many of the breeds produced by man have to a large extent the character of natural species, is shown by the inextricable doubts whether very many of them are varieties or aboriginal species.

There is no obvious reason why the principles which have acted so efficiently under domestication should not have acted under nature. In the preservation of favoured individuals and races, during the constantly-recurrent Struggle for Existence, we see the most powerful and ever-acting means of selection. The struggle for existence inevitably follows from the high geometrical ratio of increase which is common to all organic beings. This high rate of increase is proved by calculation, by the effects of a succession of peculiar seasons, and by the results of naturalisation, as explained in the third chapter. More individuals are born than can possibly survive. A grain in the balance will determine which individual shall live and which shall die,—which variety or species shall increase in number, and which shall decrease, or finally become extinct. As the individuals of the same species come in all respects into the closest competition with each other, the struggle will generally be most severe between them; it will be almost equally severe between the varieties of the same species, and next in severity between the species of the same genus. But the struggle will often be very severe between beings most remote in the scale of nature. The slightest advantage in one being, at any age or during any season, over those with which it comes into competition, or better adaptation in however slight a degree to the surrounding physical conditions, will turn the balance.

With animals having separated sexes there will in most cases be a struggle between the males for possession of the females. The most vigorous individuals, or those which have most successfully struggled with their conditions of life, will generally leave most progeny. But success will often depend on having special weapons or means of defence, or on the charms of the males; and the slightest advantage will lead to victory.

As geology plainly proclaims that each land has undergone great physical changes, we might have expected that organic beings would have varied under nature, in the same way as they generally have varied under the changed conditions of domestication. And if there be any variability under nature, it would be an unaccountable fact if natural selection had not come into play. It has often been asserted, but the assertion is quite incapable of proof, that the amount of variation under nature is a strictly limited quantity. Man, though acting on external characters alone and often capriciously, can produce within a short period a great result by adding up mere individual differences in his domestic productions; and every one admits that there are at least individual differences in species under nature. But, besides such differences, all naturalists have admitted the existence of varieties, which they think sufficiently distinct to be worthy of record in systematic works. No one can draw any clear distinction between individual differences and slight varieties; or between more plainly marked varieties and sub-species, and species. Let it be observed how naturalists differ in the rank which they assign to the many representative forms in Europe and North America.

If then we have under nature variability and a powerful agent always ready to act and select, why should we doubt that variations in any way useful to beings, under their excessively complex relations of life, would be preserved, accumulated, and inherited? Why, if man can by patience select variations most useful to himself, should nature fail in selecting variations useful, under changing conditions of life, to her living products? What limit can be put to this power, acting during long ages and rigidly scrutinising the whole constitution, structure, and habits of each creature,—favouring the good and rejecting the bad? I can see no limit to this power, in slowly and beautifully adapting each form to the most complex relations of life. The theory of natural selection, even if we looked no further than this, seems to me to be in itself probable. I have already [1] recapitulated, as fairly as I could, the opposed difficulties and objections: now let us turn to the special facts and arguments in favour of the theory.

On the view that species are only strongly marked and permanent varieties, and that each species first existed as a variety, we can see why it is that no line of demarcation can be drawn between species, commonly supposed to have been produced by special acts of creation, and varieties which are acknowledged to have been produced by secondary laws. On this same view we can understand how it is that in each region

1 Notice that the overview on the previous few pages (pp. 466–469) constitutes the analogy and mechanism components of Darwin's argument. The remainder of this discussion, through p. 480, constitutes the "consilience" component. Here his overarching point, to quote again the July 1857 letter to Gray, is that the theory "explain[s] too much, otherwise inexplicable, to be false." Darwin's point about the agency of secondary laws is central. He believed that if theists who saw divine guidance behind everything followed their reasoning to its logical conclusion, they would come to see that that view stretches credulity to its breaking point. He expressed this sentiment in a letter to Gray written in 1860: "Hensleigh Wedgwood [Darwin's brother-in-law] . . . is a very strong Theist, & I put it to him, whether he thought that each time a fly was snapped up by a swallow, its death was designed; & he admitted he did not believe so, only that God ordered general laws & left the result to what may be so far called chance, that there was no design in the death of each individual Fly" (*Correspondence* 8: 349).

where many species of a genus have been produced, and where they now flourish, these same species should present many varieties; for where the manufactory of species has been active, we might expect, as a general rule, to find it still in action; and this is the case if varieties be incipient species. Moreover, the species of the larger genera, which afford the greater number of varieties or incipient species, retain to a certain degree the character of varieties; for they differ from each other by a less amount of difference than do the species of smaller genera. The closely allied species also of the larger genera apparently have restricted ranges, and they are clustered in little groups round other species—in which respects they resemble varieties. These are strange relations on the view of each species having been independently created, but are intelligible if all species first existed as varieties.

As each species tends by its geometrical ratio of reproduction to increase inordinately in number; and as the modified descendants of each species will be enabled to increase by so much the more as they become more diversified in habits and structure, so as to be enabled to seize on many and widely different places in the economy of nature, there will be a constant tendency in natural selection to preserve the most divergent offspring of any one species. Hence during a long-continued course of modification, the slight differences, characteristic of varieties of the same species, tend to be augmented into the greater differences characteristic of species of the same genus. New and improved varieties will inevitably supplant and exterminate the older, less improved and intermediate varieties; and thus species are rendered to a large extent defined and distinct objects. Dominant species belonging to the larger groups tend to give birth to new and dominant forms; so that each large group tends to become still larger, and at the same time more divergent in character. But as all groups cannot thus succeed in increasing in size, for the world would not hold them, the more dominant groups beat the less dominant. This tendency in the large groups to go on increasing in size and diverging in character, together with the almost inevitable contingency of much extinction, explains the arrangement of all the forms of life, in groups subordinate to groups, all within a few great classes, which we now see everywhere around us, and which has prevailed throughout all time. This grand fact of the grouping of all organic beings seems to me utterly inexplicable on the theory of creation.

As natural selection acts solely by accumulating slight, successive, favourable variations, it can produce no great or sudden modification; it can act only by very short and slow steps. Hence the canon of "Natura non facit saltum," which every fresh addition to our knowledge tends to make more strictly correct, is on this theory simply intelligible. We can plainly see why nature is prodigal in variety, though niggard in innovation. But why this should be a law of nature if each species has been independently created, no man can explain.

Many other facts are, as it seems to me, explicable on this theory. How strange it is that a bird, under the form of woodpecker, should have been created to prey on insects on the ground; that upland geese, which never or rarely swim, should have been created with webbed feet; that a thrush should have been created to dive and feed on sub-aquatic insects; and that a petrel should have been created with habits and structure fitting it for the life of an auk or grebe! and so on in endless other cases. But on the view of each

1 This is a nice turn of phrase. Nature is "prodigal in variety" by virtue of the astounding diversity of life—a diversity now understood to be even greater than Darwin imagined. It is "niggard in innovation," however, in that much of this diversity is centered around a relatively small number of basic body plans. Beetles, for example, may be the single largest animal taxon on earth, but nearly all beetles are the same, just minor variations on a theme. The same is true of the basic quadruped body plan of mammals. This basic "sameness" of whole groups of species is consistent with descent of those species from a common ancestor, from whom they inherited the same body plan. What pattern would be inconsistent with Darwin's theory? Why, one in which groups of species were not variations on a theme but each a unique, created entity. The typologists thought this was true at some level—their four archetypes were stand-alone, with no supposed links between them.

2 Darwin realized the significance of such oddities early on; for example, here we see him musing in the B notebook, from 1837: "There certainly appears attempt in each dominant structure to accommodate itself to as many situations as possible.—Why should we have in open country a ground . . . woodpecker—a desert. Kingfisher.—mountain tringas.—Upland goose.—water chionis water rat with land structures" [sic] (Barrett et al. 1987, p. 185). Beyond simply being able to *account* for species strangely adapted to lifestyles that do not seem to agree with their structure, Darwin suggests that they might even be predicted or expected.

species constantly trying to increase in number, with natural selection always ready to adapt the slowly varying descendants of each to any unoccupied or ill-occupied place in nature, these facts cease to be strange, or perhaps might even have been anticipated.

[1] As natural selection acts by competition, it adapts the inhabitants of each country only in relation to the degree of perfection of their associates; so that we need feel no surprise at the inhabitants of any one country, although on the ordinary view supposed to have been specially created and adapted for that country, being beaten and supplanted by the naturalised productions from another land. [2] Nor ought we to marvel if all the contrivances in nature be not, as far as we can judge, absolutely perfect; and if some of them be abhorrent to our ideas of fitness. We need not marvel at the sting of the bee causing the bee's own death; at drones being produced in such vast numbers for one single act, and being then slaughtered by their sterile sisters; at the astonishing waste of pollen by our fir-trees; at the instinctive hatred of the queen bee for her own fertile daughters; at ichneumonidæ feeding within the live bodies of caterpillars; and at other such cases. The wonder indeed is, on the theory of natural selection, that more cases of the want of absolute perfection have not been observed.

The complex and little known laws governing variation are the same, as far as we can see, with the laws which have governed the production of so-called specific forms. In both cases physical conditions seem to have produced but little direct effect; yet when varieties enter any zone, they occasionally assume some of the characters of the species proper to that zone. In both varieties and species, use and disuse seem to have produced some effect; for it is difficult to resist this con-

1 Beginning with the fourth edition (1866), Darwin expanded his discussion of beauty in chapter VI (see p. 199). Here he inserted a paragraph on beauty immediately before this one, asserting that "we can understand how it is that such harmonious beauty generally prevails throughout nature" (Peckham 1959, p. 736). He then touched on colorful plumage, birdsong, flower and fruit coloration, and symmetry, arguing that the features we consider beautiful are adaptive to the species that possess them and are not directed at humankind. Darwin made this point in his orchid book (1862), arguing that the beauty of orchids stems from the contrivances evolved for attracting pollinators. The Duke of Argyll attacked the notion that beauty is merely adaptive, holding the natural theology view that beauty was especially designed by the creator for our benefit. Darwin wrote to Lyell in 1865: "The Duke who knows my orchid book so well might have learnt a lesson of caution from it, with respect to his doctrine of differences for mere variety or beauty. It may be confidently said that no tribe of plants presents such grotesque & beautiful differences which no one until lately conjectured were of any use; but now in almost every case, I have been able to shew their important service" (*Correspondence* 13: 34).

2 Want of perfection, or seemingly abhorrent traits, are harder to understand under special creation (by a benevolent creator) than under the naturalistic process Darwin proposes. "I had no intention to write atheistically," he wrote in 1860. "But I own that I cannot see, as plainly as others do . . . evidence of design & beneficence on all sides of us. There seems to me too much misery in the world . . . On the other hand I cannot anyhow be contented to view this wonderful universe & especially the nature of man, & to conclude that everything is the result of brute force. I am inclined to look at everything as resulting from designed laws, with the details, whether good or bad, left to the working out of what we may call chance" (*Correspondence* 8: 223).

clusion when we look, for instance, at the logger-headed duck, which has wings incapable of flight, in nearly the same condition as in the domestic duck; or when we look at the burrowing tucutucu, which is occasionally blind, and then at certain moles, which are habitually blind and have their eyes covered with skin; or when we look at the blind animals inhabiting the dark caves of America and Europe. In both varieties and species correlation of growth seems to have played a most important part, so that when one part has been modified other parts are necessarily modified. In both varieties and species reversions to long-lost characters occur. How inexplicable on the theory of creation is the occasional appearance of stripes on the shoulder and legs of the several species of the horse-genus and in their hybrids! How simply is this fact explained if we believe that these species have descended from a striped progenitor, in the same manner as the several domestic breeds of pigeon have descended from the blue and barred rock-pigeon!

On the ordinary view of each species having been independently created, why should the specific characters, or those by which the species of the same genus differ from each other, be more variable than the generic characters in which they all agree? Why, for instance, should the colour of a flower be more likely to vary in any one species of a genus, if the other species, supposed to have been created independently, have differently coloured flowers, than if all the species of the genus have the same coloured flowers? If species are only well-marked varieties, of which the characters have become in a high degree permanent, we can understand this fact; for they have already varied since they branched off from a common progenitor in certain characters, by which they have come to be specifically distinct from each other; ◀1

1 Darwin does not play these observations up, but they could be seen as another "smoking gun" for transmutation. The point, first made back in chapter V, is that traits that stand out as well-developed or distinct also happen to be highly variable, as compared with the less developed expression of the trait in closely related species. If the different species were all specially created and independent, there is no reason to think they should differ in degree of variation in any trait. Darwin thinks the explanation is that selection has been acting on the varying trait, which is why it has been rendered well developed, and that its variations will thus continue for a time—it is the tell-tale sign of selection acting, in other words. That said, this observation has not been tested, to my knowledge. We understand the nature of variation to be quite different from what Darwin imagined, and there is no reason that traits which have been notably developed by selection should be inherently more variable than the same trait, albeit less modified, in other species.

and therefore these same characters would be more likely still to be variable than the generic characters which have been inherited without change for an enormous period. It is inexplicable on the theory of creation why a part developed in a very unusual manner in any one species of a genus, and therefore, as we may naturally infer, of great importance to the species, should be eminently liable to variation; but, on my view, this part has undergone, since the several species branched off from a common progenitor, an unusual amount of variability and modification, and therefore we might expect this part generally to be still variable. But a part may be developed in the most unusual manner, like the wing of a bat, and yet not be more variable than any other structure, if the part be common to many subordinate forms, that is, if it has been inherited for a very long period; for in this case it will have been rendered constant by long-continued natural selection.

Glancing at instincts, marvellous as some are, they offer no greater difficulty than does corporeal structure on the theory of the natural selection of successive, slight, but profitable modifications. We can thus understand why nature moves by graduated steps in endowing different animals of the same class with their several instincts. I have attempted to show how much light the principle of gradation throws on the admirable architectural powers of the hive-bee. Habit no doubt sometimes comes into play in modifying instincts; but it certainly is not indispensable, as we see, in the case of neuter insects, which leave no progeny to inherit the effects of long-continued habit. On the view of all the species of the same genus having descended from a common parent, and having inherited much in common, we can understand how it is that allied species, when placed under considerably different conditions of life, yet should follow nearly the same instincts; why the thrush of South America, for instance, lines her nest with mud like our British species. On the view of instincts having been slowly acquired through natural selection we need not marvel at some instincts being apparently not perfect and liable to mistakes, and at many instincts causing other animals to suffer.

If species be only well-marked and permanent varieties, we can at once see why their crossed offspring should follow the same complex laws in their degrees and kinds of resemblance to their parents,—in being absorbed into each other by successive crosses, and in other such points,—as do the crossed offspring of acknowledged varieties. On the other hand, these would be strange facts if species have been independently created, and varieties have been produced by secondary laws.

If we admit that the geological record is imperfect in an extreme degree, then such facts as the record gives, support the theory of descent with modification. New species have come on the stage slowly and at successive intervals; and the amount of change, after equal intervals of time, is widely different in different groups. The extinction of species and of whole groups of species, which has played so conspicuous a part in the history of the organic world, almost inevitably follows on the principle of natural selection; for old forms will be supplanted by new and improved forms. Neither single species nor groups of species reappear when the chain of ordinary generation has once been broken. The gradual diffusion of dominant forms, with the slow modification of their descendants, causes the forms of life, after long intervals of time, to appear as if they had changed simultaneously throughout the world. The fact of the fossil remains of each formation being in some degree intermediate in character between the

1 Here is a good example of Darwin's making the best of a bad situation: the fossil record may be fragmentary, but such evidence as it gives supports the descent theory, he argues. If we think about it, under divine creation virtually any pattern could be in evidence. The ones we see all seem consistent with transmutation over time, with no obvious signature of supernatural intervention. Such a signature could be easily imagined: consider Darwin's point about the "chain of ordinary generation"; once extinct, a species is gone for good, but under special creation, could not the very same species appear and disappear many times through the eons, if it so pleased the creator? Why would the creator render any extinct at all? Indeed, many naturalists of the Enlightenment and well into the nineteenth century embraced the plenitude concept, which held that the creator would not render any creations extinct. Even Lyell briefly entertained the notion that extinct species could be re-created (see p. 121). Describing his concept of the "great year" of cyclical earth climatic history in the first volume of *Principles*, he fancied that when the right conditions returned, "then might those genera of animals return, of which the memorials are preserved in the ancient rocks of our continents. The huge iguanodon might reappear in the woods, and the ichthyosaur in the sea, while the pterodactyle might flit again from umbrageous groves of tree ferns" (Lyell 1830, p. 123). He was evidently serious about this: Ospovat (1977) quotes Lyell in a letter to the doctor and fossil hunter Gideon Mantell: "All these changes are to happen in future again, and iguanodons and their congeners must as assuredly live again in the latitude of Cuckfield [Mantell's home town in west Sussex] as they have done so."

fossils in the formations above and below, is simply explained by their intermediate position in the chain of descent. The grand fact that all extinct organic beings belong to the same system with recent beings, falling either into the same or into intermediate groups, follows from the living and the extinct being the offspring of common parents. As the groups which have descended from an ancient progenitor have generally diverged in character, the progenitor with its early descendants will often be intermediate in character in comparison with its later descendants; and thus we can see why the more ancient a fossil is, the oftener it stands in some degree intermediate between existing and allied groups. Recent forms are generally looked at as being, in some vague sense, higher than ancient and extinct forms; and they are in so far higher as the later and more improved forms have conquered the older and less improved organic beings in the struggle for life. Lastly, the law of the long endurance of allied forms on the same continent,—of marsupials in Australia, of edentata in America, and other such cases,—is intelligible, for within a confined country, the recent and the extinct will naturally be allied by descent.

Looking to geographical distribution, if we admit that there has been during the long course of ages much migration from one part of the world to another, owing to former climatal and geographical changes and to the many occasional and unknown means of dispersal, then we can understand, on the theory of descent with modification, most of the great leading facts in Distribution. [1] We can see why there should be so striking a parallelism in the distribution of organic beings throughout space, and in their geological succession throughout time; for in both cases the beings have been connected by the bond of ordinary generation, and the means of

1 Just a reminder that Wallace first expressed this parallelism most fully and eloquently in his 1855 Sarawak Law paper. Wallace was whiling away the time during the Sarawak wet season, enjoying the hospitality of the self-styled "Rajah" Sir James Brooke. "Having always been interested in the geographical distribution of animals and plants," he later wrote in his autobiography, "having studied Swainson and Humboldt . . . it occurred to me that these facts [of biogeography] had never been properly utilized as indications of the way in which species had come into existence. The great work of Lyell had furnished me with the main features of the succession of species in time, and by combining the two I thought that some valuable conclusions might be reached. I accordingly put my facts and ideas on paper, and the result seeming me to be of some importance, I sent it to the *Annals and Magazine of Natural History*" (see account in Berry 2002). This paper is entitled "On the law which has regulated the introduction of new species."

modification have been the same. We see the full meaning of the wonderful fact, which must have struck every traveller, namely, that on the same continent, under the most diverse conditions, under heat and cold, on mountain and lowland, on deserts and marshes, most of the inhabitants within each great class are plainly related; for they will generally be descendants of the same progenitors and early colonists. On this same principle of former migration, combined in most cases with modification, we can understand, by the aid of the Glacial period, the identity of some few plants, and the close alliance of many others, on the most distant mountains, under the most different climates; and likewise the close alliance of some of the inhabitants of the sea in the northern and southern temperate zones, though separated by the whole intertropical ocean. Although two areas may present the same physical conditions of life, we need feel no surprise at their inhabitants being widely different, if they have been for a long period completely separated from each other; for as the relation of organism to organism is the most important of all relations, and as the two areas will have received colonists from some third source or from each other, at various periods and in different proportions, the course of modification in the two areas will inevitably be different.

On this view of migration, with subsequent modification, we can see why oceanic islands should be inhabited by few species, but of these, that many should be peculiar. We can clearly see why those animals which cannot cross wide spaces of ocean, as frogs and terrestrial mammals, should not inhabit oceanic islands; and why, on the other hand, new and peculiar species of bats, which can traverse the ocean, should so often be found on islands far distant from any continent. Such facts ◀1

1 This is a clear statement of the significance of isolation, though we have seen how, in Darwin's thinking, the importance of islands and the isolation they afford eventually assumed a secondary importance in species formation. But let us not forget that observations such as these provided powerful circumstantial evidence of transmutation to minds prepared to see the world through those glasses, and were crucial as Darwin formulated his theory early on. The first mention of the role of island isolation in species formation is found in the Red notebook of 1836 and early 1837: "Mem: my idea of [volcanic] islands. elevated. then peculiar plants created." An early entry in the B notebook of mid-1837 to early 1838 speculates that species like cats, dogs, and ibises that look to have remained unchanged for thousands of years (on the basis of a comparison of mummified remains in Egyptian tombs with present individuals) would transmutate in isolation: "Aegyptian cats & dogs ibis same as formerly but separate a pair & place them on fresh [island] it is very doubtful whether they would remain constant" (Barrett et al. 1987, pp. 61, 172). Isolation also takes a leading role in Darwin's 1842 *Sketch:* "change of external conditions, and isolation either by chance landing [of] a form on an island, or subsidence dividing a continent, or great chain of mountains . . . will best favour variation and selection" (F. Darwin 1909, p. 32); there follows a detailed discussion of species formation under conditions of isolation.

as the presence of peculiar species of bats, and the absence of all other mammals, on oceanic islands, are utterly inexplicable on the theory of independent acts of creation.

The existence of closely allied or representative species in any two areas, implies, on the theory of descent with modification, that the same parents formerly inhabited both areas; and we almost invariably find that wherever many closely allied species inhabit two areas, some identical species common to both still exist. Wherever many closely allied yet distinct species occur, many doubtful forms and varieties of the same species likewise occur. It is a rule of high generality that the inhabitants of each area are related to the inhabitants of the nearest source whence immigrants might have been derived. We see this in nearly all the plants and animals of the Galapagos archipelago, of Juan Fernandez, and of the other American islands being related in the most striking manner to the plants and animals of the neighbouring American mainland; and those of the Cape de Verde archipelago and other African islands to the African mainland. It must be admitted that these facts receive no explanation on the theory of creation.

The fact, as we have seen, that all past and present organic beings constitute one grand natural system, with group subordinate to group, and with extinct groups often falling in between recent groups, is intelligible on the theory of natural selection with its contingencies of extinction and divergence of character. On these same principles we see how it is, that the mutual affinities of the species and genera within each class are so complex and circuitous. We see why certain characters are far more serviceable than others for classification; — why adaptive characters, though of paramount importance to the being, are of hardly any

1 The *Beagle* landed at the Galápagos in mid-September 1835, and Darwin's first impression suggested an inauspicious introduction to the archipelago that was later to prove formative in his transmutationist thinking. "Nothing could be less inviting than the first appearance," he wrote of landing at Chatham Island (now San Cristóbal). "A broken field of black basaltic lava, thrown into the most rugged waves, and crossed by great fissures, is everywhere covered by stunted, sunburnt brushwood, which shows little signs of life." He soon grew appreciative of the marvels of the place: "The natural history of these islands is eminently curious, and well deserves attention. Most of the organic productions are aboriginal creations found nowhere else; there is even a difference between the inhabitants of the different islands; yet all show a marked relationship with those of America, though separated from that continent by an open space of ocean, between 500 and 600 miles in width. The archipelago is a little world within itself" (Darwin 1845, chapter XVII). By 1837–1838 Darwin was collecting data on island flora and fauna; for example, he took notes on von Buch's 1836 treatment of the Canary Islands in the B notebook: "Flora of [islands] very poor . . . analogous to nearest continent: poorness in exact proportion to distance (?)" (Barrett et al. 1987, p. 209).

importance in classification; why characters derived from rudimentary parts, though of no service to the being, are often of high classificatory value; and why embryological characters are the most valuable of all. The real affinities of all organic beings are due to inheritance or community of descent. The natural system is a genealogical arrangement, in which we have to discover the lines of descent by the most permanent characters, however slight their vital importance may be.

The framework of bones being the same in the hand of a man, wing of a bat, fin of the porpoise, and leg of the horse,—the same number of vertebræ forming the neck of the giraffe and of the elephant,—and innumerable other such facts, at once explain themselves on the theory of descent with slow and slight successive modifications. The similarity of pattern in the wing and leg of a bat, though used for such different purpose,—in the jaws and legs of a crab,—in the petals, stamens, and pistils of a flower, is likewise intelligible on the view of the gradual modification of parts or organs, which were alike in the early progenitor of each class. On the principle of successive variations not always supervening at an early age, and being inherited at a corresponding not early period of life, we can clearly see why the embryos of mammals, birds, reptiles, and fishes should be so closely alike, and should be so unlike the adult forms. We may cease marvelling at the embryo of an air-breathing mammal or bird having branchial slits and arteries running in loops, like those in a fish which has to breathe the air dissolved in water, by the aid of well-developed branchiæ. ◀1

Disuse, aided sometimes by natural selection, will often tend to reduce an organ, when it has become useless by changed habits or under changed conditions

1 Darwin is passionate about this point; his theory nicely explains such oddities as embryonic structures that make no sense for the adult organism but seem to speak of history. This is emphasized on the next page.

1 Structures that "bear the plain stamp of inutility" speak of history. Such structures are not only explained by his theory but expected. Why should ruminant juveniles have been created with useless teeth that never break through the gums? Why should flightless beetles bear flight wing remnants?

In *On the Nature of Limbs* (1849) Owen acknowledged that some structures seem to be "made in vain," but he held that they somehow reflected the archetype. Typologists presumably saw such structures as a vestige of history too, insofar as they adhered to the idea of recapitulation. If there is recapitulation within a type, the order of appearance of traits must reflect the stepwise order of creation within the type. Yet it somehow seems unsatisfying that useless structures would be designedly made. Notice the passion of the last sentence in this section: the evidence for modification is plain as day, Darwin urges, but we will not see it; we remain willfully ignorant.

2 These lines open the concluding section of the chapter, and the book. Darwin is at his most lyrical in this section, and it is instructive to notice the rhetorical devices that he uses as well as the content of his statements. The sixth edition saw a significant insertion immediately following the opening sentence. Two of his points are philosophical. First, he defends his method: "It can hardly be supposed that a false theory would explain, in so satisfactory a manner . . . the several large classes of facts above specified. It has recently been objected that this is an unsafe method of arguing; but it is a method used in judging of the common events of life, and has often been used by the greatest philosophers." Second, he attempts to defuse religious objections: "I see no good reason why the views given in this volume should shock the religious feelings of any one. It is satisfactory, as showing how transient such impressions are, to remember that the greatest discovery ever made by man, namely, the law of the attraction of gravity, was also attacked by Leibnitz, 'as subversive of natural, and inferentially of revealed, religion.'"

of life; and we can clearly understand on this view the meaning of rudimentary organs. But disuse and selection will generally act on each creature, when it has come to maturity and has to play its full part in the struggle for existence, and will thus have little power of acting on an organ during early life; hence the organ will not be much reduced or rendered rudimentary at this early age. The calf, for instance, has inherited teeth, which never cut through the gums of the upper jaw, from an early progenitor having well-developed teeth; and we may believe, that the teeth in the mature animal were reduced, during successive generations, by disuse or by the tongue and palate having been fitted by natural selection to browse without their aid; whereas in the calf, the teeth have been left untouched by selection or disuse, and on the principle of inheritance at corresponding ages have been inherited from a remote period to the present day. On the view of each organic being and each separate organ having been specially created, how utterly inexplicable it is that parts, like the teeth in the embryonic calf or like the shrivelled wings under the soldered wing-covers of some beetles, should thus so frequently bear the plain stamp of inutility! Nature may be said to have taken pains to reveal, by rudimentary organs and by homologous structures, her scheme of modification, which it seems that we wilfully will not understand.

I have now recapitulated the chief facts and considerations which have thoroughly convinced me that species have changed, and are still slowly changing by the preservation and accumulation of successive slight favourable variations. Why, it may be asked, have all the most eminent living naturalists and geologists rejected this view of the mutability of species? It cannot be asserted that organic beings in a state of nature are subject to no variation; it cannot be proved that the amount of variation in the course of long ages is a limited quantity; no clear distinction has been, or can be, drawn between species and well-marked varieties. It cannot be maintained that species when intercrossed are invariably sterile, and varieties invariably fertile; or that sterility is a special endowment and sign of creation. The belief that species were immutable productions was almost unavoidable as long as the history of the world was thought to be of short duration; and now that we have acquired some idea of the lapse of time, we are too apt to assume, without proof, that the geological record is so perfect that it would have afforded us plain evidence of the mutation of species, if they had undergone mutation.

But the chief cause of our natural unwillingness to admit that one species has given birth to other and distinct species, is that we are always slow in admitting any great change of which we do not see the intermediate steps. The difficulty is the same as that felt by so many geologists, when Lyell first insisted that long lines of inland cliffs had been formed, and great valleys excavated, by the slow action of the coast-waves. The mind cannot possibly grasp the full meaning of the term of a hundred million years; it cannot add up and perceive the full effects of many slight variations, accumulated during an almost infinite number of generations. ◀1

Although I am fully convinced of the truth of the views given in this volume under the form of an abstract, I by no means expect to convince experienced naturalists whose minds are stocked with a multitude of facts all viewed, during a long course of years, from a point of view directly opposite to mine. It is so easy ◀2

1 There is probably much truth to this psychological argument. The notion that species as widely divergent as whales and ungulates are related is mind-bending, and without the existence of intermediate linking forms to literally "connect the dots," many people are incapable of or unwilling to see the transition. In the whale-ungulate case it happens that several fossil linking forms have been discovered, but even so, today most opposition to Darwin's views comes not from "eminent naturalists" but from the general public, a group not likely to be well informed about such discoveries. Perhaps an intuitive construction of the world is in fact typological, and insofar as that is a common feature of human psychology the need for a full complement of connecting dots will always be a common stumbling block preventing the acceptance of evolution.

Note the rhetorical strategy of directly comparing himself with Lyell. Indeed, there are important parallels between the geological and biological processes: both are gradual, uniformitarian processes that occur over an immense expanse of time.

2 Could Darwin be using the embarrassment ploy, inducing a hostile reader to give his theory a second look to avoid giving the appearance of *inflexibility* of mind, or of hiding his ignorance? We can imagine that such statements had the opposite effect on the likes of Owen and Agassiz, like a red cape to a bull. Be that as it may, Darwin was right to look to "young and rising naturalists"—it would be exceedingly difficult to convince most older naturalists, much of whose life work and writing had been informed by the existing worldview, that their labors had been misguided. There is something to be said for youth and flexibility of mind when it comes to so far-reaching a doctrine as Darwin's, especially given that young naturalists do not yet have much invested or at stake. This is not to say that the elder statesmen of natural history were willfully ignorant; these brilliant scientists had a well-developed and internally consistent worldview that in some respects was more firmly anchored in strict empiricism than was Darwin's. Consider that Agassiz took the fossil record seriously, seeing the gaps and abrupt appearances that Darwin lamented. Darwin argued that these are mislead-

(continued)

(continued)
ing, while Agassiz considered them data that undermined Darwin's argument.

The tide quickly turned in Darwin's favor in the decades following the *Origin*'s first appearance (though most people were convinced of the reality of transmutation and common descent, not natural selection). The old guard like Agassiz remained entrenched; shortly before his death in 1873 Agassiz even visited the Galápagos, those enchanted islands that provided Darwin with compelling examples of species divergence. Agassiz was unmoved, however, and in fact cleverly argued that since the islands were of recent origin (as was then believed), Darwin's supposed mechanism of transmutation did not require nearly as much time as he claimed (Gould 1983, Lurie 1988).

1 The lack of clear distinction between species and varieties bolsters Darwin's argument that the two represent degrees of difference; his key point is that it makes no sense for one to be specially created while the other is supposedly formed by natural law.

to hide our ignorance under such expressions as the "plan of creation," "unity of design," &c., and to think that we give an explanation when we only restate a fact. Any one whose disposition leads him to attach more weight to unexplained difficulties than to the explanation of a certain number of facts will certainly reject my theory. A few naturalists, endowed with much flexibility of mind, and who have already begun to doubt on the immutability of species, may be influenced by this volume; but I look with confidence to the future, to young and rising naturalists, who will be able to view both sides of the question with impartiality. Whoever is led to believe that species are mutable will do good service by conscientiously expressing his conviction; for only thus can the load of prejudice by which this subject is overwhelmed be removed.

Several eminent naturalists have of late published their belief that a multitude of reputed species in each genus are not real species; but that other species are real, that is, have been independently created. This seems to me a strange conclusion to arrive at. They admit that a multitude of forms, which till lately they themselves thought were special creations, and which are still thus looked at by the majority of naturalists, and which consequently have every external characteristic feature of true species,—they admit that these have been produced by variation, but they refuse to extend the same view to other and very slightly different forms. Nevertheless they do not pretend that they can define, or even conjecture, which are the created forms of life, and which are those produced by secondary laws. They admit variation as a *vera causa* in one case, they arbitrarily reject it in another, without assigning any distinction in the two cases. The day will come when this will be given as a curious illustration of the blindness of preconceived opinion. These authors seem no more startled at a miraculous act of creation than at an ordinary birth. But do they really believe that at innumerable periods in the earth's history certain elemental atoms have been commanded suddenly to flash into living tissues? Do they believe that at each supposed act of creation one individual or many were produced? Were all the infinitely numerous kinds of animals and plants created as eggs or seed, or as full grown? and in the case of mammals, were they created bearing the false marks of nourishment from the mother's womb? Although naturalists very properly demand a full explanation of every difficulty from those who believe in the mutability of species, on their own side they ignore the whole subject of the first appearance of species in what they consider reverent silence.

It may be asked how far I extend the doctrine of the modification of species. The question is difficult to answer, because the more distinct the forms are which we may consider, by so much the arguments fall away in force. But some arguments of the greatest weight extend very far. All the members of whole classes can be connected together by chains of affinities, and all can be classified on the same principle, in groups subordinate to groups. Fossil remains sometimes tend to fill up very wide intervals between existing orders. Organs in a rudimentary condition plainly show that an early progenitor had the organ in a fully developed state; and this in some instances necessarily implies an enormous amount of modification in the descendants. Throughout whole classes various structures are formed on the same pattern, and at an embryonic age the species closely resemble each other. Therefore I cannot doubt that the theory of descent with modification

1 Darwin voices frustration over what amounts to a double standard: adherents of special creation dismiss the notion of transmutation as fantastical and unlikely, while shamelessly invoking miraculous intervention on the authority of the Bible. But is miraculous fiat any more likely or less fantastical? It has only the "authority" of historical precedent, having been unquestioned for centuries. Darwin's sarcasm about whether the creator supplied his creations with the false signs of a "normal" birth touches on a theological matter. Gosse's *Omphalos* (1857) argued that the creator imbued his creations with all the (false) indications of historical continuity—trees with rings giving the impression of an age they could not have, navels suggesting birth, etc. This *prochronism*, or false historicity, went over like a lead balloon with just about everyone, chuchgoer or not. The serious point underlying Darwin's rhetorical questions is that to assert sweepingly that species were simply created begs questions that are usually not raised. Deep thought is not given to the matter, just as many people faithfully repeat what they have been taught about inerrancy of Genesis, say, or even the narrative of Jesus' birth, without realizing the troublesome contradictions inherent in these Biblical accounts.

2 Darwin inserted a paragraph here in the sixth edition that opens: "As a record of a former state of things, I have retained in the foregoing paragraphs . . . several sentences which imply that naturalists believe in the separate creation of each species; and I have been much censured for having thus expressed myself. But undoubtedly this was the general belief when the first edition of [*Origin*] appeared." By 1872 it appears that many naturalists were embarrassed at having espoused special creation! Many were transmutation converts, but often they supported neither gradualism nor natural selection. The new paragraph argues against these views as well (Peckham 1959, p. 751).

1 Darwin is willing to follow his model of common descent to its logical conclusion: tracing the tree of life backward, it is likely that all of life is descended from "some one prototype" or "one primordial form." This is termed the "universal common ancestor" today. It is interesting to note that in the very next edition of the *Origin,* which came out a mere month after this one, Darwin modified the paragraph's last sentence to end "into which life was first breathed by the Creator." This may have been an attempt to show readers that he did not reject the deity altogether, and indeed the question of *ultimate* origins is not one that is readily assessed scientifically.

2 Beginning here Darwin looks to the future. His forecast of a "considerable revolution in natural history" was borne out, but one wonders if he realized that a revolution in society was to accompany it.

embraces all the members of the same class. I believe that animals have descended from at most only four or five progenitors, and plants from an equal or lesser number.

Analogy would lead me one step further, namely, to the belief that all animals and plants have descended from some one prototype. But analogy may be a deceitful guide. Nevertheless all living things have much in common, in their chemical composition, their germinal vesicles, their cellular structure, and their laws of growth and reproduction. We see this even in so trifling a circumstance as that the same poison often similarly affects plants and animals; or that the poison secreted by the gall-fly produces monstrous growths on the wild rose or oak-tree. Therefore I should infer from analogy that probably all the organic beings which have ever lived on this earth have descended from some one primordial form, into which life was first breathed.

When the views entertained in this volume on the origin of species, or when analogous views are generally admitted, we can dimly foresee that there will be a considerable revolution in natural history. Systematists will be able to pursue their labours as at present; but they will not be incessantly haunted by the shadowy doubt whether this or that form be in essence a species. This I feel sure, and I speak after experience, will be no slight relief. The endless disputes whether or not some fifty species of British brambles are true species will cease. Systematists will have only to decide (not that this will be easy) whether any form be sufficiently constant and distinct from other forms, to be capable of definition; and if definable, whether the differences be sufficiently important to deserve a specific name. This latter point will become a far more essential consideration than it is at present; for differences, however slight, between any two forms, if not blended by intermediate gradations, are looked at by most naturalists as sufficient to raise both forms to the rank of species. Hereafter we shall be compelled to acknowledge that the only distinction between species and well-marked varieties is, that the latter are known, or believed, to be connected at the present day by intermediate gradations, whereas species were formerly thus connected. Hence, without quite rejecting the consideration of the present existence of intermediate gradations between any two forms, we shall be led to weigh more carefully and to value higher the actual amount of difference between them. It is quite possible that forms now generally acknowledged to be merely varieties may hereafter be thought worthy of specific names, as with the primrose and cowslip; and in this case scientific and common language will come into accordance. In short, we shall have to treat species in the same manner as those naturalists treat genera, who admit that genera are merely artificial combinations made for convenience. This may not be a cheering prospect; but we shall at least be freed from the vain search for the undiscovered and undiscoverable essence of the term species.

The other and more general departments of natural history will rise greatly in interest. The terms used by naturalists of affinity, relationship, community of type, paternity, morphology, adaptive characters, rudimentary and aborted organs, &c., will cease to be metaphorical, and will have a plain signification. When we no longer look at an organic being as a savage looks at a ship, as at something wholly beyond his comprehension; when we regard every production of nature as one which has had a history; when we contemplate every complex structure ◀1

1 This understated sentence is an unexpected lead-in to some of the most prophetic and beautiful language in the entire book. The ramifying and grand revolution in understanding that Darwin foresees stems from one seemingly trivial change in the way we view organisms: when species are seen to have a *history,* he says—when we see that they are genealogically related—only then will we see them comprehendingly.

1 How exciting it must have been for Darwin to realize that he alone among scientists understood this; that he was the first to behold this "grand and untrodden field of inquiry." The C notebook shows him suddenly following this line of thought in almost stream-of-consciousness adrenalin-filled passages: "Once grant that [species] may pass into each other.—grant that one instinct to be acquired (if the medullary point in ovum. has such organization as to force in one man the development of a brain capable of producing more glowing imagining or more profound reasoning than other—if this be granted!!) & whole fabric totters and falls.—look abroad, study gradation. study unity of type—study geographical distribution study relation of fossil with recent. the fabric falls!" (Barrett et al. 1987, p. 263).

and instinct as the summing up of many contrivances, each useful to the possessor, nearly in the same way as when we look at any great mechanical invention as the summing up of the labour, the experience, the reason, and even the blunders of numerous workmen; when we thus view each organic being, how far more interesting, I speak from experience, will the study of natural history become!

A grand and almost untrodden field of inquiry will be opened, on the causes and laws of variation, on correlation of growth, on the effects of use and disuse, on the direct action of external conditions, and so forth. The study of domestic productions will rise immensely in value. A new variety raised by man will be a far more important and interesting subject for study than one more species added to the infinitude of already recorded species. Our classifications will come to be, as far as they can be so made, genealogies; and will then truly give what may be called the plan of creation. The rules for classifying will no doubt become simpler when we have a definite object in view. We possess no pedigrees or armorial bearings; and we have to discover and trace the many diverging lines of descent in our natural genealogies, by characters of any kind which have long been inherited. Rudimentary organs will speak infallibly with respect to the nature of long-lost structures. Species and groups of species, which are called aberrant, and which may fancifully be called living fossils, will aid us in forming a picture of the ancient forms of life. Embryology will reveal to us the structure, in some degree obscured, of the prototypes of each great class.

When we can feel assured that all the individuals of the same species, and all the closely allied species of most genera, have within a not very remote period descended from one parent, and have migrated from some one birthplace; and when we better know the many means of migration, then, by the light which geology now throws, and will continue to throw, on former changes of climate and of the level of the land, we shall surely be enabled to trace in an admirable manner the former migrations of the inhabitants of the whole world. Even at present, by comparing the differences of the inhabitants of the sea on the opposite sides of a continent, and the nature of the various inhabitants of that continent in relation to their apparent means of immigration, some light can be thrown on ancient geography.

The noble science of Geology loses glory from the extreme imperfection of the record. The crust of the earth with its embedded remains must not be looked at as a well-filled museum, but as a poor collection made at hazard and at rare intervals. The accumulation of each great fossiliferous formation will be recognised as having depended on an unusual concurrence of circumstances, and the blank intervals between the successive stages as having been of vast duration. But we shall be able to gauge with some security the duration of these intervals by a comparison of the preceding and succeeding organic forms. We must be cautious in attempting to correlate as strictly contemporaneous two formations, which include few identical species, by the general succession of their forms of life. As [1] species are produced and exterminated by slowly acting and still existing causes, and not by miraculous acts of creation and by catastrophes; and as the most important of all causes of organic change is one which is almost independent of altered and perhaps suddenly altered physical conditions, namely, the mutual relation of organism to organism,—the improvement of one being entailing the improvement or the extermination of

1 Note that Darwin periodically reasserts, as here, that change is gradual and continuous; "slowly acting and still existing causes" is a line that could have come right out of one of Lyell's geological treatises.

1 Geological time is beyond the comprehension of most people. Happily, scientists have developed a number of analogs on a scale that humans can comprehend (e.g., Ritger and Cummins 1991). What was probably the first of these aids was devised by the physicist James Croll, who worked on the theory of glacial cycles. In an 1868 paper on geological time, Croll scaled a million years into an 83-foot, 4-inch strip of paper, such that 1/10 inch equals 100 years. (Darwin appreciated a good teaching aid when he saw it, and he cited Croll's device in the fifth edition of the *Origin;* see p. 287.) Another approach compresses the geological timescale into a calendar year: reckoning from an earth origin at midnight on January 1, a simple calculation shows that all of human existence, from the earliest appearance of *Homo sapiens,* comes late on December 31, beginning about 11:49 PM.

2 Darwin clearly saw that exciting new fields would blossom, among them researches into human evolution. He himself explored this in two subsequent books: *The Descent of Man* (1871) and *The Expression of Emotions in Man and Animals* (1872). An important thread running through these works is the precise point stated here: the development of human cognition by gradualistic evolution. Thriving subfields of psychology later grew, such as evolutionary psychology, built on the theory of sociobiology articulated in the 1960s and 1970s. Yet Darwin seems coy about the implications of his theory for humans. He came forth with his treatises on the subject only after others, like Huxley and Haeckel, argued forcefully that humans, too, had evolved. Darwin realized this very early, but in the *Origin* the furthest he would go is intensifying—and only in the last edition—the last sentence here to read: "Much light will be thrown . . ."

3 Would that most readers had shared Darwin's perspective. I agree with him that there is something exhilarating in contemplating the grand sweep of time, and the idea that there is organic continuity: we trace our ancestry in an unbroken chain that reaches back to the earliest glimmerings of life on earth.

others; it follows, that the amount of organic change in the fossils of consecutive formations probably serves as a fair measure of the lapse of actual time. A number of species, however, keeping in a body might remain for a long period unchanged, whilst within this same period, several of these species, by migrating into new countries and coming into competition with foreign associates, might become modified; so that we must not overrate the accuracy of organic change as a measure of time. During early periods of the earth's history, when the forms of life were probably fewer and simpler, the rate of change was probably slower; and at the first dawn of life, when very few forms of the simplest structure existed, the rate of change may have been slow in an
1 extreme degree. The whole history of the world, as at present known, although of a length quite incomprehensible by us, will hereafter be recognised as a mere fragment of time, compared with the ages which have elapsed since the first creature, the progenitor of innumerable extinct and living descendants, was created.

2 In the distant future I see open fields for far more important researches. Psychology will be based on a new foundation, that of the necessary acquirement of each mental power and capacity by gradation. Light will be thrown on the origin of man and his history.

Authors of the highest eminence seem to be fully satisfied with the view that each species has been independently created. To my mind it accords better with what we know of the laws impressed on matter by the Creator, that the production and extinction of the past and present inhabitants of the world should have been due to secondary causes, like those determining the
3 birth and death of the individual. When I view all beings not as special creations, but as the lineal descendants of some few beings which lived long before the

first bed of the Silurian system was deposited, they seem to me to become ennobled. Judging from the past, we may safely infer that not one living species will transmit its unaltered likeness to a distant futurity. And of the species now living very few will transmit progeny of any kind to a far distant futurity; for the manner in which all organic beings are grouped, shows that the greater number of species of each genus, and all the species of many genera, have left no descendants, but have become utterly extinct. We can so far take a prophetic glance into futurity as to foretel that it will be the common and widely-spread species, belonging to the larger and dominant groups, which will ultimately prevail and procreate new and dominant species. As all the living forms of life are the lineal descendants of those which lived long before the Silurian epoch, we may feel certain that the ordinary succession by generation has never once been broken, and that no cataclysm has desolated the whole world. Hence we may look with some confidence to a secure future of equally inappreciable length. And as natural selection works solely by and for the good of each being, all corporeal and mental endowments will tend to progress towards perfection. ◀1

It is interesting to contemplate an entangled bank, ◀2 clothed with many plants of many kinds, with birds singing on the bushes, with various insects flitting about, and with worms crawling through the damp earth, and to reflect that these elaborately constructed forms, so different from each other, and dependent on each other in so complex a manner, have all been produced by laws acting around us. These laws, taken in the largest sense, being Growth with Reproduction; Inheritance which is almost implied by reproduction; Variability from the indirect and direct action of the external con-

1 It is worth noting that although Darwin is using the word "perfection" in a general sense in this sentence (recall that on p. 82 he argues that species are *not* perfect), he does seem to see the evolutionary process as progressive. There is a fine line, perhaps, between progressive and directional; he did not see evolution as directional or purposive per se, but insofar as he thought the adaptation of organisms increases over time, he saw this as progress.

2 In the sixth edition of the *Origin* "an entangled bank" was changed to simply "a tangled bank." This is one of Darwin's most famous metaphors, and a lovely one at that. We can just *see* that sun-dappled bank, teeming with life above and below. What Darwin is driving at here is that a certain predictability underlies the apparent "tangle"—the seemingly random, chaotic melee is in fact a vibrant and structured biological community, shaped by fully explicable natural law.

ditions of life, and from use and disuse; a Ratio of Increase so high as to lead to a Struggle for Life, and as a consequence to Natural Selection, entailing Divergence of Character and the Extinction of less-improved forms. Thus, from the war of nature, from famine and death, the most exalted object which we are capable of conceiving, namely, the production of the higher animals, directly follows. There is grandeur in this view of life, with its several powers, having been originally breathed into a few forms or into one; and that, whilst this planet has gone cycling on according to the fixed law of gravity, from so simple a beginning endless forms most beautiful and most wonderful have been, and are being, evolved.

1 I hesitate to intrude on the beauty of these lines. By necessity I will simply point out a few interesting features and later changes. Note, first, that Darwin seems to speak to those perhaps reluctant to let go of their natural theology worldview: despite the reality of the "war of nature"—famine and death—exquisite beauty arises, he urges. His tone is not consoling, yet there is an air of reassurance about the statement. Note, too, the juxtaposition of Darwin's natural law of descent with modification with the law of gravitation. Even the divines of Darwin's time would have granted that the planets cycle on by Newtonian natural law, albeit set in motion by the creator. So, too, Darwin is saying, do life forms continually change—not in a cycle, he would argue, but *in response to* cycles of geological and biological change, in a grand interrelated system that spins "endless forms most beautiful and most wonderful" from perhaps but a single common ancestor. In the second edition Darwin added "by the Creator" to "originally breathed," intimating that a creator may have set this grand system in motion, just as the physicists held for the clockwork universe. Note, finally, that the very last word is the *only* use of the word "evolve" or its cognates in the book—ironic, given that "evolution" is now synonymous with Darwin's model of common descent by natural selection. In his day the word was more closely associated with embryological development, and indeed Darwin's usage in this last sentence may be invoking an image of the embryo's unfolding developmental complexity, as natural selection endlessly spins out those forms most beautiful and most wonderful.

Coda: The *Origin* Evolving

I concluded the Introduction with the observation that human understanding of the organic world and our species underwent a quantum leap with the arrival of the *Origin.* I also pointed out that few knew it at the time. In fact, the significance of Darwin's discoveries was not widely appreciated for some time. The book provoked a steady stream of editorial letters, reviews, and, in due time, even whole books both in defense of the theory and in bitter opposition to it. Darwin fought with his pen, patiently replying, often at great length, to innumerable correspondents taking issue with his arguments. As spirited a defender as he was on paper, however, Darwin had no stomach for public jousting. Fortunately he had close friends like Lyell, Hooker, Huxley, and Gray only too willing to take up his cudgels for him. Fierce battles ensued, on paper, in the lecture halls, and in discussions and debates before the learned societies. There were public debates such as the legendary (and in some respects apocryphal) face-off between Huxley and Bishop Wilberforce at the 1860 British Association meeting in Oxford; spirited reviews for and against the theory; open letters to newspapers, journals, and magazines; and privately published pamphlets. Huxley was so tenacious and formidable in his spirited verbal blocks, parries, and counter-attacks that he earned the nickname "Darwin's bulldog." Other defenders, like Gray in America, were more measured but equally tenacious. Wallace, as co-discoverer of the theory, ably defended his and Darwin's ideas with a direct and penetrating argumentative style.

Many of Darwin's critics were insightful and in some cases brilliant thinkers, people for whom Darwin had great respect and admiration. He repeatedly acknowledged in the *Origin* that his theory faced grave difficulties, and he tried to address them as best he could, arguing that most of the problems were only apparent when you got right down to it (for example, the problem of sterility and morphological differentiation in social insect colonies, or the absence of innumerable transitional forms linking all living species), or cases where we simply lacked enough information to draw any definitive conclusion as yet (for example, the seemingly sudden appearance of whole suites of species in the fossil record, or questions about the age of the earth). Nonetheless, many of the criticisms were legitimate, and Darwin took them seriously indeed.

Like a presidential hopeful on the campaign trail, Darwin kept a running tally of friends and colleagues who accepted or rejected his theory. Each new addition to his corner was met with a gleeful hurrah, while he expressed disappointment—at times bitter—over those who remained unconvinced. It pained him that the latter camp included some of the naturalists he most respected: Sir Richard Owen, Adam Sedgwick, and John Stevens Henslow. He felt that not all gave him a fair hearing, and Owen's caustic pen, turned against Darwin in an anonymously published review, especially rankled. The book's important and original observations were, in Owen's estimation, "few indeed and far apart, and leaving the determination of the origin of species very nearly where the author found it." Owen also took a jab at the "abstract" nature of the book—precisely what Darwin feared: "a rich mine of such researches is alluded to and promised by Mr. Darwin, in a more voluminous collection of his researches, extending over a period of eighteen years; and to these every naturalist now looks forward with keen interest" (Owen 1860).

To Darwin's dismay even the philosophical savants John Herschel and William Whewell were dismissive of his theory, despite his conviction that he emulated good philosophical methodology in the *Origin*. Nothing seemed to pain him more than to be accused of being unphilosophical, or to have used improper method. Darwin expressed exasperation with Sedgwick, for example, for accusing him of the sin of inventing hypotheses (natural selection) rather than following the proper scientific practice of pure induction. Incensed, Darwin wrote in a May 1860 letter to Henslow: "I can perfectly understand Sedgwick or any one saying that nat. selection does not explain large classes of facts; but that is very different from saying that I depart from right principles of scientific investigation" (*Correspondence* 8: 194). He was appreciative that the kindly Henslow, at least, defended his right to such inquiries, arguing that Darwin should be given a fair hearing.

This, then, was the lay of the philosophical land in the years immediately following the publication of the *Origin*: a scientific community divided, struggling to understand and evaluate Darwin and Wallace's bold new theory. Before long Darwin succeeded in convincing most of his colleagues of the reality of transmutation, but they were far more skeptical when it came to natural selection. Indeed, to Darwin's chagrin even some of his staunchest supporters like Huxley and Lyell were lukewarm over natural selection. The *Origin* went through six editions between 1859 and 1872, and in certain respects Darwin's message became obscured as he tried to clarify, defend, and qualify in response to criticisms leveled against the theory.

To the end, Darwin remained committed to gradual evolutionary change driven by the incessant action of natural selection, yet we see him refining and struggling to shore up elements of his theory. This vast and fascinating subject has been thoroughly treated by scholars (e.g., Peckham 1959, Vorzimmer 1963, 1972; Hull 1973, Burchfield 1974, Sulloway 1979, Liepman 1981, Rhoades 1987, Browne 2002, Hodge and Radick 2003), and it is not my intention to summarize their work in detail. Rather, here I offer a brief overview of the most significant changes Darwin made to subsequent *Origin* editions, with the aim of providing insight into the major lines of criticism he faced, and how they precipitated modifications to his ideas over time in response to contemporary debate, discussion, and discovery. Think of this as the *Origin* evolving. In short, by the time of Darwin's death in 1882, evolution was on firm footing, but natural selection and the nature, origin, limits, and heritability of variation continued to be debated—and thus also did the very nature of evolutionary change as Darwin conceived of it. The efficacy of natural selection as the main agent of change became a central question, and this interplayed with at least four general issues: the age of the earth (and so time available for natural selection), the importance of isolation, the nature of inheritance, and the mode and tempo of evolutionary change over time. Let us briefly consider the successive editions of the *Origin* in turn to see the ways these ideas played out.

The second edition of the *Origin* appeared barely a month after the first, in early 1860. Its changes were minor, consisting mainly of corrected printing errors. Yet as 1860 progressed a stream of reviews appeared, some favorable and some hostile, along with letters conveying the judgment of friends and colleagues near and far. By the end of the year Darwin was deep into revisions for a third edition, and his exasperation with the reception he was getting in some quarters came through loud and clear in a December 1860 letter to Murray, his publisher: "I hope never again to have to make so many corrections or rather additions which I have made in hopes of making my many rather stupid reviewers at least understand what is meant" (*Correspondence* 8: 515). Alas, this was not to be; Darwin was compelled to make even greater changes to later editions, and the modifications to the third, which appeared in April 1861, are minor in comparison.

The most obvious change in the third edition is a new prefatory section, a "Historical Sketch of the Recent Progress of Opinion on the Origin of Species." There may have been a strategy behind this addition: even as Darwin's ideas were being hotly debated, showing that many naturalists had come around to his view was potentially useful in legitimizing his theory and recruiting new converts. Among the less obvious but more important changes in this edition, Darwin dropped his calculation of the time needed for the erosion of the Weald in southern England in response to severe criticism (see p. 287), but he reaffirmed his commitment to natural selection. The years 1860–1861 saw much skepticism concerning the efficacy of natural selection to generate the diversity of life. Many religiously minded naturalists who were otherwise receptive to Darwin's arguments sought to reconcile transmutation with their belief in the divine. Some urged that transmutation was driven or guided by some kind of "complicating force" or law, which would direct transmutation along particular lines (unsurprisingly, usually leading to humanity). Darwin would have no part of this, and in chapter IV of the third edition he stepped up his argument that natural selection had no need for any such "necessary and universal law of advancement or development."

The fourth edition was published in 1866. One important change here was a retreat from Darwin's earlier assertion that hybrid sterility arises by natural selection. This revision may have been precipitated by critics who pointed out that sterility is often seen in experimental crosses of plants from widely separated areas. Since such species had never naturally encountered one another, sterility could not be a trait that natural selection shaped. Darwin conceded that sterility seen in species and hybrid crosses likely arises incidentally, from differences in the reproductive system. Another important change in this edition concerns the nature of variation and the rate of evolutionary change over time, stemming from an 1863 study of fossil elephants published by Darwin's friend Hugh Falconer. Falconer's paper addressed the apparent lack of change of some elephant species for long time periods—most significantly, these species remained largely unchanged even through the glacial periods. In view of the presumed severity of the climate during glacial times, the lack of change in key morphological characters suggested that natural selection was inoperative or ineffective, and therefore it became questionable what it could do in less trying times. Falconer was not a severe critic—he concluded in his article that Darwin "laid the foundation of a great edifice, but he need not be surprised if, in the progress of the erection, the superstructure is altered by his successors." Yet he forced Darwin to reconsider whether evolutionary change was indeed gradual and constant. Darwin replied in the fourth edition with an argument suggesting that some species may undergo rapid bursts of change, and that the periods in which they do undergo change tend to be short compared with the periods during which they are unchanged. This also helped respond to a related criticism Falconer made, namely that he could find no transitional forms between extinct and living elephant species. Rapid bursts of change could mean that transitional forms are short-lived and thus less likely to be picked up in the fossil record.

The fifth edition, published in August 1869—the one that saw the introduction of Herbert Spencer's phrase "survival of the fittest"—contained the most radical changes in the *Origin* to date. All through the 1860s criticisms were mounted against Darwin's views on several fronts, and among the gravest problems he faced was the matter of time. Sir William Thompson, Lord Kelvin, had estimated an earth age of some 98–200 million years, far less than Darwin thought necessary for his model of transmutation. The geologist James Croll, who accepted the upper end of Kelvin's estimate, convinced Darwin that no one bandying about figures like millions of years really had any idea what a vast period of time even one million years was. Croll had sent Darwin his papers on the subject in fall 1868; Darwin wrote back: "I have never, I think, in my life been so deeply interested by any geological discussion. I now first begin to see what a million means, and I feel quite ashamed of myself at the silly way in which I have spoken of millions of years" (Darwin and Seward 1903b, letter 544). Darwin was somewhat reassured that perhaps he did not need nearly as much time as he thought for the gradual evolution of the diversity of life. He discussed Croll's ideas at some length in chapter IX of the fifth edition in this vein, and suggested that the rate of evolution might have been greater early in earth's history: "It is . . . probable, as Sir William Thompson insists, that the world at a very early period was subjected to more rapid and violent changes in its physical conditions than those now occurring; and such changes would have tended to induce changes at a corresponding rate in the organisms which then existed" (*Origin,* 5th edition, p. 354).

Related issues dealt with in the fifth edition pertained to the origin and fate of variation. In response to persistent concerns that there is not enough variability for selection to act upon, we see Darwin getting increasingly Lamarckian. He expands the role of "conditions of life" in generating variation, tying this in with continually changing geological and climatic conditions, which would thus provide a continual engine for new variations. This dovetailed nicely with an important criticism concerning the fate of variation. In 1867, the Scottish engineer Fleeming Jenkin published an incisive critique arguing that new favorable variations no sooner arise than they become swamped by blending during reproduction, in effect stalling the action of natural selection. How, then, could favorable new variants spread, or at least hang on long enough for selection to act on them? An environmental cause for variation was helpful in countering Jenkin: perhaps multiple individuals in a population generate similar variations in response to similar environmental conditions, Darwin suggested.

A related line of criticism came into play here as well. In 1868 the German naturalist Moritz Wagner wrote to Darwin to argue that spatial isolation is necessary for the formation of new species, urging that evolutionary change is not possible in large continuous populations because individual differences would be blended together. Darwin agreed in large part with what was later called the "separation theory." In the fifth edition of the *Origin* he cited Wagner in connection with the importance of isolation in speciation, but he also declared that he could "by no means agree with this naturalist, that migration and isolation are necessary for the formation of new species." In the earlier editions Darwin seemed to downplay the importance of isolation in favor of his principle of divergence, which he thought could drive speciation even in continuous populations (sympatric as opposed to allopatric speciation). Sulloway (1979) suggested that, later, Darwin did not so much de-emphasize the role of isolation as refine it by incorporating ecological as well as geographical barriers. It is true that isolation was more implicitly than explicitly treated in the *Origin* up until the fifth edition, when Darwin came out endorsing the importance of isolation more clearly in response to Wagner. He recognized that this also helped him argue against Jenkin's protest about blending: isolation would facilitate the establishment of beneficial variations in two ways: first, the isolated individuals would likely vary in the same way, and so there would be multiple copies of the new environmentally induced variant; and second, by virtue of isolation the individuals bearing new favorable variants would interbreed with each other, and so avoid losing the favorable trait by the swamping effects of blending. One final note about Wagner and isolation: despite their initially cordial correspondence, a rift steadily grew between Darwin and Wagner over Wagner's insistence that Darwin completely neglected isolation in the initial formulation of his theory, and that it was Wagner himself who came up with the more correct theory of transmutation— a suggestion that irritated Darwin to no end. Wagner's views were first published in 1868 and later translated into English as *The Darwinian Theory and the Law of the Migration of Organisms* (1873).

February 1872 saw the publication of the sixth edition of the *Origin*. This is the edition from which "On" was dropped from the title, and the declarative tone which that change gave to the title seems to be echoed in the tone Darwin took elsewhere in the book. For example, rather than asserting that intermediate forms *may be* discovered in the future (chapter II), in this edition Darwin states confidently that they *will* be discovered. Still beleaguered over the occurrence of sufficient levels of variation, on the other hand, he relied more heavily in this edition on the direct action of the environment in inducing similar variations in groups of individuals. Time, too, was still of the essence, and Kelvin was becoming more adamant in his attacks: "The limitation of geological periods, imposed by physical science, cannot," Kelvin declared in an 1869 address, "disprove the hypothesis of transmutation of species; but it does seem sufficient to disprove the doctrine that transmutation has taken place through 'descent with modification by natural selection.'" Adding insult to injury, Darwin's own son George became a pupil of Kelvin's at Cambridge and had the thankless task of trying to explain his professor's mathematics to his father. The biologists and geologists had little option but to defer to the expertise of the physicists on such matters as the age of the earth, so many took the only recourse open to them: a compromise by figuring out ways to accelerate evolution. Wallace, for example, performed his own calculations and came up with an earth age of about 100 million years, then emphasized the supposed accelerating effects of glacial periods to give periodic spurts to evolutionary change. The idea was that during periods of intense cold more variations might be induced, and selection is more severe and thus acts on this variation more quickly. Huxley liked the idea; Darwin's response was nearly as cold as the glaciers Wallace and Huxley invoked to speed evolution along.

A larger issue connected to the age of the earth arose in the meantime, compelling the single largest alteration to the *Origin*. St. George Mivart, an early Darwin supporter, followed the geological time arguments made by Kelvin and others and became convinced that the brevity of geological time plus the supposed absence of transitional forms pointed to a sudden appearance of new species, most likely by divine agency. Mivart first published his ideas in a series of essays in 1869, then collected these into a book entitled *The Genesis of Species* in 1871. He did not discount Darwin's theory completely, arguing instead that natural selection was simply inadequate to explain the evolution of all species, including humans: natural selection occurred, but its importance was limited. The history of life was, in Mivart's view, marked by saltational jumps between forms as well as by such remarkable convergences as the vertebrate and cephalopod eye, patterns that to Mivart bespoke the "hand of God" guiding evolution. Darwin found Mivart's book to be the most serious challenge that had yet appeared, and he decided to reply with a whole new chapter dedicated to refuting Mivart in the sixth edition. This became the new chapter VII, "Miscellaneous Objections to the Theory of Natural Selection," consisting of sections from the chapter on natural selection along with a long critique of Mivart's attack. In this way Darwin addressed in one place what he took to be all the persistent misconceptions of his views. In the new chapter he reiterated and defended his position that natural selection can account for the gradual evolution of useful structures, refuted Mivart's claims for inexplicable convergence, expanded his ideas on constraint and history, and restated his position that evolution is gradual, not saltational.

It is worth noting, too, that Darwin was more explicit in acknowledging nonselective processes in evolution in the sixth edition. Natural selection was the chief means of change, but it was aided by such nonselective processes as use and disuse, effects of the environment, and spontaneous variations. In his defense, Darwin pointed out that he never said natural selection was the *only* agent of evolutionary change—it was all a matter of emphasis: "It appears that I formerly underrated the frequency and value of these latter forms of variation as leading to permanent modifications of structure independently of natural selection," he wrote in the closing pages of the sixth edition.

> But as my conclusions have lately been much misrepresented, and it has been stated that I attribute the modification of species exclusively to natural selection, I may be permitted to remark that in the first edition of this work, and subsequently, I placed in a most conspicuous position—namely, at the close of the Introduction—the following words: "I am convinced that natural selection has been the main but not the exclusive means of modification." This has been of no avail. Great is the power of steady misrepresentation.

Darwin did feel himself deliberately misrepresented by Mivart and others, and he decided that the sixth edition would stand as his final word on the subject. "I have resolved to waste no more time in reading reviews of my works or on evolution," he wrote to his staunch supporter the American philosopher Chauncy Wright. He remained as productive as ever in his final decade of life—publishing another five books!—and his commitment to natural law, as well as to his vision of an ever-ramifying tree of life largely driven by the (mostly) slow, steady accumulation of variations by natural selection, never wavered. Most of Darwin's contemporaries never accepted so central a role for natural selection, though by the century's end the reality of transmutation was almost universally accepted. The controversy, debate, and scrutiny that Darwin's ideas engendered were the sign of a healthy science, and so it is fitting that scientists' understanding of the evolutionary process so lucidly laid out by Darwin in 1859 itself continued to evolve.

Modern biologists have more properly been labeled "neo-Darwinians" than "Darwinians" ever since that productive marriage of Mendel's discoveries with empirical and theoretical population genetics, paleontology, embryology, and other fields of inquiry—the unified fields of the Modern Synthesis in the early twentieth century. There are some key differences between the modern understanding of the evolutionary process and that propounded by Darwin, perhaps paramount among them the modern conviction that isolating mechanisms are of critical importance in promoting divergence. But for all the breathtaking new insights, discoveries, and continuing puzzles, evolutionary biology today is both a science and a way of understanding the organic world that Darwin would very much have recognized.

References

Online Sources

The Alfred Russel Wallace Page (*http://www.wku.edu/~smithch/index1.htm*), edited and maintained by Charles H. Smith of Western Kentucky University. The site offers a vast amount of invaluable material relating to Alfred Russel Wallace, including an exhaustive bibliography and many of Wallace's writings in full text.

The Complete Works of Charles Darwin Online (*http://darwin-online.org.uk/*), a collaborative project of numerous Darwin scholars initiated in 2002 and currently directed by John van Wyhe at the University of Cambridge. This database contains some 40,000 pages of searchable text and 130,000 electronic images, bringing together for the first time all of Darwin's publications as well as transcriptions and images of handwritten manuscripts and private papers in their entirety.

The Darwin Correspondence Project (*http://www.darwinproject.ac.uk/*), initiated in 1974 by Frederick Burkhardt and Sydney Smith. This searchable database includes annotated text for the first 5,000 of the more than 14,500 letters Darwin exchanged with over 2,000 correspondents. The online letters are reproduced from the first thirteen volumes of the *Correspondence of Charles Darwin* (Burkhardt and Smith, Cambridge University Press, 1985–), with later material added as the print editions are published. Cited here as *Correspondence* with volume and beginning page number.

Darwin Digital Library of Evolution (*http://darwinlibrary.amnh.org/*), site dedicated to presenting a full evolutionary bibliography from the seventeenth century to the present; based at the American Museum of Natural History. Led by David Kohn, Niles Eldredge, Adam Goldstein, and Christie Stephenson, a team of advisory and contributing scholars present more than 3,400 references from the evolutionary literature in four chronological sections, as well as digital versions of Stauffer's *Natural Selection* and Darwin's natural selection note portfolios, notebooks, and more.

General References

Agassiz, L. 1841. On glaciers, and the evidence of their once having existed in Scotland, Ireland and England. *Proceedings of the Geological Society of London* 3: 327–332.

———. 1847. Plan for an investigation of the embryology, anatomy and effect of light on the blind-fish of the Mammoth Cave, *Amblyopsis spelaeus. Proceedings of the American Academy of Arts and Sciences* 1: 1–180.

———. 1851. Observations on the blind fish of the Mammoth cave. *American Journal of Science* 11: 127–128.

———. 1857 (2004). *Essay on Classification,* ed. E. Lurie. Cambridge, Mass.: Harvard University Press (first published in 1851; reprinted by Dover Publications, Mineola, N.Y.).

Allen, R. H. 1963. *Star Names: Their Lore and Meaning.* New York: Dover Publications [reprint of R. H. Allen, *Star-names and Their Meanings,* published in 1899 by G. E. Stechert].

Allwood, A. C., et al. 2006. Stromatolite reef from the Early Archaean era of Australia. *Nature* 441: 714–718.

Anderson, T. R., and T. Rice. 2006. Deserts on the sea floor: Edward Forbes and his azoic hypothesis for a lifeless deep ocean. *Endeavour* 30: 131–137.

Appel, T. A. 1987. *The Cuvier-Geoffroy Debate: French Biology in the Decades before Darwin.* New York: Oxford University Press.

Arrhenius, O. 1921. Species and area. *Journal of Ecology* 9: 95–99.

Audubon, M. R. 1898. *Audubon and His Journals,* vol. I. London: John C. Nimmo.

Baldwin, J. M. 1896. A new factor in evolution. *American Naturalist* 30: 441–451.

Ballard, H. E., Jr., and K. J. Sytsma. 2000. Evolution and biogeography of the woody Hawaiian violets (*Viola,* Violaceae): Arctic origins, herbaceous ancestry and bird dispersal. *Evolution* 54: 1521–1532.

Barlow, N., ed. 1963. Darwin's ornithological notes. *Bulletin of the British Museum (Natural History) Historical Series* 2: 201–278.

Barney, D. L., and K. E. Hummer. 2005. *Currants, Gooseberries, and Jostaberries: A Guide for Growers, Marketers, and Researchers in North America.* New York: Food Products Press.

Barrett, P. H., et al., eds. 1987. *Charles Darwin's Notebooks, 1836–1844.* Ithaca, N.Y.: Cornell University Press.

Barrett, S. C. H. 2003. Mating strategies in flowering plants: The outcrossing-selfing paradigm and beyond. *Proceedings of the Royal Society of London B* 358: 991–1004.

Barrow, M. V. 2000. *A Passion for Birds: American Ornithology after Audubon.* Princeton, N.J.: Princeton University Press.

Behe, M. J. 1996. *Darwin's Black Box: The Biochemical Challenge to Evolution.* New York: The Free Press.

Belon, P. 1555. *L'histoire de la Nature des Oyseavx.* Paris.

Bentley, B. L. 1977. Extrafloral nectaries and protection by pugnacious bodyguards. *Annual Review of Ecology and Systematics* 8: 407–427.

Bergsma, D. R., and K. S. Brown. 1971. White fur, blue eyes, and deafness in the domestic cat. *Journal of Heredity* 62: 171–185.

Berry, A., ed. 2002. *Infinite Tropics: An Alfred Russel Wallace Anthology.* London and New York: Verso.

Berta, A., J. L. Sumich, and K. M. Kovacs. 2005. *Marine Mammals: Evolutionary Biology* (2nd ed.). Burlington, Mass.: Academic Press.

Birky, C. W., Jr. 2004. Bdelloid rotifers revisited. *Proceedings of the National Academy of Sciences USA* 101: 2651–2652.

Blechman, A. D. 2006. *Pigeons: The Fascinating Saga of the World's Most Revered and Reviled Bird.* New York: Grove Press.

Bollinger, R. R., et al. 2007. Biofilms in the large bowel suggest an apparent function of the human vermiform appendix. *Journal of Theoretical Biology* 249: 826–831.

Boufford, D. E., and S. A. Spongberg. 1983. Eastern Asian–Eastern North American phytogeographical relationships: A history from the time of Linnaeus to the twentieth century. *Annals of the Missouri Botanical Garden* 70: 423–439.

Bouvier, E. L. 1916. The life and work of J. H. Fabre. *Annual Report of the Smithsonian Institution* 1916: 587–597.

Bowler, P. J. 2003. *Evolution: The History of an Idea,* 3rd ed. Berkeley: University of California Press.

Boyden, A. 1943. Homology and analogy: A century after the definitions of "homologue" and "analogue" of Richard Owen. *Quarterly Review of Biology* 18: 228–241.

Brady, S. G. 2003. Evolution of the army ant syndrome: The origin and long-term evolutionary stasis of a complex of behavioral and reproductive adaptations. *Proceedings of the National Academy of Sciences USA* 100: 6575–6579.

Bredekamp, H. 2005. *Darwins Korallen: Frühe Evolutionsmodelle und die Tradition der Naturgeschichte.* Berlin: Klaus Wagenbach Verlag.

Browne, J. 1980. Darwin's botanical arithmetic and the "Principle of Divergence," 1854–1858. *Journal of the History of Biology* 13: 53–89.

———. 1983. *The Secular Ark: Studies in the History of Biogeography.* New Haven, Conn.: Yale University Press.

———. 2002. *Charles Darwin: The Power of Place.* New York: Alfred A. Knopf.

Buckland, W. 1841. On the evidences of glaciers in Scotland and the north of England. *Proceedings of the Geological Society of London* 3: 333–337, 345–348.

———. 1842. On diluvio-glacial phaenomena in Snowdonia and adjacent parts of North Wales. *Proceedings of the Geological Society of London* 3: 579–584.

Burchfield, J. D. 1974. Darwin and the dilemma of geological time. *Isis* 65: 300–321.

———. 1990. *Lord Kelvin and the Age of the Earth.* Chicago: University of Chicago Press.

Burkhardt, R. W., Jr. 1979. Closing the door on Lord Morton's mare: The rise and fall of telegony. *Studies in the History of Biology* 3: 1–21.

Burkhardt, F., and S. Smith, eds. 1985–2008. *The Correspondence of Charles Darwin,* vols. I–XVI. Cambridge; New York: Cambridge University Press.

Calsbeek, R., and T. B. Smith. 2003. Ocean currents mediate evolution in island lizards. *Nature* 426: 552–555.

Camerini, J. 1993. Evolution, biogeography, and maps: An early history of Wallace's Line. *Isis* 84: 700–727.

Candolle, A. P. de. 1820. *Essai Élémentaire de géographie botanique.* Paris and Strasbourg.

———. 1855. *Géographie botanique raisonnée ou exposition des faits principaux et des lois concernant la distribution géographique des plantes de l'époque actuelle.* Paris: Victor Mason.

Cannon, W. F. 1960. The Uniformitarian-Catastrophist debate. *Isis* 51: 38–55.

Causier, B., et al. 2005. Evolution in action: Following function in duplicated floral homeotic genes. *Current Biology* 15: 1508–1512.

Censky, E. J., K. Hodge, and J. Dudley. 1998. Over-water dispersal of lizards due to hurricanes. *Nature* 395: 556.

Czajkowski, M. 2002. A geological tour of the islands of Madeira and Porto Santo. *Geology Today* 18: 26–34.

d'Archiac, E. J. A., and E. P. de Verneuil. 1842. On the fossils of the older deposits in the Rhenish Provinces, preceded by a general survey of the fauna of Paleozoic rocks, and followed by a tabular list of the organic remains of the Devonian system in Europe. *Transactions of the Geological Society of London* 6: 303–410.

Darlington, P. J., Jr. 1943. Carabidae of mountains and islands: Data on the evolution of isolated faunas, and on atrophy of wings. *Ecological Monographs* 13: 37–61.

———. 1957. *Zoogeography: The Geographical Distribution of Animals.* New York: John Wiley and Sons.

———. 1970. Carabidae on tropical islands, especially the West Indies. *Biotropica* 2: 7–15.

Darwin, C. R. 1835. 'Galapagos. Otaheite Lima' (1835). Beagle field notebook. EH1.17. Transcribed by Gordon Chancellor; available at http://darwin-online.org.uk.

———. 1839. *Journal of Researches into the Natural History and Geology of the Countries Visited During the Voyage of H.M.S. Beagle Round the World.* London: John Murray.

———. 1841a. *Birds.* Part 3, No. 5, of *The Zoology of the Voyage of H.M.S.* Beagle, by John Gould. London: Smith Elder and Co.

———. 1841b. Humble-bees. *Gardeners' Chronicle* 34 (August 21): 550.

———. 1842. Notes on the effects produced by the ancient glaciers of Caernarvonshire, and on the boulders transported by floating ice. *Philosophical Magazine* 21: 180–188.

———. 1845. *Journal of Researches into the Natural History and Geology of the Countries Visited During the Voyage of H.M.S. Beagle Round the World,* 2nd ed. [*Voyage of the Beagle*] London: John Murray.

———. 1846. *Geological Observations on South America. Being the Third Part of The Geology of the Voyage of the Beagle.* London: Smith, Elder and Co.

———. 1855a. Does sea-water kill seeds? *Gardeners' Chronicle and Agricultural Gazette* 15 (April 14): 242.

———. 1855b. Does sea-water kill seeds? *Gardeners' Chronicle and Agricultural Gazette* 21 (May 26): 356–357.

———. 1855c. Effect of salt-water on the germination of seeds. *Gardeners' Chronicle and Agricultural Gazette* 47 (November 24): 773.

———. 1855d. Effect of salt-water on the germination of seeds. *Gardeners' Chronicle and Agricultural Gazette* 48 (December 1): 789.

———. 1857a. Bees and fertilisation of kidney beans. *Gardeners' Chronicle* 43 (October 24): 725.

———. 1857b. On the action of sea-water on the germination of seeds. *Journal of the Proceedings of the Linnean Society of London* (Botany) 1: 130–140.

———. 1858. On the agency of bees in the fertilisation of papilionaceous flowers, and on the crossing of kidney beans. *Gardeners' Chronicle* 46 (November 13): 828–829.

———. 1861a. Is the female bombus fertilised in the air? *Journal of Horticulture* (October 22): 76.

———. 1861b. On the three remarkable sexual forms of *Catasetum tridentatum,* an orchid in the possession of the Linnean Society. *Proceedings of the Linnean Society of London (Botany)* 6: 151–157.

———. 1862. *On the Various Contrivances by which British and Foreign Orchids are Fertilized by Insects, and on the Good Effects of Intercrossing.* London: John Murray.

———. 1868a. On the specific difference between *Primula veris,* Brit. F. (var. *officinalis* of Linn.), *P. vulgaris,* Brit. Fl. (var. *acaulis,* Linn.), and *P. elatior,* Jacq.; and on the hybrid nature of the common oxlip. With supplementary remarks on naturally-produced hybrids in the genus *Verbascum. Journal of the Linnean Society of London (Botany)* 10: 437–454.

———. 1868b. *The Variation of Animals and Plants under Domestication.* London: John Murray.

———. 1870. Note on the habits of the pampas woodpecker *(Colaptes campestris). Proceedings of the Zoological Society of London* 47: 705–706.

———. 1871. *The Descent of Man, and Selection in Relation to Sex.* London: John Murray.

———. 1877. *The Different Forms of Flowers on Plants of the Same Species.* London: John Murray.

———. 1878. Transplantation of shells. *Nature* 18: 120–121.

———. 1882. On the dispersal of freshwater bivalves. *Nature* 25: 529–530.

———. 1883. *The Variation of Animals and Plants under Domestication,* 2nd ed. New York: Appleton [reprinted by Johns Hopkins University Press, 1998].

———. 1958. *The Autobiography of Charles Darwin, 1809–1882,* ed. N. Barlow. London: Collins.

Darwin, F., ed. 1896. *The Life and Letters of Charles Darwin, Including an Autobiographical Chapter,* vols. I and II. New York: D. Appleton and Co.

———. 1909. *The Foundations of the Origin of Species. Two Essays Written in 1842 and 1844.* Cambridge: Cambridge University Press.

Darwin, F., and A. C. Seward, eds. 1903a. *More Letters of Charles Darwin: A Record of his Work in a Series of Hitherto Unpublished Letters,* vol. I. London: John Murray.

———. 1903b. *More Letters of Charles Darwin: A Record of his Work in a Series of Hitherto Unpublished Letters,* vol. II. London: John Murray.

Dawkins, R. 1996. *The Blind Watchmaker.* New York: W. W. Norton.

De Beer, G. 1959. Some unpublished letters of Charles Darwin. *Notes and Records of the Royal Society of London* 14: 12–66.

Dennett, D. 1995. *Darwin's Dangerous Idea.* New York: Simon and Schuster.

de Queiroz, A. 2005. The resurrection of oceanic dispersal in historical biogeography. *Trends in Ecology and Evolution* 20: 68–73.

Dixon, E. S. 1851. *The Dovecot and the Aviary: Being Sketches of the Natural History of Pigeons and Other Domestic Birds in a Captive State, with Hints for their Management.* London: John Murray.

Doolittle, F., and E. Bapteste. 2007. Pattern pluralism and the tree of life hypothesis. *Proceedings of the National Academy of Sciences USA* 104: 2043–2049.

Dressler, R. L. 1990. *The Orchids: Natural History and Classification.* Cambridge, Mass.: Harvard University Press.

Dwyer, J. D. 1955. The Botanical Catalogues of Auguste De St. Hilaire. *Annals of the Missouri Botanical Garden* 42: 153–170.

Earl, G. W. 1845. On the physical structure and arrangement of the islands in the Indian archipelago. *Journal of the Royal Geographical Society* 15: 358–365.

Egerton, F. N. 2003. *Hewett Cottrell Watson: Victorian Plant Ecologist and Evolutionist.* Aldershot, U.K., and Burlington, Vt.: Ashgate Publishing.

Eisner, T. 2003. *For Love of Insects.* Cambridge, Mass.: Harvard University Press.

Emerson, R. W., with an introduction by N. H. Dole. 1899. *Early Poems of Ralph Waldo Emerson.* New York, Boston: Thomas Y. Crowell and Company.

Emig, C. C. 2003. Proof that *Lingula* (Brachiopoda) is not a living-fossil, and emended diagnoses of the Family Lingulidae. *Carnets de Géologie / Notebooks on Geology* Letter 2003/01. [http://paleopolis.rediris.es/cg/CG2003_L01_CCE/CG2003_L01_CCE.pdf]

Entrikin, R. K., and L. C. Erway. 1972. A genetic investigation of roller and tumbler pigeons. *Journal of Heredity* 63: 351–354.

Evans, L. T. 1984. Darwin's use of the analogy between artificial and natural selection. *Journal of the History of Biology* 17: 113–140.

Fabre, A. 1921. *The Life of Jean Henri Fabre,* trans. B. Miall. New York: Dodd, Mead, and Company.

Falconer, H. 1859. *Descriptive Catalogue of the Fossil Remains of Vertebrata from the Sewalik Hills, the Nerbudda, Perim Island, etc. in the Museum of the Asiatic Society of Bengal.* Calcutta: C. B. Lewis.

———. 1863. On the American fossil elephant of the regions bordering the Gulf of Mexico (*E. Columbi,* Falc.); with general observations on the living and extinct species. *Natural History Review,* new series 3: 43–114.

Fessl, B., and S. Tebbich. 2002. *Philornis downsi*—a recently discovered parasite on the Galápagos archipelago—a threat for Darwin's finches? *Ibis* 144: 445–451.

Fessl, B., B. J. Sinclair, and S. Kleindorfer. 2006. The life-cycle of *Philornis downsi* (Diptera: Muscidae) parasitizing Darwin's finches and its impacts on nestling survival. *Parasitology* 133: 739–747.

Fleming, J. 1829. On systems and methods in natural history. *Quarterly Review* 41: 302–327.

Forbes, E. 1846. On the connexion between the distribution of the existing fauna and flora of the British Isles and the geological changes which have affected their area, especially during the epoch of the Northern Drift. *Memoirs of the Geological Survey of Great Britain* 1: 336–432.

———. 1854. On the manifestation of polarity in the distribution of organized beings in time. *Notices of the Proceedings of the Meetings of the Members of the Royal Institution* 1 (1854): 428–433.

Forey, P. L., and P. Janvier. 1994. Evolution of the early vertebrates. *American Scientist* 82: 554–565.

Fries, E. M. 1850. A monograph of the Hieracia; being an abstract of Prof. Fries's *"Symbolae ad Historiam Hieraciorum." Botanical Gazette* 2: 85–92, 185–188, 203–219.

Galperin, M. Y., D. R. Walker, and E. V. Koonin. 1998. Analogous enzymes: Independent inventions in enzyme evolution. *Genome Research* 8: 779–790.

Gans, C. 1993. Evolutionary origin of the vertebrate skull. Pp. 1–35 in J. Hanken and B. K. Hall, eds., *The Skull,* vol. 2. Chicago: University of Chicago Press.

Gärtner, C. F. von. 1849. *Versuche und Beobachtungen uber die Bastarderzeugung im Pflanzenreich.* Stuttgart.

Gehring, W. J. 1998. *Master Control Genes in Development and Evolution.* New Haven, Conn.: Yale University Press.

Gingerich, P. D., et al. 2001. Origin of whales from early artiodactyls: Hands and feet of Eocene Protocetidae from Pakistan. *Science* 293: 2239–2242.

Gleason, H. A. 1922. On the relation between species and area. *Ecology* 3: 158–162.

Goldschmidt, R. 1940. *The Material Basis of Evolution.* New Haven, Conn.: Yale University Press.

Gould, J. 1837a. Remarks on a group of ground finches from Mr Darwin's collection, with characters of the new species. *Proceedings of the Zoological Society of London* 5: 4–7.

———. 1837b. [Two new species of the genus *Sterna,* a species of cormorant, and three species of the genus *Orpheus,* from the Galapagos, in the collection of Mr. Darwin.]. *Proceedings of the Zoological Society of London* 5: 26–27.

———. 1837c. [Observations on the raptorial birds in Mr. Darwin's

collection, with characters of the new species.] *Proceedings of the Zoological Society of London* 5: 9–11.
———. 1848. *The Birds of Australia*, vols. I–VII. London: Printed by R. and J. E. Taylor.
Gould, S. J. 1977. *Ontogeny and Phylogeny.* Cambridge, Mass.: Belknap Press of Harvard University Press.
———. 1983. *Hen's Teeth and Horse's Toes: Further Reflections in Natural History.* New York: W. W. Norton and Company.
———. 1993. *Eight Little Piggies: Reflections in Natural History.* New York: W. W. Norton and Company.
———. 2002. *The Structure of Evolutionary Theory.* Cambridge, Mass.: Harvard University Press.
Gray, A. 1846. Analogy between the flora of Japan and that of the United States. *American Journal of Science and Arts*, 2nd series, 2: 135–136.
———. 1856–1857. Statistics of the flora of the northern United States. *American Journal of Science and Arts*, 2nd series, 22 (1856): 204–232; 23 (1857): 62–84, 369–403.
———. 1859. Diagnostic characters of new species of phaenogamous plants, collected in Japan by Charles Wright, Botanist of the U.S. North Pacific Exploring Expedition. With observations upon the relations of the Japanese flora to that of North America, and of other parts of the Northern Temperate Zone. *Memoirs of the American Academy of Arts and Science* 6, pt. II (new series): 377–452.
———. 1876. *Darwiniana: Essays and Reviews Pertaining to Darwinism.* New York: D. Appleton and Company.
Gray, P. H. 1968. The early animal behaviorists: Prolegomenon to Ethology. *Isis* 59: 372–383.
Green, C. H. 1990. The effect of colour on the numbers, age and nutritional status of *Glossina tachinoides* (Diptera: Glossinidae) attracted to targets. *Physiological Entomology* 15: 317–329.
Gross, L. 2006. Scientific illiteracy and the partisan takeover of biology. *PLoS Biology* 4(5): e167.
Haeckel, E. 1880. *The History of Creation, Or, The Development of the Earth and Its Inhabitants by the Action of Natural Causes*, vol. I, trans. E. R. Lankester. New York: D. Appleton and Co.
Haldane, J. B. S. 1990. *The Causes of Evolution.* [Princeton Science Library reprint of the 1932 edition.] Princeton, N.J.: Princeton University Press.
Hales, T. C. 2001. The Honeycomb Conjecture. *Discrete and Computational Geometry* 25: 1–22.
Hamilton, W. D. 1964. The genetical evolution of social behaviour I and II. *Journal of Theoretical Biology* 7: 1–16 and 17–52.
Hamilton Smith, C. 1853. Experiments on the comparative effect of rifle and musketry fire on different colours. *Aide Memoire to the Military Sciences*, R. E. 1: 257–259.
Hankins, T. L. 1985. *Science and the Enlightenment.* Cambridge, England: Cambridge University Press.
Hanotte, O., et al. 2002. African pastoralism: Genetic imprints of origins and migrations. *Science* 296: 336–339.
Harris, D. J., et al. 2000. Phylogeny of the thoracician barnacles based on 18S rDNA sequences. *Journal of Crustacean Biology* 20: 393–398.
Hazlewood, N. 2000. *Savage: The Life and Times of Jemmy Button.* New York: St. Martin's Press.
Hearne, S. 1795. *A Journey from Prince of Wales's Fort in Hudson's Bay, to the Northern Ocean.* London: Strahan and Cadell. [Edited by J. B. Tyrrell, 1911, Champlain Society editions.]
Hector, A., and R. Hooper. 2002. Darwin and the first ecological experiment. *Science* 295: 639–640.
Heinicke, M. P., W. E. Duellman, and S. B. Hedges. 2007. Major Caribbean and Central American frog faunas originated by ancient oceanic dispersal. *Proceedings of the National Academy of Sciences USA* 104: 10092–10097.
Helferich, F. 2004. *Humboldt's Cosmos: Alexander von Humboldt and the Latin American Journey That Changed the Way We See the World.* New York: Gotham Books.
Herbers, J. M. 2007. Watch your language! Racially loaded metaphors in scientific research. *BioScience* 57: 104–105.
Herbert, S. 2005. *Charles Darwin, Geologist.* Ithaca, N.Y.: Cornell University Press.
Herbert, W. 1837. *Amaryllidaceae, an attempt to arrange the Monocotyledonous Orders.* London: J. Ridgway.
Heron, R. 1835. Notes on the habits of the pea-fowl. *Proceedings of the Zoological Society of London* (1833–1835), pt. 3: 54.
Hodge, M. J. S. 1977. The structure and strategy of Darwin's 'long argument.' *British Journal of the History of Science* 10: 237–246.
———. 1991. Darwin, Whewell, and natural selection. *Biology and Philosophy* 6: 457–460.
———. 1992. Darwin's argument in the *Origin. Philosophy of Science* 59: 461–464.
Holden, C., ed. 2007. Random samples: Forest primeval. *Science* 317: 877.
Hollingsworth, M. J. 1960. Studies on the polymorphic workers of the army ant *Dorylus (Anomma) nigricans* Illiger. *Insectes Sociaux* 7: 17–37.
Hooker, J. D. 1844–1860. *The Botany of the Antarctic Voyage of H. M.*

Discovery Ships 'Erebus' and 'Terror' in the Years 1839–1843. Part I: *Flora Antarctica,* 2 vols., 1844–1847. Part II: *Flora Novae-Zelandiae,* 2 vols., 1853–1855. Part III: *Flora Tasmaniae,* 2 vols., 1855–1860. London: Lovell Reeve.

———. 1846. On the vegetation of the Galapagos Archipelago, as compared with that of some other tropical islands and of the continent of America. Read December 1 and 15, 1846; published in *Proceedings of the Linnean Society of London* 1 (1849): 312–314; *Transactions of the Linnean Society of London* 20 (1851): 235–262.

———. 1853. *Introductory Essay to the Flora of New Zealand.* London: Lovell Reeve.

———. 1853–1855. *Flora Novae-Zelandiae,* 2 vols. Pt. II of *The Botany of the Antarctic Voyage of H. M. Discovery Ships 'Erebus' and 'Terror' in the Years 1839–1843, under the command of Captain Sir James Clark Ross.* London: Lovell Reeve.

———. 1856. Notice of Alphonse de Candolle's *Géographie Botanique Raisonée. Hooker's Journal of Botany and Kew Garden Miscellany* 8: 54–64, 82–88, 112–121, 151–157, 181–191, 214–219, 248–256.

Hopson, J. A. 1987. The mammal-like reptiles: A study of transitional fossils. *American Biology Teacher* 49: 16–26.

Huber, P. 1810. *Recherche sur les Moeurs des Fourmis Indigènes.* Paris and Geneva: J. J. Paschoud.

———. 1836. Mémoire pour servir a l'histoire de la chenille du hamac, *Tinea Harisella* Linnaei; oecophore de Latreille. *Mémoires de la Société de Physique et d'Histoire Naturelle de Genève* 7: 121–160, plus 2 plates.

———. 1839. Notice sur la Mélipone domestique, abeille domestique Mexicaine. *Mémoires de la Société de Physique et d'Histoire Naturelle de Genève* 8: 1–26, plus 3 plates.

Hull, D. 1973. *Darwin and His Critics.* Chicago: University of Chicago Press.

———. 1988. *Science as a Process: An Evolutionary Account of the Social and Conceptual Development of Science.* Chicago: University of Chicago Press.

———. 2003. Darwin's science and Victorian philosophy of science. In J. Hodge and G. Radick, eds., *The Cambridge Companion to Darwin,* pp. 168–191. Cambridge, England: Cambridge University Press.

Huxley, T. H. 1858. The Croonian Lecture: On the theory of the vertebrate skull. *Proceedings of the Royal Society of London* 9: 381–457.

———. 1970. *Darwiniana Essays* [reprint of 1896 edition]. New York: AMS Press.

Irmscher, C., ed. 1999. *John James Audubon: Writings and Drawings.* New York: The Library of America.

Jameson, R. 1851. The theory of successive development in the scale of being, both animal and vegetable, from the earliest periods to our own time, as deduced from Paleontological evidence. *Edinburgh New Philosophical Journal* 51: 1–31.

Jenkin, F. 1867. The Origin of Species. *North British Review* 46: 276–318.

Jockusch, E. L., T. A. Williams, and L. M. Nagy. 2004. The evolution of patterning of serially homologous appendages in insects. *Development, Genes and Evolution* 214: 324–338.

Jones, D. K. C. 1999. On the uplift and denudation of the Weald. *Special Publications of the Geological Society of London.* 162: 25–43.

Jones, J. S., and N. Barton. 1985. Sex chromosomes and evolution. Haldane's Rule OK. *Nature* 314: 668–669.

Just, T. 1953. Review of Gray's Manual of Botany. *Quarterly Review of Biology* 28: 294.

Kaplan, M. R., et al. 2004. Cosmogenic nuclide chronology of millennial-scale glacial advances during O-isotope stage 2 in Patagonia. *Geological Society of America Bulletin* 116: 308–321.

Kappmeier, K., and E. M. Nevill. 1999. Evaluation of coloured targets for the attraction of *Glossina brevipalpis* and *Glossina austeni* (Diptera: Glossinidae) in South Africa. *Onderstepoort Journal of Veterinary Research* 66: 291–305.

Kemp, T. 1982. The reptiles that became mammals. *New Scientist* 93: 583.

Keynes, R. D. 2001. *Charles Darwin's* Beagle *Diary.* Cambridge, England: Cambridge University Press.

Kitchell, K. F., Jr., and I. M. Resnick, trans. 1999. *Albertus Magnus, On Animals: A Medieval* Summa Zoologica, vol. II. Baltimore and London: Johns Hopkins University Press.

Kitching, I., et al. 1998. *Cladistics: Theory and Practice of Parsimony Analysis.* Oxford, England: Oxford University Press.

Knight, A. 1799. An account of some experiments on the fecundation of vegetables. *Philosophical Transactions of the Royal Society of London* 89: 195–204.

Kölreuter, J. G. 1806. Some further observations and experiments on the irritability of the stamens of the barberry. *Annals of Botany* 2: 1–10.

Kríz, J., and J. Pojeta, Jr. 1974. Barrande's colonies concept and a comparison of his stratigraphy with the modern stratigraphy of the middle Bohemian lower Paleozoic rocks (Barrandian) of Czechoslovakia. *Journal of Paleontology* 48: 489–494.

Kullberg, M., et al. 2006. Housekeeping genes for phylogenetic analysis of Eutherian relationships. *Molecular Biology and Evolution* 23: 1493–1503.

Land, M. F., and R. D. Fernald. 1992. The evolution of eyes. *Annual Review of Neuroscience* 15: 1–29.
Lever, C. 2003. *Naturalized Reptiles and Amphibians of the World.* New York: Oxford University Press.
Lindblad-Toh, K., et al. 2005. Genome sequence, comparative analysis and haplotype structure of the domestic dog. *Nature* 438: 803–819.
Liu, R., and H. Ochman. 2007. Stepwise formation of the bacterial flagellar system. *Proceedings of the National Academy of Sciences USA* 104: 7116–7121.
Lonsdale, W. 1840. Notes on the age of the limestones of south Devonshire. *Transactions of the Geological Society of London,* series 2, 5: 721–738.
Löve, D. 1970. Subarctic and subalpine: Where and what? *Arctic and Alpine Research* 2: 63–73.
Lovejoy, A. O. 1964. *The Great Chain of Being.* Cambridge, Mass.: Harvard University Press.
Lubbock, J. 1858. On the digestive and nervous systems of *Coccus hesperidum. Proceedings of the Royal Society of London* 9 (1857–1859): 480–486.
———. 1884. *Ants, Bees, and Wasps. A Record of Observations on the Habits of the Social Hymenoptera.* New York: D. Appleton and Company.
Luo, D. R., et al. 1996. Origin of floral asymmetry in *Antirrhinum. Nature* 383: 794–799.
Lurie, E. 1988. *Louis Agassiz: A Life in Science.* Baltimore: Johns Hopkins University Press.
Lyell, C. 1830–1833. *Principles of Geology,* vols. I–III. London: John Murray. [Facsimile edition published by the University of Chicago Press, 1990.]
———. 1837. *Principles of Geology,* 5th edition, vol. III. London: John Murray.
———. 1849. *A Second Visit to the United States of North America,* vol. II. New York: Harper and Brothers; London: John Murray.
———. 1865. *Elements of Geology; Or, The Ancient Changes of the Earth and Its Inhabitants as Illustrated by Geological Monuments,* 6th ed. London: John Murray.
Mabberley, D. J. 1975. The giant lobelias: Toxicity, inflorescence and tree-building in the Campanulaceae. *New Phytologist* 75: 289–295.
———. 1985. *Jupiter Botanicus: Robert Brown of the British Museum.* London: Wheldon and Wesley, the British Museum (Natural History), and J. Cramer.
MacLeod, R. M. 1965. Evolutionism and Richard Owen, 1830–1868: An episode in Darwin's century. *Isis* 56: 259–280.
MacWhirter, R. B. 1989. On the rarity of intraspecific brood parasitism. *The Condor* 41: 485–492.
Maderspacher, F. 2006. The captivating coral: The origins of early evolutionary imagery. *Current Biology* 16: R476–R478.
———. 2007. All the queen's men. *Current Biology* 17: R191–R195.
Maherali, H., and J. N. Klironomos. 2007. Influence of phylogeny on fungal community assembly and ecosystem functioning. *Science* 316: 1746–1748.
Malthus, T. R. 1798. *An Essay on the Principle of Population.* London.
Marshall, W. 1788. *The Rural Economy of Yorkshire.* London: T. Cadell.
———. 1796. *The Rural Economy of Yorkshire: Comprising the Management of Landed Estates, and the Present Practice of Husbandry in the Agricultural Districts of that County,* vol. II, 2nd edition. London: T. Cadell.
Martens, M. C. 1857. Expériences sur la persistence de la vitalité des graines flottant à la surface de la mer. *Bulletin de la Société Botanique de France* 4: 324–337.
Martin, R. A. 2004. *Missing Links: Evolutionary Concepts and Transitions through Time.* Sudbury, Mass.: Jones and Bartlett.
Mayr, E. 1942. *Systematics and the Origin of Species.* Cambridge, Mass.: Harvard University Press.
———. 1961. Cause and effect in biology: Kinds of causes, predictability, and teleology are viewed by a practicing biologist. *Science* 134: 1501–1506.
———. 1991. *One Long Argument: Charles Darwin and the Genesis of Modern Evolutionary Thought.* Cambridge, Mass.: Harvard University Press.
McCullough, D. 1992. *Brave Companions: Portraits in History.* New York: Prentice Hall Press.
Mills, W. 1983. Darwin and the iceberg theory. *Notes and Records of the Royal Society of London* 38: 109–127.
Milne-Edwards, H. 1844. Considérations sur quelques principes relatifs à la clasification naturelle des animaux, et plus particulièrement sur la distribution méthodique des mammifères. *Annales des Sciences Naturelles (Zoologie),* 3rd series, 1: 65–99.
Mokady, O., et al. 1999. Speciation versus phenotypic plasticity in coral inhabiting barnacles: Darwin's observations in an ecological context. *Journal of Molecular Evolution* 49: 367–375.
Moreno, P. I., et al. 2001. Interhemispheric climate links revealed by a late-glacial cooling episode in southern Chile. *Nature* 409: 804–808.
Morton, The Earl of. 1821. A communication of a singular fact in natural history. *Philosophical Transactions of the Royal Society of London* 111: 20–22.
Muir, J. 1901. *The Mountains of California.* New York: The Century Co.

Müller-Wille, S. 2007. The love of plants. *Nature* 446: 268.
Müller-Wille, S., and V. Orel. 2007. From Linnaean species to Mendelian factors: Elements of hybridism, 1751–1870. *Annals of Science* 64: 171–215.
Nagy, L., et al., eds. 2003. *Alpine Biodiversity in Europe.* Berlin: Springer-Verlag.
Nelson, G. 1979. From Candolle to Croizat: Comments on the history of biogeography. *Journal of the History of Biology* 11: 269–305.
Newman, H. W. 1851. On the habits of the Bombinatrices. *Transactions of the Entomological Society of London* (new series) 1: 86–92, 109–112, 116–118.
Ni, X., et al. 2004. A euprimate skull from the early Eocene of China. *Nature* 427: 65–68.
Nilsson, D.-E. 2005. Photoreceptor evolution: Ancient siblings serve different tasks. *Current Biology* 15: R94–R96.
Oakley, T. H. 2003. On homology of arthropod compound eyes. *Integrative and Comparative Biology* 43: 522–530.
Oakley, T. H., and C. W. Cunningham. 2002. Molecular phylogenetic evidence for the independent evolutionary origin of an arthropod compound eye. *Proceedings of the National Academy of Sciences USA* 99: 1426–1430.
O'Brien, C. F. 1970. *Eozoön Canadense,* "The Dawn Animal of Canada." *Isis* 61: 206–223.
Ogawa, M. 2001. The mysterious Mr. Collins: Living for 140 years in *Origin of Species. Journal of the History of Biology* 34: 461–479.
O'Hara, R. J. 1991. Representations of the Natural System in the Nineteenth Century. *Biology and Philosophy* 6: 255–274.
Ohta, T. 2003. Gene families: Multigene families and superfamilies. In *Encyclopedia of the Human Genome.* London: Nature Publishing Group.
Opitz, J. M. 2004. Goethe's bone and the beginnings of morphology. *American Journal of Medical Genetics,* pt. A, 126A: 1–8.
Ospovat, D. 1976. The influence of Karl Ernst von Baer's embryology, 1828–1859: A reappraisal in light of Richard Owen's and William B. Carpenter's "Palaeontological Application of 'Von Baer's Law.'" *Journal of the History of Biology* 9: 1–28.
———. 1977. Lyell's theory of climate. *Journal of the History of Biology* 10: 317–339.
Owen, R. 1830–1831. On the anatomy of the Ourang-outang (*Simia satyrus,* L.). *Proceedings of the Zoological Society of London* 1: 4–5, 9–10, 28–29, 66–72.
———. 1837. A description of the cranium of the *Toxodon Platensis,* a gigantic extinct mammiferous species, referrible by its dentition to the *Rodentia,* but with affinities to the *Pachydermata* and the *Herbivorous Cetacea. Proceedings of the Geological Society of London* 2 [Read 19 April]: 541–542.
———. 1839. [Notes on the dugong]. *Annals of Natural History* 2: 300–307.
——— 1843. *Lectures on the Comparative Anatomy and Physiology of the Invertebrate Animals, Delivered at the Royal College of Surgeons, in 1843.* London: Longman, Brown, Green, and Longmans.
———. 1846. *Lectures on the Comparative Anatomy and Physiology of the Vertebrate Animals, Delivered at The Royal College of Surgeons of England, in 1844 and 1846.* Part I: Fishes. London: Longman, Brown, Green, and Longmans.
———. 1848a. Description of teeth and portions of jaws of two extinct Anthracotheroid Quadrupeds *(Hyopotamus vectianus* and *Hyop. bovinus)* discovered by the Marchioness of Hastings in the Eocene deposits on the N. W. coast of the Isle of Wight: with an attempt to develop Cuvier's idea of the classification of Pachyderms by the number of their toes. *Quarterly Journal of the Geological Society of London* 4: 103–141.
———. 1848b. *On the Archetype and Homologies of the Vertebrate Skeleton.* London: John van Voorst.
———. 1849. *On the Nature of Limbs: A Discourse Delivered on Friday February 9, at an Evening Meeting of the Royal Institution of Great Britain* (London). [Reprinted by University of Chicago Press, R. Amundson, ed., 2007.]
———. 1860. Darwin on the Origin of Species. *Edinburgh Review* 3: 487–532.
———. 1868. *On the Anatomy of Vertebrates,* vol. III. London: Longman, Green, and Co.
Padian, K. 1997. The rehabilitation of Sir Richard Owen. *BioScience* 47: 446–453.
———. 1999. Charles Darwin's views of classification in theory and practice. *Systematic Biology* 48: 352–364.
Paley, W. 1802. *Natural Theology: Or, Evidences of the Existence and Attributes of the Deity, Collected from the Appearances of Nature.* Boston: Gould and Lincoln.
Pallen, M. J., and N. J. Matzke. 2006. From *The Origin of Species* to the origin of bacterial flagella. *Nature Reviews Microbiology* 4: 784–790.
Panchen, A. L. 1992. *Classification, Evolution, and the Nature of Biology.* Cambridge, England: Cambridge University Press.
Paul, D. B. 1988. The selection of the "survival of the fittest." *Journal of the History of Biology* 21: 411–424.
Peckham, M., ed. 1959. *The Origin of Species: A Variorum Text.* Philadelphia: University of Pennsylvania Press.

Perry, S. F., et al. 2001. Which came first, the lung or the breath? *Comparative Biochemistry and Physiology*, pt. A, 129: 37–47.

Pontin, A. J. 1978. The number and distribution of subterranean aphids and their exploitation by the ant *Lasius flavus* (Fabr.). *Ecological Entomology* 3: 203–207.

Preston, F. W. 1962. The canonical distribution of commonness and rarity: Part I. *Ecology* 43: 185–215.

Pumphrey, R. J. 1958. The Forgotten Man: Sir John Lubbock, F.R.S. *Notes and Records of the Royal Society of London* 13: 49–58.

Ren, N., and M. P. Timko. 2001. AFLP analysis of genetic polymorphism and evolutionary relationships among cultivated and wild *Nicotiana* species. *Genome* 44: 559–571.

Rhoades, F. H. T. 1987. Darwinian gradualism and its limits: The development of Darwin's views on the rate and pattern of evolutionary change. *Journal of the History of Biology* 20: 139–157.

Richards, R. J. 1981. Instinct and intelligence in British natural theology: Some contributions to Darwin's theory of the evolution of behavior. *Journal of the History of Biology* 14: 193–230.

Riffenburgh, B., ed. 2007. *Encyclopedia of the Antarctic.* New York: Routledge.

Riley, J. R., et al. 2005. The flight paths of honeybees recruited by the waggle dance. *Nature* 435: 205–207.

Rinderknecht, A., and R. E. Blanco. 2008. The largest fossil rodent. *Proceedings of the Royal Society B* 275: 923–928.

Ritger, S. D., and R. H. Cummins. 1991. Using student-created metaphors to comprehend geologic time. *Journal of Geological Education* 9: 9–11.

Romer, A. S., and T. S. Parsons. 1986. *The Vertebrate Body,* 6th edition. Philadelphia: Saunders.

Rudwick, M. J. S. 1985. *The Great Devonian Controversy: The Shaping of Scientific Knowledge among Gentlemanly Specialists.* Chicago and London: University of Chicago Press.

———. 1992. *Scenes from Deep Time: Early Pictorial Representations of the Prehistoric World.* Chicago: University of Chicago Press.

———. 1997. *Georges Cuvier, Fossil Bones, and Geological Catastrophes: New Translations and Interpretations of the Primary Texts.* Chicago: University of Chicago Press.

Ruse, M. 1979. *The Darwinian Revolution.* Chicago: University of Chicago Press.

Sabine, J. 1820. Observations on, and account of, the species and varieties of the genus *Dahlia;* with instructions for their cultivation and treatment. *Transactions of the Royal Horticultural Society of London* 3: 217–243.

Sachs, A. 2006. *The Humboldt Current: Nineteenth Century Exploration and the Roots of American Environmentalism.* New York: Viking.

Sagan, C. 1996. *The Demon-Haunted World: Science as a Candle in the Dark.* New York: Random House.

Salvini-Plawen, L. V., and E. Mayr. 1977. On the evolution of photoreceptors and eyes. *Evolutionary Biology* 10: 207–263.

Sánchez-Villagro, M. R., O. Aguilera, and I. Horovitz. 2003. The anatomy of the world's largest extinct rodent. *Science* 301: 1708–1710.

Sanford, W. F., Jr. 1965. Dana and Darwinism. *Journal of the History of Ideas* 26: 531–546.

Savolainen, P., et al. 2002. Genetic evidence for an east Asian origin of dogs. *Science* 298: 1610–1613.

Schmitt, M. 2003. Willi Hennig and the rise of cladistics. Pp. 369–379 in A. Legakis et al., eds., *The New Panorama of Animal Evolution: Proceedings of the 18th International Congress of Zoology.* Sofia and Moscow: Pensoft Publishers.

Secord, J. A. 1981. Nature's fancy: Charles Darwin and the breeding of pigeons. *Isis* 72: 162–186.

———. 1986. *Controversy in Victorian Geology: The Cambrian-Silurian Dispute.* Princeton, N.J.: Princeton University Press.

———. 2000. *Victorian Sensation: The Extraordinary Publication, Reception, and Secret Authorship of "Vestiges of the Natural History of Creation."* Chicago: University of Chicago Press.

Shaffer, M. L. 1981. Minimum population sizes for species conservation. *BioScience* 31: 131–134.

Simmons, N. B., et al. 2008. Primitive early Eocene bat from Wyoming and the evolution of flight and echolocation. *Nature* 451: 818–821.

Simpson, G. G. 1950. History of the fauna of Latin America. *American Scientist* 38: 361–389.

———. 1953. The Baldwin effect. *Evolution* 7: 110–117.

Smith, F. 1853. Monograph of the genus *Cryptocerus,* belonging to the group Cryptoceridae—family Myrmicidae—division Hymenoptera Heterogyna. *Transactions of the Entomological Society of London* 2 (1854): 213–228.

Smith, K. G. V. 1987. Darwin's insects: Charles Darwin's entomological notes, with an introduction and comments by Kenneth G. V. Smith. *Bulletin of the British Museum (Natural History) Historical Series* 14: 1–143.

Somes, R. G., Jr., and R. E. Burger. 1993. Inheritance of the white and pied plumage color patterns in the Indian peafowl *(Pavo cristatus). Journal of Heredity* 84: 57–62.

Somkin, F. 1962. The contributions of Sir John Lubbock, F.R.S. to the

"Origin of Species": Some annotations to Darwin. *Notes and Records of the Royal Society of London* 17: 183–191.

Southward, A. J. 1983. A new look at variation in Darwin's species of acorn barnacles. *Biological Journal of the Linnean Society* 20: 59–72.

Stauffer, R. C. 1960. Ecology in the long manuscript version of Darwin's "Origin of Species" and Linnaeus' "Oeconomy of nature." *Proceedings of the American Philosophical Society* 104: 235–241.

———. 1975. *Charles Darwin's Natural Selection: Being the Second Part of His Big Species Book Written from 1856 to 1858.* Cambridge, England: Cambridge University Press.

Stevens, P. F. 1984. Metaphors and typology in the development of botanical systematics, 1690–1960, or the art of putting new wine in old bottles. *Taxon* 33: 169–211.

Sulloway, F. J. 1979. Geographic isolation in Darwin's thinking: The vicissitudes of a crucial idea. *Studies in the History of Biology* 3: 23–65.

———. 1982a. Darwin and his finches: The evolution of a legend. *Journal of the History of Biology* 15: 1–53.

———. 1982b. Darwin's conversion: The *Beagle* voyage and its aftermath. *Journal of the History of Biology* 15: 325–396.

———. 1984. Darwin and the Galápagos. *Biological Journal of the Linnean Society* 21: 29–59.

Tammone, W. 1995. Competition, the division of labor, and Darwin's principle of divergence. *Journal of the History of Biology* 28: 109–131.

Tegetmeier, W. B. 1869. *Pigeons: Their Structure, Varieties, Habits, and Management.* London: George Routledge and Sons.

Thewissen, J. G. M., S. I. Madar, and S. T. Hussain. 1996. *Ambulocetus natans,* an Eocene cetacean (Mammalia) from Pakistan. *Courier Forschungsinstitut Senckenberg* 191: 1–86.

Thompson, D. 1959. *On Growth and Form,* 2nd ed., vol. II. Cambridge, England: Cambridge University Press.

Thomson, J. A. 1932. *Riddles of Science.* New York: Liveright, Inc.

Thomson, W. 1862. On the age of the sun's heat. *Macmillan's Magazine* 5: 288–393.

———. 1863. On the secular cooling of the earth. *Philosophical Magazine* 25: 1–14.

Tomarev, S. I., et al. 1997. Squid *Pax-6* and eye development. *Proceedings of the National Academy of Sciences USA* 94: 2421–2426.

Torrey, B., ed. 1968. *The Writings of Henry David Thoreau. Journal,* vol. X. New York: AMS Press.

Vences, M., et al. 2003. Multiple overseas dispersal in amphibians. *Proceedings of the Royal Society of London B* 270: 2435–2442.

Vilà, C., et al. 1997. Multiple and ancient origins of the domestic dog. *Science* 276: 1687–1689.

von Frisch, K. 1967. *The Dance-Language and Orientation of Bees.* Cambridge, Mass.: Harvard University Press.

Vorzimmer, P. J. 1970. *Charles Darwin: The Years of Controversy; The* Origin of Species *and Its Critics, 1859–1882.* Philadelphia: Temple University Press.

———. 1977. The Darwin reading notebooks (1838–1860). *Journal of the History of Biology* 10: 107–153.

Voss, J. 2007. Darwin's diagrams: Images of the discovery of disorder. Max Planck Institute for the History of Science, Berlin, Germany.

Vrba, E. S., and N. Eldredge, eds. 2005. *Macroevolution: Diversity, Disparity, Contingency.* Lawrence, Kans.: The Paleontology Society and Allen Press.

Wagner, M. 1868. *Die Darwinische Theorie, und das Migrationsgesetz der Organismus.* Leipzig.

Wallace, A. R. 1855. On the law which has regulated the introduction of new species. *Annals and Magazine of Natural History* 16 (2nd series): 184–196.

———. 1858. On the tendency of varieties to depart indefinitely from the parental type. *Proceedings of the Linnean Society of London* 3: 53–62.

———. 1863. Remarks on the Rev. S. Haughton's paper on the bee's cell, and on the Origin of Species. *Annals and Magazine of Natural History* 12 (3rd series): 303–309.

———. 1889. *Darwinism: An Exposition of the Theory of Natural Selection with Some of its Applications.* London and New York: Macmillian and Co.

———. 1892. The permanence of ocean basins. *Natural Science* (November 1892): 717–718.

———. 1905. *My Life: A Record of Events and Opinions,* vols. I and II. London: Chapman and Hall, Ltd.

Wallace, D. R. 1997. *The Monkey's Bridge: Mysteries of Evolution in Central America.* San Francisco: Sierra Club Books.

Waterhouse, G. R. 1843. Observations on the classification of the Mammalia. *Annals and Magazine of Natural History* 12: 399–412.

Waters, C. K. 2003. The arguments in the *Origin of Species.* Pp. 116–139 in J. Hodge and G. Radick, eds., *The Cambridge Companion to Darwin.* Cambridge, England: Cambridge University Press.

Weissmann, G. 1998. *Darwin's Audubon: Science and the Liberal Imagination.* New York: Basic Books.

Wen, J. 1999. Evolution of eastern Asian and eastern North American disjunct distributions in flowering plants. *Annual Review of Ecology and Systematics* 30: 421–455.

Westwood, J. O. 1874. *Thesaurus Entomologicus Oxoniensis; or, Illustrations of New, Rare, and Interesting Insects.* Oxford: Clarendon Press.

Wilkins, J. F., and D. Haig. 2003. What good is genomic imprinting: The function of parent-specific gene expression. *Nature Reviews Genetics* 4: 359–368.

Williams, G. C. 1966. *Adaptation and Natural Selection.* Princeton, N.J.: Princeton University Press.

Wilson, E. O., and D. L. Perlman. 2000. *Conserving Earth's Biodiversity.* Washington, D.C.: Island Press.

Wilson, L. G., ed. 1970. *Sir Charles Lyell's Scientific Journals on the Species Question.* New Haven: Yale University Press.

Withers, T. H. 1928. *Catalog of Fossil Cirripedia in the Department of Geology, 1 (Triassic and Jurassic).* London: British Museum (Natural History).

Wollaston, T. V. 1854. *Insecta Maderensia; Being an Account of the Insects of the Islands of the Madeiran Group.* London.

Wood, R. J. 1973. Robert Bakewell (1725–1795) Pioneer animal breeder and his influence on Charles Darwin. *Folia Mendeliana* 8: 231–242.

Woodson, T. 1975. Notes for the annotation of Walden. *American Literature* 46: 550–555.

Woodward, H. 1901. 'Pyrgoma cretacea,' a cirripede from the upper chalk of Norwich and Margate. *Geological Magazine* (New Series) 8: 145–151.

Biographical Notes

The parenthetical page numbers following each entry indicate where the individual is mentioned in the *Origin.*

Agassiz, Jean Louis Rodolphe (1807–1873), Swiss-born systematist, comparative anatomist, and paleontologist who became professor of natural history at Neuchâtel and later (1846) moved to the United States, where he founded the Museum of Comparative Zoology at Harvard. One of the leading naturalists of his day, Agassiz produced monumental works on fossil fish, systematics and classification, and natural history; he also championed the theory of glacial periods. Agassiz never accepted Darwin's theory of descent with modification, resisting it to his dying day. A rich record of debate and discussion on the issue survives at the American Academy of Arts and Sciences in Cambridge, Mass., of which he and Asa Gray, Darwin's staunch defender, were members. Agassiz's final essay on the subject, "Evolution and the permanence of type," was published in *The Atlantic Monthly* in 1874, shortly after his death. *(pp. 139, 302, 305, 310, 338, 366, 418, 439, 449)*

Audubon, John James (1785–1851), celebrated ornithologist and painter best known for his work *Birds of America* (first published in London in 1827–1838 in four huge volumes, then issued in seven volumes in Philadelphia in 1840–1844). Audubon's *Ornithological Biography* (published in Edinburgh in five volumes 1831–1839) was the only other work he published in his lifetime. Audubon traveled to London, Cambridge, and Oxford in the spring of 1828 to drum up subscriptions for his planned *Birds of America* folios. Audubon met John Stevens Henslow, Adam Sedgwick, William Whewell, and other notables in Cambridge, and he may have crossed paths with the young Darwin, who arrived in town in January of that year but was not yet known as "the man who walks with Henslow." Audubon was unimpressed with the Thames upriver at Oxford; writing from there to his wife, Lucy, back in the United States, Audubon commented, "We crossed the Thames twice, near its head; it does not look like the Ohio, I assure thee; a Sand-hill Crane could easily wade across it without damping its feathers" (Audubon 1898, p. 292). *(pp. 185, 212, 387)*

Babington, Charles Cardale (1808–1895), English botanist, entomologist, and archaeologist. Babington became editor of the *Annals and Magazine of Natural History* in 1842, served as chair of the Cambrian Archaeological Association between 1855 and 1885, and was professor of botany at Cambridge University from 1861 to his death in 1895. The 1851 edition of Babington's *Manual of British Botany containing the Flowering Plants and Ferns arranged according to the Natural Order* was included in Darwin's "botanical arithmetic" analysis of species and varieties in large and small genera (see p. 55). *(p. 48)*

Bakewell, Robert (1725–1795), highly influential English cattle and sheep breeder who developed or improved such popular varieties as longhorn and shorthorn cattle and the New Leicester sheep, developed from the Lincolnshire. Bakewell began running his family's Leicestershire farm in 1760. He pioneered methods of irrigation and use of experimental plots to test different water and fertilization regimes on crop productivity. He also developed an innovative approach to cattle and sheep breeding, which he called breeding "in and in"—essentially controlled inbreeding to effect directional change in characteristics of interest. *(pp. 35, 36)*

Barrande, Joachim (1799–1883), French geologist and paleontologist who did extensive work on the Paleozoic fossils of Bohemia. Barrande, who was a strong supporter of Cuvier's theory of catastrophes, referred to fossiliferous rocks underlying Silurian strata near Prague as the "primordial zone." After 1852 Barrande produced a remarkable twenty-three volumes describing the "primordial" fossil fauna of Bohemia (two of these were published posthumously), numerous scientific papers, plus a series of volumes on his theory of "colonies" (see p. 313). He was awarded the Wollaston Medal in 1857 in recognition of his efforts. *(pp. 307, 310, 313, 317, 325, 328, 330)*

Bentham, George (1800–1884), distinguished English botanist and prolific researcher who published extensively on plant taxonomy, including contributions to Augustin Pyramus de Candolle's seventeen-volume work *Prodromus Systematis Naturalis Regni Vegetabilis,* published between 1824 and 1873. Bentham undertook preparation of comprehensive floras of the British colonies, producing *Flora Hongkongensis* (1862) and *Flora Australiensis* (seven volumes, 1863–1878). With Sir Joseph Hooker, Bentham authored the three-volume work *Genera Plantarum,* in which the two introduced a new system of botanical classification and nomenclature. Bentham was awarded the Royal Society's Royal Medal in 1859 and served as president of the Linnean Society from 1861 to 1874. *(pp. 48, 419)*

Berkeley, Miles Joseph (1803–1889), English clergyman, botanist and mycologist (he described the fungi from the *Beagle* voyage), and editor of the *Journal of the Royal Horticultural Society* from 1866 to 1877. Berkeley was awarded the Royal Medal of the Royal Society in 1863, and elected a fellow of the Royal Society in 1879. In the 1850s Berkeley conducted experiments on the immersion of seeds in salt water to determine how long they would remain viable—a topic of great interest to Darwin in terms of dispersal. Berkeley provided Darwin with results of his seed-immersion experiments for Darwin's paper on the subject. *(p. 358)*

Birch, Samuel (1813–1885), English archaeologist and Egyptologist who worked in the British Museum, first as assistant keeper of the Department of Antiquities and later as keeper of the Oriental, British, and medieval antiquities. *(p. 27)*

Blyth, Edward (1810–1873), English naturalist and curator of the museum of the Asiatic Society of Bengal in Calcutta, India, from 1841 to 1862. Blyth wrote prolifically under the pseudonym "Zoophilus," and in the years just prior to the *Origin*'s publication he had an extensive correspondence with Darwin regarding Indian plants and animals. He was also an early theorist on species and selection. *(pp. 18, 163, 253, 254)*

Borrow, George Henry (1803–1881), English traveler and writer famed for his books on Romany (Gypsy) culture and language. *(p. 35)*

Bory de Saint-Vincent, Jean Baptiste (1778–1846), French naturalist and soldier. As a young man Bory de Saint-Vincent traveled abroad as naturalist on an expedition bound for Australia, but he abandoned the expedition at Mauritius and spent the next several years exploring the islands of the west Indian Ocean. He later edited the seventeen-volume *Dictionnaire Classique d'Histoire Naturelle* (Paris, 1822–1831), which Darwin owned, led collecting expeditions to north Africa, and published several books recounting his travels. Bory de Saint-Vincent was perhaps the earliest commentator on the flora and fauna of oceanic islands and their relationship with continental land masses. Some of these observations were given in his book *Voyage dans les Quatre Principales Îles des Mers d'Afrique* (Paris, 1804). *(p. 393)*

Bosquet, Joseph Augustin Hubert (1814–1880), Belgian invertebrate paleontologist who specialized on Crustacea. *(p. 304)*

Brent, Bernard Peirce (d. 1867), English pigeon fancier and author. Brent was a regular contributor to the poultry section of the *Cottage Gardener* and provided Darwin with much information pertaining to varieties and breeding of poultry and pigeons. *(pp. 214, 362)*

Brewer, Thomas Mayo (1814–1880), American physician, journalist, editor, and ornithologist. Brewer's interest in birds led him to contribute to John James Audubon's *Ornithological Biography,* and in 1874 Brewer published his three-volume *History of North American Birds* (Boston). Brewer ignominiously played a role in the "sparrow wars" of the 1870s, supporting the movement to introduce English sparrows such as the house sparrow into the United States, ostensibly to control insect populations (see Barrow 2000). *(p. 217)*

Bronn, Heinrich Georg (1800–1862), leading German paleontologist and professor of natural history at the University of Heidelberg. Bronn authored several important geological works, including a three-volume treatise (*Letkaea Geognostica,* 1851–1856) that established German stratigraphy. Bronn provided the first German translation of the *Origin* in 1860. *(pp. 293, 312)*

Brown, Robert (1773–1858), Scottish botanist. Early on Brown studied medicine and became an assistant surgeon in the British navy. Also a self-taught naturalist, in 1800 Joseph Banks recommended Brown as naturalist on Matthew Flinders's Australian expedition on HMS *Investigator.* Brown returned to England in 1805, and five years later published the first (and only) volume of his projected series of volumes treating the botany of Australia and Tasmania: *Prodromus Florae Novae Hollandiae et Insulae Van-Diemen* (London, 1810). Brown later became librarian and then officer of the Linnean Society and keeper of the British Museum's herbarium, which he inherited from Banks and donated to the museum. Given the affectionate nickname "Jupiter Botanicus," he became a prolific author of papers and one of the leading botanists of his day. As an aside, Brown was also interested in microscopical studies of pollen, in the course of which he discovered the stochastic movement of particles in fluids now called Brownian motion (see biography by Mabberley 1985). *(pp. 415, 416)*

Buckland, William (1784–1856), celebrated English geologist and paleontologist, canon of Christ Church College professor of mineralogy at Oxford, and president of the Geological Society of London and British Association for the Advancement of Science. Buckland was the last of the adherents of "flood geology" in England, but he also became an early champion of Agassiz's theory of glaciation and made important contributions to paleontology (describing the first dinosaur, *Megalosaurus,* in 1824). He worked to establish the Geological Survey of Great Britain and authored several well-known works, including the 1823 book *Reliquiae Diluvianae* (Relicts of the Flood) and *Geology and Mineralogy Considered with Reference to Natural Theology* (1836), one of the eight Bridgewater Treatises. *(p. 329)*

Richard Whatley's *Elegy intended for Professor Buckland* (1820) captures his essence:

Where shall we our great Professor inter
That in peace may rest his bones?
If we hew him a rocky sepulchre
He'll rise and break the stones
And examine each stratum that lies around
For he's quite in his element underground

Buckley, John (fl. 1770s–1780s), English sheep breeder and member of the Dishley Society (named for leading breeder Robert Bakewell's home of Dishley, Leicestershire), an influential group of breeders who helped popularize the New Leicester sheep developed by Bakewell. *(p. 36)*

Buckman, James (1814–1884), professor in the Royal Agricultural College in Cirencester, England, and author of books and papers on horticulture, geology, and other subjects. *(p. 10)*

Burgess, Joseph (fl. 1780s–1800s), English sheep breeder who, like John Buckley, was influential in the popularization of the New Leicester breed of sheep. *(p. 36)*

Candolle, Alphonse Louis Pierre Pyrame de (1806–1893), noted French-Swiss botanist and son of the even better known Augustin-Pyramus de Candolle (1778–1841). Alphonse succeeded his father in the botany chair and as director of the Botanical Garden in Geneva, but he retired early from both to devote himself full time to botanical research. *(pp. 53, 115, 146, 175, 360, 379, 386, 387, 389, 392, 402, 406)*

Candolle, Augustin Pyramus de (1778–1841), renowned Swiss botanist and plant biogeographer, professor of botany at the University of Montpellier and later of natural history at the Academy of Geneva. He was perhaps best known for his multivolume work *Prodromus Systematis Naturalis Regni Vegetabilis,* a comprehensive treatment of plant families published between 1824 and 1873 (the posthumous volumes continued by his son Alphonse), but his 1820 *Essai Élémentaire de Géographie Botanique* is an important founding document of biogeography; Lyell referred to this essay as "luminous" in the second volume of his *Principles of Geology,* recognizing its importance for the "species question." *(pp. 62, 146, 430)*

Cassini, Alexandre Henri Gabriel de (1781–1832), French botanist known for his considerable work on Asteraceae, particularly the North American flora. *(p. 145)*

Cautley, Proby Thomas (1802–1871), English engineer and naturalist who emigrated to India at the age of seventeen. He served in the army for many years, becoming an army engineer (and notably designed and oversaw construction of the Ganga Canal; in 1836 he became Superintendent-General of Canals in India. Perhaps, like William Smith, Cautley became interested in geology and paleontology as a result of his canal excavations. He collaborated with Hugh Falconer on expeditions to the Siwálik Hills, and the two co-authored a treatise on the fossils of the region (*Fauna Antiqua Sivalensis,* 1846), as well as several papers for the Bengal Asiatic Society and the Geological Society of London; they were jointly awarded the Wollaston Medal in 1837. *(p. 340)*

Clausen, Peter (?1804–1855), Danish collector of natural history artifacts, briefly served as an officer in the Brazilian army. He and Peter Wilhelm Lund explored many limestone caves around Lagoa Santa, in central Brazil, unearthing a remarkable number of fossil mammals and birds. *(p. 339)*

Clift, William (1775–1849), naturalist, anatomist, artist, and curator of the Hunterian Museum of the Royal College of Surgeons. *(p. 339)*

Collins [Colling], Charles (1751–1836), English cattle breeder, apprentice of the noted breeder Robert Bakewell. Instrumental in developing the shorthorn breed of cattle in the late eighteenth century. Ogawa (2001) showed that Darwin erroneously referred to Colling as "Collins" through all six editions of the *Origin. (p. 35)*

Cuvier, Frédéric (1773–1838), younger brother of the acclaimed comparative anatomist and naturalist Baron Georges Cuvier. Frédéric was also a naturalist, overseeing the menagerie at the *Muséum d'Histoire Naturelle* in Paris from 1804 to 1838, and was appointed professor of comparative physiology at the Muséum in 1837. Frédéric co-authored, with Étienne Geoffroy Saint-Hilaire, the lavishly illustrated four-volume *Histoire Naturelle des Mammifères* (Paris, 1819–1842). One wonders if the fierce rivalry that developed between Geoffroy Saint-Hilaire and Frédéric's elder brother Georges strained the relationship between the co-authors. *(p. 208)*

Cuvier, Baron Georges (1769–1832), eminent French naturalist who became one of the leading comparative anatomists and paleontologists of his day. In 1795, at the age of twenty-six, Cuvier was appointed (at the invitation of Geoffroy Saint-Hilaire) assistant to the professor of comparative anatomy at the Muséum National d'Histoire Naturelle in Paris. Three years later, he was appointed professor of natural history at the Collège de France. Cuvier's contributions to physiology, anatomy, classification, paleontology, and zoology are too numerous to recount, but several proved to be especially influential. Cuvier's paper on fossil and living elephants, published in 1796, put forth his idea of successive catastrophes in earth history. Between 1800 and 1805 he published a treatise on comparative anatomy, *Leçons d'Anatomie Comparée* (Paris), and developed a theory of "correlation of parts" to explain structure and function. In 1812 Cuvier included a "preliminary discourse" prefatory to his work *Recherches sur les Ossemens Fossiles de Quadrupèdes,* later (1826) republished as *Discours sur les Révolutions de la Surface du Globe* (Discourse on the Revolutions of the Surface of the Globe). Cuvier published the first edition of his magisterial work on the comparative anatomy of fossil and living animals, the four-volume *Règne Animal Distribué d'après son Organisation,* in 1817, and the five-volume second edition in 1829–1830. He originated the idea of classifying animals into four *"embranchements,"* and his commitment to typology led him to vigorously oppose and marginalize his colleague Jean-Baptiste Lamarck and all transmutationist ideas. He debated Lamarck over extinction and morphological change and Geoffroy Saint-Hilaire over typological classification. Cuvier influenced a generation of French naturalists, among them Louis Agassiz, and was made a baron and a peer of France just a year before his death. See Rudwick (1997) for translations and discussions of Cuvier's important geological and paleontological writings. *(pp. 206, 303, 310, 329, 440)*

Dana, James Dwight (1813–1895), leading American geologist and zoologist; professor of natural history and later geology and mineralogy at Yale University (where he inherited the chair of his father-in-law, Benjamin Silliman). Dana published many articles in "Silliman's Journal," the *American Journal of Science and Arts,* which he co-edited with Benjamin Silliman, Jr., and later edited alone. He also authored several books on mineralogy (his book *Systems of Mineralogy* is still in print, now in its eighth edition), volcanoes, and coral reefs. (Dana independently came up with the same theory of coral reefs as Darwin, but he was beaten to the punch by Darwin's 1842 book on the subject.) Dana was geologist for the U.S. Exploring Expedition under Charles Wilkes from 1838 to 1842, during which he made a special study of the geology of California; in 1849 he published a lengthy treatment of Mount Shasta and environs, which is thought to have contributed to the gold rush of that year. Darwin sent Dana, who was ill for a few years, a presentation copy of the *Origin,* writing: "Whenever you are strong enough to read it, I know you will be dead against me, but I know equally well that your opposition will be liberal & philosophical. And this is a good deal more than I can say of all my opponents in this country" (*Correspondence* 8: 303). Indeed, Dana was initially opposed to Darwin's theory but eventually came to accept it (Sanford 1965). *(pp. 139, 372, 376)*

d'Archiac, Adolphe [Etienne Jules Adolphe Desmier de Saint-Simon, Vicomte d'Archiac] (1802–1868), came to geology and paleontology relatively late in life, after retiring from a brief military career in 1830. He is best known for his work *Histoire des Progrès de la Géologie de 1834 à 1859,* published in eight volumes between 1847 and 1860, followed by the three-volume work *Paléontologie Stratigraphique* (1864–1865) and *Géologie et Paléontologie* (1866). The Vicomte d'Archiac was appointed professor of paleontology at the Muséum National d'Histoire Naturelle, in Paris. *(p. 325)*

Dawson, Sir John William (1820–1899), Canadian paleontologist and geologist who presided over McGill University in Montreal for many years. Dawson founded the Royal Society of Canada and served as president of both the British and the American Associations for the Advancement of Science. Darwin likely drew on Dawson's book *Acadian Geology: An Account of the Geological Structure and Mineral Resources of Nova Scotia* (1855). *(p. 296)*

d'Azara, Félix (1746–1821), Spanish army officer and naturalist. D'Azara spent more than twenty years in South America, where he was initially sent to mediate border disputes between the Spanish and Portuguese colonies. While in South America he worked at map making and made numerous zoological observations that he sent to his brother, the Spanish ambassador to France. These were published the year d'Azara returned to Europe as *Essais sur l'histoire naturelle des quadrupèdes de la Province de Paraguay* (Paris, 1801). The following year Don Félix, as he was known, published a similar work in Spain: *Apuntamientos para la Historia natural de los cuadrúpedos del Paraguay y Río de la Plata* (Madrid, 1802). A four-volume account of d'Azara's travels in South America, with notes by George Cuvier, was published in Paris in 1809. *(p. 72)*

De Verneuil, Edouard (1805–1873), French geologist and paleontologist and three-time president of the Geological Society of France who made contributions to the study of Paleozoic formations. De Verneuil collaborated with Sedgwick and Murchison in geological research in Belgium and Russia and traveled extensively in Spain and the United States analyzing Paleozoic rocks. *(p. 325)*

d'Orbigny, Alcide Dessalines (1802–1857), French naturalist who made important contributions to zoology, geology, and paleontology. Early on he was especially interested in marine biology and described and named the foraminifera. Between 1826 and 1833 d'Orbigny made extensive collections in South America, overlapping in time with Darwin's *Beagle* voyage. He published his findings as part of *La Relation du Voyage dan l'Amérique Méridionale* (1835–1847). A protégé of Georges Cuvier, d'Orbigny was appointed professor of paleontology at the Muséum National d'Histoire Naturelle in Paris. His major works included the eight-volume *La Paléontologie Française* (1840–1860) and *Prodrome de Paléontologie Stratigraphique* (1849). *(p. 297)*

Downing, Andrew Jackson (1815–1852), American landscape designer and horticulturist. Darwin got his information on smooth- and downy-skinned fruits from Downing's 1845 manual *Fruits and Fruit Trees of America,* which enjoyed wide readership in both England and the United States and went through a remarkable thirteen editions in Downing's lifetime. *(p. 85)*

Earl (Earle), George Windsor (d. 1865), traveler in southeast Asia who published numerous accounts of the geography and ethnography of the region. Earl published several books, the best known of which is *The Eastern Seas, or, Voyages and Adventures in the Indian Archipelago in 1832–33–34* (London, 1837); Darwin recorded reading this book in October 1838 (Vorzimmer 1977). In 1853 Earl published *The Native Races of the Indian Archipelago. Papuans.* (London). *(p. 395)*

Edwards, W. W. (?b–?d), acquaintance of Darwin's, knowledgeable about racehorses, judging from the context in which he is cited in the *Origin* as well as in *Variation of Animals and Plants Under Domestication.* In the latter book, Darwin mentions that "Mr. W. W. Edwards examined for me twenty-two foals of race-horses, and twelve had the spinal stripe more or less plain" (vol. I, p. 62). A note from 1870 in the Darwin collection of the Cambridge University Library reads, "Mr W. W Edwards has observed himself & Mr Fox with violent retching"—a rather odd observation to report, perhaps, but this might be related to Darwin's research for his book *Expressions of the Emotions in Man and Animals* (1872). *(p. 164)*

Élie de Beaumont, Jean-Baptiste (1798–1874), French geologist who served as engineer-in-chief and then inspector-general of mines in France, overseeing production of the first comprehensive geological map of France. In 1852 Élie de Beaumont published an ill-fated theory of mountain ranges *(Notice sur le systeme des montagnes)* that generated much discussion. *(p. 317)*

Elliot, Walter (1803–1887), English sportsman, writer, and councilman for the government of Madras in northern India. Hooker first told Darwin about Elliot, and the two eventually met in 1855, at the British Association for the Advancement of Science meeting in Scotland. Darwin wrote to Elliot soon after, seeking skins of pigeons and poultry from India (and, incidentally, some tiger measurements!). Elliot obligingly sent Darwin a sizable shipment of material. Darwin later (1867) donated much of this collection to the British Museum (Natural History). *(p. 20)*

Eyton, Thomas Campbell (1809–1880), naturalist friend of Darwin who lived in Shropshire. Eyton had a special interest in birds; in 1855 he built a museum at his family estate featuring a collection of skins and skeletons of European birds. *(pp. 182, 253)*

Fabre, Jean-Henri Casimir (1823–1915), renowned French naturalist and prolific author of popular books on insect natural history (collectively known as the *Souvenirs Entomologiques,* ten volumes published between 1879 and 1907), many of which were translated into English by Alexander Teixiera de Mattos. Fabre, who never accepted Darwin's theories, was nonetheless hailed by Darwin as an "incomparable observer," and indeed he was known for his ingenious experiments in insect behavior and physiology. Fabre was a teacher by profession for many years, but he later dedicated himself full time to his studies at his home in Provence, his Harmas de Sérignan, as he called it. See Bouvier (1916) for an early biographical sketch, and Fabre (1921) for a biography by distant relative Augustin Fabre. *(p. 218)*

Falconer, Hugh (1808–1865), Scottish botanist and paleontologist who served as superintendent of the botanical gardens in Saharanpur and Calcutta, India, and professor of botany at the Calcutta Medical College. Falconer co-authored, with Proby Thomas Cautley, the *Fauna Antiqua*

Sivalensis (1846), a treatise on the fossils of the Siwálik Hills, India, as well as a number of contributions to the Bengal Asiatic Society and the Geological Society of London. The two men were awarded the Wollaston Medal by the Geological Society of London in 1837. *(pp. 65, 310, 313, 334, 340, 378)*

Forbes, Edward (1815–1854), English naturalist with far-ranging interests, but especially known for his work in marine biology, and as originator of the idea that vast continental land masses once linked Europe, Africa, and America. He championed the "continental extension" hypothesis to explain the biogeography of some plant and animal groups. The theory had its adherents, including in some measure Hooker and Lyell, but Darwin vehemently opposed it, favoring oversea dispersal and migration facilitated by periods of climate change and sea-level fluctuation. In the 1850s Forbes also proposed his "polarity theory" of two supposed great epochs of creation, the Paleozoic and the Neozoic; this paper is notable today only because it provoked Alfred Russel Wallace to write his landmark Sarawak Law paper in 1855. Forbes became the Regius Professor of Natural History at Edinburgh shortly before his untimely death. Forbes was Darwin's second pick, after Lyell, for a candidate editor to publish his 1844 *Essay* in the event of his *own* untimely death, as Darwin directed in a letter to his wife, Emma, dated July 5 of that year (*Correspondence* 3: 43). *(pp. 132, 175, 287, 291, 292, 307, 310, 316, 357, 358, 366, 372, 389, 409)*

Fries, Elias Magnus (1794–1878), Swedish mycologist and botanist who held professorships at the University of Lund and, later, the University of Uppsala. Considered the father of mycology, Fries originated the classification system used for lichens and fungi today. *(p. 57)*

Gardner, George (1812–1849), English botanical explorer who traveled in Brazil from 1836 to 1841 and published an account of his expedition in 1846 *(Travels in the Interior of Brazil).* On p. 374 of *Origin* Darwin refers to Gardner's account of the botany of the Organ Mountains. In 1844 Gardner was appointed superintendent of the botanic barden at Peradeniya, Sri Lanka (formerly known as Ceylon). *(p. 374)*

Gärtner, Karl Friedrich (1772–1850), German botanist and physician. Gärtner left medicine to pursue botany in 1800 and later became a leading authority on plant hybridization. Darwin made much use of Gärtner's 1844 and 1849 treatises on hybridization. He later wrote to Gärtner's daughter Emma, "I have long venerated your Father and estimated his two great works most highly" (*Correspondence* 8: 250). *(pp. 50, 98, 246–250, 252, 255, 257, 258, 259, 262, 268, 270, 272, 273, 274)*

Geoffroy Saint-Hilaire, Étienne (1772–1844), celebrated French embryologist, comparative anatomist, and paleontologist; professor of vertebrate zoology at the Muséum National d'Histoire Naturelle. He embraced the idea of unity of underlying body plan for all animal groups and held a quasi-materialist view of species transmutation. In 1830 he engaged Georges Cuvier in a famous debate over the possibility of homology across animal phyla, and the nature of form and function (see p. 206). Geoffroy's views were likely communicated to Darwin at the University of Edinburgh through Saint-Hilaire's friend Robert Edmund Grant, who was receptive to transmutationism and whom Darwin assisted there in 1826 and 1827. (Grant also introduced Darwin to the uglier side of the scientific pursuit by appropriating a discovery made by Darwin and reporting it as his own.) Geoffroy Saint-Hilaire was perhaps best known for his two-volume treatise *Philosophie Anatomique* (1818–1822). *(pp. 8, 147, 434)*

Geoffroy Saint-Hilaire, Isidore (1805–1861), French zoologist noted for his work in functional anatomy and physiological acclimatization in animals. He was the son of Étienne Geoffroy Saint-Hilaire and eventually succeeded his father as professor of zoology in the faculty of sciences at the University of Paris. *(pp. 11, 144, 149, 155)*

Girou de Buzareingues, Louis François Charles (1773–1856), French agronomist and plant physiologist. Girou de Buzareingues authored two 1828 works read by Darwin: *De la génération* and *Philosophie physiologique, politique et morale. (p. 270)*

Gmelin, Johann Georg (1709–1755), German naturalist and professor of chemistry and natural history at the University of St. Petersburg and, later, the University of Tübingen. In 1733 Gmelin took part in an expedition to Siberia at the request of the empress Anna Ivanovna, returning in 1743. This experience led to his works *Flora Sibirica* (1747–1769) and *Reisen durch Siberien* (1751). *(p. 365)*

Godwin-Austen, Robert Alfred Cloyne (1808–1884), English geologist who studied under William Buckland at Oxford. Godwin-Austen made contributions to what is now called paleoecology. When he was awarded the Geological Society of London's Wollaston Medal in 1862, the president, Roderick Murchison, famously hailed Godwin-Austen as "pre-eminently the physical geographer of bygone periods." *(p. 299)*

Goethe, Johann Wolfgang von (1749–1832), celebrated German poet, playwright, novelist, painter, philosopher, and natural scientist. Goethe's scientific work alone was expansive; he published a significant analysis of light and color in 1810 entitled *Zur Farbenlehre* (Theory of Colors), and in the life sciences he worked on plant morphology (inspiring in 1790 his work *Metamorphose der Pflanzen,* a poetic vision of transmutation) and comparative anatomy. One of Goethe's zoological contributions was his theory of the skull, and the independent discovery of the human intermaxillary bone in 1784 (Opitz 2004). It is not widely appreciated that Goethe helped establish a new philosophy of organic relationships in the Enlightenment period; consider this assessment by the zoologist Ernst Haeckel, writing in the late nineteenth century: "Jean Lamarck and Wolfgang Goethe stand at the head of all the great philosophers of nature who first established a theory of organic development, and who are the illustrious fellow-workers of Darwin" (Haeckel 1880, p. 80). *(p. 147)*

Gould, Augustus Addison (1805–1866), American physician and noted invertebrate zoologist. Gould's specialty was mollusks. Among his many publications, Gould described the mollusk specimens of the United States Exploring Expedition of 1838–1841 (volume XII of the expedition, 1852: *Mollusca and Shells*) and edited a book titled *The Terrestrial and Airbreathing Mollusks of the United States* (1851–1855). *(p. 397)*

Gould, John (1804–1881), distinguished British ornithologist and curator and preserver at the museum of the Zoological Society of London. Gould was renowned as an artist and as a taxonomist. He published thousands of hand-colored lithographs depicting birds and mammals from all over the world in several books, including the five-volume *Birds of Europe,* the seven-volume *Birds of Australia,* as well as volumes on kangaroos and their relatives, Australian mammals generally, hummingbirds, parrots, and birds of the Himalayas. Gould was given Darwin's mammal and bird specimens from the *Beagle* voyage for identification, and it was he who noted that Darwin's indifferently collected Galápagos blackbirds, "gross-bills," and finches actually constituted "a series of ground Finches which are so peculiar" as to form "an entirely new group, containing 12 species." Recall that Darwin had neglected to identify the islands of origin of his finches, but fortunately he was able to reconstruct this information thanks to the greater care taken by his assistant Syms Covington and Captain Fitzroy. Gould also informed Darwin that the small rhea specimen he collected in South America (the one that almost became Christmas dinner) was a separate species distinct from the northern rheas; he named it *Rhea darwinii* in 1837. (Unfortunately for Darwin, this name did not stick; French zoologist Alcide d'Orbigny, who collected in South America between 1825 and 1833, published a description of this rhea under the name *Pterocnemia pennata* three years earlier, in 1834.) In any case, both the finches and the rheas were significant in the development of Darwin's evolutionary insights. *(pp. 132, 133, 398, 404)*

Gray, Asa (1810–1888), noted American botanist and longtime Darwin supporter and correspondent. Gray's scientific career was spent at Harvard University, where he became the leading American botanist of the nineteenth century and authored several botanical manuals on North American flora. In addition to providing Darwin with a great deal of botanical information, Gray was a strong and vocal champion of Darwin's theory of transmutation by natural selection. Their mutual sympathies are abundantly evident in their rich correspondence; many of their letters touch on family matters and politics (especially the American Civil War, which was raging in the years just after the *Origin* was published). Gray is remarkable, too, for his commitment to both his personal religious beliefs and Darwin's ideas. An orthodox Presbyterian, Gray argued that Darwin's transmutation theory and Christian belief were not mutually exclusive. In 1876 Gray produced his book *Darwiniana: Essays and Reviews Pertaining to Darwinism,* in which he staunchly defended Darwin. Gray and his wife traveled to England in late summer 1868, and he and Darwin met at last soon after their arrival that September: "We reached Kew last evening, Mrs. Gray somewhat dazed by the voyage and fatigued by the travel. When she gets a little stronger I shall be delighted to come and see you" (Darwin Correspondence Project, Letter 6370). *(pp. 100, 115, 176, 365, 372)*

Gray, John Edward (1800–1875), English zoologist who worked for many years at the British Museum (eventually becoming keeper of zoology) and founded the Entomological Society of London. Darwin drew on Gray's book *Gleanings from the Menagerie and Aviary at Knowsley Hall: Hoofed Quadrupeds* (Liverpool, 1850) for illustrations of hybrid equines, calling this a "splendid work" in *Natural Selection. (pp. 163, 165, 218)*

Hamilton Smith, Charles (1776–1859), colonel in the British army and author of several zoological books. Hamilton Smith is also known to military historians for his experiment with the color of soldiers' uniforms; in field tests conducted in 1800 with targets of different colors, Hamilton Smith showed that the conspicuous red uniform of the British regulars was more than twice as likely to receive a bullet-hole as gray uniforms. He submitted a report concluding, "The general result is . . . of so important a nature, that it appears exceedingly desirable they should be repeated . . . the question arises whether all riflemen and light infantry should not take the field in some grey unostentatious uniform, leaving the parade dress for peace and garrison duty" (Hamilton Smith 1853). Thus was conducted perhaps the first experiment in relative apparency and its effects on "predation"—oddly appropriate, given the evolutionary context of this book. *(p. 164)*

Harcourt, Edward William Vernon (1825–1891), Oxfordshire parliamentarian and sometime sheriff. Harcourt was also a well-regarded amateur naturalist who made a special study of the natural history of Madeira, off the coast of north Africa. In 1851 he published a guide to the archipelago entitled *A Sketch of Madeira,* and in 1855 a paper on Madeiran ornithology. *(p. 391)*

Hartung, Georg (1821–1891), German geologist and author who made a geological study of the Azores and, with Lyell, the Canary Islands and Madeira in the 1850s. In 1860 Hartung published several papers and books on the geology of these islands, including a volume on the geology of the Azores with Heinrich Georg Bronn: *Die Azoren in ihrer äusseren Erscheinung und nach ihrer geognostischen Natur* (Leipzig). *(p. 363)*

Hearne, Samuel (1745–1792), English explorer who, in the employ of Hudson's Bay Company, undertook three expeditions to North America between 1769 and 1772 in which he surveyed the Coppermine River, discovered the "Northern" (Arctic) Ocean, and searched for the Northwest Passage. Hearne became best known for his *Journey from Prince of Wales's Fort in Hudson's Bay, to the Northern Ocean* (1795), a book that inspired poets William Wordsworth and Samuel Taylor Coleridge (Hearne is the model for Coleridge's *Rime of the Ancient Mariner*). It was in Hearne's *Journey* that Darwin read the account of the insect-eating black bears that led to the brouhaha with Richard Owen. *(p. 184)*

Heer, Oswald (1809–1883), Swiss paleontologist and botanist, professor of botany and entomology at the University of Zürich. Heer's paleobotanical work was admired by Lyell and Darwin (though Darwin was not impressed with Heer's fondness for continental extensions as explanations for plant distribution). Heer was especially interested in the flora of the Tertiary period, and between 1855 and 1859 he produced his three-volume work *Flora Tertiaria Helvetiae. (p. 107)*

Herbert, William (1778–1847), naturalist, classicist, and clergyman who served as rector of Spofforth, Yorkshire, from 1814 to 1840, and dean of Manchester from 1840 to his death in 1847. Herbert's scientific investigations focused on plant hybridization, and he made lilies his specialty. The journal of the International Bulb Society, *Herbertia,* was named for him. *(pp. 62, 250, 251)*

Heron, Sir Robert (1765–1854), second baronet and politician who served as member of Parliament for Grimsby and Peterborough. Heron was something of a gentleman farmer who experimented with animal breeding. *(p. 89)*

Heusinger von Waldegg, Karl Friedrich (1792–1883), German physician; professor of practical medicine and director of the Marburg Clinic from 1829 until his death in 1883. Heusinger von Waldegg studied comparative pathology; Darwin consulted his 1846 paper "On the diverse effects of certain external influences on different colored animals" (*Wochenschrift fürdie gesammte Heilkunde* 18: 277–283) for information on constitutional differences among animal breeds in response to toxins and nutritional factors. *(p. 12)*

Hewitt, Edward (fl. 1850s–1860s), Birmingham poultry enthusiast, judge at poultry shows, and an apparently avid poultry breeder. Hewitt corresponded with William Bernhard Tegetmeier, providing him with information relating to poultry breeding. Tegetmeier put Darwin in touch with Hewitt concerning Hewitt's hybridization experiments. *(p. 264)*

Hooker, Joseph Dalton (1817–1911), English botanist and director of Royal Botanic Gardens, Kew (succeeding his father, Sir William Jackson Hooker, the first director). Hooker voyaged around the world from 1839 to 1843 as assistant surgeon aboard HMS *Erebus* on James Clark Ross's Antarctic expedition (with its sister ship HMS *Terror*). He later explored the botany of the Himalaya region (1847–1850). Hooker was perhaps the leading botanist of his day, producing as a result of his explorations several volumes on the biogeography of southern hemisphere and Himalayan flora. Hooker described Darwin's botanical collections from the Falkland Islands, Tierra del Fuego, and the Galápagos. Darwin cautiously approached Hooker in 1845 about his species theory: "as you say . . . geographical [distribution] will be the key which will unlock the mystery of species," he wrote in November of that year. Darwin closed this letter saying: "I wish I could get you sometime hence to look over a rough sketch (well copied) on this subject, but it is too impudent a request" (*Correspondence* 3: 263). Hooker visited Darwin at Down House not long afterward and likely took home with him a copy of the 1844 *Essay. (pp. 53, 100, 140, 145, 373, 374, 375, 376, 378, 379, 381, 387, 391, 429)*

Horner, Leonard (1785–1864), Scottish geologist and strong promoter of science education; president of the Geological Society of London on two occasions. Darwin had a long correspondence with Horner, who also happened to be Sir Charles Lyell's father-in-law. *(p. 18)*

Huber, François (1750–1831), Swiss entomologist and apiculturist honored as the father of modern beekeeping. (He is also the father of entomologist Pierre Huber.) Blind from about the age of fifteen, he worked closely with his assistant, François Burnens, researching aspects of bee biology, in particular helping to solve the mystery of bee reproduction (see Maderspacher 2007 for a review of this topic and Huber's contributions). François Huber's treatise *Nouvelles Observations sur les Abeilles* (New Observations of Bees; Geneva, 1792) was very influential and is often considered the origin of modern beekeeping. *(pp. 230, 231)*

Huber, Jean Pierre (1777–1840), Swiss entomologist known largely for his work on Hymenoptera, and son of the renowned blind Swiss apiculturist François Huber, the father of modern beekeeping. Pierre continued his father's work, helping him produce later editions of his landmark book on bees, but he also conducted his own entomological research, primarily with ants and insect behavior. In 1804 Pierre discovered the phenomenon of "slavery," or dulosis, in ants. His book *Recherches sur les mour des Fourmis Indigènes* (Paris and Geneva, 1810) was highly acclaimed. Darwin acquired an English translation, which came out in 1820 under the title *The Natural History of Ants. (pp. 208, 219, 220, 221, 225, 226)*

Humboldt, Alexander von (1769–1859), celebrated Prussian naturalist, explorer, diplomat, and author. Humboldt traveled to Mexico, Central America, and South America between 1799 and 1804 with French botanist Aimé Bonpland, producing their monumental *Personal Narrative of a Journey to the Equinoctial Regions of the New Continent,* published in thirty volumes over a twenty-year period. Humboldt's acclaim and influence cannot be over-emphasized: his legacy is evident in the numerous species, geographical features, and even towns, counties, parks, and high schools and universities named in his honor in the United States and abroad. Darwin and Wallace each acknowledged their debt to Humboldt, whose writings inspired in them a burning passion for travel and exploration. Darwin famously wrote of Humboldt: "He was the greatest travelling scientist who ever lived.—I have always admired him; now I worship him" (See McCullough 1992, Helferich 2004, and Sachs 2006, for treatments of Humboldt's importance.) *(p. 374)*

Hunter, John (1728–1793), Scottish anatomist and surgeon who authored in 1786 the treatise *Observations on Certain Parts of the Animal Oeconomy.* John Hunter's collections, acquired by the Royal College of Surgeons of England in 1799, form the core of the College's Hunterian Collection. These spectacular collections can be viewed in the newly renovated Hunterian Museum. John's brother William (1718–1783) was also a physician of note, and he left funds and his collections to establish what is now the Hunterian Museum in Glasgow, which opened in 1807. *(p. 150)*

Hutton, Thomas (?1807–1874), captain in the Bengal army (invalided in 1841) and the author of various papers on natural history and scriptural geology in the *Calcutta Journal of Natural History.* Darwin contacted Hutton at the suggestion of Edward Blyth, and Hutton provided Darwin with information regarding pigeons and geese. *(p. 253)*

Huxley, Thomas Henry (1825–1895), English zoologist and one of Darwin's staunchest defenders after the *Origin* was published. Huxley was known as "Darwin's bulldog" for his pugnacious style in a profusion of lectures, articles, and debates on the subject. He was largely self-educated, and like Hooker and Darwin embarked on a lengthy voyage to the southern hemisphere, serving as assistant surgeon on HMS *Rattlesnake* (1846–1850). During this voyage Huxley conducted research on marine invertebrates and embryology. He later became a prolific writer, addressing such far-ranging topics as human evolution, fossil fishes, and educational reform. *(pp. 101, 338, 438, 442)*

Jones, John Matthew (1828–1888), Canadian zoologist and army captain; president of the Nova Scotian Institute of Science, 1863–1873. Jones published several studies on the geology and flora and fauna of Bermuda. *(p. 391)*

de Jussieu, Adrien-Henri Laurent (1797–1853), French botanist of the distinguished botanical de Jussieu family. He succeeded his father, Antoine Laurent de Jussieu, as professor of botany at the Muséum National d'Histoire Naturelle in Paris. Among his many works, he published treatises on euphorbs and on classification (*Taxonomie,* Paris, 1848). *(p. 417)*

Kirby, William (1759–1850), English naturalist and author, founding member of the Linnean Society. Kirby's first entomological work of note was *Monographia Apum Angliae* (Monograph on the Bees of England) of 1802. He co-authored, with his friend William Spence (1783–1860), the celebrated *Introduction to Entomology,* which came out in several editions between 1815 and 1826 and became the standard reference in general entomology for many years. Among other works, Kirby also contributed one of the Bridgewater Treatises commissioned by the Earl of Bridgewater (*History, Habits, and Instincts of Animals,* 1835). Kirby came close to being appointed to the botany chair at Cambridge in 1815, but complications arising from his political affiliations led to the appointment of John Stevens Henslow instead—a fateful event in the history of biology, to be sure. *(p. 135)*

Knight, Thomas Andrew (1759–1838), English botanist and horticulturist, president of the Horticultural Society of London from 1811 to 1838. Knight carried out an extensive experimental breeding program on various fruits and vegetables and developed an important strawberry variety that was widely planted. In addition to many papers in the Transactions of the Horticultural Society of London, he authored the popular *Treatise on the Culture of the Apple and Pear* (1797), which Darwin recorded reading in 1844, and *Pomona Herefordieneis* (1809), admired for its illustrations. *(pp. 7, 96)*

Kölreuter, Joseph Gottlieb (1733–1806), German botanist who served as assistant curator at the Imperial Academy of Sciences in St. Petersburg from 1756 to 1761, and professor of natural history and director of the botanical gardens in Karlsruhe from 1763 to 1786. Kölreuter conducted extensive research on plant hybridization and authored a treatise on pollination that Darwin heavily annotated. Kölreuter's most important work was a series of four reports (the *Vorläufige Nachricht*) presenting his plant hybridization experiments, published between 1761 and 1766; Darwin discusses this work in chapter VIII. *(pp. 98, 246, 247, 248, 250, 258, 271, 274, 451)*

Lamarck, Jean-Baptiste (1744–1829), French naturalist and professor of invertebrates at the newly (post-Revolution) reorganized Muséum National d'Histoire Naturelle in Paris. Lamarck's early work was in botany, but in his new position he studied paleontology and invertebrate zoology (he coined the term "invertebrate"). He authored the first detailed evolutionary theory in his work *Philosophie Zoologique* (1809) and later published his seven-volume *Histoire Naturelle des Animaux sans Vertebras* between 1815 and 1822. Darwin paid tribute to Lamarck in the introduction to later editions of *Origin:* "Lamarck was the first man whose conclusions on the subject excited much attention," he wrote. "This justly celebrated naturalist first published his views in 1801 . . . he first did the eminent service of arousing attention to the probability of all changes in the organic, as well as in the inorganic world, being the result of law, and not of miraculous interposition." *(pp. 242, 427)*

Lepsius, Karl Richard (1810–1884), Prussian Egyptologist; professor of Egyptology at the University of Berlin, where he helped establish the Egyptian Museum. Lepsius led an expedition to the Nile Valley and Sudan (Nubia) between 1842 and 1845, sponsored by King Wilhelm IV of Prussia, an undertaking that led to his twelve-volume work *Denkmaler aus Aegypten und Aethiopien. (p. 27)*

LeRoy, Charles Georges (1723–1789), French naturalist and gamekeeper who succeeded his father as ranger of Versailles and Marley. In 1764 Leroy published a collection of papers on animal intelligence entitled *Lettres Philosophiques sur l'intelligence et la perfectibilité des animaux, avec quelques letters sur l'homme,* which Darwin consulted. *(p. 214)*

Linnaeus, Carolus [Carl von Linné] (1707–1778), the great Swedish botanist. Linnaeus studied medicine in Sweden and the Netherlands but devoted most of his time to botany. He mounted plant-collecting expeditions to Lapland in 1731 and central Sweden in 1734, and in 1735 he published the first edition of the *Systema Naturae,* in which he set forth his method for classifying all living things and introduced consistent use of binomials in naming species. Linnaeus practiced medicine in Sweden until he was awarded a professorship at the University of Uppsala in 1741. He was extremely influential, sending many students on voyages around the world for plant exploration (Linneaus's "apostles"); these included Peter Kalm, who traveled to northeastern North America, Daniel Solander, who accompanied the first *Endeavor* voyage around the world under Captain Cook, and Carl Peter Thunberg, who traveled to India, Japan, and southern Africa. *(pp. 64, 413, 417, 427)*

Livingstone, David (1813–1873), Scottish missionary and renowned explorer of central and southern Africa; outspoken critic of the African slave trade. His first book, *Missionary Travels and Researches in South Africa* (1857), written back in England, was a sensation. It was followed by *Narrative of an Expedition to the Zambezi and Its Tributaries* (1865) after his second expedition, and *The Last Journals of David Livingstone in Central Africa* (1874), published posthumously following Livingstone's death in 1873 on this third expedition. Darwin commented of Livingston's *Travels* in his reading notebooks: "the best travels I ever read"—high praise coming from one who read many a travelogue! (Vorzimmer 1977, p. 151). *(p. 34)*

Lubbock, Sir John (1834–1913), the first Lord Avebury, was a politician, banker, and entomologist. Lubbock was Darwin's neighbor in Downe, and the two became friends and frequent correspondents. Lubbock's scientific pursuits were many and varied, from paleontology to experiments on bee vision to insect anatomy. He was especially interested in social insects, one result of which was his book *Ants, Bees, and Wasps* (London, 1882). Lubbock was an ardent supporter of Darwin and his theory, and he assisted Darwin by providing advice and information before and after the publication of the *Origin*—notably setting Darwin straight on the most robust way to analyze the species-to-variety ratio in large and small genera, and providing Darwin with data on variation in flower structure and characters of the insect nervous system (see Pumphrey 1958, Somkin 1962). One of Lubbock's best-known legislative initiatives was the introduction of the bank holiday, adopted in England in 1871 ("St. Lubbock's Day" was its tongue-in-cheek name).

The magazine *Punch* lampooned this curious combination of interests in an 1882 cartoon with the verse caption:

> *How doth the Banking Busy Bee*
> *Improve his shining Hours?*
> *By studying on Bank Holidays*
> *Strange insects and Wild Flowers!*
> *(pp. 46, 241)*

Lucas, Prosper (1805–1885), French physician who studied heredity in people; Lucas authored the two-volume treatise *Traité Philosophique et Physiologique de L'Hérédité Naturelle* (Paris, 1847 and 1850). *(pp. 12, 275)*

Lund, Peter Wilhelm (1801–1880), Danish naturalist, often called the "father of Brazilian paleontology." He explored with fellow Dane Peter Clausen the limestone caves around the central Brazilian town of Lagoa Santa in the modern state of Minas Gerais, achieving renown for the fossil mammals, birds, and other species they unearthed. Early on Lund interpreted his cave discoveries in much the same way Buckland interpreted his: a window onto an antediluvial fauna. This was the thesis of his book on the topic, the title of which translates as *A View of the Fauna of Brazil before the Last Global Revolution.* He later adopted a more uniformitarian view. *(p. 339)*

Lyell, Sir Charles (1797–1875), English geologist and naturalist, and Darwin's close personal friend. Lyell had become a lawyer by profession, but this career was short-lived. While he was a student at Oxford, his interest in geology was piqued by William Buckland, and in the 1820s he was already honored for his contributions to geology by being elected fellow of the Linnean, Geological, and Royal societies. Lyell gave up law in 1827 to devote himself to geology and was appointed professor of geology at King's College, London, 1831–1833. This was the period that his greatest work appeared: his three-volume *Principles of Geology* was published between 1830 and 1833, the first volume of which was auspiciously given to Darwin by Captain Fitzroy at the start of the *Beagle*'s voyage. Lyell became the leading English geologist and was knighted in 1848 for his contributions. Among his later books, the most notable is perhaps *The Antiquity of Man* (1863), in which he argued for an ancient human history and first embraced, in some measure, Darwin's ideas on species origins. *(pp. 62, 95, 282, 283, 289, 292, 296, 304, 307, 310, 312, 313, 323, 328, 356, 363, 381, 382, 385, 402, 481)*

Macleay, William Sharpe (1792–1865), English entomologist and originator of the Quinarian school of systematics, which he presented in the second volume of his work *Horae Entomologicae* (London, 1819, 1821). Macleay held that a natural system of classification involved symmetry and numerical regularity, a system of circles within circles all based on the number 5. Macleay's system also incorporated the concepts of homology (what he termed "affinity") and analogy (parallelism), and some people have erroneously attributed the origin of these terms to him (see Panchen 1992). *(p. 427)*

Malthus, Thomas Robert (1766–1834), famous (to some infamous) social and political philosopher best known for his *Essay on the Principle of Population* (1798) and several works on economics. Malthus was an Anglican clergyman from 1796 until his marriage in 1804, which required forfeiture of his fellowship from Jesus College, Cambridge. He was then appointed professor of history and political economy (the first such appointment in England) at the East India Company College in Hertfordshire. In the wake of revolution in France, many English intellectuals debated whether society could be improved. The ideas on this subject of such leading thinkers as William Godwin and the Marquis de Condorcet prompted Malthus to write his *Essay,* which went through six editions between 1798 and 1826. Malthus also authored several important works on economics, treating (among others) such topics as regulation of corn prices, rent, and a theory of supply-and-demand mismatches. In 1820 Malthus published *Principles of Economics,* one of the earliest economics textbooks. The "Malthusian" formulation of population's outstripping the means of subsistence came to be known as the "dismal theorem." Malthus was pessimistic about the perfectibility of society, but his humane recommendations to mitigate the problems of overpopulation (free public education, universal suffrage) are often overlooked. The revised and expanded second edition of his *Essay on Population* (1803) includes in its subtitle "an Inquiry into our Prospects respecting the Removal or Mitigation of the Evils which it occasions." *(p. 63)*

Marshall, William (1745–1818), English agriculturist and philologist best known for his *General Survey* series of volumes on the rural economy of England in the 1780s and 1790s and a book on the management of estate properties in 1804. *(pp. 41, 423)*

Martens, Martin Charles (1797–1863), professor of botany and chemistry at the University of Louvain in Belgium. Darwin was interested in an experiment on seed germination following long-term immersion in sea water that Martens published in 1857 in the *Bulletin de la Société Botanique de France. (p. 360)*

Martin, William Charles Linnaeus (1798–1864), naturalist and curator of the Zoological Society of London's museum. Martin was a prolific author of natural history articles and books, including volumes on dogs, horses, and quadrupeds in general; his 1845 work *The History of the Horse: Its Origin, Physical and Moral Characteristics, its Principal Varieties and Domestic Allies* was a standard reference. *(p. 165)*

Matteuchi [Matteucci], Carlo (1811–1868), Italian physicist, neurophysiologist, and politician who conducted important work on the electrical properties of nerves and muscle tissue. Matteucci was appointed professor of physics at the University of Pisa in 1840, but he later became a senator and eventually minister of education in Italy. Darwin probably consulted some of Matteucci's many papers on neurophysiology, which were later collected under the title *Electro-Physiological Researches* in the Philosophical Transactions of the Royal Society of London. *(p. 193)*

Miller, Hugh (1802–1856), Scottish stonemason, lay theologian, and geologist. Miller was a self-taught geologist and achieved much acclaim for his several popular geological works, including *The Old Red Sandstone* (1841) and the posthumously published *Testimony of the Rocks* (1857). *(p. 283)*

Miller, William Hallowes (1801–1880), Professor of Mineralogy at Cambridge, succeeding William Whewell in that post in 1832; Fellow of the Royal Society and its Foreign Secretary from 1856 to 1873. Miller provided Darwin with a mathematician's assessment of honeybee cell geometry. *(p. 226)*

Milne-Edwards, Henri (1800–1885), French invertebrate zoologist and physiologist. Initially trained in medicine, Milne-Edwards later devoted himself to science, studying under Georges Cuvier. He became a professor at the *Collège Central des Arts et Manufactures* in 1823, chair of entomology at the Muséum National d'Histoire Naturelle in 1841, and eventually (1862) chair of zoology at the Muséum. Milne-Edwards served for many years as editor of the journal *Annales des Sciences Naturelles* and published several works on crustaceans and corals (*Histoire naturelle des Crustacés,* 1837–1841, and *Histoire Naturelle des Coralliaires,* 1858–1860). *(pp. 116, 194, 418, 433)*

Moquin-Tandon, Christian Horace Benedict Alfred (1804–1863), French naturalist and physician. Moquin-Tandon was professor of zoology at Marseilles from 1829 to 1833, when he became professor of botany and director of the botanical gardens at Toulouse. He later was appointed director of the Jardin des Plantes and Académie des Sciences in Paris. The fleshy-leaved seashore plants Darwin mentions were listed in Moquin-Tandon's 1841 book *Eléments de tératologie végétale, ou histoire abrégée des anomalies de l'organisation dans les végétaux*—Darwin's library included a heavily annotated copy of this volume. *(p. 132)*

Morton, Lord [George Douglas, the sixteenth Earl of Morton] (1761–1827), Scottish naturalist and fellow of the Royal Society. The Earl of Morton (misspelled Moreton in this *Origin* edition) is perhaps best known in the history of biology for his 1821 paper "A communication of a singular fact in natural history," submitted as a letter to the Royal Society. The "singular fact" was the astonishing observation that a mare once mated to a quagga apparently produced offspring from later sires that continued to bear quagga-like markings. This suggested that some heritable element from the quagga had become a part of the mare's reproductive constitution; the term "telegony" described this purported lingering reproductive influence—a long-discredited idea. *(p. 165)*

Mozart, Wolfgang Amadeus (1756–1791), renowned Austrian composer and musician, a child prodigy who learned his first musical piece, and created his first musical composition, at the age of five. *(p. 209)*

von Müller, Ferdinand (1825–1896), German physician and botanist who emigrated to Australia in 1847, becoming government botanist for Victoria beginning in 1853 and director of the Royal Botanic Gardens in Melbourne from 1857 to 1873. Von Müller was a prolific botanical explorer of Australia, producing several botanical works, among them *Fragmenta Phytographica Australiae* (eleven volumes, 1862–1881) and *Plants of Victoria* (two volumes, 1860–1865). *(p. 375)*

Müller, Johannes Peter (1801–1858), German physiologist, comparative anatomist, and ichthyologist. Müller was professor of physiology at Bonn University for many years, then chair of anatomy and physiology at Humboldt University in Berlin. He edited the journal *Archiv fur Anatomie und Physiologie* and authored more than half a dozen treatises, the best known of which is probably his physiological work *Handbuch der Physiologie des Menschen* 1833–1840), translated into English in 1842 as *Elements of Physiology. (p. 10)*

Murchison, Sir Roderick Impey (1792–1871), noted Scottish geologist, had an early career as a soldier and turned to science relatively late. Murchison served as president of the Geological and Royal Geographical societies and was appointed director-general of the Geological Survey of the United Kingdom and director of the Royal School of Mines and the Museum of Practical Geology. He authored several important geological works, the best known of which is *The Silurian System* (1839). *(pp. 289, 307, 310, 317)*

Murray, Sir Charles Augustus (1806–1895), English diplomat and writer. Murray served as consul-general in Egypt from 1846 to 1853 and as envoy and minister plenipotentiary to the court of Persia (now Iran) from 1854 to 1859. Darwin first contacted Murray in December 1855, while Murray was employed in the Persian court. Introducing himself as a friend of Charles Lyell, whom Murray met on his travels in North America, Darwin went on to ask Murray's assistance in obtaining specimens: "Skins of any domestic breed of Pigeons, Poultry, Ducks Rabbits or even Cats (or skeletons of Dogs) which have been bred for many generations in domestication would be of great value" (*Correspondence* 5: 530). *(p. 20)*

Newman, Henry Wenman (1788–1865), captain and later lieutenant-colonel commandant in the British army. Newman inherited a family estate at Thornbury Park, Gloucestershire, in 1829, where he pursued his interest in bees. Newman authored several articles on bees and corresponded with Darwin on bee biology through the medium of gardening journals such as the *Cottage Gardener* and the *Journal of Horticulture. (p. 74)*

Noble, Charles (1817–1898), English nurseryman. Noble was a partner with John Standish in a nursery at Bagshot Park, Surrey, from 1848 to 1853, after which he ran his own nursery (Sunningdale) in Windlesham specializing in Asian rhododendrons. Noble co-authored, with Standish, the best-selling book *Practical Hints on Planting Ornamental Trees* (London, 1852). *(p. 251)*

Owen, Sir Richard (1804–1892), leading British comparative anatomist, first at the Hunterian Museum of the Royal College of Surgeons and later as superintendent of the natural history departments of the British Museum. Owen was instrumental in the establishment of the Natural History Museum at its current location in South Kensington, London. He published innumerable papers and books on comparative anatomy and paleontology, including important early studies of the platypus, extinct giant moas, *Archaeopteryx* (which he named), and fossil mammals and reptiles (coining the word "dinosaur" in 1842). Famously, Owen determined in 1842 that an anomalous bone discovered in New Zealand belonged to a giant flightless bird, and he was subsequently proven correct when complete skeletons were found the following year; he dubbed this bird *Diornis.* Owen was also a strong supporter of the "archetype" concept and developed the concepts of homology and analogy in morphology. Owen described Darwin's fossil mammals from the *Beagle* voyage but was a lifelong opponent of Darwin's views; his 1860 review of the *Origin* was scathing, and Darwin and Owen's relationship never recovered. Owen's personal copy of the *Origin of Species* now resides at Shrewbury School in Shropshire—ironically, the school Darwin attended as a child. *(pp. 134, 149, 150, 191, 192, 310, 319, 329, 339, 414, 416, 435, 437, 442, 452)*

Paley, Rev. William (1743–1805), English clergyman, theologian, and antislavery activist who studied and taught for a time at Christ's College, Cambridge—Darwin's college some years later. Paley was famous for his teaching and writing; among his many books, *Natural Theology* (1802) was perhaps the most influential. In that book he developed his watch analogy for inferring design in the universe. Paley's *View of the Evidences of Christianity* (1794) was required reading at Cambridge, and Darwin wholeheartedly subscribed to the "Evidences," as they were called, when a student there: "In order to pass the B.A. examination, it was, also, necessary to get up Paley's *Evidences of Christianity,* and his *Moral Philosophy* . . . The logic of this book and as I may add of his *Natural Theology* gave me as much delight as did Euclid. The careful study of these works, without attempting to learn any part by rote, was the only part of the Academical Course which, as I then felt and as I still believe, was of the least use to me in the education of my mind. I did not at that time trouble myself about Paley's premises; and taking these on trust I was charmed and convinced of the long line of argumentation" (Darwin 1958). With a perhaps intentionally ironic nod to history, portraits of Darwin and Paley now hang side-by-side in the grand dining hall of Christ's College, just behind the high table. *(p. 201)*

Pallas, Peter Simon (1741–1811), German zoologist who worked in Russia for many years under the patronage of Catherine II. Pallas was professor of natural history at the St. Petersburg Academy of Sciences, and he led two expeditions of exploration to remote Russian provinces, resulting in several zoological and botanical volumes. *(pp. 163, 253)*

Philippi, Rudolph Amandus (1808–1904), German-born geologist and paleontologist. Philippi was professor of natural history and geography in a technical school in Cassel from 1835 to 1851, when he emigrated to Chile to join his brother. Two years later, Philippi was appointed professor of zoology and botany at the University of Chile and director of the National Museum and Botanical Garden. Darwin is referencing Philippi's 1836 work on the mollusks of Sicily. *(p. 312)*

Pictet de la Rive, François Jules (1809–1872), Swiss zoologist and paleontologist; succeeded Augustin de Candolle as professor of zoology and comparative anatomy at the University of Geneva. In books such as *Traité élémentaire de paléontologie* (1844–1846), Pictet argued for successive bouts of creation and, in the second edition of 1853–1857, the absence of continuity of faunas in the fossil record. *(pp. 302, 305, 313, 316, 335, 338)*

Pierce, James (fl. early nineteenth century), American geologist and naturalist; founding member of the Lyceum of Natural History in New York in 1818, the forerunner institution of the extant New York Academy of Sciences. Pierce published several articles in the *American Journal of Science and Arts* ("Silliman's Journal") treating the geology, mineralogy, and to a lesser extent the zoology of the New York, New Jersey, and middle Atlantic regions. Darwin read Pierce's "Memoir on the Catskill Mountains" published in Silliman's journal in 1823. *(p. 91)*

Pliny the Elder [Gaius Plinius Secundus] (23–79 CE), soldier, philosopher, and encyclopedist of ancient Rome who famously came to an untimely end while attempting to observe an eruption of Mount Vesuvius. *(pp. 28, 34, 37)*

Poole, Skeffington (fl. early nineteenth century), English soldier who entered the East India Company service in 1820 and retired from military service as lieutenant-colonel in 1850. In 1858 Darwin and Poole exchanged several letters regarding Indian horses. *(pp. 163, 164, 166)*

Prestwich, Sir Joseph (1812–1896), chair of geology at Oxford from 1874 to 1888. Prestwich specialized on Tertiary formations of Britain and Europe and made contributions to archaeology. He was best known for his two-volume work *Geology: Chemical, Physical, and Stratigraphical* (1886, 1888). Prestwich was awarded the Wollaston Medal of the Geological Society of London for his work on the Eocene strata of the London and Hampshire basins, published between 1846 and 1857. *(p. 328)*

Ramond, Louis François Élisabeth (1753–1827), the Baron de Carbonnières, accomplished in many areas, including literature, politics, and natural history. He achieved renown as a passionate explorer of the high Pyrenees, publishing several books on the subject between 1789 and 1825. The Société Ramond (Ramond Society), formed in 1865, is still in existence. An especially appropriate honor bestowed upon him is the alpine plant genus *Ramonda* (Gesneriaceae), one of which *(R. pyrenaica)* is a rare endemic of the Pyrenees. *(p. 368)*

Ramsay, Sir Andrew Crombie (1814–1891), distinguished Scottish geologist who taught at University College, London, and later the Royal School of Mines (becoming director-general of Britain's Geological Survey in 1871). Ramsay authored numerous papers and books, among them *Geology of the Island of Arran* (Glasgow, 1841), *Geology of North Wales* (London, 1866), and *Physical Geology and Geography of Great Britain* (London, 1894). He also produced an updated geological map: *Very Large Geological Map of England & Wales* (London, 1872). In 1858 Ramsay provided Darwin with information on the thickness of geological formations in the British Isles, which Darwin incorporated into chapter IX. *(pp. 284, 285, 286)*

Rengger, Johann Rudolf (1795–1832), Swiss physician and naturalist who traveled in South America between 1818 and 1826. Rengger was imprisoned for six years (1819–1825) in Paraguay with colleague Marcelin Longchamp by the dictator José Gaspar Rodríguez Francia. Once back in Switzerland, Rengger and Longchamp published an account of their imprisonment in 1827 under the title *Essai historique sur la révolution du Paraguay, et le gouvernement dictatorial du docteur Francia,* which was translated into several languages. Darwin referred to Rengger's later work on the natural history of the mammals of Paraguay: *Naturgeschichte der Säugethiere von Paraguay* (Basel, 1830). *(p. 72)*

Richard, Louis-Claude Marie (1754–1821), French botanist who studied under Bernard de Jussieu; collected extensively in the French colonies of the Americas and was appointed professor of botany at the Ecole de Médecine in 1794. Son Achille Richard (1794–1852) became a prominent botanist as well. L.-C. Richard made a special study of orchids, coining the term "pollinia" for the pollen packet produced by these plants. *(p. 417)*

Richardson, Sir John (1787–1865), Scottish naturalist, arctic explorer, and surgeon in the Royal Navy. Richardson served as surgeon and naturalist on the arctic expeditions of John Franklin between 1819 and 1827 and led the fruitless search expedition for the lost Franklin in 1847–1849. Between 1829 and 1837 he published *Fauna Boreali-Americana,* an important four-volume work on the zoology of northern North America. Richardson's ground squirrel *(Spermophilus richardsonii)* is named in his honor. *(pp. 180, 376)*

Rollin [Roulin, François Désiré] (1796–1874), French naturalist who lived in Central America from 1821 to 1828, then worked as a librarian and editor for the *Académie des sciences* in Paris. Darwin erroneously cites "Rollin" as the authority for the observation of striping and coat coloration in mules. Although this entry was never changed in the *Origin,* the citation for this same observation in *Variation* (vol. II, p. 16) is changed to Roulin and references observations made in South America. Darwin is likely citing F. D. Roulin's paper "Recherches sur quelques changemens observés dans les animaux domestiques transportés de l'ancien dans le nouveau continent," which appeared in the *Mémoires présentés par divers Savans a l'Académie Royale des Sciences de l'Institut de France* in 1835. *(p. 165)*

Sagaret [Sageret], Augustin (1763–1851), French botanist and horticulturist, founding member of the Société d'horticulture de Paris and a member of the Société royale d'horticulture. Sagaret became well known for his work with melon hybridization, published in 1826, and in 1830 he published *Pomologie physiologique,* which treated the biology of fruit trees. Darwin cited this work several times in *Natural Selection* and *Variation of Animals and Plants under Domestication.* Sagaret is credited with elucidating the principle of genetic dominance in the early nineteenth century. *(pp. 262, 270)*

Saint-Hilaire, Augustin François César Prouvençal de (1799–1853), French botanist who traveled extensively in South America and made a special study of Brazilian flora, collecting some 30,000 specimens in Brazil between 1816 and 1822. See Dwyer (1955) for a treatment of Saint-Hilaire's extensive botanical catalogs in the Muséum National d'Histoire Naturelle in Paris. Saint-Hilaire's best-known works include the three-volume *Flora Brasiliae Meridionalis* (1825–1832), published with A. de Jussieu and J. Cambessdes; *Histoire des Plantes les plus Remarquables du Brésil et de Paraguay* (1824); and a treatment of plant morphology: *Leçons de Botanique, Comprénant Principalement la Morphologie Végetale* (1841). *(p. 418)*

Saint John, Charles George William (1809–1856), nature writer and sportsman. Darwin recorded having read Saint John's *Wild Sports and Natural History of the Highlands of Scotland* (1845). *(p. 91)*

Schiødte, Jørgen Matthias Christian (1815–1884), Danish zoologist and professor and curator at the Copenhagen Museum. Schiødte published several books and papers on the zoology of Denmark and Greenland; he had a special interest in cave fauna. *(p. 138)*

Schlegel, Hermann (1804–1884), well-known German-born herpetologist and ornithologist whose career was spent in the Netherlands. Schlegel worked for many years at the Natural History Museum in Leiden, first as assistant to Director Coenraad Jacob Temminck and then as his successor in 1858. Schlegel co-authored, with Temminck and naturalist Philipp Franz von Siebold, a monograph series on the zoology of Japan, the acclaimed work *Fauna Japonica* (1833–1850). In 1855 Edward Blyth recommended that Darwin read Schlegel for his discussion of species and varieties in the *Essai sur la physiognomie des serpents:* "But what constitutes a species? You should read Schlegel's introductory letter on the subject, addressed to Temminck, & prefatory to his work on serpents" (*Correspondence* 5: 425). *(p. 144)*

Sebright, John Saunders (1767–1846), agriculturist and Whig politician noted for his books on animal breeding. Among Sebright's works are two important contributions to animal breeding, both of which were read by Darwin: *The Art of Improving the Breeds of Domestic Animals* (London, 1809, written in the form of a letter to Royal Society president Sir Joseph Banks); and *Observations Upon the Instinct of Animals* (London, 1836). *(pp. 20, 31)*

Sedgwick, Adam (1785–1873), English geologist, Woodwardian Professor of Geology at Cambridge. Sedgwick, who remained a catastrophist and never accepted Lyell's gradualistic model of geological processes, made extensive contributions to the stratigraphy of the early Paleozoic era. His 1835 paper with Roderick Murchison *(On the Silurian and Cambrian Systems)* led to first disagreement and then a rift between Sedgwick and Murchison over the question of Sedgwick's Cambrian strata (detailed in Secord 1986). Darwin, who had attended Sedgwick's geology lectures at Cambridge, assisted Sedgwick in some of the fieldwork leading up to this episode. At Henslow's request Darwin accompanied Sedgwick on a geological trip to Wales in the summer of 1831. Darwin returned home from this very trip to find a letter inviting him to join the *Beagle* voyage. Sedgwick admired Darwin's geological work on the voyage, but later he was intensely critical of his transmutation theory. *(pp. 302, 310)*

Silliman, Benjamin, Jr. (1816–1885), professor of chemistry at the University of Louisville and later Yale University, in the latter post succeeding his father, Benjamin Silliman, Sr., upon his death in 1864. For many years Silliman helped his father edit the *American Journal of Science and Arts,* a popular periodical founded by the elder Silliman in 1818 and known informally as "Silliman's Journal." Now the *American Journal of Science,* it has the distinction of being the longest continuously published scientific periodical in the United States. *(p. 137)*

Smith, Frederick (1805–1879), entomologist at the British Museum, and president of the Entomological Society of London in 1862–1863. Smith authored several works on British insects, especially ants and bees. In 1854 (revised and expanded in 1857) Smith authored an essay on British ants, and it was here that Darwin was first made aware of the habits of the slave-making ant *Formica sanguinea.* Darwin's heavily annotated copy of Smith's paper is now in Cambridge University Library. *(pp. 219, 220, 222, 239, 240)*

Smith, James (1782–1867), the second Smith of the famed Jordanhill estate in Glasgow, Scotland. Smith was an antiquary, yachtsman, and man of letters and science, elected to the Royal Society for his geological contributions. *(p. 283)*

Somerville, Lord John Southey (1765–1819), fifteenth Lord Somerville and second president of the Board of Agriculture, an organization founded in 1793 dedicated to agricultural research of all kinds. Lord Somerville was a pioneer sheep breeder especially well known for his work with Merino sheep. *(p. 31)*

Spencer, Lord John Charles (1782–1845), the third Earl Spencer and a president of the Royal Agricultural Society. Lord Spencer was also a leading Whig politician, serving as chancellor of the Exchequer and leader of the House of Commons. *(p. 35)*

Sprengel, Christian Konrad (1750–1816), German botanist who published an important early treatise on plant pollination in Berlin in 1793: *Das entdeckte Geheimniss der Natur im Bau und in der Befruchtung der Blumen* (Nature's Secret in the Structure and Fertilization of Flowers Unveiled). Botanist Robert Brown called Darwin's attention to Sprengel's work in spring 1841, and that August Darwin heavily annotated a copy of Sprengel's book. *(pp. 98, 99, 145)*

Steenstrup, Johannes Japetus Smith (1813–1897), Danish naturalist and professor of zoology at the University of Copenhagen, discoverer of the phenomenon of alternation of generations. Steenstrup is best known for his pioneering book *On the Alternation of Generations; or The Propagation and Development of Animals Through Alternate Generations* (London, 1845). *(p. 424)*

Strickland, Hugh Edwin (1811–1853), English geologist and zoologist who worked on zoological nomenclatural reform. In 1842 Strickland published the report of a committee, of which Darwin was a member, that was established by the Council of the British Association to consider various proposals for nomenclatural reform. Among other important positions, the committee advocated the rule of priority in naming species and other rules of zoological nomenclature in practice today. Strickland was among those Darwin recommended to Emma as a possible editor for the 1844 *Essay* in the event of her husband's death. *(p. 25)*

Tausch, Ignaz Friedrich (1793–1848), Czech professor of botany and director of the garden of the duke of Canal de Malabaillas in Prague. Tausch published a botanical manual of this area, the *Hortus Canalius,* in 1823. *(p. 146)*

Tegetmeier, William Bernhard (1816–1912), English pigeon fancier and poultry expert who served as secretary of the Apiarian and Philoperisteron societies of London, and pigeon and poultry editor for the *Field* from 1864 to 1907. He also had an interest in apiculture and contributed papers on this subject to the *Cottage Gardener* magazine. *(pp. 228, 233)*

Temminck, Coenraad Jacob (1778–1858), distinguished Dutch ornithologist and director of the National Natural History Museum in Leiden. Temminck authored a manual of European birds, the *Manuel d'Ornithologie* (Paris, 1815), as well as other ornithological works and a monograph on mammals. Darwin paid special attention to Temminck's natural history of pigeons and their relatives (*Histoire Naturelle Générale des Pigeons et des Gallinacés,* Amsterdam and Paris, 1813–1815), which he recorded reading in 1847. *(p. 419)*

Thouin, André (1747–1824), French horticulturist, chair of horticulture at the Muséum National d'Histoire Naturelle in Paris. Thouin made important contributions in agronomy and grafting and published several treatises on arboriculture and grafting techniques, including *Monographie des greffes* (Paris, 1821). *(p. 262)*

Thuret, Gustave Adolphe (1817–1875), French botanist who made considerable contributions to the study of marine algae. With colleague Edouard Bornet, for many years Thuret conducted detailed studies of development and reproduction of seaweeds, culminating in the posthumously published work *Etudes phycologiques* (1878). *(pp. 258, 264)*

Thwaites, George Henry Kendrick (1811–1882), botanist and entomologist; director of the botanical gardens at Peradeniya, Sri Lanka (formerly known as Ceylon). *(p. 140)*

Tomes, Robert Fisher (1823–1904), Gloucestershire farmer and zoologist who specialized on bats. Tomes worked on birds early on but became increasingly interested in bats and described several species. He published many papers on members of the widespread and diverse vesper or evening bat family (Vespertilionidae) and contributed sections on bats and "insectivora" for Thomas Bell's *History of British Quadrupeds,* second edition (London, 1874). *(p. 394)*

Valenciennes, Achille (1794–1865), French zoologist who studied under Baron Georges Cuvier. Valenciennes is best known as a founder of parasitology, but he was also an accomplished ichthyologist. He worked with Cuvier to produce *Histoire Naturelle des Poissons* (Natural History of Fish) in the 1830s and 1840s, a monumental work of twenty-two volumes, treating more than 4,500 species of fish. *(p. 384)*

Van Mons, Jean Baptiste (1765–1842), Belgian pharmacist and physician, professor at the University of Louvain, and well known for his pioneering efforts in pear and apple breeding and propagation. In 1836 Van Mons published the important treatise *Arbres Fruitiers ou Pomologie Belge Expérimentale et Raisonnée. (p. 29)*

Wallace, Alfred Russel (1823–1913), celebrated Welsh-born English naturalist, geographer, and traveler; co-discoverer of the principle of natural selection and founder of modern biogeography. A largely self-educated man, Wallace developed a passion for natural history at an early age. In 1848 he traveled to the Amazon with Henry Walter Bates and spent the next four years collecting and observing. Nearly all of Wallace's collections and field notes were subsequently lost in a ship fire en route home in 1852, yet he still managed to produce six papers and two books (*Palm Trees of the Amazon and Their Uses* and *Travels on the Amazon*). Undeterred from traveling, Wallace then left on a remarkable eight-year journey of exploration in the Malay Archipelago (1854 to 1862). It was there that Wallace produced, among many other papers, his insightful works on evolution: the Sarawak Law paper (Wallace 1855) and the paper on the principle of natural selection that he fatefully sent to Darwin for comment (Wallace 1858). Wallace collected hundreds of thousands of specimens in his southeast Asian travels, well over a thousand of them new to science. His eight years of adventures and observations were published in 1869 as *The Malay Archipelago,* which he dedicated to Darwin—perhaps the finest book of its kind. He subsequently drew on his vast store of notes and observations to produce the seminal books *The Geographical Distribution of Animals* (1876) and *Island Life* (1880), considered founding works of the modern era of biogeography. His definitive work on evolution by natural selection, *Darwinism,* came out in 1889. *(pp. 355, 395)*

Waterhouse, George Robert (1810–1888), curator for the Zoological Society of London from 1836–1843, and later assistant keeper and then keeper of mineralogy and geology at the British Museum. Waterhouse studied the beetle and mammal specimens from Darwin's *Beagle* voyage. In 1846–1848 Waterhouse published his two-volume *Natural History of Mammals. (pp. 116, 150, 151, 225, 429, 430)*

Watson, Hewett Cottrell (1804–1881), English botanist who wrote extensively on plant biogeography, including the monumental treatise *Cybele Britannica, or British Plants and Their Geographical Relations* (London, 1847–1859). Watson had a long and extensive correspondence with Darwin on botanical matters. *(pp. 48, 53, 58, 140, 176, 363, 367, 376)*

Westwood, John Obadiah (1805–1893), English entomologist who became curator of the University Museum and then professor at the University of Oxford. Westwood was a talented artist, and in 1874 he produced his acclaimed *Thesaurus Entomologicus Oxoniensis,* an illustrated guide to insects in the Hope entomological collection at the Oxford University Museum. The Rev. Frederick William Hope, Westwood's patron, donated this remarkable collection to the museum in 1849 and nominated Westwood as the first Hope Professor. *(pp. 57, 157, 415)*

Wollaston, Thomas Vernon (1822–1878), entomologist and malacologist who conducted research in Madeira. Among Wollaston's entomological works are a treatise on the insects of the Madeira island group, *Insecta Maderensia* (London, 1854), *On the Variation of Species, with Especial Reference to the Insecta* (London, 1856), and *Coleoptera Atlantidum* (London, 1865). In 1860 Wollaston published one of the harsher reviews of the *Origin,* and he never accepted Darwin's transmutation argument. *(pp. 48, 52, 132, 135, 136, 176, 389, 402)*

Woodward, Henry (1832–1921), British paleontologist and geologist appointed keeper of the geology and paleontology department of the British Museum in 1880. He worked on fossil crustaceans and for many years edited the *Geological Magazine,* which he also co-founded. *(pp. 293, 316, 339)*

Youatt, William (1776–1847), English veterinarian and author of several important books on farm animal breeds and husbandry, several of which, including *The Dog* (1845) and *The Horse* (1831), are still in print. The first appearance of the term "natural selection" is a marginal note Darwin made in his copy of *The Horse. (pp. 31, 454)*

Acknowledgments

In the Introduction to his 1868 book *Variation of Animals and Plants under Domestication,* Darwin concludes a concise summary of his transmutation theory with a heartfelt acknowledgment of assistance: "I have continually been led to ask for information from many zoologists, botanists, geologists, breeders of animals, and horticulturists, and I have invariably received from them the most generous assistance. Without such aid I could have effected little . . . I cannot express too strongly my obligations to the many persons who have assisted me, and who, I am convinced, would be equally willing to assist others in any scientific investigation." His words often echoed in my mind as I worked on this annotated *Origin,* and I have long appreciated that research is fundamentally a collaborative undertaking.

This book has its origins in an observation the late Stephen Jay Gould made to me, back when I served as a teaching fellow in his legendary History of Life course in the early 1990s. There was but one required text in that course, Harvard's wonderful facsimile first edition *Origin* edited by Ernst Mayr. Gould lamented that few students of biology or even professional biologists ever read that founding document of their science in its entirety, let alone understood it. I privately reflected with embarrassment that I was among the guilty, having read the odd chapter and passage but never undertaking the whole text. I resolved to remedy that situation, and though I was a confirmed evolutionary biologist already, the exercise inspired something akin to the zeal of the newly converted to spread the word. Soon after arriving at Western Carolina University (WCU) in 1996, I developed a course entitled Darwin's *Origin of Species,* a detailed study of the book focusing on Darwin's arguments and their historical context, and a modern take on his examples and assertions. Every new reading brought fresh insights, and I am grateful to the cohorts of students who have taken the course and shared their perspectives.

The present book is directly descended from the commentaries that grew out of my *Origin* course, and I was fortunate to be able to share earlier drafts with students in my Spring 2008 *Origin* courses at Western Carolina University and with the CLE community group in Highlands, North Carolina (thank you, Mark Whitehead, for enthusiastically endorsing the course!), as well as in Harvard's Darwin program in the summers of 2007 and 2008. The critical feedback and suggestions of these students, too numerous to name individually here, were invaluable indeed, and I am grateful for their help. I am equally grateful to my colleagues in WCU's department of biology, who supported my developing and teaching an *Origin* course to begin with, untraditional offering that it is, with as much history and philosophy as science.

Since 2005 I have had the good fortune to teach with Andrew Berry and Naomi Pierce in Harvard's summer Darwin program based at Oxford University, a fabulous full immersion in Darwiniana and the history of evolutionary thinking. This annual adventure influenced this book in significant ways. I continue to learn a great deal from Naomi and Andrew, as I have from my days as a postdoctoral fellow in Naomi's lab at Harvard, and I am deeply grateful for the opportunity to spend time with them each summer. I have gained much from innumerable casual and not-so-casual discussions with the scientists, historians, and other scholars who meet with our Darwin group in the United Kingdom each year, especially George Beccaloni of the Natural History Museum, London; Janet Browne of Harvard's History of Science Department, who helps us see Down House through Darwin's eyes; Richard Dawkins of Oxford University; Gina

Douglas of the Linnean Society; Tom Kemp and Malgosia Nowak-Kemp at the University Museum, Oxford; Ian Lacey and Michael Morrogh of Shrewsbury School; Jim Moore of the Open University; and the Rev. Michael Roberts, geologist and vicar of Cockerham, England. A special thank you to Janet for generously offering use of her home in Oxford when my family comes to visit, and to Naomi for hosting us at St. Frideswide's in a pinch. Many an enjoyable and instructive evening was spent at St. Frideswide's farmhouse just outside Oxford, where the inimitable and gracious Sally Craddock holds forth with panache and the Harvard students inexplicably find the sheep just beyond the ha-ha irresistible. Thanks, too, to my friend Wenfei Tong, an integral part of our Oxford course team, whose love of Darwiniana may be rivaled only by her love of good ice cream.

For unstinting support in my annual Oxford endeavor I thank Western Carolina University's provost, Kyle Carter, former University of North Carolina system vice president for research Russ Lea and current vice president for research Steve Leath, and the Highlands Biological Station Board of Directors. Needless to say I could not undertake this project without outstanding staff on the home front either, and I appreciate the professionalism of Guy Cook, Cyndi Banks, Patrick Brannon, and especially office manager Leslie Davis and my wonderful associate director at the Highlands Biological Station, Anya Hinkle.

Several friends and colleagues at WCU share an interest in evolutionary biology, Darwin, and the broader cultural or philosophical context of evolution; over the years I have enjoyed stimulating discussions with Tim Carstens, Kefyn Catley, Daryl Hale, Hal Herzog, Seán O'Connell, Brian Railsback, and Steve Yurkovich. I am also fortunate to work with talented librarians at Western Carolina University and Harvard University, who are ever ready, willing, and able to assist. I thank WCU's Brenda Moore, Kitty Taylor, and Terry Ensley for all their help, and I especially thank Krista Schmidt and Bart Voskuil for coming to my rescue, often at the drop of a hat, to ferret out obscure articles or books. I am grateful to Bart Voskuil, too, for translating French and German literature for me, and to Alessia Zanin-Yost for Italian translations. Thanks, finally, to former university librarian Bil Stahl for providing me with office space in the library. I was in good hands up North, too—many thanks to Ronnie Broadfoot, Mary Sears, and especially Dana Fisher at Harvard's Ernst Mayr Library for all manner of assistance, from tracking down references to scanning manuscripts.

George Beccaloni, Gaden Robinson, and Doug Yanega kindly assisted with taxonomic questions. Naomi Pierce, Wenfei Tong, Howie Neufeld, Anya Hinkle, Leslie Costa, and especially Janet Browne and Andrew Berry improved the manuscript with comments, criticisms, and suggestions; all have been champions of this project from its inception, and I am deeply grateful for their help. Harvard University Press has been a great champion, too, and I appreciate the enthusiasm and support of my editor, Ann Downer-Hazell, as well as editorial assistant Vanessa Hayes and production team Christine Thorsteinsson, Annamarie Why, and David Foss for editorial and design assistance.

I had the good fortune to accompany Frank Sulloway in the Galápagos. I can think of no better companion with whom to follow in Darwin's footsteps. Thank you, Frank, for so freely sharing your knowledge of Galápagos ecology, history, and lore.

I wish to acknowledge and thank the veritable army of dedicated Darwin and Wallace scholars who did much of my work for me; I am humbled and awed by the collective efforts of these scholars, past and present, in making available to the world such priceless jewels as Darwin's transmutation notebooks, print and online correspondence projects, published books and articles, and the *Beagle* diary, as well as annotated print and online collections of Wallace's diverse writings, and more. Beyond merely useful—and they are invaluable resources—these magisterial labors of love are deeply moving.

And speaking of labors of love, last but certainly not least I am ever grateful to my wife, Leslie Costa, for supporting me in myriad ways during the writing of this book, especially in undertakings like the Oxford program, which require extended periods away from a home made quite lively by our boys, Addison and Eli. Museum curatorial associate until school lets out, then taxi driver, chef, finance minister, referee, homework monitor, soccer cheerleader, judge, gardener, violin coach, and more, Leslie is a force of nature.

Subject Index

This index is reproduced unaltered, except for minor corrections, from Harvard University Press's facsimile edition of 1964. It consists of Darwin's entries for the first edition of the *Origin*, along with additions from the renowned evolutionary biologist Ernst Mayr. Darwin's terms reflect concepts and observations important to his time and place, and as such have historical value to us. Mayr's additions, italicized to distinguish them from Darwin's, include many modern evolutionary terms and concepts. These are equally valuable in underscoring the continuity between the modern and the historical, illustrating the continued currency of many of Darwin's ideas. JAMES T. COSTA